# LOGIC
# A Foundation
# for
# Computer Science

# INTERNATIONAL COMPUTER SCIENCE SERIES

*Consulting editors*   **A D McGettrick**   University of Strathclyde

**J van Leeuwen**   University of Utrecht

## SELECTED TITLES IN THE SERIES

# LOGIC
# A Foundation
# for
# Computer Science

## V. Sperschneider & G. Antoniou

University of Osnabrück

ADDISON-WESLEY
PUBLISHING
COMPANY

Wokingham, England · Reading, Massachusetts · Menlo Park, California · New York
Don Mills, Ontario · Amsterdam · Bonn · Sydney · Singapore
Tokyo · Madrid · San Juan · Milan · Paris · Mexico City · Seoul · Taipei

Cover designed by Crayon Design of Henley-on-Thames and
printed by The Riverside Printing Co. (Reading) Ltd.
Printed in Malta.

First printed 1991.

**British Library Cataloguing in Publication Data**
Sperschneider, V.
   Logic : a foundation for computer science.
   I. Title   II. Antoniou, G.
   005.131

   ISBN 0201565145

**Library of Congress Cataloging in Publication Data**
Sperschneider, V.
      Logic : a foundation for computer science / V. Sperschneider and
   G. Antoniou.
         p.    cm.
      Includes bibliographical references (p.    ) and index.
      ISBN 0-201-56514-5
      1. Computer science.  2. Artificial intelligence.  3. Logic,
   Symbolic and mathematical.  I. Antoniou, G. (Grigorios)  II. Title.
   QA76.S6787   1991                                           91-2885
   004--dc20                                                   CIP

# Preface

This book is based on a number of experiences and convictions. The first concerns the dominant role that logical foundations play in present-day computer science and artificial intelligence. It is best described by quoting John McCarthy, one of the fathers of computer science:

> It is reasonable to hope that the relationship between computation and mathematical logic will be as fruitful in the next century as that between analysis and physics in the last. The development of this relationship demands a concern for both applications and mathematical elegance. (1967)

These prophetic sentences are surely no longer controversial today. Lots of fields such as automatic theorem proving, logic programming, specification of abstract data types, program verification, program synthesis, non-monotonic reasoning, and so on, have emerged since 1967, which could not possibly have been successfully handled without a broad and solid knowledge of logic.

The second conviction results from experience the first author has had in the last 10 years with courses on logical foundations of computer science, recently with the assistance of the second author. This concerns the sometimes broad gap between the practicability and applicability of logical results, and their theoretical treatment. Often, books either address only versatile mathematicians and logicians ignoring the question of how a practitioner might learn to understand and apply the subject matter, or they are restricted to a superficial presentation of examples and methods without offering the reader a chance to obtain a profound understanding of the material. Such a deeper understanding is by no means a luxury. It is our experience that reasonable practical application of logical concepts is achieved mainly by a solid teaching of the foundations.

We believe that is is possible to teach the logical foundations of computer science in the required mathematically rigorous way, without intimidating people who are mainly interested in the practical applications. This is not to say that the material is trivial, or even easy. Certain basic mathematical prerequisites, as well as the willingness to make some effort, are certainly necessary for a beginner in logical foundations.

This leads us to our third experience. Once having worked through some basic concepts, theorems, methods, examples, and so on, the reader will find it more and more easy to make his/her way into new fields of application such as logic programming, specification, verification and AI-languages. Each of these fields is represented in the current literature by specialist text books.

But there are only few books that cover a broad range of logical foundations with a uniform and homogeneous presentation. Thus, the reader who is interested in several fields is forced to work through several books with different notations and methods of presentation, sometimes even with incompatible concepts. One of the purposes of this book is to overcome this difficulty. It is self-contained and presupposes nothing but some familiarity with elementary mathematical notation and reasoning; also, a basic understanding of computer science may be helpful for some parts. We think that this book will be suitable both for self-study and as a source of additional material for university courses.

## Organization of the book

The book is in two volumes, the first one dealing with logics that are sublogics or simple extensions of predicate logic. Such logics have been used in the past in more classical parts of computer science. Recent developments show the importance of logics that are considerably removed from predicate logic (for example dynamic logic), or are even incompatible with predicate logic (non-monotonic logic). This first volume (Parts I-IV) discusses quite standard material although at some points it reaches into current research (such as verification of logic programs or verification of modules). The second volume (Parts V-IX) introduces special logics for distinguished applications.

Part I introduces *predicate logic*, which is the basis of all subsequent parts. We discuss syntax, semantics, proof theory and foundations of automated theorem proving. Part II is concerned with *logic programming*. We discuss declarative and procedural interpretation of logic programs, discuss verification aspects and give a short introduction to basic features of Prolog. Part III is devoted to *equational logic and algebraic specification*. It is shown how equational logic may be used to specify abstract data types as well as serving as an operational model. Part IV discusses *program verification*. We introduce *Hoare's logic* and show how it can be used to verify programs written in a quite realistic programming language. Part V introduces *modal logic*, an extension of predicate logic that builds the basis for several program logics as well as non-standard logics. Part VI is concerned with *dynamic logic*, a logic that is well suited to reasoning about programs. Part VII discusses *temporal logic* which addresses reasoning about time. It can be used for proving properties of parallel processes. Part VIII introduces logics related to *program synthesis*. Finally, Part IX discusses the broad field of *non-monotonic logics* and its application to artificial intelligence.

## How to use this volume

This volume may be used by many readers with different interests. Naturally, it may be read in a sequential way chapter by chapter. There are, however, some alternative paths through this volume, if the reader has a particular interest.

A course in *predicate logic* should consist of Chapters 1-6.

A course in *logic programming* should consist of Chapters 1 and 2 introducing predicate logic, Sections 3.1 and 3.4 about resolution, and Part II.

A course in *specification techniques* should consist of Chapters 1 and 2 on predicate logic, Chapters 4 and 6 on predicate logic with equality and many-sorted logic, and Part III. Sections 18.1-18.3 could also be of interest in seeing how a module concept may be developed based on the specification methods introduced.

A course in *program verification* should consist of Chapters 1, 2, 4, 5 and 6, as well as Chapters 14-17.

Sections that are devoted to specialists and that may be omitted in a first reading are indicated with a ✂.

## Acknowledgements

We want to thank numerous colleagues and students for hints, helpful criticism and corrections during the development of this book.

Above all, thanks to all the staff of Addison-Wesley and several anonymous referees who participated in the development of this book. They all gave us enormous support, concerning scientific advice as well as the difficult task of language editing.

Finally, special thanks to Jana, Jörg and Ursula for their patience during the writing of this volume.

Volker Sperschneider and Grigorios Antoniou
Universität Osnabrück
Germany
Osnabrück, March 1991

# Contents

# Introduction

## About philosophical and mathematical logic

Logic is an ancient discipline: having roots that range back to Ancient Greece with Aristotle as prominent representative; being constantly present in the ideal world of the Middle Ages; playing a dominant role in Leibniz's idea of an *ars magna*; exhibiting its first highlight in its role in the well-known foundational crisis of mathematics at the end of the last and beginning of this century with Frege, Hilbert and Gödel as three of many important contributors; and recently receiving fresh impetus with the dawn of computer science and artificial intelligence.

The primary purpose of logic was to investigate the objective laws of human thought. Objectivity, of course, means that arguments must be communicable to and verifiable by other people. Since its beginnings, it was undoubted that objectivity of thinking, or strictness of reasoning and arguing, is related to formalizability. Only within a formal language is it possible to communicate arguments unambiguously to others, and to verify arguments. What formalizability might mean became more and more clear during the history of logic.

It was the application of logical concepts to the foundations of mathematics, and, conversely, the use of mathematical methods within logic, that led to the first highlight of logic, which was from then called mathematical logic or formal logic. It was the strict separation of pragmatical, syntactical and semantical aspects that made mathematical logic a clean mathematical discipline, with major contributions to problems in the foundations of mathematics.

In a field of application, be it a mathematical discipline or an everyday life application, statements are translated into formulas of a formal language, with semantics as the bridge between the world of mathematical or real objects, and the world of syntactic representatives. Thus, whereas *semantics* is interested in the *meaning* (or *content* or *truth* or *validity*) of formulas, *syntax* deals only with the *form* of formulas.

Not only the 'static' concern of laying down declarative knowledge in a formal language that is less susceptible to misunderstandings than natural language, but also its 'dynamic' counterpart, to mechanize human reasoning, was a constant demand in the development of logic (Leibniz's *ars magna* as a machine able to answer arbitrary questions about the world). It is at this point where finitary games, called logic calculi, whose purpose is to derive (or deduce or prove) formulas on a purely syntactical level, come in. Such calculi are closely related to semantics. Inevitably, they must be sound, meaning that nothing may be provable that is not true. Preferably, they should be also complete, meaning that everything that is true is also provable. It is these calculi that provide a basis for the mechanization of reasoning.

Concerning its role in mathematics, at least the following four facets of mathematical logic, in particular in its prototypical form called predicate logic, must be mentioned: as a formalization language to express and make precise mathematical statements; as a calculus, clarifying the notion of a mathematical proof and mechanizing the process of proving; as a mathematical subdiscipline, like analysis or algebra; as a mathematical tool, applicable to classical mathematical problems.

The first facet does not need further clarification. The second facet is, strictly speaking, of greater importance for computer science than it ever was for mathematics. Usually, mathematicians do not like to write down proofs up to the finest details within a formal calculus. Though this is in many cases possible in principle, it would soon obstruct the productivity of a mathematician, owing to the enormous overhead imposed by the formalities of a calculus. In contrast, a computer likes to deal with formal proofs, while it cannot handle proofs which are not totally formalized. But note that formalization of provability, even combined with the power of present-day computers, should not lead to the erroneous idea that the proof of statements is trivial. A look at chess might be helpful at this point. The game of chess is totally described by a starting position and a handful of rules. With this knowledge it is possible both to create every possible game and to check whether a series of moves is legal. However, a great portion of creativity and ingenuity is necessary in order to produce the games of Fischer or Kasparov.

The third facet has already been addressed, by mentioning that the great success of mathematical logic was because it was treated as a mathematical discipline. Finally, concerning the fourth facet, it is worth mentioning that there is a non-vanishing number of classical mathematical theorems (without any relation to logic) that were discovered or first proved using logical methods.

## About logic and computer science

What are currently the special concerns of computer science with mathematical logic? First of all it is important to mention the increased importance of pragmatical aspects. Concepts and methods must fit well into the demands of everyday experiences in computer science. Efficiency considerations play a dominant role.

Needless to say, because computers understand nothing more than formalized statements, the first facet mentioned above, logic as a formalization language, becomes even more important. As the use of computers generates lots of new challenging problems (formal specification languages, program verification) with the evidence that these may be solved only by using computers, theorem proving receives special attention, too. Only by efficient implementation of appropriate logic calculi will it be possible to manage the flood of lemmas which usually must be derived in order to establish the correctness of a program.

Looking at logic from the computer science point of view, a new aspect emerges, namely logic as an execution or programming language, referring to fields like logic programming (specifying by logical formulas *what* a program should do and leaving it to the computer to decide *how* it is done), abstract interpretation of specifications (providing mechanisms that allow the execution, and thus observing and testing of systems at a very early stage of development, also called rapid prototyping), or program synthesis (from so-called constructive proofs of statements extracting algorithmic information, finally leading to efficient algorithms solving the problem behind the considered statement).

These concerns are attacked using various logical languages, the most prominent being predicate logic. There are other languages, some being sublanguages of predicate logic with more efficient proof procedures tailored to the peculiarities of the considered language (Horn logic, equational logic, term rewriting systems), and some being extensions of predicate logic with more comfortable features adapted to specific applications (Hoare logic, dynamic logic).

## About logic and artificial intelligence

Artificial intelligence (AI) is a scientific discipline dealing with reproducing what is called human intelligence. The starting point of AI is an intelligent being communicating with its environment. The AI approach assumes that the being has an internal model of the surrounding world; it makes decisions using this model, whereupon these decisions lead to actions with effects in the real world.

The logical approach to AI assumes that the being's model consists essentially of declarative knowledge. This is the point where logic comes in. Logic is namely the natural knowledge representation language. Furthermore, intelligent beings must be able to process their knowledge and make deductions on its basis. Logic supports this aspect of intelligence, since it provides us with powerful mechanisms for manipulating knowledge.

# Part I

# PREDICATE LOGIC

We begin this book by treating a standard logic which is fundamental for almost all work in logic, namely **predicate logic**. As we have already stated in the introduction, there are two different views of logic, a syntactical and a semantical one. So we start by presenting the formal language of predicate logic, followed by its semantics. This is the basis for what is called **knowledge representation**.

We then treat calculi for predicate logic, some types of syntactical game being able to derive exactly those formulas that are semantically valid. We present a Hilbert-type calculus, two sequent calculi, the resolution calculus and the calculus of analytic tableaux. Each of these calculi is shown to be **sound** and **complete**, that is the working of these calculi is related in a specific way (that varies from calculus to calculus) to semantics. As calculi allow the derivation of new facts from other known ones, they can be regarded as a means for **knowledge extraction**.

We also discuss some widely used extensions of predicate logic, namely predicate logic with equality and many-sorted logic.

# Chapter 1
# Syntax of Predicate Logic

## 1.1  Predicate logic formulas

Let us consider some sentences from various fields.

- Either it rains or it does not rain.

- Every student who studies computer science must study a further field which may not be archaeology.

- Every number different from 0 is the successor of another number.

- In a domain with an associative multiplication, a left-neutral element e and a left-inverse element for every domain element, e is always a right-neutral element, too.

- A two-person game with stones is played according to the rule that alternately the two players may take between one and three stones. The player taking the last stone wins.

What is required to formalize these natural language sentences adequately in a formal language? Of course, 'adequately' means that the fine structure of the sentences should be faithfully mirrored in the corresponding formal sentences. Let us take a first look at the ingredients of the sentences above:

(1)     There are distinguished objects of the domains of discourse, like: computer science, archaeology, e, 0, 1, 2, 3, ... These objects are represented in our formal language by names cs, arch, 0, e, 1, 2, 3, ... that we call **constants**.

(2)     As the example '1, 2, 3, ...' already shows, it is not always economical to refer to distinguished objects only via constants. A better way may be to use **function symbols** which generate names of objects from constants. Examples are succ for successor function, ∘ for associative multiplication and i for the inverse function in a group. Function symbols have a certain **arity**. For example, succ is of arity 1 and ∘ is 2-ary. To briefly denote arities, we will use a special sort symbol s and write ∘:$s^2$→s to indicate that ∘ is 2-ary. Using constants and function symbols, we may generate composed names of objects like succ(succ(succ(0))) and ∘(i(e),∘(e,e)). Strings like these are called **terms**.

(3)     Our sentences above make statements about terms. Such statements are constructed by using **relation** (or predicate) **symbols**, like student (1-ary), field (1-ary), studies (2-ary), equal (2-ary), wins (1-ary). Arities of predicate symbols are indicated by for example, equal:$s^2$.

The symbols introduced so far form an application-dependent vocabulary used to talk about the domain of discourse. It will be called a **signature**.

(4)     Given a signature as above, we may construct simple sentences about the objects of the domain of discourse, like field(cs), equal(0,succ(0)), wins(3). Such basic statements are called **atomic formulas**.

(5)     The sentences mentioned above consist of several atomic formulas, combined in different ways: ...is *not* archaeology..., ...is associative *and* has a left-neutral element..., ...may take one *or* two *or* three stones..., ...*if* different from 0 *then*... . Let us complete the list of used **propositional connectives** (not, and, or, if then) by an often used one, namely 'if and only if', also written 'iff'. There are standard abbreviations for these connectives, namely ¬ for 'not', ∧ for 'and', ∨ for 'or', → for 'if then', ↔ for 'if and only if'.

(6)     The atomic formulas constructible so far are of a quite simple form. In particular, they express properties of single objects, but not properties of a whole (perhaps infinite) class of objects. In other words, we may express single facts but not general rules stating something about 'all objects of the domain of discourse'. To express more general sentences we need **variables** X,Y,Z ..., that is 'place-holders' for the concrete objects of the domain of discourse. Using variables similarly as constants, we may build terms and atomic formulas containing variables, like student(X), equal(X,X), equal(∘(i(e),∘(e,e)),X).

(7)    Finally, there are two modes of variable use: **existential** ('...there is a left-inverse element...') and **universal** ('...every student...'). Thus, we introduce quantifiers ∀ (universal) and ∃ (existential).

Now we are prepared to express the examples presented above as **formulas** (the different sizes of brackets may help at this point to identify the structure of the formulas):

(φ∨¬φ), where φ stands for the sentence 'It rains'.

$$\forall X\big((\text{student}(X)\wedge\text{studies}(X,\text{cs}))\rightarrow$$
$$\exists Y(\text{field}(Y)\wedge(\text{studies}(X,Y)\wedge\neg\text{equal}(Y,\text{arch})))\big)$$

$$\forall X\big(\neg\text{equal}(X,0)\rightarrow\exists Y\ \text{equal}(X,\text{succ}(Y))\big)$$

Let φ be the formula ∀X∀Y∀Z equal(∘(X,∘(Y,Z)),∘(∘(X,Y),Z)), ψ the formula ∀X equal(∘(e,X),X), χ the formula ∀X equal(∘(i(X),X),e), and ζ the formula ∀X equal(∘(X,e),X). Then, our fourth sentence is formalized as follows:

((φ∧(ψ∧χ))→ζ)

Here we can make more concrete what was meant by 'adequate formalization'. Simply introducing names φ,ψ,χ,ζ for the sentences above without going into the fine structure of these sentences would mirror the structure of our third example only in a very rudimentary way. Indeed, it would mirror only the 'propositional' structure of our sentence. So keep in mind that there are, in general, many ways of representing knowledge in predicate logic.

The four auxiliary formulas look much better if we use an infix equality symbol = and an infix multiplication ∘, and omit the outermost brackets:

(∀X∀Y∀Z X∘(Y∘Z)=(X∘Y)∘Z ∧ ∀X e∘X=X ∧ ∀X i(X)∘X=e)
→ ∀X X∘e=X

Finally, our fifth example is formalized as follows (using a minimal amount of built-in arithmetic and φ←ψ as an abbreviation for ψ→φ):

wins(1)
wins(2)
wins(3)
loses(0)

$$\forall X\big(\text{wins}(X)\leftarrow\big(X\geq4\wedge(\text{loses}(X\text{-}1)\vee(\text{loses}(X\text{-}2)\vee\text{loses}(X\text{-}3)))\big)\big)$$

$$\forall X\big(\text{loses}(X)\leftarrow\big(X\geq4\wedge(\text{wins}(X\text{-}1)\wedge(\text{wins}(X\text{-}2)\wedge\text{wins}(X\text{-}3)))\big)\big)$$

**Definition 1.1:** The following special characters are used to build predicate logic formulas:

| Character | Called | Spelled |
|---|---|---|
| ¬ | negation symbol | not |
| ∧ | conjunction symbol | and |
| ∨ | disjunction symbol | or |
| → | implication symbol | if...then, or implies |
| ↔ | bi-implication symbol | iff (equivalent) |
| ∀ | universal quantifier | for all |
| ∃ | existential quantifier | there exists a |
| s | the sort symbol | |
| ( ) , | brackets and comma | |
| $V_0, V_1, ..., V_n, ...$ | variables | |

The first five characters are called **propositional connectives**. ∀ and ∃ are called **quantifiers**. Propositional connectives and quantifiers together form the set of **logical symbols**. Variables are always denoted by strings with an upper-case letter as leftmost symbol.

Besides these special characters we will need a set of application-dependent symbols to model specific domains of discourse. These are introduced by the notion of a signature.

**Definition 1.2:** A **signature** is a denumerable set $\Sigma$ of strings of the form $f:s^n \rightarrow s$ or $p:s^n$, with a natural number n and our sort symbol s. If $f:s^n \rightarrow s$ is an element of $\Sigma$ we call f a **function symbol** of $\Sigma$ of **arity** n. Abusing language, we sometimes write $f \in \Sigma$. In the case that n=0 we call f a **constant**. If $p:s^n$ is an element of $\Sigma$ we call p a **relation symbol** or **predicate symbol** of $\Sigma$ of arity n. Abusing language, we sometimes write $p \in \Sigma$. In the case that n=0 we call p an **atom**. It is required that different strings in a signature have different leftmost symbols (that is we do not allow the same symbol to be simultaneously used as a function symbol and a predicate symbol, or used twice as a function symbol with different arities).

**Definition 1.3:** For a signature $\Sigma$, we define $\Sigma$-**terms** inductively as follows:

(a)     Every variable is a $\Sigma$-term.

(b)     If f is a function symbol of $\Sigma$ of arity n and $t_1, ..., t_n$ are $\Sigma$-terms, then $f(t_1, ..., t_n)$ is a $\Sigma$-term, too.

In particular, constants of $\Sigma$ are $\Sigma$-terms. Let $T(\Sigma)$ denote the set of all $\Sigma$-terms. A $\Sigma$-term is called **ground $\Sigma$-term** iff it does not contain variables. Let $T_0(\Sigma)$ denote the set of all ground $\Sigma$-terms.

**Definition 1.4:**  For a signature $\Sigma$, we define an **atomic $\Sigma$-formula** to be a string of the form $p(t_1,\ldots,t_n)$, with a predicate symbol p of arity n in $\Sigma$ and $\Sigma$-terms $t_1,\ldots,t_n$. $p(t_1,\ldots,t_n)$ is called a **ground atomic $\Sigma$-formula** iff it does not contain variables. Let $At(\Sigma)$ and $At_0(\Sigma)$ denote the set of all atomic and ground atomic $\Sigma$-formulas, respectively.

**Definition 1.5:**  For a signature $\Sigma$, we define **$\Sigma$-formulas** inductively as follows:

(a)     Every atomic $\Sigma$-formula $p(t_1,\ldots,t_n)$ is a $\Sigma$-formula.

(b)     If $\varphi$ and $\psi$ are $\Sigma$-formulas and X is a variable, then the following strings are also $\Sigma$-formulas:
$\neg\varphi$, $(\varphi\wedge\psi)$, $(\varphi\vee\psi)$, $(\varphi\to\psi)$, $(\varphi\leftrightarrow\psi)$, $\forall X\varphi$ and $\exists X\varphi$

We denote by $F(\Sigma)$ the set of all $\Sigma$-formulas. $\forall X$ and $\exists X$ are called **quantifications**.

---

**Exercise 1.1**   Formalize in predicate logic the following conclusion: Some patients like every doctor. No patient likes a quack. Therefore no doctor is a quack.

---

## 1.2  Induction principles for terms and formulas

Let us assume that we wish to show that every term or formula exhibits a certain property P. A common method for showing such statements is **induction**. There are several variants of induction in the context of term and formula syntax, all of which will be used later in appropriate situations.

- **Structural induction on the term syntax**

  Show that P is true for every variable.

  Show that P is true for every term $f(t_1,...,t_n)$, provided that P is true for each of $t_1,...,t_n$.

- **Structural induction on the formula syntax**

  Show that P is true for every atomic formula.
  Show that P is true for $\neg\varphi$, $(\varphi\wedge\psi)$, $(\varphi\vee\psi)$, $(\varphi\rightarrow\psi)$, $(\varphi\leftrightarrow\psi)$, $\forall X\varphi$ and $\exists X\varphi$, provided that P is true for $\varphi$ and $\psi$.

- **Induction on term length**

  Show that P is true for a term t provided that it is true for every term s consisting of fewer symbols than t.

- **Induction on the length of a formula**

  See the corresponding induction principle for terms.

- **Induction on the number of logical symbols in a formula**

  Show that P is true for a formula $\varphi$ provided that it is true for every formula $\psi$ containing fewer logical symbols than $\varphi$.

This latter induction principle has the following advantage over the former ones: in proving P for a quantified formula $QX\varphi$, it may be necessary to know that P has already been shown for every formula $\varphi\{X/t\}$ which results from $\varphi$ by replacing certain occurrences of X by a term t. Note that $\varphi\{X/t\}$ contains fewer logical symbols than $QX\varphi$, although it may be longer than $QX\varphi$.

## 1.3   Uniqueness of predicate logic syntax

*Theorem 1.1*

(1)   The decomposition of a term t into its constituents is unique. The same holds for formulas.

(2)   Given a string x that is a proper prefix of some term, x uniquely determines what type of symbol follows x in any term s having x as a proper prefix, namely:

   (a)   an opening bracket
   (b)   a comma
   (c)   a closing bracket
   (d)   a variable or a function symbol

If the symbol following x is a variable or a function symbol, then every term s with prefix x uniquely determines the term r following prefix x in s (later called the **subterm** of s starting after prefix x).

| s = | proper prefix x | subterm r | ... |
|-----|-----------------|-----------|-----|

---

**Exercise 1.2**     Use induction on the length of terms to show that no term can be a proper prefix of another term. Show a corresponding statement for formulas. Then show that this proves part (1) of Theorem 1.1.

**Exercise 1.3**     Use structural induction on term s to show that each proper prefix x of s uniquely determines which one of the four cases of part (2) of Theorem 1.1 applies to the symbol following x.

**Exercise 1.4**     Use structural induction on term s to show that at each occurrence of a function symbol in s, a unique subterm r of s starts.

---

The uniqueness of the syntax of terms and formulas allows functions on terms and formulas to be defined by structural induction.

**Definition 1.6:** We may define functions h on $T(\Sigma)$ as follows:

- First, define h(X), for every variable X.
- Assuming that $h(t_1),...,h(t_n)$ are already defined for $\Sigma$-terms $t_1,...,t_n$, define $h(f(t_1,...,t_n))$ using $h(t_1),...,h(t_n)$, for every n-ary function symbol f in $\Sigma$.

Similarly, functions g on $F(\Sigma)$ may be defined as follows:

- First, define g(φ), for atomic $\Sigma$-formulas φ.
- Assuming that g(φ) and g(ψ) are already defined for two $\Sigma$-formulas φ and ψ, define g(¬φ), g((φ op ψ)) and (QXφ) using g(φ) and g(ψ), for every op∈ {∧,∨,→,↔} and every quantifier Q.

Structural induction on φ and t tells us that g(φ) and h(t) are defined for every formula $\varphi$ and term t. Uniqueness of the syntax tells us that g and h are indeed well-defined functions. As an example, consider the following definition of a function ls counting the number of logical symbols occurring in a formula:

- $ls(p(t_1,...,t_n))=0$
- $ls(\neg\varphi)=ls(\varphi)+1$
- $ls((\varphi \text{ op } \psi))=ls(\varphi)+ls(\psi)+1$, for every op$\in \{\wedge,\vee,\rightarrow,\leftrightarrow\}$
- $ls(QX\varphi)=ls(\varphi)+1$, for $Q\in \{\forall,\exists\}$

## 1.4 Some important syntactical concepts

**Definition 1.7:** Given a term t and a substring s of t that is itself a term, we call s a **subterm of term** t. In the same way, the concepts **subterm of a formula** and **subformula of a formula** are defined.

As an example, consider the formula $\forall X(p(X)\wedge\exists Yq(Y))$. Its subterms are X and Y, its subformulas $\forall X(p(X)\wedge\exists Yq(Y))$, $(p(X)\wedge\exists Yq(Y))$, $p(X)$, $\exists Yq(Y)$ and q(Y).

**Definition 1.8:** For a $\Sigma$-formula ...$QX\varphi$... with subformula $QX\varphi$, we define $\varphi$ to be the **scope** of the considered quantification QX. Next we consider occurrences of variables X in a formula $\varphi$ that are not part of a substring QX (proper occurrences of X). Such an occurrence is called a **bound occurrence** of X in $\varphi$ if it lies within the scope of a quantification QX; otherwise it is called a **free occurrence** of X in $\varphi$. We define $fv(\varphi)$, the **set of free variables** of $\varphi$, as the set of all variables X such that there exists a free occurrence of X in $\varphi$.

Consider formula $\forall X(p(X)\vee(\exists Yq(Y)\wedge\exists Zr(X,Y,Z)))$. The only free variable of this formula is Y: its second occurrence is free. The scope of the first quantifier is $(p(X)\vee(\exists Yq(Y)\wedge\exists Zr(X,Y,Z)))$, of the second one q(Y) and of the third one r(X,Y,Z).

Finally, we introduce some important syntactical operations on formulas and special syntactical categories of formulas.

**Definition 1.9:** A $\Sigma$-formula is called a **closed** formula (or a **sentence**) iff it does not contain free variables. Let $F_0(\Sigma)$ denote the set of all closed $\Sigma$-formulas. For a formula $\varphi$ whose free variables are $X_1,\ldots,X_n$ (in any order) let the **universal** and **existential closure** of $\varphi$, denoted $\forall(\varphi)$ and $\exists(\varphi)$, be defined as the formulas $\forall X_1\ldots\forall X_n\varphi$ and $\exists X_1\ldots\exists X_n\varphi$. A **universal formula** is a formula of the form $\forall(\varphi)$, with a quantifier-free formula $\varphi$. An **existential formula** is a formula of the form $\exists(\varphi)$, with a quantifier-free formula $\varphi$. A **literal** is either an atomic formula $p(t_1,\ldots,t_n)$ (also called a **positive literal**) or a negated atomic formula $\neg p(t_1,\ldots,t_n)$ (also called a **negative literal**). For a literal L, let its **denial** $\sim$L be defined as follows: $\sim p(t_1,\ldots,t_n)$ is $\neg p(t_1,\ldots,t_n)$, and $\sim\neg p(t_1,\ldots,t_n)$ is $p(t_1,\ldots,t_n)$. That is, $\sim$ either introduces a new, or removes a present, negation symbol.

---

**Exercise 1.5**   Give an inductive definition of the notions subterm of a term, subterm of a formula and subformula of a formula. Define fv($\varphi$) by structural induction on $\varphi$.

---

## 1.5  Substitutions

**Definition 1.10:** A **substitution** $\sigma$ (over a signature $\Sigma$) is a finite set $\{X_1/t_1,\ldots,X_n/t_n\}$ of pairs X/t with the following properties:

- $X_1,\ldots,X_n$ are different variables
- $t_1,\ldots,t_n$ are $\Sigma$-terms
- $X_i$ is different from $t_i$, for i=1,...,n

For $\sigma$ as above, we say that $\sigma$ **acts on variables** $X_1,\ldots,X_n$. Furthermore, if the terms $t_1,\ldots,t_n$ are ground terms, we call $\sigma$ a **ground substitution**. For a substitution $\sigma$ and a variable X, let $\sigma$-$\{X/\_\}$ be the substitution which results from $\sigma$ by deleting an eventually occurring pair X/t.

**Definition 1.11:** For a substitution $\sigma=\{X_1/t_1,...,X_n/t_n\}$ and a term s, we define $s\sigma$, called the result of the **application of substitution** $\sigma$ to s, as the term that is obtained from s by simultaneously replacing all occurrences of $X_i$ in s by $t_i$, for i=1,...,n ('simultaneously' refers to all i=1,...,n as well as to all occurrences of one of the variables $X_i$). Thus:

- $X_i\sigma=t_i$, for i=1,...,n
- $Y\sigma=Y$, for every $Y\in V-\{X_1,...,X_n\}$
- $f(s_1,...,s_m)\sigma=f(s_1\sigma,...,s_m\sigma)$

For a set T of terms, define $T\sigma=\{s\sigma|\ s\in T\}$.

**Definition 1.12:** For a substitution $\sigma=\{X_1/t_1,...,X_n/t_n\}$ and a formula $\varphi$, we define $\varphi\sigma$, called the result of the **application of substitution** $\sigma$ to $\varphi$, as the formula that is obtained from $\varphi$ by simultaneously replacing all *free* occurrences of $X_i$ in $\varphi$ by $t_i$, for i=1,...,n. We call $\varphi\sigma$ a **ground instance** of $\varphi$ if $\varphi\sigma$ contains no free variables. For a set M of formulas, define $M\sigma=\{\varphi\sigma|\ \varphi\in M\}$.

As an example, consider the formula $\varphi=(\forall Xp(X)\lor\exists Zq(X,Y,Z))$ and the substitution $\sigma=\{X/Y,Y/c\}$. Then $\varphi\sigma$ is the formula $(\forall Xp(X)\lor\exists Zq(Y,a,Z))$.

---

**Exercise 1.6**    Define $\varphi\sigma$ by structural induction on $\varphi$.

---

**Definition 1.13:** For a substitution $\sigma=\{X_1/t_1,...,X_n/t_n\}$ and a formula $\varphi$, we say that $\varphi\sigma$ is **admissible** iff none of the variables of $t_i$ becomes bound after substitution of $t_i$ for the free occurrences of $X_i$, for i=1,...,n.

**Exercise 1.7**    Define admissibility of $\varphi\sigma$ by structural induction on $\varphi$.

Let us look at why there is a restriction in applying a substitution to a formula. The next chapter on semantics of predicate logic will introduce a formal understanding of all the syntactic notions introduced so far. In particular, the universal quantifier will be interpreted as a construct expressing something like 'for all objects of the domain of discourse'. This has, of course, already been anticipated here by reading the symbol $\forall$ as 'for all'. As a consequence, we intend to interpret $\forall$ in such a way that we may always conclude from $\forall X\varphi$ any formula $\varphi\{X/t\}$, for a term t. This means that we conclude any concrete instance of a formula $\varphi$ from its universal closure. Now let us consider the formula $\forall X \exists Y \neg equal(X,Y)$, with a predicate equal whose intended meaning is equality. This formula certainly expresses a true statement over every domain of discourse with at least two elements. Let us instantiate X by Y. We obtain the formula $\exists Y \neg equal(Y,Y)$ which is never true, if equal is interpreted as equality. What went wrong with the instantiation above? The free occurrence of X in the formula $\exists Y \neg equal(X,Y)$ was replaced by a term t, namely Y, one of whose variables became bound after the replacement of X by t (in our case Y itself). This is just equivalent to saying that $(\exists Y \neg equal(X,Y)\{X/Y\})$ is not admissible.

Hence, the requirement that the application of a substitution must be admissible prevents us from the above erroneous situation. Nevertheless, at the present stage of discussion we have not shown that admissibility of substitution application prevents errors in any case, nor have we shown that omission of the admissibility requirement always leads to erroneous situations. This will require a formal treatment in the next chapter. Here we are more interested in the syntactical properties of substitutions.

**Notational convention**

A term whose variables are contained in $\{X_1,...,X_n\}$ is sometimes denoted $t(X_1,...,X_n)$. Likewise, a formula whose free variables are contained in $\{X_1,...,X_n\}$ is sometimes denoted $\varphi(X_1,...,X_n)$. Given a term $t(X_1,...,X_n)$ and a substitution $\sigma=\{X_1/s_1,...,X_n/s_n\}$ we sometimes use the more comfortable notion $t(s_1,...,s_n)$ instead of the notion $t(X_1,...,X_n)\{X_1/s_1,...,X_n/s_n\}$. Likewise, for a formula $\varphi(X_1,...,X_n)$, we use the notion $\varphi(s_1,...,s_n)$ instead of the more lengthy notion $\varphi(X_1,...,X_n)\{X_1/s_1,...,X_n/s_n\}$.

We may look at a substitution $\sigma$ from different points of view. First, it is a finite set of pairs $X/t$. Second, it is a function assigning to every term $t$ a new term $t\sigma$. Third, it is a function assigning to every variable $Y$ a term $Y\sigma$. The following lemma states that these three views are equivalent.

### Lemma 1.1

For two substitutions $\sigma$ and $\rho$ over the same signature $\Sigma$ the following are equivalent:

(1) $\sigma=\rho$ (as finite sets of pairs)

(2) $t\sigma=t\rho$, for every $\Sigma$-term $t$

(3) $Y\sigma=Y\rho$, for every variable $Y$

Hence, substitutions may be considered as concrete finite representations for functions subst assigning to every variable $Y$ a term subst($Y$) with the property that for almost all variables $Y$, subst($Y$)=$Y$.

*Proof* Trivial implications are (1)$\Rightarrow$(2) and (2)$\Rightarrow$(3). Let us assume (3). We show that $\sigma\subseteq\rho$. So let a pair $X/t$ from $\sigma$ be given. Then, by definition, $X$ and $t$ are different terms and $X\sigma=t$. By (3), it follows $X\rho=t$. This implies that $X/t$ is also a pair in $\rho$ (otherwise, $X\rho$ would be $X$, that is, different from $t$). Symmetrically we obtain also $\rho\subseteq\sigma$, hence $\sigma=\rho$.    ■

**Definition 1.14:**    Let substitutions $\sigma=\{X_1/t_1,...,\ X_n/t_n\}$ and $\rho=\{Y_1/s_1,...,Y_m/s_m\}$ over the same signature be given. We define the **composition** ($\sigma\rho$) of $\sigma$ and $\rho$ as the following substitution:

$$\{X_i/t_i\rho \mid 1\leq i\leq n \text{ and } X_i\neq t_i\rho\}\cup\{Y_j/s_j \mid 1\leq j\leq m \text{ and } Y_j\notin \{X_1,...,X_n\}\}$$

### Compositionality Lemma 1.2

Let $\sigma,\rho,\tau$ be substitutions over the same signature $\Sigma$, $t$ a $\Sigma$-term and $\varphi$ a $\Sigma$-formula. Then:

(1) $t(\sigma\rho)=t\sigma\rho$

(2) If $\varphi\sigma$ and $\varphi\sigma\rho$ are admissible then $\varphi(\sigma\rho)$ is admissible.

(3) If $\varphi\sigma$ and $\varphi\sigma\rho$ are admissible then $\varphi(\sigma\rho)=\varphi\sigma\rho$.

(4) $((\sigma\rho)\tau)=(\sigma(\rho\tau))$

Take care to read (1) correctly. $t(\sigma\rho)$ is the application of the single substitution $(\sigma\rho)$ to t. t$\sigma\rho$ is the application of $\rho$ to the term t$\sigma$. Thus, (1) states that the construct $(\sigma\rho)$ indeed represents the composition of the application of $\sigma$ and $\rho$ (first $\sigma$, then $\rho$).

It should also be remarked that $\varphi(\sigma\rho)=\varphi\sigma\rho$ does not hold if we omit the requirement that $\varphi\sigma$ and $\varphi\sigma\rho$ are admissible. Furthermore, the assumption that $\varphi(\sigma\rho)$ is admissible neither suffices to conclude that $\varphi\sigma$ and $\varphi\sigma\rho$ are admissible, nor that $\varphi(\sigma\rho)=\varphi\sigma\rho$. As an example consider $\varphi=\forall Xp(X,Y)$, $\sigma=\{Y/X\}$, $\rho=\{X/a\}$. Then, $(\sigma\rho)=\{Y/a, X/a\}$, so $\varphi(\sigma\rho)=\forall X\ p(X,a)$, $\varphi\sigma=\forall X$ $p(X,X)$, $\varphi\sigma\rho=\forall X\ p(X,X)$, $\varphi(\sigma\rho)$ is admissible, but $\varphi\sigma$ is not admissible.

*Proof* Let $\sigma$ be $\{X_1/t_1,...,X_n/t_n\}$ and $\rho$ be $\{Y_1/s_1,...,Y_m/s_m\}$.
To show (1) it suffices to prove the claim for variables Z. We distinguish three cases.

| Case | $Z(\sigma\rho)$ | $Z\sigma$ | $Z\sigma\rho$ |
|---|---|---|---|
| $Z=X_i$, with $1\leq i\leq n$ | $t_i\rho$ | $t_i$ | $t_i\rho$ |
| $Z=Y_j$ with $Y_j\notin\{X_1,...,X_n\}$ | $s_j$ | $Y_j$ | $s_j$ |
| $Z\notin\{X_1,...,X_n\}\cup\{Y_1,...,Y_m\}$ | $Z$ | $Z$ | $Z$ |

To show (2) and (3) assume that $\varphi\sigma$ and $\varphi\sigma\rho$ are admissible. We show that $\varphi(\sigma\rho)$ is admissible, too, and $\varphi(\sigma\rho)=\varphi\sigma\rho$. Let us consider the pairs Z/t occurring in $(\sigma\rho)$.

We start with pairs $Y_j/s_j$ with $1\leq j\leq m$ and $Y_j\notin\{X_1,...,X_n\}$. Assume that there is a free occurrence of $Y_j$ in $\varphi$. Since $Y_j\notin\{X_1,...,X_n\}$, the considered free occurrence of $Y_j$ in $\varphi$ is also a free occurrence in $\varphi\sigma$. Since $\varphi\sigma\rho$ is admissible, no variable Z of $s_j$ is put into the scope of a quantifier QZ when replacing the considered occurrence of $Y_j$ in $\varphi\sigma$ by $s_j$. This implies that no variable Z of $s_j$ is put into the scope of a quantifier QZ when replacing the considered occurrence of $Y_j$ in $\varphi$ by $s_j$. Furthermore, application of $(\sigma\rho)$ replaces $Y_j$ by $s_j$, application of $\sigma$ does not affect $Y_j$, and the succeeding application of $\rho$ replaces $Y_j$ by $s_j$.

Next we consider pairs $X_i/t_i\rho$ with $1\leq i\leq n$ and $X_i\neq t_i\rho$. Assume that there is a free occurrence of $X_i$ in $\varphi$. Since $\varphi\sigma$ is admissible this occurrence of $X_i$ in $\varphi$ is not within the scope of a quantifier QZ, where Z is a variable of $t_i$.

Hence, the subsequent application of $\rho$ on $\varphi\sigma$ replaces $t_i$ by $t_i\rho$. Since $\varphi\sigma\rho$ is admissible, none of the variables Z in the substituted term $t_i\rho$ which are not already present in $t_i$ lies within the scope of a quantifier QZ of $\varphi$.

(4) follows immediately from Lemma 1.2 and the trivial observation that

$$Z((\sigma\rho)\tau)=Z(\sigma\rho)\tau=Z\sigma\rho\tau=Z\sigma(\rho\tau)=Z(\sigma(\rho\tau)).$$

(Are you sure that you correctly understood each one of the five expressions above?) Do not try to prove (3) by directly computing $((\sigma\rho)\tau)$ and $(\sigma(\rho\tau))$. ∎

---

**Exercise 1.8**    Show that $fv(\varphi\{X/c\})=fv(\varphi)-\{X\}$, for every constant c.

**Exercise 1.9**    Show that $\varphi\{X/t\}\{Z/c\}=\varphi\{Z/c\}\{X/t\}$, provided that all occurring substitutions are admissible, X and Z are different variables, c is a constant, and Z does not occur in term t.

**Exercise 1.10**    Show that $\varphi\{X/t\}\sigma=\varphi\{X/t\sigma\}$, provided that $\sigma$ does not act on the free variables of $\varphi$.

---

## 1.6   Unification

**Definition 1.15:**    Let T be a non-empty set of $\Sigma$-terms. A **unifier** of T is a substitution $\sigma$ such that T$\sigma$ consists of exactly one term. T is called **unifiable**. Likewise, a unifier of a non-empty set M of formulas is a substitution $\sigma$ such that $\varphi\sigma$ is admissible, for every $\varphi\in M$, and M$\sigma$ consists of exactly one formula. A unifier of two terms s and t is a unifier of the set $\{s,t\}$. Likewise, a unifier of two formulas $\varphi$ and $\psi$ is defined to be a unifier of the set $\{\varphi,\psi\}$.

**Definition 1.16:**    A unifier $\mu$ of a set of terms T or a set of formulas M is called a **most general unifier** (**mgu**) of T or M iff every other unifier $\theta$ of T or M can be obtained from $\mu$ by subsequent application of a substitution $\tau$, that is, $\theta=\mu\tau$. Most general unifiers of two terms or two formulas are defined as mgus of the corresponding sets.

As an example consider the terms $f(g(a,X),g(Y,b))$ and $f(Z,g(U,V))$. Then a unifier of these two terms is the substitution $\theta=\{X/a,Z/g(a,a),Y/U,V/b\}$. Another one is $\mu=\{Z/g(a,X),Y/U,V/b\}$. The latter is more general than the former one in the sense that $\theta$ can be obtained from $\mu$ by applying a subsequent substitution. In fact, $\mu$ is an mgu. $\theta$ can be obtained from $\mu$ as follows: $\theta=\mu\{X/a\}$. There are more mgus of both our terms. For example, $\mu'=\{Z/g(a,X),U/Y,V/b\}$. This shows that mgus are not uniquely determined. But note that $\mu$ and $\mu'$ are related as follows: $\mu=\mu'\{Y/U\}$ and $\mu'=\mu\{U/Y\}$. $\{Y/U\}$ and $\{U/Y\}$ are called renaming substitutions.

The following unification algorithm shows that mgus always exist for unifiable sets of terms.


## Unification algorithm for sets of terms

Let a finite set $T$ of $\Sigma$-terms be given. We compute a finite sequence $\mu_0,\mu_1,...,\mu_m$ of substitutions as follows (with increasing i, $\mu_i$ will unify increasingly longer prefixes of the terms in $T$, until either $T$ is unified by $\mu_m$ or it becomes clear that $T$ is not unifiable):

$\mu_0:=\varnothing$

Assume that $\mu_k$ is already computed.

Compute $T\mu_k$.

> Case 1:     $T\mu_k$ is a singleton.
>
> Stop and pass back the message '$\mu_k$ is a mgu of $T$'.
>
> Case 2:     $T\mu_k$ is not a singleton.
>
> Compute the maximal common prefix x of all the terms in $T\mu_k$.
>
> Compute the set $\Omega(T\mu_k,x)$ of all subterms of terms in $T\mu_k$ starting after the common prefix x.
>
> > Subcase 2a: $\Omega(T\mu_k,x)$ contains two terms which begin with different function symbols, or $\Omega(T\mu_k,x)$ contains a variable X and a term $r{\neq}X$ which contains X.
> >
> > Then stop with message 'T is not unifiable'.
> >
> > Subcase 2b: Not subcase 2a.
> >
> > Choose a variable X in $\Omega(T\mu_k,x)$ and a term $r{\neq}X$ in $\Omega(T\mu_k,x)$.
> >
> > Define $\mu_{k+1}=(\mu_k\{X/r\})$.

Let us consider case 2. We may conclude that x is a proper prefix of all the terms in $T\mu_k$, since otherwise (due to Theorem 1.1) $T\mu_k$ was a singleton. Furthermore, there are at least two terms in $T\mu_k$ which differ by the symbol following x. According to Theorem 1.1, the symbols following x in the terms in $T\mu_k$ are variables or function symbols. (Brackets or commas would be uniquely determined by the common prefix x, hence the same for all terms in $T\mu_k$, contradicting the maximality of x.) Finally, $\Omega(T\mu_k,x)$ consists of at least two terms.

Let us consider subcase 2b. $\Omega(T\mu_k,x)$ must contain a variable X, since otherwise, because of the definition of case 2b, all the terms in $\Omega(T\mu_k,x)$ would begin with the same function symbol. Hence, a variable X may be chosen. Since $\Omega(T\mu_k,x)$ consists of at least two terms, a term $r \neq X$ can be chosen from $\Omega(T\mu_k,x)$, too. As a result of the definition of subcase 2b, r does not contain the variable X.

### Lemma 1.3 (Termination of the unification algorithm)
If m is the number of variables occurring in T, then the maximum k such that $\mu_k$ is computed is less than or equal to m.

*Proof* First observe that all the variables occurring in $\mu_k$, $T\mu_k$ and $\Omega(T\mu_k,x)$ are variables from T. Thus, the unification algorithm does not introduce further variables besides the ones occurring in T. In the computation of $\mu_{k+1}$ from $\mu_k$, a variable X occurring in T is substituted by a term r which does not contain X. Thus, every computation of $\mu_k$ reduces the number of variables occurring in $T\mu_k$ by one. This can happen at most m times. ∎

### Lemma 1.4 (Partial correctness for non-unifiable set T)
If T is not unifiable, the unification algorithm stops with message 'T is not unifiable'.

*Proof* As shown above, the unification algorithm always stops. The result cannot come from case 1, since then T would be unifiable (by construction, $\mu_k$ would be a unifier of T). Hence termination must be in case 2a, where the correct answer is passed back. ∎

### Lemma 1.5 (The intermediate substitutions are most general)
If $\mu_k$ is computed by the unification algorithm, then every unifier θ of T

can be written as $(\mu_k \tau_k)$, for a suitable substitution $\tau_k$.

*Proof* Let a unifier $\theta$ of T be given. The lemma is proved by induction on k. The case k=0 is trivial since $\theta$ is $(\varnothing\theta)$ (take $\theta$ as $\tau_0$). Assume that $\mu_{k+1}$ is computed by the unification algorithm. By the induction hypothesis, we choose $\tau_k$ such that $\theta=(\mu_k \tau_k)$. Since $\mu_{k+1}$ was computed via subcase 2b, we know that $\mu_{k+1}=(\mu_k\{X/r\})$, for a variable X and a term r which does not contain X. Since $\theta$ is a unifier of T, we know that $(\mu_k \tau_k)$ is a unifier of T, hence $\tau_k$ is a unifier of $T\mu_k$, so $\tau_k$ is a unifier of $\Omega(T\mu_k,x)$, if x is the common prefix computed in case 2 of the k+1th step. As a consequence, $\tau_k$ unifies X and r: $X\tau_k=r\tau_k$. We define $\tau_{k+1}=\tau_k-\{X/\_\}$ and show that $\tau_k=(\{X/r\}\tau_{k+1})$. This requires the distinction of two cases.

Case 1: $X\tau_k \neq X$. Then, $\tau_k$ contains the pair $X/X\tau_k$, that is the pair $X/r\tau_k$. It follows that

$$\tau_k=$$
$$\tau_{k+1}\cup\{X/r\tau_k\}=$$
$$\{X/r\tau_k\}\cup\tau_{k+1}= \text{(since r does not contain X)}$$
$$\{X/r\tau_{k+1}\}\cup\tau_{k+1}= \text{(composition of substitutions and } X\tau_k \neq X)$$
$$(\{X/r\}\tau_{k+1})$$

Case 2: $X\tau_k=X$. Then

$$\tau_k= \text{(since } \tau_k \text{ does not contain a pair } X/\_)$$
$$\tau_{k+1}= \text{(composition of substitutions and } r\tau_{k+1}=r\tau_k=X\tau_k=X)$$
$$(\{X/r\}\tau_{k+1})$$

Using $\tau_k=(\{X/r\}\tau_{k+1})$, we conclude that :

$$\theta=(\mu_k \tau_k)=$$
$$(\mu_k(\{X/r\}\tau_{k+1}))=$$
$$((\mu_k\{X/r\})\tau_{k+1})=$$
$$(\mu_{k+1}\tau_{k+1}).$$

■

### Lemma 1.6 (Partial correctness for unifiable set T)
Let T be a unifiable set. Then the unification algorithm stops with case 1 and passes back a most general unifier $\mu_m$ of T.

*Proof* Let $\theta$ be a unifier of T. Let $\mu_m$ be the last substitution computed by the unification algorithm. By the previous lemma there is a substitution $\tau_m$ such that $\theta=(\mu_m\tau_m)$. This implies that $\tau_m$ is a unifier of $T\mu_m$, so $\tau_m$ is also a unifier of $\Omega(T\mu_m,x)$, if x is the computed common prefix of all the terms in $T\mu_m$. Hence, the unification algorithm cannot stop with case 2a, since the facts of case 2a imply that $\Omega(T\mu_m,x)$ is not unifiable. Thus, the algorithm stops with case 1 and passes back $\mu_m$ as a unifier of T. That $\mu_m$ is an mgu of T follows directly from Lemma 1.5. ∎

**Total correctness of the unification algorithm**

Given a finite set T of terms, the unification algorithm terminates and outputs the message 'T is not unifiable' in the case that T is not unifiable, and a most general unifier of T in the case that T is unifiable.

---

**Exercise 1.11** Determine the result of the unification algorithm for the following sets of terms (a and c are constants, X, Y, Z, U, V and W are variables):

(1)  $\{ f(X,g(a,Y),Z), f(h(U,c),g(U,W),k(X)), f(h(U,V),g(a,Y),W) \}$

(2)  $\{ f(X,h(Z),h(X)), f(g(a,Y),h(b),h(Y)) \}$

---

Next we treat the question of the uniqueness of most general unifiers.

**Definition 1.17:** A substitution $\sigma$ is a **variant** of a substitution $\rho$ iff there are substitutions $\tau_1$ and $\tau_2$ such that $\sigma=\rho\tau_1$ and $\rho=\sigma\tau_2$. A term s is called a variant of a term t iff there are substitutions $\tau_1$ and $\tau_2$ such that $s\tau_1=t$ and $t\tau_2=s$.

*Theorem 1.2*

Let $\mu_1$ and $\mu_2$ be most general unifiers of a non-empty set T of terms. Then $\mu_1$ is a variant of $\mu_2$. In particular, if $T\mu_1=\{t_1\}$ and $T\mu_2=\{t_2\}$, then $t_1$ is a variant of $t_2$.

*Proof*    Since $\mu_1$ is an mgu of T and $\mu_2$ a unifier of T, we can choose a substitution $\tau_1$ with $\mu_2=\mu_1\tau_1$. Likewise, we obtain a substitution $\tau_2$ with $\mu_1=\mu_2\tau_2$.    ■

The idea of variants of terms can be made more explicit. (As an exercise the reader might discuss similar questions for variants of substitutions.)

**Definition 1.18:** A **renaming** of a term t is a substitution $\sigma=\{X_1/Y_1,\ldots,X_n/Y_n\}$ with the following properties:

- $Y_1,\ldots,Y_n$ are different variables
- $X_1,\ldots,X_n$ are variables occurring in t
- For all i=1,...,n either $Y_i\in\{X_1,\ldots,X_n\}$ or $Y_i$ does not occur in t.

Thus, a renaming allows us to permute some of the variables occurring in t and give some of the other variables of t new names (variables not occurring in t so far). The intention is to preserve the syntactic structure of t. Thus it is not allowed to rename two variables occurring in t by the same name, nor is it allowed to replace a variable of t by a non-variable, nor is it allowed to give a variable $X_i$ of t a name $Y_i$ with $Y_i\notin\{X_1,\ldots,X_n\}$ and $Y_i$ occurring in t. In the last case, two variables of t would be identified in the renamed version of t which were different before.

### Theorem 1.3

(1)    If $\sigma=\{X_1/Y_1,\ldots,X_n/Y_n\}$ is a renaming of a term t and we define $\rho=\{Y_1/X_1,\ldots,Y_n/X_n\}$, then $t\sigma\rho=t$. Hence, $t\sigma$ is a variant of t.

(2)    If a term s is a variant of a term t then there exists a renaming $\sigma$ of t such that $s=t\sigma$. Likewise, there is a renaming $\rho$ of s such that $t=s\rho$.

*Proof*    To show (1) we prove that $Z\sigma\rho=Z$, for every variable Z occurring in t. This is clear for $Z\in\{X_1,\ldots,X_n\}$. $Z\in\{Y_1,\ldots,Y_n\}-\{X_1,\ldots,X_n\}$ and Z in t is impossible because of the definition renaming substitutions. The case $Z\notin\{Y_1,\ldots,Y_n\}\cup\{X_1,\ldots,X_n\}$ is trivial.

To show (2) consider a variant s of t. Choose substitutions $\sigma$ and $\rho$ such that $s=t\sigma$ and $t=s\rho$. We may assume that for all pairs X/r in $\sigma$, X is a variable of t, and for all pairs Y/r in $\rho$, Y is a variable in s. Note that $t=s\rho=t\sigma\rho$. Thus, the application of $\rho$ renders null and void the effect of $\sigma$ on t. Since for every pair $X/r$ in $\sigma$ we know that $X$ occurs in $t$, $r$ must be a variable (replacing $X$ by a constant or composed term cannot be rendered null and void). Thus, $\sigma$ is $\{X_1/Y_1,...,X_n/Y_n\}$ with variables $Y_1,...,Y_n$. Of course, $Y_1,...,Y_n$ must be different variables. If not, say if $Y_i=Y_j=Y$, for some $i{\neq}j$, we would conclude that $\sigma$ replaces the different variables $X_i$ and $X_j$ by the same variable $Y$, an effect which cannot be undone by $\rho$. For the same reasons, it is impossible that, for some $j$, $Y_j$ occurs in $t$ and $Y_j{\notin}\{Y_1,...,Y_n\}$, because the different variables $Y_j$ and $Y_j$ in $t$ would be mapped by $\sigma$ to the same variable $Y_j$. Hence, $\sigma$ is a renaming of $t$.    ∎

---

**Exercise 1.12: Unification of sets of literals**  Prove the following:

(a)  Let M be a set of atomic formulas. If M is unifiable then all formulas in M must have the same leftmost predicate symbol p. Furthermore, a most general unifier of M can be computed by the unification algorithm applied to the set T of all terms, which are obtained from the formulas from M by replacing p by a new n-ary function symbol f of the same arity as p.

(b)  Let M be a set of negated atomic formulas. Then M is unifiable iff the corresponding set M' of unnegated formulas is unifiable. Then, a most general unifier of M is a most general unifier of M', and vice versa.

---

# Chapter 2
# Semantics of Predicate Logic

In Chapter 1 we considered formulas from a *syntactical* point of view as character strings, that is, only the *form* of these strings was investigated. In this chapter we are concerned with the *meaning* of strings, so we introduce several *semantical* concepts.

Let us look at the formula $\forall X p(X,X)$. From a syntactical point of view it is only a sequence of characters. But we may as well assign to it a meaning. Anticipating the semantics of the constituents of this formula (which has already led to the suggested spelling of the different logical symbols), the formula should state the following: 'For all objects X, p(X,X) is true'. To give a formal meaning of the formula, two things are necessary:

(1)  Lay down a domain, over which variable X varies.

(2)  Lay down the meaning of predicate symbol p as a relation between elements of this domain.

We can, for example, take the set $\mathbb{N}$ of natural numbers as the domain and interpret p(X,Y) as the relation 'X is a divisor of Y'. Under this **interpretation** (sometimes called an **algebra**) our formula is obviously true. However, there are further interpretations. Consider, for example, one with the same domain, which interprets p(X,Y) as 'X is less than Y'. The above formula does not hold under this interpretation.

Of course, we may also take a domain other than $\mathbb{N}$. We could, for example, use domain {a,b,...,y,z} and interpret p(X,Y) as the relation 'letter X occurs in the alphabet before Y'.

In the following sections we shall give a formal definition of interpretations and say what it means that a formula is **valid** in some interpretation. Primarily, we are interested in **sentences**, that is, closed formulas. But since the definition of validity will be by structural induction on the syntax of formulas, we must unavoidably deal with formulas with free variables, too.

Having defined the validity of formulas within an interpretation, we may use formulas as a formal language to represent knowledge about a given domain of discourse, that is, an interpretation. From this point of view, it becomes important to know what it means to extract further knowledge from a set of formulas. This will be encapsulated by the concept of logical conclusion. So we will define what it means when a formula $\varphi$ **follows** from a set of formulas M.

Finally, we will study **normal forms** and show why such normal forms exhibit nice properties.

## 2.1    Interpretations and states

**Definition 2.1:**   For a signature $\Sigma$, a **$\Sigma$-interpretation** or **$\Sigma$-algebra** $A$ consists of:

- a non-empty set dom($A$), the so-called **domain** of $A$
- a function $f_A : \text{dom}(A)^n \rightarrow \text{dom}(A)$, for each n-ary function symbol $f \in \Sigma$
- an element $c_A \in \text{dom}(A)$, for each constant $c \in \Sigma$
- a relation $r_A \subseteq \text{dom}(A)^n$, for each n-ary predicate symbol $r \in \Sigma$
- a truth value $r_A \in \{\text{true,false}\}$, for each atom $r \in \Sigma$

**Definition 2.2:**   Let signature $\Sigma$ be a subset of signature $\Omega$. A $\Sigma$-interpretation $A$ is called the **reduct** to signature $\Sigma$ of an $\Omega$-interpretation $B$ iff dom($A$)=dom($B$), $f_A = f_B$ and $p_A = p_B$, for all function symbols f and predicate symbols p from $\Sigma$.

Then, **B** is also called an **expansion** to signature $\Omega$ of **A**. This property of interpretations is denoted by $A \triangleleft B$.

**Notational conventions**

$f_A$ and $r_A$ are called the **interpretation of** f and r within algebra **A** respectively. Sometimes, it is usual to write a 2-ary function or predicate in **infix notation** (t f t') or t f t' instead of the prescribed prefix notation f(t,t'), for example addition on natural numbers or the relation < on an ordered domain. In this case, we allow the corresponding function or predicate symbol to be written in infix notation, too.

If a signature $\Sigma$ is given by a listing ..., $f:s^n \rightarrow s$, ..., $r:s^m$, ... of its function and predicate symbols then we usually denote a $\Sigma$-interpretation **A** by its domain and corresponding listing of interpretations of the symbols: $(\text{dom}(A),...,f_A,...,r_A,...)$.

Usually, we will be interested in a fixed signature $\Sigma$ and lots of $\Sigma$-algebras **A**. In some cases, there will be only one fixed $\Sigma$-algebra **A** that is investigated. In this case we sometimes use the same symbol f to denote a function symbol as well as its interpretation within **A**.

An expansion of **A** by functions f,g,h... and relations p,q,r,... is often denoted $(A,f,g,h, ...,p,q,r, ...)$.

> **Definition 2.3:** Let **A** be a $\Sigma$-interpretation. A **state** over **A** is a function sta from the set of all variables into dom(**A**). If sta is a state over **A**, $X_1,...,X_n$ are different variables and $a_1,...,a_n \in \text{dom}(A)$, then $\text{sta}(X_1/a_1,...,X_n/a_n)$ is the **modified state** sta' over **A** defined by $\text{sta}'(X_k)=a_k$, for k=1,...,n, and sta'(Y)=sta(Y), for $Y \notin \{X_1,...,X_n\}$ (that is, $\text{sta}(X_1/a_1,...,X_n/a_n)$ is the state obtained from sta by assigning to variables $X_1,...,X_n$ new values $a_1,...,a_n$ and leaving the rest unchanged).

## 2.2    Evaluation of terms

States provide the necessary information for assigning terms a value.

> **Definition 2.4:**    Given a $\Sigma$-interpretation **A**, a state sta over **A** and a $\Sigma$-term t, then we define the value of t in algebra **A** and state sta, briefly denoted by $\text{val}_{A,\text{sta}}(t)$, by structural induction on t as follows:

- $\mathrm{val}_{A,sta}(X)=sta(X)$
- $\mathrm{val}_{A,sta}(f(t_1,\ldots,t_n))=f_A(\mathrm{val}_{A,sta}(t_1),\ldots,\mathrm{val}_{A,sta}(t_n))$

for every variable X, n-ary function symbol f and $\Sigma$-terms $t_1,\ldots,t_n$. In particular, for a constant c we obtain $\mathrm{val}_{A,sta}(c)=c_A$.

As an example, consider the signature $\Sigma_{nat}=\{0{:}{\to}s,\ succ{:}s{\to}s,\ +{:}s^2{\to}s,\ *{:}s^2{\to}s\}$ and $\Sigma_{nat}$-algebra $A{=}Nat$ with the set of natural numbers $\mathbb{N}$ as domain, number 0 as interpretation of constant 0, and successor function, addition and multiplication as interpretation of function symbols succ, + and *. Let sta be a state over $Nat$ with sta(X)=3 and sta(Y)=2, and t be the term $(succ(X)+Y)*succ(succ(0))$. Then:

$$\mathrm{val}_{A,sta}(t)$$
$$=\mathrm{val}_{A,sta}((succ(X)+Y)*succ(succ(0)))$$
$$=\mathrm{val}_{A,sta}((succ(X)+Y)) *_A \mathrm{val}_{A,sta}(succ(succ(0)))$$
$$=(\mathrm{val}_{A,sta}((succ(X)) +_A \mathrm{val}_{A,sta}(Y)) *_A succ_A(\mathrm{val}_{A,sta}(succ(0)))$$
$$=(succ_A(\mathrm{val}_{A,sta}(X)) +_A sta(Y)) *_A succ_A(succ_A(\mathrm{val}_{A,sta}(0)))$$
$$=(succ_A(sta(X)) +_A sta(Y)) *_A succ_A(succ_A(0_A))$$
$$=(succ_A(3) +_A 2) *_A succ_A(succ_A(0))$$
$$=(4+_A 2) *_A 2$$
$$=6 *_A 2$$
$$=12$$

In this example, $\mathrm{val}_{A,sta}(t)$ depended only on sta(X) and sta(Y), that is, the values of sta for the variables occurring in t. Obviously, this can be generalized as follows.

### Coincidence Lemma 2.1 (for terms)

Given a $\Sigma$-interpretation $A$, states sta and sta' over $A$ and a $\Sigma$-term t, if sta(X)=sta'(X), for every variable X occurring in t, then $\mathrm{val}_{A,sta}(t)$ $=\mathrm{val}_{A,sta'}(t)$.

*Proof* Simple structural induction on t.                                   ■

## Notational convention

By $t(X_1,...,X_m)$ we denote a term with at most variables $X_1,...,X_m$ occurring in it. Given such a term $t(X_1,...,X_m)$, an algebra $A$ and $a_1,...,a_m \in \text{dom}(A)$, we write $t_A(a_1,...,a_m)$ instead of $\text{val}_{A,\text{sta}}(t(X_1,...,X_m))$, where sta is an arbitrary state over $A$ with $\text{sta}(X_1)=a_1,...,\text{sta}(X_m)=a_m$. By the Coincidence Lemma 2.1, $\text{val}_{A,\text{sta}}(t(X_1,...,X_m))$ depends indeed only on $\text{sta}(X_1),...,\text{sta}(X_m)$, hence the introduced notion makes sense. If t is a ground term, then we write $t_A$ instead of $\text{val}_{A,\text{sta}}(t)$ for an arbitrary state sta, whose choice has no effect on $\text{val}_{A,\text{sta}}(t)$.

Next, we deal with a technical lemma relating term evaluation with the notion of substitution. To motivate this lemma, consider a term t with lots of occurrences of a variable X. Now evaluate $t\{X/r\}$, for a further term r, in an algebra $A$ and state sta. What happens? The inductive definition of term evaluation is applied and, whenever we arrive at a subterm r of $t\{X/r\}$, the value of term r in algebra $A$ and state sta must be computed. This procedure is rather uneconomical (think of thousands of occurrences of X in t and a long subterm r). A better way to evaluate $t\{X/r\}$ would be to evaluate term r in algebra $A$ and state sta, leading to value $a$, storing this value $a$ within the state by changing sta into $\text{sta}(X/a)$, and then evaluating the original term t in algebra $A$ and the modified state $\text{sta}(X/a)$. Obviously, the result will be the same as the one obtained above.

### Substitution Lemma 2.2 (for terms)

Let $A$ be a $\Sigma$-interpretation, sta a state over $A$, X a variable and  t and r $\Sigma$-terms. If $\text{val}_{A,\text{sta}}(r)=a$ then $\text{val}_{A,\text{sta}}(t\{X/r\})= \text{val}_{A,\text{sta}(X/a)}(t)$.

*Proof*  By structural induction on term t, if t coincides with variable X, then

$$\text{val}_{A,\text{sta}}(t\{X/r\})=\text{val}_{A,\text{sta}}(r)=a=\text{val}_{A,\text{sta}(X/a)}(X).$$

If t is a variable Y different from X, then

$$\text{val}_{A,\text{sta}}(t\{X/r\})=\text{val}_{A,\text{sta}}(Y)=\text{sta}(Y)=\text{sta}(X/a)(Y)=\text{val}_{A,\text{sta}(X/a)}(Y).$$

If t is a composite term $f(t_1,...,t_n)$, then

$$\text{val}_{A,\text{sta}}(t\{X/r\})=\text{val}_{A,\text{sta}}(f(t_1\{X/r\},...,t_n\{X/r\}))$$
$$=f_A(\text{val}_{A,\text{sta}}(t_1\{X/r\}),...,\text{val}_{A,\text{sta}}(t_n\{X/r\}))$$
$$=f_A(\text{val}_{A,\text{sta}(X/a)}(t_1),...,\text{val}_{A,\text{sta}(X/a)}(t_n))=\text{val}_{A,\text{sta}(X/a)}(f(t_1,...,t_n))$$

∎

## 2.3    Validity of formulas

**Definition 2.5:**    Let $A$ be a $\Sigma$-interpretation. For a $\Sigma$-formula $\varphi$ and a state sta over $A$ we define, when $\varphi$ is **valid in algebra $A$ in state** sta, denoted by $A \models_{sta}\varphi$:

- $A \models_{sta}p(t_1,...,t_n)$ iff $(val_{A,sta}(t_1),...,val_{A,sta}(t_n))\in p_A$, for every n-ary predicate symbol p in $\Sigma$ with n>0

- $A \models_{sta}p$ iff $p_A$=true, for every atom p in $\Sigma$

- $A \models_{sta}\neg\varphi$ iff $A \not\models_{sta}\varphi$

- $A \models_{sta}(\varphi\wedge\psi)$ iff $A \models_{sta}\varphi$ and $A \models_{sta}\psi$

- $A \models_{sta}(\varphi\vee\psi)$ iff $A \models_{sta}\varphi$ or $A \models_{sta}\psi$

- $A \models_{sta}(\varphi\rightarrow\psi)$ iff $A \not\models_{sta}\varphi$ or $A \models_{sta}\psi$

- $A \models_{sta}(\varphi\leftrightarrow\psi)$ iff

    $A \models_{sta}\varphi$ and $A \models_{sta}\psi$, or $A \not\models_{sta}\varphi$ and $A \not\models_{sta}\psi$

- $A \models_{sta}\forall X\varphi$ iff $A \models_{sta(X/a)}\varphi$, for all $a\in dom(A)$

- $A \models_{sta}\exists X\varphi$ iff there is an $a\in dom(A)$ such that $A \models_{sta(X/a)}\varphi$

For a set M of $\Sigma$-formulas we define $A \models_{sta}M$ iff $A \models_{sta}\varphi$, for every formula $\varphi$ in M.

As an example, consider the expansion $A=(Nat,<,=)$ of the algebra *Nat* used in Section 2.2 by the less-than relation $<$ and the equality relation $=$ on $\mathbb{N}$ (in an extended signature with 2-ary, infixedly written relation symbols $<$ and $=$). Consider a state sta over $A$ and the formula $\varphi=\forall X\exists Y((Y*Z<X\vee Y*Z=X)\wedge X<succ(Y)*Z)$. What does $A \models_{sta}\varphi$ mean? We expect that validity of $\varphi$ in algebra $A$ and state sta depends only on sta(Z), since Z is the only free variable in $\varphi$. Let us see that this is indeed so.

$A \models_{sta}\varphi$

$\Leftrightarrow$ *(interpretation of $\forall X$)*

*For every* $x\in \mathbb{N}, A \models_{sta(X/x)}\exists Y ((Y*Z<X \vee Y*Z=X) \wedge X<succ(Y)*Z)$

$\Leftrightarrow$ *(interpretation of $\exists Y$)*

For every $x\in \mathbb{N}$ *there is some* $y\in \mathbb{N}$ with

$A \models_{\text{sta}(X/x)(Y/y)} ((Y*Z<X \lor Y*Z=X) \land X<\text{succ}(Y)*Z)$

$\Leftrightarrow$ (*interpretation of* $\land$)

For every $x \in \mathbb{N}$ there is some $y \in \mathbb{N}$ with $A \models_{\text{sta}(X/x)(Y/y)}(Y*Z<X \lor$

$Y*Z=X)$ *and* $A \models_{\text{sta}(X/x)(Y/y)} X<\text{succ}(Y)*Z$

$\Leftrightarrow$ (*interpretation of* $\lor$)

For every $x \in \mathbb{N}$ there is some $y \in \mathbb{N}$ with $A \models_{\text{sta}(X/x)(Y/y)} Y*Z<X$ *or*

$A \models_{\text{sta}(X/x)(Y/y)} Y*Z=X$, *and* $A \models_{\text{sta}(X/x)(Y/y)} X<\text{succ}(Y)*Z$

(Note that, concerning the bracketing of 'and' and 'or', we are already having some difficulty in stating clearly how the informal statement above is to be understood. This difficulty does not exist within our formal sentences.)

$\Leftrightarrow$ (*evaluation of terms and predicate symbols*)

For every $x \in \mathbb{N}$ there is some $y \in \mathbb{N}$ with $y*sta(Z)<x$ *or* $y*sta(Z)=x$, and $x<(y+1)*sta(Z)$

$\Leftrightarrow$

For every $x \in \mathbb{N}$ there is some $y \in \mathbb{N}$ with $y*sta(Z) \leq x < (y+1)*sta(Z)$

$\Leftrightarrow$

For every $x \in \mathbb{N}$ there is a least $y \in \mathbb{N}$ with $x<(y+1)*sta(Z)$.

So, finally we arrive at a number theoretic statement about sta(Z). Note that, after the first five equivalences, an informal statement was reached which mirrored the given predicate logic sentence directly. As was predicted, $A \models_{\text{sta}} \varphi$ depends only on sta(Z). For sta(Z)=0 the statement is false, for sta(Z)>0 it is true.

### Coincidence Lemma 2.3 (*for formulas*)

Given a $\Sigma$-interpretation $A$, states sta and sta' over $A$ and a $\Sigma$-formula $\varphi$, if sta(X)=sta'(X) for every free variable X of $\varphi$, then $A \models_{\text{sta}} \varphi$ iff $A \models_{\text{sta}'} \varphi$.

*Proof*   The statement of the lemma is shown by structural induction on $\varphi$. The induction base concerning atomic formulas follows immediately from the semantics of atomic formulas and Lemma 2.1. Induction passes trivially over the propositional connectives (this is left as an exercise for the reader). We are left with the quantifiers. Assume that $\varphi$ is $\forall Z \psi$. Then we conclude:

$A \models_{\text{sta}} \varphi$

$\Leftrightarrow$ (*interpretation of* $\forall Z$)

For every $a \in dom(A)$, $A \models_{sta(Z/a)} \psi$

$\Leftrightarrow$ (*induction hypothesis*: note that $sta(Z/a)(X)=sta'(Z/a)(X)$, for all free variables X of $\psi$)

For every $a \in dom(A)$, $A \models_{sta'(Z/a)} \psi$

$\Leftrightarrow$

$A \models_{sta'} \varphi$

The existential quantifier is treated analogously.    ∎

## Notational convention

By $\varphi(X_1,...,X_m)$ we denote a formula all of whose free variables are contained in $\{X_1,...,X_m\}$. By Lemma 2.3, this notation makes sense. Given such a formula $\varphi(X_1,...,X_m)$, an algebra $A$ and $a_1,...,a_m \in dom(A)$, we write $A \models \varphi(a_1,...,a_m)$ instead of $A \models_{sta}\varphi$, where sta is an arbitrary state over $A$ with $sta(X_1)=a_1,...,sta(X_m)=a_m$.

### Substitution Lemma 2.4 (for formulas)

Given a $\Sigma$-interpretation $A$, a state sta over $A$, a variable X, a $\Sigma$-term r and a formula $\varphi$ such that $\varphi\{X/r\}$ is admissible, define $a=val_{A,sta}(r)$.

Then $A \models_{sta}\varphi\{X/r\}$ iff $A \models_{sta(X/a)}\varphi$.

*Proof*    First of all, we deal with the special case that variable X does not occur free in $\varphi$. Since in this case $\varphi\{X/t\}$ coincides with $\varphi$ and $A \models_{sta(X/a)}\varphi$ is equivalent to $A \models_{sta}\varphi$ by Lemma 2.3, the statement of Lemma 2.4 follows. So, let us assume that X occurs free in $\varphi$. Lemma 2.4 is shown by structural induction on $\varphi$. The induction base, $\varphi$ is an atomic formula, immediately follows from Lemma 2.2. Induction then passes over the propositional connectives in a trivial way (as it usually does for these connectives). The only interesting cases are the quantifiers. We treat the universal quantifier. Let $\varphi$ be the formula $\forall Z\psi$. Since it was assumed that X occurs free in $\varphi$, it follows that X is different from Z and occurs free in $\psi$. Also, as it was assumed that $\varphi\{X/r\}$ is admissible, it follows that Z does not occur in term r (see Definition 1.13). This implies the following three facts:

(1)    $\psi\{X/r\}$ is admissible, too (see Definition 1.13, use $Z \neq X$).

(2)    $\varphi\{X/r\}=(\forall Z\psi\{X/r\})=\forall Z(\psi\{X/r\})$ (because $Z \neq X$).

(3)    $a=val_{A,sta}(r)=val_{A,sta(Z/b)}(r)$, for arbitrary $b \in dom(A)$.

Now we may conclude:

$A \models_{sta} \varphi\{X/r\}$

$\Leftrightarrow$ (use fact (2))

$A \models_{sta} \forall Z(\psi\{X/r\})$

$\Leftrightarrow$ (interpretation of $\forall Z$)

For every $b \in dom(A)$, $A \models_{sta(Z/b)} \psi\{X/r\}$

$\Leftrightarrow$ (use (3) and induction hypothesis with $sta(Z/b)$ instead of $sta$)

For every $b \in dom(A)$, $A \models_{sta(Z/b)(X/a)} \psi$

$\Leftrightarrow$ (since $Z \neq X$)

For every $b \in dom(A)$, $A \models_{sta(X/a)(Z/b)} \psi$

$\Leftrightarrow$

$A \models_{sta(X/a)} \forall Z \psi$     ∎

### Corollary

If $\varphi\{X/r\}$ is admissible then, for every $\Sigma$-interpretation A and state sta over $A$, $A \models_{sta} (\forall X \varphi \rightarrow \varphi\{X/r\})$. Likewise, $A \models_{sta} (\varphi\{X/r\} \rightarrow \exists X \varphi)$. If $\varphi$ is a universal formula then the conclusion of the corollary is true without the admissibility assumption on $\varphi\{X/r\}$. (This latter statement will be of importance in later chapters on resolution and logic programming.)

*Proof* Assume that $A \models_{sta} \forall X \varphi$. This means that $A \models_{sta(X/b)} \varphi$ for every $b \in dom(A)$. Define $a = val_{A,sta}(r)$. Thus, $A \models_{sta(X/a)} \varphi$. By Lemma 2.4 we may conclude that $A \models_{sta} \varphi\{X/r\}$. The second validity follows from the fact that $(\varphi\{X/r\} \rightarrow \exists X \varphi)$ is equivalent to $(\neg \exists X \varphi \rightarrow \neg \varphi\{X/r\})$, and hence to the formula $(\forall X \neg \varphi \rightarrow \neg \varphi\{X/r\})$.

For a universal formula $\varphi = \forall X_1 ... \forall X_n \psi$, with quantifier-free formula $\psi$, we argue as follows (not using Lemma 2.4 which requires admissibility of the occurring substitution): we may assume that X is different from all $X_1, ..., X_n$.

Evaluating all universal quantifiers, $A \models_{sta} \forall X \varphi$ implies:

$A \models_{sta(X/b)(X_1/b_1)...(X_n/b_n)} \psi$, for all elements $b, b_1, ..., b_n$ of $dom(A)$

$\Rightarrow$

$A \models_{sta(X_1/b_1)...(X_n/b_n)(X/b)} \psi$, for $b = val_{A,sta}(r)$ and all $b_1, ..., b_n \in dom(A)$

$\Rightarrow$

$A \models_{sta(X_1/b_1)...(X_n/b_n)} \psi\{X/r\}$ for all $b_1,...,b_n \in dom(A)$ (note that $\psi\{X/r\}$ is admissible since $\psi$ is quantifier-free)

$\Rightarrow$

$A \models_{sta} \varphi\{X/r\}$                                                                                     ∎

The corollary above is not true if we leave out the admissibility condition for the substitution $\{X/r\}$, so Lemma 2.4 is not true either. An example is provided by the formula $\varphi = \exists Y p(Y,X)$ and the substitution $\{X/Y\}$. Note that $\varphi\{X/Y\}$ is not admissible. The corresponding implication considered in the corollary is the formula $(\forall X \exists Y p(X,Y) \rightarrow \exists Y p(Y,Y))$. Now consider the interpretation $(\mathbb{N},<)$, with the less-than relation $<$ on $\mathbb{N}$, and an arbitrary state sta (which has no effect on the validity of our formula as it is a closed one). Then $A \models_{sta} (\forall X \varphi \rightarrow \varphi\{X/r\})$ means: if for every natural number x there is a greater number y, then there is a number y which is greater than itself. Certainly a false statement.

---

**Exercise 2.1**    Generalize the substitution lemma to arbitrary substitutions. Conclude that $A \models_{sta} (\forall X \varphi \rightarrow \varphi \sigma)$ whenever $\varphi \sigma$ is admissible.

---

## 2.4    Models and semantical conclusion

**Definition 2.6:**    Let $A$ be a $\Sigma$-interpretation, $\varphi$ a $\Sigma$-formula and M a set of $\Sigma$-formulas. We say that $A$ is a **model** of $\varphi$ (denoted by $A \models \varphi$) iff $A \models_{sta} \varphi$ for all states sta over $A$. $A$ is called a model of M, denoted $A \models M$, if it is a model of every formula in M.

*Lemma 2.5*

$A$ is a model of a formula $\varphi$ iff $A$ is a model of $\forall(\varphi)$. $A$ is a model of a set of formulas M iff $A$ is a model of $\forall(M)$.

*Proof* Let $X_1,...,X_n$ be the variables occurring free in $\varphi$. Then, $\forall(\varphi)$ is the formula $\forall X_1...\forall X_n\varphi$, and we obtain: $A \models \forall(\varphi)$ iff (see Definition 2.6) for every state sta over $A$, $A \models_{sta} \forall X_1...\forall X_n\varphi$ iff for every state sta over $A$ and all $a_1,...,a_n \in dom(A)$, $A \models_{sta(X_1/a_1)...(X_n/a_n)}\varphi$ iff for every state sta over $A$, $A \models_{sta}\varphi$ iff $A \models \varphi$. ∎

It should be remembered that the definition of $A \models \varphi$ involves an implicit universal quantification of the free variables of $\varphi$. As a consequence, several properties true for the validity relation $\models_{sta}$ do not hold for the model relation $\models$. For example, $A \models_{sta}(\varphi\vee\psi)$ is equivalent, by definition, to $A \models_{sta}\varphi$ or $A \models_{sta}\psi$. In contrast, $A \models (\varphi\vee\psi)$ does not always imply that $A \models \varphi$ or $A \models \psi$. As an example, consider $A \models (p(X)\vee\neg p(X))$ which is true for arbitrary interpretation $A$, whereas neither $A \models p(X)$ nor $A \models \neg p(X)$ are true, for the interpretation $A$ with $dom(A)=\mathbb{N}$ and $p_A=$the set of all even numbers. Likewise, $A \models_{sta}\neg\varphi$ is equivalent to $A \not\models_{sta}\varphi$, whereas $A \not\models \varphi$ does not imply that $A \models \neg\varphi$. Simply consider $\varphi=p(X)$.

### Lemma 2.6

If $A \models \varphi$ and $A \models (\varphi\rightarrow\psi)$ then $A \models \psi$.

*Proof*    Assume that $A \models \varphi$ and $A \models (\varphi\rightarrow\psi)$. This means that $A \models_{sta}\varphi$ and $A \models_{sta}(\varphi\rightarrow\psi)$, for every state sta over $A$. By definition of $\models_{sta}$ (subcase $\rightarrow$) it follows that $A \models_{sta}\psi$, for every state sta over $A$, so $A \models \psi$. ∎

Let us now discuss a couple of axiom systems M which, via the model relation $\models$, encode knowledge about several domains of discourse, that is, about interpretations.

(1)    The first algebra we consider is the expansion $(Nat,<,=)$ introduced in Section 2.3.

| Formula φ | What $(Nat,<,=) \models \varphi$ expresses |
|---|---|
| $\neg succ(X)=0$ | 0 is not the successor of any natural number |
| $succ(X)=succ(Y)\rightarrow X=Y$ | succ is injective |
| $\forall X(\neg X=0 \rightarrow \exists Y\ succ(Y)=X)$ | every number $\neq 0$ has a predecessor |
| $X+0=X$<br>$X+succ(Y)=succ(X+Y)$ | usual recursive definition of + |
| $X*0=0$<br>$X*succ(Y)=(X*Y)+X$ | usual recursive definition of * |
| $\forall X\ \exists Y\ Y*Y=X$ | every natural number is a square, a false statement |
| $\forall X\ \exists Y\ X*X=Y$ | squares of natural numbers are natural numbers |
| find a formula | the gaps between squares become arbitrarily large |
| find a formula | every natural number is the sum of four squares |

(2)    Next we consider a signature $\Sigma_{group}=\{e:\rightarrow s,\ \circ:s^2\rightarrow s$ (infix notation), $i:s\rightarrow s,\ =:s^2\}$. Let $G=(G,e,\circ,i,=_G)$ be a $\Sigma_{group}$-algebra with equality $=_G$ as interpretation of relation symbol =.

| Formula φ | What $G \models \varphi$ expresses |
|---|---|
| $X\circ(Y\circ Z)=(X\circ Y)\circ Z$<br>$e\circ X=X$ | $\circ$ is associative<br>e is a left neutral element w.r.t. $\circ$ |

| | |
|---|---|
| $i(X) \circ X = e$ | $i(X)$ is a left inverse of $X$ w.r.t. $\circ$ |
| the 3 formulas above together | $G$ is a group |

(3)     Now consider the signature $\Sigma_{ord} = \{<:s^2, =:s^2\}$ with infix relation symbols $<$ and $+$. Let $Ord = (A, <, =_A)$ be a $\Sigma_{ord}$-algebra with equality $=_A$ as interpretation of relation symbol $=$.

| *Formula* $\varphi$ | *What Ord* $\models \varphi$ *expresses* |
|---|---|
| $(X<Y \wedge Y<Z) \rightarrow X<Z$ | $<$ is a transitive relation |
| $\neg X<X$ | $<$ is an irreflexive relation |
| the 2 formulas above and | |
| $X<Y \vee (X=Y \vee Y<X)$ | *Ord* is a linear ordering |

(4)     Finally, we consider the signature $\Sigma_{real} = \{+:s^2 \rightarrow s, \; -:s^2 \rightarrow s, \; *:s^2 \rightarrow s, \; <:s^2, \; =:s^2\}$. Let *Real* be the $\Sigma_{real}$-algebra with the set of real numbers as domain, addition, subtraction and multiplication as interpretations of $+,-$ and $*$, and less-than and equality relations on real numbers as interpretations of $<$ and $=$.

| *Formula* $\varphi$ | *What Real* $\models \varphi$ *expresses* |
|---|---|
| $(\exists X \; t(X)<0 \wedge \exists X \; 0<t(X)) \rightarrow \exists X \; t(X)=0$ for a term $t$ with exactly one variable $X$ | a polynomial (represented by term t) which takes both positive and negative values has a zero |

Now, we come to one of the most important notions in logic, the concept of **semantical conclusion**.

**Definition 2.7:**   A formula φ **follows from a set of formulas** M (or φ is a **logical consequence** of M, or M **entails** φ), briefly denoted by M $\models$ φ, iff every model $A$ of M is also a model of φ.

Concerning the definition of **logical conclusion**, several comments should be made.

(1)    As usual in the literature, the symbol $\models$ is used to denote two different concepts, namely the model relation $A \models$ φ, for an interpretation $A$, as well as logical conclusion M $\models$ φ, for a set of formulas M.

(2)    Drawing conclusions from an axiom set M intuitively means to *extract further knowledge* implicitly present in the knowledge that is *represented* by the axiom system M. This view of logical conclusion as knowledge extraction gives a hint as to why this notion plays such a central role. Extracting knowledge from an axiom system may be simple in some cases, but rather sophisticated in others. The former call for automatization of the process of knowledge extraction (as will be achieved later with logic programming), the latter call for clarification of the difficulty of knowledge extraction in general (this will be answered by Church's undecidability result for logical conclusion). Below we will treat some examples. The reader must always be cautious and use only the information present in an axiom system Ax to draw conclusions from Ax, and not information that is erroneously suggested by some mnemotechnical notations.

(3)    From the definition of the model relation we obtain the following simple, but important fact: M $\models$ φ iff $\forall$(M) $\models$ $\forall$(φ), that is, in the process of knowledge extraction both the axiom system and the considered formula must be thought of as being implicitly universally quantified.

(4)    There is an alternative notion of semantical conclusion sometimes used in the literature. This reads as follows. A formula φ follows from a set of formulas M (briefly denoted by M $\models_{alt}$ φ) iff, for every interpretation $A$ and state sta over $A$, the following is true:

$$\text{If } A \models_{sta} \psi, \text{ for every } \psi \in M, \text{ then } A \models_{sta} \varphi.$$

Note the difference in our concept of semantical conclusion in the quantification over states. It is mainly a matter of taste which notion is preferred. Since it is usual in several fields to save universal quantifiers in axiom systems (in particular in larger specifications of abstract data types to be studied later) we adopt our proposed notion. A second reason is that the proof theory (discussed in Chapter 3) is somehow more easily developed with our notion of semantical conclusion. There are obvious correlations between these two concepts, showing that they can be mutually simulated.

---

**Exercise 2.2**     Prove the following statements:

(a)     If $M \models_{\text{alt}} \varphi$ then $M \models \varphi$.

(b)     $\{p(X)\} \models \forall X\, p(X)$, but not $\{p(X)\} \models_{\text{alt}} \forall X\, p(X)$.

(c)     If M consists of closed formulas, then $M \models_{\text{alt}} \varphi$ iff $M \models \varphi$.

(d)     $M \models \varphi$ iff $\forall(M) \models_{\text{alt}} \varphi$.

(e)     $\{\psi_1, \psi_2, \ldots, \psi_n\} \models_{\text{alt}} \varphi$ iff $\models (\psi_1 \to (\psi_2 \to (\ldots \to (\psi_n \to \varphi)) \ldots))$.

---

We now discuss some examples for logical conclusion.

(1)     First, we formalize addition on natural numbers. To avoid the use of equality we use a 3-ary **predicate symbol** add instead of a 2-ary function symbol +. Besides add, we use a constant 0 and an 1-ary function symbol succ. The intended meaning of add should be: $\text{add}(\text{succ}^n(0), \text{succ}^m(0), \text{succ}^k(0))$ iff k=n+m.
        We propose the following axiom system Add:

> (Add1)          add(X,0,X)
> (Add2)          $(\text{add}(X,Y,Z) \to \text{add}(X,\text{succ}(Y),\text{succ}(Z)))$

This axiom system formalizes the usual recursive definition of addition. The following formulas follow from Add, for all numbers n and m:

> $\text{add}(\text{succ}^n(0), \text{succ}^m(0), \text{succ}^{n+m}(0))$
> $\text{add}(X, \text{succ}^m(0), \text{succ}^m(X))$

This is easily shown by induction on m. Up to now we could have also used our intuition on numbers. But the next examples also show that we must be very careful not to conclude - tempted by our intentions and a suggestive notation - information which is not present in an axiom system. Consider the formulas:

> $\text{add}(\text{succ}^n(0), Y, \text{succ}^n(Y))$
> $(\text{add}(X,Y,Z) \to \text{add}(Y,X,Z))$

Neither one of them follows from Add. To see this, consider the following interpretation $A$:

> $\text{dom}(A) = \mathbb{N} \cup \{-1\}$

$0_A$=the natural number 0

$\text{succ}_A$=the successor function on $\mathbb{N}\cup\{-1\}$

$\text{add}_A=\{\text{add}(\text{succ}^n(0),\text{succ}^m(0),\text{succ}^{n+m}(0)) \mid m\in \mathbb{N} \text{ and } n\in \mathbb{N}\}$

$A$ is a model of Add, but not a model of the proposed two formulas.

(2)    Secondly, consider the axiom system M describing the rules of the game introduced in Section 1.1. It consists of the following five formulas:

loses(0)

(loses(X)→wins(succ(X)))

(loses(X)→wins($\text{succ}^2$(X)))

(loses(X)→wins($\text{succ}^3$(X)))

((wins(X)∧wins(succ(X))∧wins($\text{succ}^2$(X)))
        →loses($\text{succ}^3$(X)) )

Starting with the formula loses(0) and cyclically applying the four implications, we may successively conclude from M the formulas loses(0), wins(succ(0)), wins($\text{succ}^2$(0)), wins($\text{succ}^3$(0)), loses($\text{succ}^4$(0)), and so on. Let us consider the formula ¬wins(0). Despite our intuition that winning means that the opponent loses, the formula ¬wins(0) does not follow from M. This is easily seen by taking the interpretation $A$ with dom($A$)=$\mathbb{N}$, $0_A$=the number 0, $\text{succ}_A$=the successor function, $\text{wins}_A=\mathbb{N}$ and $\text{loses}_A=\mathbb{N}$. This provides us with a model of M which is not a model of ¬wins(0). In the algebra $A$ just constructed, we defined dom($A$), $0_A$ and $\text{succ}_A$ according to the suggestive reading of the symbols 0 and succ. But note that M does not contain information that the domain of a model of M and the interpretations of 0 and succ are in any way related to the algebra of natural numbers.

Extracting information of the form wins($\text{succ}^n$(0)) and loses($\text{succ}^n$(0)), for formulas which indeed follow from the considered axiom system, is a quite simple task, since a complete goal-oriented (top-down) **reduction procedure** starting with wins($\text{succ}^n$(0)) can be readily realized (see Figure 2.1). Though the search space for the proposed reduction procedure is conceptually simple, it is exponentially growing with n. A better solution, applicable for our given problem, would be a forward-oriented (bottom-up) **deduction procedure** generating all conclusions from the given axiom loses(0) via applications of the implications. Since the numbers i occurring in the subterms $\text{succ}^i$(0) grow monotonically, we may stop the procedure when i=n is reached.

The situation is illustrated in Figure 2.1. In the figure, we abbreviate $\text{succ}^n$(0) by n, wins(X) by w(X) and loses(X) by l(X).

*deduction tree (bottom-up derivation, forward chaining)*

$$1(0) \longrightarrow w(1) \longrightarrow w(2) \longrightarrow w(3) \longrightarrow 1(4) \longrightarrow w(5) \longrightarrow w(6)$$

*reduction tree:*
*(top-down reduction, goal-oriented reduction, backward chaining)*

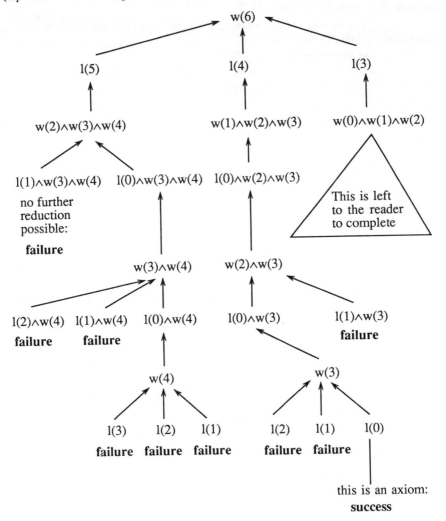

**Figure 2.1**

This final example serves the purpose of showing that the process of information retraction may be quite delicate, indicating limits of automatization. The example is taken from group theory. As already stated, the following is an adequate axiomatization of group theory (provided that the relation symbol = is always fixedly interpreted as the equality relation on the domain of the algebra under consideration):

(Gr1)$(X \circ (Y \circ Z)) = ((X \circ Y) \circ Z)$
(Gr2)$(e \circ X) = X$
(Gr3)  $(i(X) \circ X) = e$

As is well known from group theory, the left-neutral element in a group is always right-neutral, too. Thus, we might try to conclude the formula $X \circ e = X$ from (Gr1)-(Gr3). There is an obvious difficulty concerning the equality symbol. There are no formulas stating that = has to be interpreted as the equality on the domain of a given model of (Gr1)-(Gr3). Note that our concept of a model of (Gr1)-(Gr3) does not prescribe that = is to be fixedly interpreted as equality. Indeed, from (Gr1)-(Gr3) alone, our formula $X \circ e = X$ does not follow (see Exercise 2.3). What is needed are the following 'axioms for equality':

(Eq1)  $X = X$                                      (reflexivity)
(Eq2)  $(X = Y \rightarrow Y = X)$                    (symmetry)
(Eq3)  $((X = Y \wedge Y = Z) \rightarrow X = Z)$     (transitivity)
(Eq4)  $(X = Y \rightarrow i(X) = i(Y))$              (congruence w.r.t. i)
(Eq5)  $((X_1 = Y_1 \wedge X_2 = Y_2) \rightarrow X_1 \circ X_2 = Y_1 \circ Y_2)$ (congruence w.r.t. $\circ$)

Now we show that $X \circ e = X$ follows from (Gr1)-(Gr3) and (Eq1)-(Eq5). For this purpose, let $(G, \circ, i, =)$ be a model of (Gr1)-(Gr3) and (Eq1)-(Eq5). We want to show that $(G, \circ, i, =)$ is a model of $(X \circ e) = X$. Let an element x of G be given. We compute:

| | |
|---|---|
| $(x \circ e)$ | (Gr2) |
| $= (e \circ (x \circ e))$ | (Gr3)+(Eq2)+(Eq1)+(Eq5) |
| $= ((i(i(x)) \circ i(x)) \circ (x \circ e))$ | (Gr1)+(Eq2) |
| $= (i(i(x)) \circ (i(x) \circ (x \circ e)))$ | (Gr1)+(Eq2)+(Eq1)+(Eq5) |
| $= (i(i(x)) \circ ((i(x) \circ x) \circ e)))$ | (Gr3)+(Eq1)+(Eq5) |
| $= (i(i(x)) \circ (e \circ e))$ | (Gr2)+(Eq1)+(Eq5) |
| $= (i(i(x)) \circ e)$ | (Gr3)+(Eq2)+(Eq5) |
| $= (i(i(x)) \circ (i(x) \circ x))$ | (Gr1) |
| $= ((i(i(x)) \circ i(x)) \circ x)$ | (Gr3)+(Eq1)+(Eq5) |
| $= (e \circ x)$ | (Gr2) |
| $= x$ | |

Iterated application of (Eq3) allows us to conclude that $(x \circ e) = x$. Axiom (Eq4) was not used in this example.

**Exercise 2.3**    Show that X∘e=X does not follow from the set of formulas {X∘(Y∘Z)=(X∘Y)∘Z, e∘X=X and i(X)∘X=e}.

**Exercise 2.4**    Let M be the following set of formulas:

    father(nick, peter)
    father(nick,caren)
    father(peter,paul)
    mother(caren,bob)
    mother(caren,suzan)
    ∀G∀E (grandfather(G,E) ↔ ∃X (father(G,X) ∧ father(X,E)))
    ∀G∀E (grandfather(G,E) ↔ ∃X (father(G,X) ∧ mother(X,E)))

Show    that    M ⊨ ∃X    grandfather(X,suzan),    M ⊭ grandfather(peter,nick)    and M ⊭ ¬grandfather(peter,nick).

**Exercise 2.5**    Formalize and show the following conclusion: some patient likes every doctor. No patient likes any quack. Hence no doctor is a quack.

## 2.5    Further important semantical concepts and facts

**Definition 2.8:**    Let $A$ be a $\Sigma$-interpretation, $\varphi$ a $\Sigma$-formula and M a set of $\Sigma$-formulas:

- $\varphi$ is **valid in algebra** $A$ iff $A$ is a model of $\varphi$.
- $\varphi$ is **valid** iff $\varphi$ is valid in every $\Sigma$-interpretation $A$.
- $\varphi$ is **satisfiable in algebra** $A$    iff $A \models_{sta}\varphi$, for some state sta over $A$.
- $\varphi$ is **satisfiable** iff for some $\Sigma$-interpretation $A$, $\varphi$ is satisfiable in $A$.
- M is **valid in algebra** $A$ iff $A$ is a model of M.
- M is **valid** iff every $\Sigma$-interpretation $A$ is a model of M.
- M is **satisfiable in algebra** $A$    iff $A \models_{sta}M$, for some state sta over $A$.
- M is **satisfiable** iff for some $\Sigma$-interpretation $A$, M is satisfiable in $A$.
- $\varphi$ and $\psi$ are **equivalent** iff formula $(\varphi\leftrightarrow\psi)$ is valid.

## EXAMPLE 2.1

We show that the formula $\exists X(p(X) \to \forall X\ p(X))$ is valid. (Can this indeed be possible, faced with the interpretation whose domain consists of all men and p is interpreted as the property of liking football? Then our sentence reads: 'There is a man such that, if this man likes football, everyone likes football.') To prove the stated validity, we take an arbitrary interpretation $A$ and a state sta over $A$. We have to show that $A \models_{sta} \exists X(p(X) \to \forall X\ p(X))$, that is, there is an $a \in dom(A)$ such that $A \models_{sta(X/a)}(p(X) \to \forall X p(X))$. We consider two cases. If $p_A = dom(A)$, then choose a arbitrarily. Since, in this case, $A \models_{sta(X/a)} \forall X\ p(X)$, we obtain the desired conclusion by the definition of the semantics of implication. If $p_A \neq dom(A)$, then choose an element $a \in dom(A) - p_A$. Since, in this case, $A \not\models_{sta(X/a)} p(X)$ the desired conclusion again follows from the semantics of implication.

The formula $(0<X \wedge X<Y) \to \exists Z(X<(Z*Z) \wedge (Z*Z)<Y)$ is valid in the algebra *Real* (see Section 2.4), since it states the true fact that every open interval of positive real numbers contains a square. It is not valid in the algebra $(Nat,<,=)$ though. Hence it is not a valid formula. But note that it is satisfiable in $(Nat,<,=)$ (there are states sta such that the considered formula is valid in state sta).

The following exercises contain simple facts that are often used in succeeding chapters. So we advise the reader to solve them all.

---

**Exercise 2.6**    Prove that the following three statements are equivalent:

- $\varphi$ is valid in $A$
- $A \models_{sta} \varphi$, for every state sta over $A$
- $\forall(\varphi)$ is valid in $A$

**Exercise 2.7**    Prove that the following five statements are equivalent:

- $\varphi$ is valid
- $A \models_{sta} \varphi$, for every interpretation $A$ and every state sta over $A$
- Every interpretation is a model of $\varphi$
- $\varphi$ follows from the empty set of formulas
- $\forall(\varphi)$ is valid

**Exercise 2.8**    Show that the following three statements are equivalent:

- φ is satisfiable in $A$
- $A \models_{sta} φ$, for some state sta over $A$
- $\exists(φ)$ is satisfiable in $A$

**Exercise 2.9**    Show that the following three statements are equivalent:

- φ is satisfiable
- $A \models_{sta} φ$, for some interpretation $A$ and some state sta over $A$
- $\exists(φ)$ is satisfiable

**Exercise 2.10**   Prove the following statements:

- φ is satisfiable in $A$ iff $¬φ$ is not valid in $A$
- φ is valid in $A$ iff $¬φ$ is not satisfiable in $A$
- φ is satisfiable iff $¬φ$ is not valid
- φ is valid iff $¬φ$ is not satisfiable

**Exercise 2.11**   Prove the following statements:

- $A \models ¬φ$ implies $A \not\models φ$
- $A \models φ$ implies $A \models (φ∨ψ)$
- $A \models (φ∧ψ)$ is equivalent to ($A \models φ$ and $A \models ψ$)

**Exercise 2.12**   Show that the following formulas are not valid:

- $(φ\{X/t\}→∀Xφ)$
- $(\exists Xφ→φ\{X/t\})$ where $φ\{X/t\}$ is admissible
- $(\exists Xφ→∀Xφ)$
- $(∀X(φ∨ψ)→(∀Xφ∨∀Xψ))$

**Exercise 2.13**
Show that $(\exists X∀Yφ→∀Y\exists Xφ)$ is valid, but not $(∀Y\exists Xφ→\exists X∀Yφ)$.

**Exercise 2.14**   Show that the following equivalences hold between formulas

- $(φ↔ψ)$ and $((φ→ψ)∧(ψ→φ))$
- $(φ∨ψ)$ and $(¬φ→ψ)$
- $(φ∧ψ)$ and $¬(φ→¬ψ)$

These equivalences allow replacement of every formula by an equivalent one with $¬$ and $→$ as the only propositional connectives. Further equivalences exist between formulas

- $(φ→ψ)$ and $¬(φ∨ψ)$

- $(\varphi \wedge \psi)$ and $\neg(\neg\varphi \vee \neg\psi)$
- $(\varphi \vee \psi)$ and $\neg(\neg\varphi \wedge \neg\psi)$

Thus we may replace every formula by an equivalent one with $\neg$ and $\wedge$, or with $\neg$ and $\vee$ as the only propositional connectives.

**Exercise 2.15**  Show that the following equivalences hold between formulas
- $(\varphi\wedge(\psi\wedge\xi))$ and $((\varphi\wedge\psi)\wedge\xi)$  (associativity of $\wedge$)
- $(\varphi\vee(\psi\vee\xi))$ and $((\varphi\vee\psi)\vee\xi)$  (associativity of $\vee$)
- $(\varphi\wedge(\psi\vee\xi))$ and $((\varphi\wedge\psi)\vee(\varphi\wedge\xi))$  (distributivity of $\wedge$ over $\vee$)
- $(\varphi\vee(\psi\wedge\xi))$ and $((\varphi\vee\psi)\wedge(\varphi\vee\xi))$  (distributivity of $\vee$ over $\wedge$)

**Exercise 2.16**  Show that $\forall$ and $\exists$ can be mutually expressed via the equivalence of

- $\forall X\varphi$ and $\neg\exists X\neg\varphi$
- $\exists X\varphi$ and $\neg\forall X\neg\varphi$

Show that redundant quantification can be removed via the equivalence of

- $QX\varphi$ and $\varphi$, provided that X does not occur free in $\varphi$

**Exercise 2.17**  Show that equivalence of formulas passes over propositional connectives and quantifiers: If $\varphi$ is equivalent $\varphi'$ and $\psi$ is equivalent $\psi'$, then equivalent are:

$\neg\varphi$ and $\neg\varphi'$, $(\varphi \text{ op } \psi)$ and $(\varphi' \text{ op } \psi')$, $QX\varphi$ and $QX\varphi'$

for every 2-ary propositional connective op and quantifier Q. This means that the equivalences stated in the lemma above may be applied locally to subformulas.

**Exercise 2.18**  Check which of the following relations are always true:

- $\{\exists X\varphi\} \models \varphi\{X/c\}$
- $\{\exists X\varphi\} \models \varphi\{X/Y\}$, Y is a new variable
- $\{\forall X\varphi\} \models \varphi$
- $\{\varphi\} \models \forall X\varphi$
- $\models (\varphi \rightarrow \forall X\varphi)$
- $\models \forall X\varphi \Leftrightarrow \models \varphi\{X/c\}$ with a new constant c

## 2.6    Normal forms

In this section we shall introduce some normal forms for formulas. It will be shown that formulas may be transformed into these normal forms without affecting certain semantical properties. Formulas in normal form exhibit advantages over arbitrary formulas in several respects, as we will soon see.

### 2.6.1    Prenex normal form

Our first concern will be, given a formula, to move all quantifiers left to the front of the given formula, thereby obtaining an equivalent one in **prenex normal form**. The following theorem describes the way to move quantifiers from the inside of formulas out to the left.

*Lemma 2.7*

The following formulas are equivalent (for $Q \in \{\forall, \exists\}$ ): it is required that X does not occur free in $\varphi$ for (3), (5), (7) and (9), that X does not occur free in $\psi$ for (4), (6), (8) and (10), and that $\varphi\{X/Y\}$ is admissible and Y is not free in $\varphi$ for (11) and (12):

(1)    $\neg \forall X\, \varphi$ and $\exists X\, \neg\varphi$

(2)    $\neg \exists X\, \varphi$ and $\forall X\, \neg\varphi$

(3)    $(\varphi \wedge QX\psi)$ and $QX(\varphi \wedge \psi)$

(4)    $(QX\varphi \wedge \psi)$ and $QX(\varphi \wedge \psi)$

(5)    $(\varphi \vee QX\psi)$ and $QX(\varphi \vee \psi)$

(6)    $(QX\varphi \vee \psi)$ and $QX(\varphi \vee \psi)$

(7)    $(\varphi \rightarrow \forall X\psi)$ and $\forall X(\varphi \rightarrow \psi)$

(8)    $(\forall X\varphi \rightarrow \psi)$ and $\exists X(\varphi \rightarrow \psi)$

(9)    $(\varphi \rightarrow \exists X\psi)$ and $\exists X(\varphi \rightarrow \psi)$

(10)    $(\exists X\varphi \rightarrow \psi)$ and $\forall X(\varphi \rightarrow \psi)$

(11)    $\forall X\varphi$ and $\forall Y\varphi\{X/Y\}$

(12)    $\exists X\varphi$ and $\exists Y\varphi\{X/Y\}$

*Proof* (exemplary)    For all interpretations $A$ and states sta over $A$, we have to show that one formula of the equivalence to be proven is valid in $A$ in state sta if, and only if, the other side of the equivalence is valid in $A$ in state sta.

Let us consider the third equivalence, for $Q=\forall$:

$$A \models_{sta} \forall X(\varphi \wedge \psi)$$

$$\Leftrightarrow$$

for all $a \in \text{dom}(A)$, $A \vDash_{\text{sta}(X/a)} (\varphi \wedge \psi)$

$\Leftrightarrow$

for all $a \in \text{dom}(A)$, $A \vDash_{\text{sta}(X/a)} \varphi$ and $A \vDash_{\text{sta}(X/a)} \psi$

$\Leftrightarrow$ (use Lemma 2.3 and the fact that X does not occur free in $\varphi$)

for all $a \in \text{dom}(A)$, $A \vDash_{\text{sta}} \varphi$ and $A \vDash_{\text{sta}(X/a)} \psi$

$\Leftrightarrow$

for all $a \in \text{dom}(A)$, $A \vDash_{\text{sta}} \varphi$ and, for all $a \in \text{dom}(A)$, $A \vDash_{\text{sta}(X/a)} \psi$

$\Leftrightarrow$

$A \vDash_{\text{sta}} \varphi$ and $A \vDash_{\text{sta}} \forall X \, \psi$

$\Leftrightarrow$

$A \vDash_{\text{sta}} (\varphi \wedge \forall X \psi)$

Without the assumption that X does not occur free in $\varphi$, formulas $(\varphi \wedge \forall X \psi)$ and $\forall X(\varphi \wedge \psi)$ are not equivalent, in general. This is shown by the following example: $(p(X) \wedge \forall X q(X))$ and $\forall X(p(X) \wedge q(X))$, an interpretation $A$ with $\text{dom}(A)=\{0,1\}$, $p_A=\{0\}$, $q_A=\{0,1\}$ and a state sta with $\text{sta}(X)=0$.

Next, let us consider the ninth equivalence. Assume that X does not occur free in $\psi$. We show that $(\forall X \varphi \rightarrow \psi)$ and $\exists X(\varphi \rightarrow \psi)$ are equivalent by showing that the negated formulas are equivalent. Now, $\neg(\forall X \varphi \rightarrow \psi)$ is equivalent to $(\forall X \varphi \wedge \neg \psi)$, and hence, by the equivalences just shown above, also to $\forall X(\varphi \wedge \neg \psi)$, $\forall X \neg(\varphi \rightarrow \psi)$ and finally to $\neg \exists X(\varphi \rightarrow \psi)$.

Finally, we deal with the 11th equivalence. We may assume without loss of generality that $X \neq Y$. Since $\varphi\{X/Y\}$ is admissible, we know that $(\forall X \varphi \rightarrow \varphi\{X/Y\})$ is valid; therefore, the following formulas are valid, too:

$\forall Y(\forall X \varphi \rightarrow \varphi\{X/Y\})$

$(\forall X \varphi \rightarrow \forall Y \varphi\{X/Y\})$ (since Y does not occur free in $\forall X \varphi$)

Now note that $\varphi\{Y/X\}$ was assumed to be admissible, therefore X cannot occur free in $\varphi\{X/Y\}$, and $\varphi\{X/Y\}\{Y/X\}$ is $\varphi$.

| | | | |
|---|---|---|---|
| $\varphi$ with four free occurrences of X (no free occurrence of Y) | ... X ... X | ... | X ... X ... |
| $\varphi\{X/Y\}$ with four free occurrences of Y (no further Y, Y not in the scope of a quantifier $\forall X/\exists X$) | ... Y ... Y | ... | Y ... Y ... |
| $\varphi\{X/Y\}\{Y/X\}$ | ... X ... X | ... | X ... X ... |

By the same argument as above it follows that $(\forall Y\varphi\{X/Y\}\rightarrow$ $\forall X\varphi\{X/Y\}\{Y/X\})$, that is, the formula $(\forall Y\varphi\{X/Y\}\rightarrow\forall X\ \varphi)$ is valid.     ■

---

**Exercise 2.19** Complete the proof of the above lemma and show that the given assumptions are necessary. In particular give an example for $\forall X\varphi$ and $\forall Y\varphi\{X/Y\}$ not being equivalent because

- $\varphi\{X/Y\}$ is not admissible, but $Y\notin fv(\varphi)$
- $\varphi\{X/Y\}$ is admissible, but $Y\in fv(\varphi)$.

---

**Definition 2.9:** We call a formula of the form $Q_1X_1...Q_nX_n\ \chi$, where $Q_1,...,Q_n$ are quantifiers and $\chi$ is a formula containing no quantifiers, a formula in **prenex normal form**.

### *Theorem 2.1*

For each formula $\varphi$ we can effectively construct an equivalent formula in prenex normal form.

*Proof*     First, we eliminate all occurrences of $\leftrightarrow$ in $\varphi$ using the fact that $(\varphi\leftrightarrow\psi)$ and $((\varphi\rightarrow\psi)\wedge(\psi\rightarrow\varphi))$ are equivalent formulas. Then we successively apply Lemma 2.7 above to move quantifiers from inside out to the left. If at some point during this process the condition concerning variable occurrence in specific subformulas is violated, we may rename some bound variables in an appropriate way according to the last two equivalences in Lemma 2.7.     ■

Let us for example consider the formula:

$$\forall X\ \exists Y\ ((\forall X\ p(X)\rightarrow q(X,f(Y),Z))\wedge\neg\forall Z\ \exists X\ \neg r(g(X,Z),Z))$$

The following transformation steps lead to a formula in prenex normal form:

$$\forall X\ \exists Y\ ((\forall X\ p(X)\rightarrow q(X,f(Y),Z))\wedge\neg\forall Z\ \exists X\ \neg r(g(X,Z),Z))$$
$$\forall X\ \exists Y\ ((\forall U\ p(U)\rightarrow q(X,f(Y),Z))\wedge\neg\forall Z\ \exists X\ \neg r(g(X,Z),Z))$$
$$\forall X\ \exists Y\ (\exists U(\ p(U)\rightarrow q(X,f(Y),Z))\wedge\neg\forall Z\ \exists X\ \neg r(g(X,Z),Z))$$

$$\forall X \, \exists Y \, \exists U((\, p(U) \to q(X,f(Y),Z)) \wedge \neg \forall Z \, \exists X \, \neg r(g(X,Z),Z))$$
$$\forall X \, \exists Y \, \exists U((\, p(U) \to q(X,f(Y),Z)) \wedge \exists Z \, \underline{\neg} \exists X \, \neg r(g(X,Z),Z))$$
$$\forall X \, \exists Y \, \exists U((\, p(U) \to q(X,f(Y),Z)) \wedge \exists V \, \neg \exists X \, \neg r(g(X,V),V))$$
$$\forall X \, \exists Y \, \exists U \, \exists V((\, p(U) \to q(X,f(Y),Z)) \wedge \neg \exists X \, \neg r(g(X,V),V))$$
$$\forall X \, \exists Y \, \exists U \, \exists V((\, p(U) \to q(X,f(Y),Z)) \wedge \forall X \, \neg \neg r(g(X,V),V))$$
$$\forall X \, \exists Y \, \exists U \, \exists V((\, p(U) \to q(X,f(Y),Z)) \wedge \forall W \, \neg \neg r(g(W,V),V))$$
$$\forall X \, \exists Y \, \exists U \, \exists V \, \forall W((\, p(U) \to q(X,f(Y),Z)) \wedge \neg \neg r(g(W,V),V))$$

## 2.6.2    Conjunctive and disjunctive normal form

Having transformed a formula into prenex normal form, we next treat the quantifier-free kernel of the obtained prenex normal form.

First, we need two notational conventions concerning **finite conjunctions** and **disjunctions**:

$\displaystyle\bigwedge_{1 \le i \le n} \chi_i$  is the formula $(\chi_1 \wedge (\chi_2 \wedge (\ldots \wedge \chi_n)\ldots))$, called a finite conjunction

$\displaystyle\bigvee_{1 \le i \le n} \chi_i$  is the formula $(\chi_1 \vee (\chi_2 \vee (\ldots \vee \chi_n)\ldots))$, called a finite disjunction

for $n \ge 1$ and formulas $\chi_1, \chi_2, \ldots, \chi_n$. Sometimes we write $\chi_1 \wedge \chi_2 \wedge \ldots \wedge \chi_n$ and $\chi_1 \vee \chi_2 \vee \ldots \vee \chi_n$ instead of the fully bracketed versions above (since in this case the bracketing has no effect on the semantics of formulas).

### Theorem 2.2

For every quantifier-free formula $\varphi$, we can effectively construct equivalent formulas $\gamma$ and $\delta$ of the following form:

$\gamma$ is a finite conjunction of finite disjunctions of literals:

$$\gamma = \bigwedge_{1 \le i \le n} \chi_i, \text{ with } \chi_i = \bigvee_{1 \le j \le m(i)} L_{ij} \text{ with literals } L_{ij}$$

$\delta$ is a finite disjunction of finite conjunctions of literals:

$$\delta = \bigvee_{1 \le i \le n} \chi_i, \text{ with } \chi_i = \bigwedge_{1 \le j \le m(i)} L_{ij} \text{ with literals } L_{ij}$$

$\gamma$ is called a formula in **conjunctive normal form**, $\delta$ a formula in **disjunctive normal form**.

*Proof*    Using the equivalences from Section 2.5 we may assume without loss of generality that $\varphi$ contains only the logical symbols $\neg$ and $\wedge$. The proof goes by structural induction over $\varphi$. If $\varphi$ is an atomic formula then it is already in conjunctive ($\varphi$ is the conjunction over one formula, which is the disjunction over a single formula, namely $\varphi$) and disjunctive normal form. Since the negation of a conjunctive normal form is equivalent to a disjunctive normal form, and the negation of a disjunctive normal form is equivalent to a conjunctive normal form, induction passes over $\neg$ in a simple way.

Finally, consider the case that $\varphi$ is of the form $(\omega \wedge \xi)$. By induction hypothesis we have conjunctive and disjunctive normal forms for $\omega$ and $\xi$. Since the conjunction of conjunctive normal forms is readily transformed into conjunctive normal form (according to the associativity of conjunction we may rearrange the bracketing of the conjunctively connected formulas) we have a conjunctive normal form for $(\omega \wedge \xi)$. A disjunctive normal form for $(\omega \wedge \xi)$ is obtained from disjunctive normal forms $(\alpha_1 \vee \alpha_2 \vee ... \vee \alpha_n)$ and $(\beta_1 \vee \beta_2 \vee ... \vee \beta_n)$ of $\omega$ and $\xi$ by applying the following equivalence:

$$((\alpha_1 \vee \alpha_2 \vee ... \vee \alpha_n) \wedge (\beta_1 \vee \beta_2 \vee ... \vee \beta_n)) \text{ is equivalent to}$$
$$(\alpha_1 \wedge \beta_1) \vee (\alpha_1 \wedge \beta_2) \vee ... \vee (\alpha_1 \wedge \beta_n) \vee ... \vee (\alpha_n \wedge \beta_1) \vee (\alpha_n \wedge \beta_2) \vee ... \vee (\alpha_n \wedge \beta_n)$$

∎

What are the advantages of formulas in conjunctive and disjunctive normal forms? The following theorem collects some of them. It states that for quantifier-free formulas without free variables semantical properties can be reduced to simple and decidable syntactical ones.

### Theorem 2.3

Let $\Sigma$ be a signature containing at least one constant and $L_1, L_2, ..., L_k$ be ground literals (atomic or negated atomic formulas without variables) in signature $\Sigma$. The following table characterizes several semantical properties of formulas of special forms by simple syntactical properties.

| Syntactical form | Semantical property | Syntactical characterization |
|---|---|---|
| $L_1 \wedge L_2 \wedge ... \wedge L_k$ | has a model | $\{L_1, L_2, ..., L_k\}$ does not contain a complementary pair |
| $L_1 \wedge L_2 \wedge ... \wedge L_k$ | is valid | is never the case |
| $L_1 \vee L_2 \vee ... \vee L_k$ | has a model | is always the case |

| $L_1 \lor L_2 \lor \ldots \lor L_k$ is valid | $\{L_1, L_2, \ldots, L_k\}$ contains a complementary pair |
|---|---|

Recall that a complementary pair was a pair of literals $L, \sim L$.

For the proof we require the concept of a **Herbrand algebra**, which is of great importance in subsequent chapters.

**Definition 2.10:** Let $\Sigma$ be a signature with at least one constant. A $\Sigma$-algebra $A$ is called a **Herbrand algebra** iff $dom(A)$ is $T_0(\Sigma)$, that is, the set of all ground $\Sigma$-terms, and the interpretations of the function symbols in $\Sigma$ is fixed as follows: $f_A(t_1, \ldots, t_n) = f(t_1, \ldots, t_n)$, where f is an n-ary function symbol and $t_1, \ldots, t_n \in T_0(S)$. Note that $f_A$ is indeed a mathematical function on $T_0(\Sigma)^n$, whereas f is merely a symbol and $f(t_1, \ldots, t_n)$ is a ground term. Note further, that the only thing not fixed in a Herbrand algebra is the interpretation of predicate symbols. A **Herbrand model** of a formula set M is a Herbrand algebra that is a model of M.

*Lemma 2.8*

Let $\Sigma$ be a signature with at least one constant and $A$ be a $\Sigma$-Herbrand algebra. Let t be a $\Sigma$-term with variables $X_1, \ldots, X_n$ and sta be a state over $A$ with $sta(X_1) = r_1, \ldots, sta(X_n) = r_n$, where $r_1, \ldots, r_n$ are ground terms. Then $val_{A, sta}(t) = t\{X_1/r_1, \ldots, X_n/r_n\}$. In particular, if t is a ground term, then $t_A = t$.

Thus, term evaluation in a Herbrand algebra coincides with substitution, and ground terms evaluate to themselves.

*Proof*    The simple structural induction on t is left to the reader.

*Proof of Theorem 2.3*    We start with a closed formula of the form $L_1 \land L_2 \land \ldots \land L_k$. If $\{L_1, L_2, \ldots, L_k\}$ contains a complementary pair then, of course, $L_1 \land L_2 \land \ldots \land L_k$ cannot possess a model. Conversely, assume that $\{L_1, L_2, \ldots, L_k\}$ does not contain a complementary pair. We define a Herbrand algebra $A$ as follows:

$p_A = \{ (r_1,...,r_n) \in T_0(\Sigma)^n \mid \{L_1,L_2,...,L_k\}$ contains the formula $p(r_1,...,r_n) \}$

We show that $A$ is a model of $L_i$, for $i=1,...,k$. If $L_i$ is atomic, say $L_i$ is $p(r_1,...,r_n)$ with $(r_1,...,r_n) \in T_0(\Sigma)^n$, then $p(r_1,...,r_n)$ occurs among $\{L_1,L_2,...,L_k\}$. By definition of $p_A$, $(r_1,...,r_n) \in p_A$. Since the values of the ground terms $r_1,...,r_n$ in $A$ are $r_1,...,r_n$, respectively, it follows that $A$ is a model of $p(r_1,...,r_n)$, that is, a model of $L_i$. If $L_i$ is negated atomic, say $L_i$ is $\neg p(r_1,...,r_n)$ with $(r_1,...,r_n) \in T_0(\Sigma)^n$, then $p(r_1,...,r_n)$ does not occur among $\{L_1,L_2,...,L_k\}$, since otherwise $\{L_1,L_2,...,L_k\}$ would contain a complementary pair. By the definition of $p_A$, $(r_1,...,r_n) \notin p_A$. Thus, $A$ is a model of $\neg p(r_1,...,r_n)$, that is, a model of $L_i$.

Closed formulas of the form $L_1 \wedge L_2 \wedge ... \wedge L_k$ are never valid since by the result just proven $\sim L_1$ always has a model.

A closed formula of the form $L_1 \vee L_2 \vee ... \vee L_k$ is valid iff $\sim L_1 \wedge \sim L_2 \wedge ... \wedge \sim L_k$ does not possess a model iff $\{\sim L_1, \sim L_2,...,\sim L_k\}$ contains a complementary pair iff $\{L_1,L_2,...,L_k\}$ contains a complementary pair.

A closed formula of the form $L_1 \vee L_2 \vee ... \vee L_k$ always possesses a model, since $L_1$ possesses a model, as was shown above.  ∎

The proof above required the assumption that the considered formulas are variable free. Indeed, the theorem is not true for formulas with variables. As an example, consider the formula $p(c) \wedge \neg p(X)$. Although $\{p(c), \neg p(X)\}$ does not contain a complementary pair, the proposed formula does not possess a model.

## 2.6.3   Skolem normal form

Now, we turn our interest to existential quantifiers and seek a possible method of eliminating them in an effective way that preserves satisfiability. The following shows how this can be done.

### Theorem 2.4 (Skolem Normal Form)

Let $\forall X_1...\forall X_n \exists Y \varphi$ be a closed $\Sigma$-formula with different variables $X_1,...,X_n,Y$. Assume that $\varphi$ does not contain quantifiers $Q X_1,...,Q X_n$. (Thus $\varphi\{Y/g(X_1,...,X_n)\}$ is admissible.) We extend $\Sigma$ by a new n-ary function symbol $g$ to a signature $\Sigma_g = \Sigma \cup \{g:s^n \rightarrow s\}$. Then, every $\Sigma_g$-

model of $\forall X_1...\forall X_n \varphi\{Y/g(X_1,...,X_n)\}$ is a model of $\forall X_1...\forall X_n \exists Y \varphi$. Conversely, every $\Sigma$-model of $\forall X_1...\forall X_n \exists Y \varphi$ can be expanded to a $\Sigma_g$-model of $\forall X_1...\forall X_n \varphi\{Y/g(X_1,...,X_n)\}$.

As a consequence, the following statements are equivalent:

- There exists a model of $\forall X_1...\forall X_n \exists Y \varphi$.
- There exists a model of $\forall X_1...\forall X_n \varphi\{Y/g(X_1,...,X_n)\}$

*Proof* Since $(\forall X_1...\forall X_n \varphi\{Y/g(X_1,...,X_n)\} \rightarrow \forall X_1...\forall X_n \exists Y \varphi)$ is a valid formula, every $\Sigma_g$-model of $\forall X_1...\forall X_n \varphi\{Y/g(X_1,...,X_n)\}$ is a model of $\forall X_1...\forall X_n \exists Y \varphi$. Conversely, assume that $A$ is a $\Sigma$-model of the formula $\forall X_1...\forall X_n \exists Y \varphi$. This means that for every $(a_1,...,a_n) \in dom(A)^n$ there is an element $a \in dom(A)$ such that $A \models_{sta(Y/a)} \varphi$, where sta is a state over $A$ with $sta(X_1)=a_1,..., sta(X_n)=a_n$. We define a (selection) function G, which supplies us with such an element $a=G(a_1,...,a_n)$ for each tuple $(a_1,...,a_n) \in dom(A)^n$. Then $A \models_{sta(Y/G(a_1,...,a_n))} \varphi$, for all $(a_1,...,a_n) \in dom(A)^n$ and states sta over $A$ such that $sta(X_1)=a_1,...,sta(X_n)=a_n$. Now we expand $A$ to a $\Sigma_g$-interpretation $B$ by defining $g_B$ to be the chosen selection function G. Applying Lemma 2.4 we obtain that $B \models_{sta} \varphi\{Y/g(X_1,...,X_n)\}$, for every state sta over $B$. Thus, $B$ is a model of the formula $\forall X...\forall X_n \varphi\{Y/g(X_1,...,X_n)\}$.    ∎

Let us look at some examples:

- $\exists Y\ r(Y)$ has a model $\Leftrightarrow r(c)$ has a model.
- $\forall X \exists Y\ p(X,Y)$ has a model $\Leftrightarrow \forall X\ p(X,g(X))$ has a model.
- $\forall X_1 \forall X_2 \exists Y \forall U \exists V\ q(X_1,X_2,Y,U,V)$ has a model $\Leftrightarrow$

  $\forall X_1 \forall X_2 \forall U \exists V\ q(X_1,X_2,h(X_1,X_2),U,V)$ has a model $\Leftrightarrow$

  $\forall X_1 \forall X_2 \forall U\ q(X_1,X_2,h(X_1,X_2),U,k(X_1,X_2,U))$ has a model.

Here, c is a constant, g is 1-ary, h is 2-ary and k is 3-ary.

**Definition 2.11:** A closed formula of the form $\forall X_1...\forall X_n \varphi$, where $\varphi$ is a quantifier-free formula in conjunctive normal form, is called a formula in **Skolem normal form**.

### Theorem 2.5  (Skolem Normal Form )

For each formula $\varphi$ we can effectively construct a formula $\psi$ in Skolem normal form, such that $\psi$ has a model iff $\varphi$ has a model. (Emphasis is on the effective constructibility: the pure existence of such a formula $\psi$ is trivial; is it clear why?)

*Proof*   Collect the results presented above.                                    ■

---

**Exercise 2.19** Show that the Skolem normal form of a formula is not uniquely determined.

**Exercise 2.20** Show that usually a formula $\varphi$ and its Skolem normal forms are not equivalent.

**Exercise 2.21** Transform the following formulas into Skolem normal form:

$$(\forall X(p(X) \to \exists Yq(X,Y)) \wedge \forall X(\neg p(X) \to \neg \exists Yq(X,Y)))$$

$$\exists X(p(X) \to \forall Xp(X)).$$

**Exercise 2.22** Let $L_1,\ldots,L_k$ be literals containing variables. Show that $L_1 \wedge \ldots \wedge L_k$ is satisfiable iff $\{L_1,\ldots,L_k\}$ contains no complementary pair. (Hint: Skolemize!)

---

## 2.7   Herbrand's theorem

Formulas in Skolem normal form are **closed, universally quantified** formulas. The following theorem shows some of the advantages of such formulas and formula sets.

### Theorem 2.6 (Herbrand's Theorem)

Let $\Sigma$ be a signature containing at least one constant, and M be a set of closed universal $\Sigma$-formulas. We call **ground-instances(M)** the set of all formulas $\psi\{X_1/t_1,\ldots,X_n/t_n\}$, where $\psi$ is a quantifier-free formula, $\forall X_1,\ldots,\forall X_n\psi$ is an element of M, and $t_1,\ldots,t_n$ are ground $\Sigma$-terms. The following statements are equivalent:

(a)   M has a model.
(b)   M has a Herbrand model.

(c)    formula set ground-instances(M) has a model.

(d)    formula set ground-instances(M) has a Herbrand model

*Proof*    Since Herbrand models are models, and since formulas of the form $\forall X_1...\forall X_n \psi \rightarrow \psi\{X_1/t_1,...,X_n/t_n\}$ are valid, the implications of Figure 2.2 are trivial:

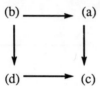

**Figure 2.2**

To prove the entire theorem, it is sufficient to show the implication (c) $\Rightarrow$ (b). For this purpose let a model $A$ of ground-instances(M) be given. We define a Herbrand interpretation $B$ as follows:

$$r_B = \{ (t_1,...,t_n) \in (dom(B)^n) \mid A \models r(t_1,...,t_n) \}, \text{ for all } r \in \Sigma \text{ of arity } n > 0$$

$$r_B = r_A, \text{ for all 0-ary } r \in R$$

An alternative description of the definition of $r_B$ is the following: for all n-ary $r \in R$ and all $t_1,...,t_n \in dom(B) = T_0(\Sigma)$:

$$A \models r(t_1,...,t_n) \Leftrightarrow B \models r(t_1,...,t_n).$$

This means that the same closed atomic formulas are valid in $A$ and $B$. This result is directly extended (simple structural induction) to arbitrary closed, quantifier-free formulas.

Finally, we show that $B$ is a model of M. Let $\forall X_1...\forall X_n \psi$ be a formula in M, with a quantifier-free formula $\psi$. We conclude that $B \models \forall X_1...\forall X_n \psi$    is true iff, for all $t_1,...,t_n \in dom(B) = T_0(S)$, $B \models \psi\{X_1/t_1,...,X_n/t_n\}$. This latter statement is equivalent to the following one: for all $t_1,...,t_n \in dom(B) = T_0(\Sigma)$, $A \models \psi\{X_1/t_1,...,X_n/t_n\}$.

But the last assertion is indeed true, because $A$ is a model of ground-instances(M), and $\psi\{X_1/t_1,...,X_n/t_n\}$ is an element of ground-instances(M), for all $t_1,...,t_n \in dom(B) = T_0(\Sigma)$.

■

## 2.8    Compactness and Löwenheim-Skolem theorem

*Compactness Theorem 2.7 (for sets of quantifier-free closed formulas)*

Let $\Sigma$ be a signature and M be a set of quantifier-free $\Sigma$-formulas without free variables (note that such formulas do not contain any variables). Then M has a model iff every finite subset of M has a model.

*Proof*    We may assume that $\Sigma$ contains a constant. (Otherwise extend $\Sigma$ by one.) If M has a model then, of course, every finite subset of M has a model, too. Conversely, assume that M does not possess a model. Fix an enumeration $A_0, A_1, A_2, \ldots$ of all ground $\Sigma$-atoms. Observe that the validity of a closed, quantifier-free formula $\varphi$ in an interpretation $A$ depends only on the validity in $A$ of the ground atoms occurring in $\varphi$.

Let us take a closer look at Herbrand algebras. Each such Herbrand algebra is uniquely determined by the set of closed formulas of the form $r(t_1, \ldots, t_n)$ which are valid in it. We may express this fact in the following way. Each path $\pi$ in the **semantic tree** of Figure 2.3 defines a unique Herbrand interpretation $B(\pi)$, namely that one in which exactly the formulas $r(t_1, \ldots, t_n)$ occurring along the path $\pi$ are valid. Conversely, every Herbrand algebra can be obtained as a $B(\pi)$, for a suitable path $\pi$.

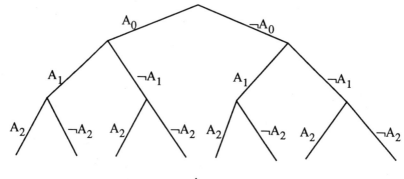

**Figure 2.3**

Since M does not possess a model, for each path $\pi$ in the tree, there exists a formula $\varphi(\pi)$ in M such that $B(\pi) \not\models \varphi(\pi)$. As $\varphi(\pi)$ contains only a finite number of ground atoms, the statement $B(\pi) \not\models \varphi(\pi)$ depends only on a finite initial segment $\varepsilon(\pi)$ of $\pi$ up to some node $k(\pi)$. More precisely, this means that for all paths $\pi'$ of the semantic tree which coincide with $\pi$ up to node $k(\pi)$, $B(\pi') \not\models \varphi(\pi)$, too.

Now, for all paths $\pi$ in the semantic tree, we cut off the part of $\pi$ below $k(\pi)$. In this way we obtain a tree with only finitely long paths. By König's lemma (see appendix), T is a finite tree. The finite set of its paths may be written as $\{\varepsilon(\pi) \mid \pi \in \prod\}$ for a finite set $\prod$ of paths in the original semantic tree. Let E be the finite subset $\{\varphi(\pi) \mid \pi \in \prod\}$ of M. We show that E does not possess a Herbrand model, and therefore by Herbrand's Theorem no model at all.

For this reason, let a Herbrand algebra $B(\nu)$ be given for a path $\nu$ of the semantic tree. Since $\varepsilon(\nu)$ is an element of $\{\varepsilon(\pi) \mid \pi \in \prod\}$, there is a path $\pi \in \prod$ such that $\varepsilon(\pi) = \varepsilon(\nu)$. But then $B(\pi) \not\models \varphi(\pi)$, so $B(\nu) \not\models \varphi(\pi)$. Because $\varphi(\pi) \in E$, it follows that $B(\nu)$ is not a model of E.    ∎

### Compactness Theorem 2.8

Let M be an arbitrary set of closed $\Sigma$-formulas. Then M has a model iff every finite subset of M has a model.

*Proof*   As above, one direction is trivial. For the other one, assume that M does not possess a model. Transform every formula $\varphi \in M$ into Skolem normal form $sko(\varphi)$, thereby using for every such formula a separate set of new function symbols required in the Skolem construction. It follows that the set $sko(M) = \{sko(\varphi) \mid \varphi \in M\}$ does not possess any model. By Herbrand's Theorem, the set ground-instances$(sko(M))$ does not possess a model, either. By Theorem 2.7 we obtain a finite subset of ground-instances$(sko(M))$ without a model. This finite subset is a subset of ground-instances$(sko(E))$, for a finite subset E of M. It follows that E cannot possess a model, either.    ∎

### Corollary

Let M be an arbitrary set of $\Sigma$-formulas (free variables allowed). Then, M is satisfiable iff every finite subset of M is satisfiable.

*Proof*   Skolemize (using new constants) and apply Theorem 2.8 for the model relation.    ∎

### Corollary

Let $\Sigma$ be a signature containing at least one constant and M be a set of closed, universal $\Sigma$-formulas. Let $\varphi$ be a closed, existential $\Sigma$-formula $\exists X \psi$, where $\psi$ is without quantifiers. Then the following statements are equivalent:

(1)    $M \models \varphi$

(2)    $M \models \psi\{X/t_1\}\vee...\vee\psi\{X/t_m\}$, for a finite number of ground terms $t_1,...,t_m$

This means that a true existential question to a knowledge base consisting of universal formulas may be answered by providing a finite number of ground answer terms.

*Proof*  If there exist ground terms $t_1,...,t_m$ with $M \models \psi\{X/t_1\}\vee...\vee \psi\{X/t_m\}$, then obviously $M \models \exists X\psi$. Conversely, suppose that $M \models \exists X\psi$. Then $M\cup\{\forall X\neg\psi\}$ does not possess a model. By Theorems 2.8 and 2.6 there is a finite subset of ground-instances($M\cup\{\forall X\neg\psi\}$) which does not possess a model. Let such a subset be $M'\cup \{\neg\psi\{X/t_1\},...,\neg\psi\{X/t_m\}\}$, where M' is a finite subset of ground-instances(M). Since all formulas in formula set ground-instances(M) follow from M, $M\cup\{\neg\psi\{X/t_1\},...,\neg\psi\{X/t_m\}\}$ does not possess a model, either. This implies $M \models \neg(\neg\psi\{X/t_1\}\wedge...\wedge\neg\psi\{X/t_m\})$, that is, $M \models \psi\{X/t_1\}\vee...\vee\psi\{X/t_m\}$.    ∎

A so-called **definite answer** to an existential question $\exists X\psi$ to a knowledge base M consisting of closed, universal formulas, is a *single* ground term t such that $M \models \psi\{X/t\}$. Definite answers are not always possible, as the following example indicates. Let M be the set $\{(p(a)\vee p(b))\}$. We consider the existential question $\exists Xp(X)$. Clearly, $M \models \exists Xp(X)$. There also exist ground terms $t_1$ and $t_2$ such that $M \models \psi\{X/t_1\}\vee\psi\{X/t_2\}$, namely a as $t_1$ and b as $t_2$. But there is no ground term t with $M \models p(t)$, because t could only be a or b. But neither $M \models p(a)$ nor $M \models p(b)$ is true. (Is it clear why?)

Theorem 2.8 does not hold for arbitrary sets of formulas M. Let M be the set consisting solely of the formula $(q(a)\wedge\exists Xp(X))$ and $S=\{a:\rightarrow s, p:s\}$. Then $M \models \exists Xp(X)$, but not $M \models p(a)$.

### Theorem 2.9 (Löwenheim-Skolem)

If a set M of formulas has a model then it has a model with a domain which is at most denumerable.

*Proof*  Assume that M has a model. Skolemize the formulas of M obtaining a set sko(M) of formulas. Then, sko(M) has a model, too. By Herbrand's Theorem, sko(M) has a Herbrand model. Since a Herbrand model of sko(M) has a domain which is at most denumerable, and since every model of sko(M) is a model of M, too, the conclusion of our theorem follows.    ∎

## 2.9    Semi-decidability of logical conclusion

For closed formulas $\varphi_1,...,\varphi_n,\varphi$, we want to find out whether $\{\varphi_1,...,\varphi_n\} \models \varphi$. We may proceed in the following way:

$\{\varphi_1,...,\varphi_n\} \models \varphi \Leftrightarrow$

$\{\varphi_1,...,\varphi_n,\neg\varphi\}$ does not possess a model $\Leftrightarrow$

$sko(\{\varphi_1,...,\varphi_n,\neg\varphi\})$ does not possess a model

(where sko indicates the operator producing Skolem normal form) $\Leftrightarrow$

there is a finite subset of ground-instances(sko($\{\varphi_1,...,\varphi_n,\neg\varphi\}$) that does not possess a model.

We check whether a finite set $\{A_1,...,A_n\}$ of quantifier free, closed formulas (note that such formulas are variable free) has a model by bringing $(A_1\wedge...\wedge A_n)$ into disjunctive normal form $(K_1\vee...\vee K_m)$ and using the syntactic characterization presented in Section 2.6.2   to determine whether $(K_1\vee...\vee K_m)$ has a model:

$(K_1\vee...\vee K_m)$ has a model $\Leftrightarrow$

there is an i=1,...,m such that $K_i$ has a model $\Leftrightarrow$

there is an i=1,...,m such that among the literals of $K_i$ there is no complementary pair.

Now we define the following effective procedure:

Enumerate all finite subsets of $sko(\{\varphi_1,...,\varphi_n,\neg\varphi\})$ until, eventually, one is found that possesses a model.

If $\{\varphi_1,...,\varphi_n\} \models \varphi$, the process terminates. If $\{\varphi_1,...,\varphi_n\} \not\models \varphi$, then the process does not terminate.

### Theorem 2.10 (Recursive enumerability of logical conclusion)

Let $\Sigma$ be a finite signature. There exists an algorithm that systematically generates all tuples $(\varphi_1,...,\varphi_n,\varphi)$ such that $\{\varphi_1,...,\varphi_n\} \models \varphi$. Thus, the set of all such tuples forms a recursively enumerable set of strings.

Next, we will prove that the enumeration algorithm presented above is the best that can be achieved from an algorithmic point of view. This means that it is impossible to develop a sharpened algorithm that also terminates in cases where $\{\varphi_1,...,\varphi_n\} \not\models \varphi$ with a message expressing this fact. This result forms Church's famous undecidability theorem for logical implication.

## 2.10    Undecidability of predicate logic

**Definition 2.12:**    The following problem is known as **Post's correspondence problem**: given a finite list $[(a_1,b_1),...,\{a_k,b_k)]$ consisting of pairs of non-empty strings over the alphabet $\{0,1\}$, find a sequence of indices $i_1,...,i_n$ such that $a_{i_1}...a_{i_n}=b_{i_1}...b_{i_n}$. Such a sequence $i_1,...,i_n$ is called a **solution** of the correspondence system $[(a_1,b_1),...,\{a_k,b_k)]$.

As an example, consider the correspondence system $[(1,101),(10,00), (011,11)]$. A possible solution for this system is the sequence 1,3,2,3 since $a_1a_3a_2a_3 = 101110011 = b_1b_3b_2b_3$.

*Theorem 2.11 (Post)*
Post's correspondence problem is undecidable, that is, there is no algorithm which, given a correspondence system $[(a_1,b_1),...,\{a_k,b_k)]$ as input, terminates with answer 'yes' if the given system has a solution, and with answer 'no' otherwise.

Now we are interested in investigating whether a formula is valid (or satisfiable). We shall use a well-known method from computability theory in order to bridge the gap between our concern and Post's correspondence problem.

**Definition 2.13:**    Let A and B be two sets of strings over a finite alphabet $\Omega$. We say that A is **reducible** to B iff there exists a total function $f:\Omega^*\to\Omega^*$ such that:

- f can be computed by an algorithm
- For every string x over $\Omega$, $x\in A$ iff $f(x)\in B$.

Function f is called a **reduction** of A to B.

Thus, if A is reducible to B via a function f, then every proposed question $x \in A$ can be effectively transformed into an equivalent question $f(x) \in B$. The following result is easy to show.

**Fact**  Let A be reducible to B. If B is decidable, then A is decidable, too. If B is recursively enumerable, then A is recursively enumerable, too.

### Undecidability Theorem 2.12 (Church)

Let $\Sigma$ be a signature with a constant c, 1-ary function symbols $f_0$ and $f_1$, and a 2-ary predicate symbol p. Then the set of all valid $\Sigma$-formulas is an undecidable set of strings.

*Proof*  The proof idea is to define an effective procedure that reduces Post's correspondence problem to validity of predicate logic formulas. First we formalize Post's correspondence problem in predicate logic, using the given signature $\Sigma$. For a non-empty bit-sequence $i_1 \ldots i_m$ and term t we use the abbreviation $f_{i_1 \ldots i_m}(t)$ for the term $f_{i_1}(\ldots(f_{i_m}(t)\ldots)$. For a correspondence problem $C=[(a_1,b_1),\ldots,(a_k,b_k)]$ we define a formula $((\varphi_1 \wedge \varphi_2) \rightarrow \varphi_3)$ and then show that:

$$C \text{ has a solution} \Leftrightarrow \ \models ((\varphi_1 \wedge \varphi_2) \rightarrow \varphi_3) \qquad (2.1)$$

The formulas are:

$$\varphi_1 = (p(f_{a_1}(c), f_{b_1}(c)) \wedge \ldots \wedge p(f_{a_k}(c), f_{b_k}(c)))$$
$$\varphi_2 = \forall X \forall Y (p(X,Y) \rightarrow (p(f_{a_1}(X), f_{b_1}(Y)) \wedge \ldots \wedge p(f_{a_k}(X), f_{b_k}(Y))))$$
$$\varphi_3 = \exists X p(X,X)$$

Further, we define an interpretation $A_C=(\{0,1\}^*, \varepsilon, F_0, F_1, P)$, where $\varepsilon$ is the empty word, $F_0(u)=0u$, $F_1(u)=1u$, and $P(u,v)$ holds iff there exist $n>0$, $i_1, \ldots, i_n \leq k$ such that $u=a_{i_1} \ldots a_{i_n}$ and $v=b_{i_1} \ldots b_{i_n}$.

It is easy to see that $A_C \models (\varphi_1 \wedge \varphi_2)$. Furthermore, $A_C \models \varphi_3$ iff K has a solution. Now we show (2.1). The direction from right to left is clear and follows from the definition of predicate P within $A_C$. To show the reversed implication, let C be solvable and $j_1, \ldots, j_n$ be a solution. Let $A=(\text{dom}(A), \varepsilon', F_0', F_1', P')$ be an interpretation such that $A \models (\varphi_1 \wedge \varphi_2)$. As for the terms above, we introduce the abbreviation $F_{i_1 \ldots i_m}'(a)$ for $F_{i_1}'(\ldots(F_{i_m}'(a)\ldots)$. (Note that $F_u'(F_v'(a)) = F_{uv}'(a)$.)

As $(\varphi_1 \wedge \varphi_2)$ is valid in $A$, $(F'_{a_{j_1}}(...(F'_{a_{j_n}}(\varepsilon')...)),F'_{b_{j_1}}(...(F'_{b_{j_n}}(\varepsilon')...)))$ $\in P'$, hence $(F'_{a_{j_1}...a_{j_n}}(\varepsilon'),F'_{b_{j_1}...b_{j_n}}(\varepsilon')) \in P'$. Then, $A \models \exists Xp(X,X)$. The proof of (2.1) and the theorem is thus completed.                                ∎

### Corollary

The set of all pairs $(\Sigma,\varphi)$, with a finite signature $\Sigma$ and a valid $\Sigma$-formula $\varphi$, is undecidable. This implies that the set of all sequences $(\varphi_1,...,\varphi_n,\varphi)$ such that $\varphi_1,...,\varphi_n,\varphi$ are $\Sigma$-formulas, for a finite signature $\Sigma$, and $\{\varphi_1,...,\varphi_n\} \models \varphi$ is undecidable, too.

*Proof*   Direct consequence of Theorem 2.12.                                ∎

### Corollary

It is undecidable whether a formula is satisfiable.

*Proof*   If this problem were decidable then so would be validity, since a formula $\varphi$ is satisfiable iff $\neg\exists(\varphi)$ is not valid (this is also a reduction between problems).                                ∎

# Chapter 3
# Proof Theory

Chapter 2 introduced the central notion of logical implication: $M \models \varphi$. A disadvantage of this notion, considered from an algorithmical point of view, is that it refers to infinite objects like interpretations. The purpose of this chapter is to replace these semantical notions completely by equivalent syntactical ones.

We shall introduce four 'games' to be played with formulas, whose purpose is to **derive** formulas via certain **rules** from a given set of initial formulas, called the **axioms**. Such games will be called **calculi**. These calculi will be **sound**, meaning for example for Hilbert's calculus H that every formula that can be derived from a set of formulas M also follows semantically from M, and **complete**, meaning conversely that every formula that follows from M is derivable from M. Soundness and completeness may link for other calculi derivability with other semantical properties.

The calculi studied here are:

- A **Hilbert calculus** H as a means for easily obtaining theoretical results.

- A **Gentzen calculus** LK as a means mimicking natural deduction.

- **Resolution calculus** RES as the basis for automated theorem proving.

- **Analytic tableaux**, a popular basis for automated theorem provers.

## 3.1    What is a calculus?

A calculus is a game with finite objects. It allows certain objects to be 'written down' or 'generated' or 'derived' or 'deduced'. The objects which may initially be written down are either elements of a given set M of objects or so-called **axioms** of the calculus. What the axioms of a calculus are is always introduced via the following notation:

$$\frac{}{\alpha}$$

Thus, an axiom like this allows us to write down $\alpha$. Having written down some objects, we may write down further ones according to so-called **rules**. We denote rules by

$$\frac{\pi_1\ \pi_2\ \dots\ \pi_n}{\gamma}$$

where $\pi_1,\pi_2,\dots,\pi_n$ are called the **premisses** of the rule and $\gamma$ its **conclusion**. A rule like that presented allows us to write down $\gamma$, provided that $\pi_1,\pi_2,\dots,\pi_n$ have already been written down (either as axioms or as results of prior rule applications). Looking at the usage of axioms and rules, axioms could be considered as rules without premisses (n=0). Usually, rules and axioms will be rule and axiom **schemata**, that is, forms with infinitely many instantiations. More formally, the notion of a calculus can be introduced as follows.

> **Definition 3.1:**    A **calculus** C over an alphabet $\Omega$ consists of a finite number of **rules**, each one of a certain finite arity n. A rule of arity n is a set of n+1 tuples $(\pi_1,\pi_2,\dots,\pi_n,\gamma)$ of strings $\pi_1,\pi_2,\dots,\pi_n,\gamma$ over $\Omega$. Each sequence $(\pi_1,\pi_2,\dots,\pi_n,\gamma)$ occurring in a rule is called an **instance of that rule**, with premisses $\pi_1,\pi_2,\dots,\pi_n$ and conclusion $\gamma$. The conclusions of 0-ary rules are called **axioms**. (Usually, rules will be decidable sets of n+1 tuples.)

> **Definition 3.2:**    Let C be a calculus over an alphabet $\Omega$ and M be a set of strings over $\Omega$. A **deduction** (or **derivation**) in C from M is a finite sequence $(\psi_1,\psi_2,\dots,\psi_m)$ of strings over $\Omega$ with the following property:
>
> For all i=1,...,m, either $\psi_i \in$ M or there exists an instance $(\pi_1,\pi_2,\dots,\pi_n,\gamma)$ of a rule of C such that $\gamma=\psi_i$ and $\pi_1,\pi_2,\dots,\pi_n$ are contained in $\{\psi_1,\dots,\psi_{i-1}\}$.

Deductions are often written vertically. Note how a deduction in C emerges from a set M. Having already constructed a deduction $(\psi_1, \psi_2, \ldots, \psi_{i-1})$ in C from M, we may extend it as follows to a sequence which again is a deduction in C from M:

- Extend $(\psi_1, \psi_2, \ldots, \psi_{i-1})$ by a formula $\psi_i \in M$.

- Choose an instance $(\pi_1, \pi_2, \ldots, \pi_n, \gamma)$ of a rule of C whose premisses are already contained within $\psi_1, \psi_2, \ldots, \psi_{i-1}$ and extend $(\psi_1, \psi_2, \ldots, \psi_{i-1})$ by the conclusion $\gamma$; this subsumes the case of rules without premisses, that is, $\psi_i$ is always allowed to be an axiom.

**Definition 3.3:** A string $\psi$ can be **derived** (**deduced**) in a calculus C from a set M of strings iff there exists a deduction in C from M with last element $\psi$. This property is denoted by $M \vdash_C \psi$.

### 3.1.1   Proofs by induction on the deduction length

Suppose we wished to show that for all strings $\psi$ with $M \vdash_C \psi$ a certain property P is fulfilled. This can be done by induction on the length of a deduction of $\psi$ as follows: Assuming that P is already shown to be true for all $\psi$ which can be derived in C from M with a derivation of length $<l$, show that P is true for all $\psi$ which can be derived in C from M with a derivation of length $=l$.

**Lemma 3.1**

If $M_1 \vdash_C \pi_1, \ldots, M_n \vdash_C \pi_n$ and $(\pi_1, \ldots, \pi_n, \gamma)$ is an instance of a rule of C, then $M_1 \cup \ldots \cup M_n \vdash_C \gamma$.

If $M_1 \vdash_C \psi$ and $M_2 \cup \{\psi\} \vdash_C \varphi$, then $M_1 \cup M_2 \vdash_C \varphi$.

*Proof*    The first statement is shown by sticking together deductions of $\pi_1, \ldots, \pi_n$ in C from $M_1, \ldots, M_n$, respectively, and extending the resulting derivation in C from $M_1 \cup \ldots \cup M_n$ by the string $\gamma$. The second statement is shown by induction on the length $l$ of a deduction of $\varphi$ from $M_2 \cup \{\psi\}$.    ∎

Sometimes it is more perspicuous to represent a deduction in C from M by a tree instead of a sequence. This leads to the concept of a **deduction tree**.

**Definition 3.4:**    Let C be a calculus over an alphabet $\Omega$. A **deduction**

tree (or **derivation tree**) in C is a finite tree T whose nodes are labelled with strings over $\Omega$ in the following way. If a non-leaf node is labelled with $\gamma$ and its predecessor nodes are labelled (from left to right) with $\pi_1,\ldots,\pi_n$, then $(\pi_1,\ldots,\pi_n,\gamma)$ is an instance of a rule of C.

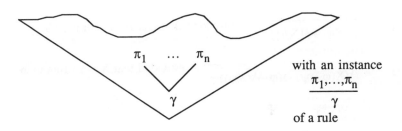

with an instance

$$\frac{\pi_1,\ldots,\pi_n}{\gamma}$$

of a rule

**Fact**    $M \vdash_C \psi$ iff there is a deduction tree in C whose root is labelled with $\psi$ and whose leaves are labelled with axioms of C or strings from M.

## 3.2    A Hilbert calculus H

The first kind of calculi developed for predicate logic were similar to the one presented in this section. All the formulas that occur in this section are so-called **restricted formulas**, that is, they contain only the logical symbols $\neg$, $\rightarrow$ and $\forall$. Faced with the results from Section 2.5 (see Exercises 2.6-2.18), restriction to this subclass of formulas does not affect the generality of our results. The advantage is that it leads to a very compact calculus well suited for theoretical investigations.

### 3.2.1    The axioms and rules of calculus H

The table below shows the axioms and rules of Hilbert's calculus H ($\varphi,\psi$ and $\chi$ are $\Sigma$-formulas, t is a $\Sigma$-term and X is a variable). Hilbert's calculus consists of five axiom schemata, $Ax_1$-$Ax_5$, and two rules, MP and GEN. MP is **modus ponens**; GEN is the **generalization rule**.

If a $\Sigma$-formula can be derived from axioms $Ax_1$-$Ax_3$ and MP as the only rule, it is called a **tautology**.

For a set V of variables, we denote by $H(\Sigma,V)$ the subcalculus of $H(\Sigma)$ with the restriction of GEN to variables from V. Usually we write H instead of $H(\Sigma)$, and $H(V)$ instead of $H(\Sigma,V)$.

$Ax_1$     $\dfrac{}{(\varphi\to(\psi\to\varphi))}$

$Ax_2$     $\dfrac{}{((\varphi\to(\psi\to\chi))\to((\varphi\to\psi)\to(\varphi\to\chi)))}$

$Ax_3$     $\dfrac{}{((\neg\varphi\to\neg\psi)\to(\psi\to\varphi))}$

$Ax_4$     $\dfrac{}{(\forall X\varphi\to\varphi\{X/t\})}$     provided that $\varphi\{X/t\}$ is admissible

$Ax_5$     $\dfrac{}{(\forall X(\varphi\to\psi)\to(\varphi\to\forall X\psi))}$     provided that $X$ isn't free in $\varphi$

MP     $\dfrac{(\varphi\to\psi)\qquad\varphi}{\psi}$

GEN     $\dfrac{\varphi}{\forall X\varphi}$

As a first example for a deduction in H, we show that $(\varphi\to\varphi)$ can be derived in H from the empty set (it is in fact a tautology). It is quite astonishing that the derivation of such a trivial formula can be so complicated.

(1) $((\varphi\to((\varphi\to\varphi)\to\varphi))\to((\varphi\to(\varphi\to\varphi))\to(\varphi\to\varphi)))$     $(Ax_1)$

(2) $(\varphi\to((\varphi\to\varphi)\to\varphi))$     $(Ax_1)$

(3) $((\varphi\to(\varphi\to\varphi))\to(\varphi\to\varphi))$     (MP)

(4) $(\varphi\to(\varphi\to\varphi))$     $(Ax_1)$

(5) $(\varphi\to\varphi)$     (MP)

## 3.2.2 Soundness of Hilbert's calculus H

### Theorem 3.1

For every formula set M and formula $\varphi$, if $M \vdash_H \varphi$ then $M \models \varphi$.

The proof of this theorem by induction on the length of a deduction is a direct consequence of Lemmas 3.2 and 3.3 which state that axioms of H are valid, and rules of H preserve validity of formulas.

### Lemma 3.2

Every instance of an axiom of H is valid.

*Proof* Given an instance $\varphi$ of an axiom of H, an interpretation $A$ and a state sta over $A$, we must show that $A \vDash_{sta}\varphi$. We start with an instance $(\varphi \to (\psi \to \varphi))$ of Ax$_1$. Assume that $A \nvDash_{sta}(\varphi \to (\psi \to \varphi))$. Successively we obtain:

$A \vDash_{sta}\varphi$

$A \nvDash_{sta}(\psi \to \varphi)$

$A \vDash_{sta}\psi$

$A \nvDash_{sta}\varphi$

Thus a contradiction is obtained.

Next we consider an instance $((\varphi \to (\psi \to \chi)) \to ((\varphi \to \psi) \to (\varphi \to \chi)))$ of Ax$_2$. Assume that $A \nvDash_{sta}((\varphi \to (\psi \to \chi)) \to ((\varphi \to \psi) \to (\varphi \to \chi)))$. Successively we obtain:

| | |
|---|---|
| $A \vDash_{sta}(\varphi \to (\psi \to \chi))$ | (a) |
| $A \nvDash_{sta}((\varphi \to \psi) \to (\varphi \to \chi)))$ | (b) |
| $A \vDash_{sta}(\varphi \to \psi)$ | (c), obtained from (b) |
| $A \nvDash_{sta}(\varphi \to \chi)$ | (d), obtained from (b) |
| $A \vDash_{sta}\varphi$ | (e), obtained from (d) |
| $A \nvDash_{sta}\chi$ | (f), obtained from (d) |
| $A \vDash_{sta}(\psi \to \chi)$ | (g), obtained from (a) and (e) |
| $A \vDash_{sta}\psi$ | (h), obtained from (c) and (e) |
| $A \vDash_{sta}\chi$ | (i), obtained from (g) and (h) |

(f) and (i) contradict each other.

Finally, consider an instance $((\neg\varphi \to \neg\psi) \to (\psi \to \varphi))$ of Ax$_3$. Assume that $A \nvDash_{sta}((\neg\varphi \to \neg\psi) \to (\psi \to \varphi))$. Successively we obtain:

$A \vDash_{sta}(\neg\varphi \to \neg\psi)$

$A \nvDash_{sta}(\psi \to \varphi)$

$A \vDash_{sta}\psi$

$A \nvDash_{sta}\varphi$

$A \vDash_{sta}\neg\varphi$

$A \vDash_{sta}\neg\psi$

which is a contradiction. The validity of an instance of $Ax_4$ or $Ax_5$ has already been shown in Sections 2.3 and 2.6.    ∎

### Lemma 3.3

MP and GEN preserve validity.

*Proof* See Section 2.4.    ∎

---

**Exercise 3.1:**    Show that the formulas in $Ax_4$ and $Ax_5$ of H are not valid, in general, if we omit the restrictions posed on the forms of these formulas.

---

## 3.2.3    Working with H

H is rather inflexible concerning the development of concrete deductions. It is far easier to use the following deduction theorem, which allows switching between the use of a formula as an axiom in an axiom set M, and its use as the left-hand side of an implication. Thus we want to establish a correlation between statements $M \cup \{\varphi\} \vdash_H \psi$ and $M \vdash_H (\varphi \rightarrow \psi)$. Of course, the latter always implies the former by an additional application of modus ponens. However, in general the reverse implication is not true.

This can be easily seen from the following example: $\{p(X)\} \vdash_H \forall X p(X)$ by a single application of the generalization rule, with generalized variable X, whereas $\vdash_H (p(X) \rightarrow \forall X p(X))$ is not true (the fact that x=3 is a prime number does not imply that every x is prime). We will see that the reason that prevented a shift of p(X) from $\{p(X)\} \vdash_H \forall X p(X)$ to $\vdash_H (p(X) \rightarrow \forall X p(X))$ was the generalization over the free variable X of p(X). If we forbid such a generalization the desired shift is possible. This can be achieved by using the restricted Hilbert calculus H(V) which allows generalization only over variables from V.

### Deduction Theorem 3.2

Let M be a set of formulas, V a set of variables, $\varphi$ and $\psi$ formulas. If $M \vdash_H (\varphi \rightarrow \psi)$ then $M \cup \{\varphi\} \vdash_H \psi$. If $M \cup \{\varphi\} \vdash_{H(V)} \psi$ and no variable from V occurs free in $\varphi$, then $M \vdash_{H(V)} (\varphi \rightarrow \psi)$. In particular, if $\varphi$ is a closed formula, then $M \cup \{\varphi\} \vdash_H \psi$ implies $M \vdash_H (\varphi \rightarrow \psi)$.

*Proof* The first statement is trivial. The third statement follows from the second one by using as V the set of all variables. The second statement is shown by induction on the length of a deduction of $\psi$ in H(V) from $M\cup\{\varphi\}$.

Assume that $\psi$ is derived in H(V) from $M\cup\{\varphi\}$ with a deduction of length 1. We distinguish whether $\psi$ is an instance of an axiom, an element of $M\cup\{\varphi\}$, obtained by an application of modus ponens, or obtained by an application of generalization with a generalized variable $X\in V$ that does not occur free in $\varphi$. If $\psi$ is an instance of an axiom or an element of M we obtain the following deduction in H from $M\cup\{\varphi\}$ without any application of generalization:

| | |
|---|---|
| $\psi$ | (as an axiom or element of M) |
| $(\psi\rightarrow(\varphi\rightarrow\psi))$ | ($Ax_1$) |
| $(\varphi\rightarrow\psi)$ | (MP) |

Since we did not use GEN at all, it follows that $M\vdash_{\overline{H(V)}} (\varphi\rightarrow\psi)$. If $\psi$ coincides with $\varphi$ (note that our derivation of $\psi$ is from axiom set $M\cup\{\varphi\}$), then a look at the example worked out in Section 3.2.1 proves the desired result: $M\vdash_{\overline{H(V)}} (\varphi\rightarrow\varphi)$. If $\psi$ is obtained from $(\chi\rightarrow\psi)$ and $\chi$ by an application of modus ponens, we may assume by induction hypothesis that $M\vdash_{\overline{H(V)}} (\varphi\rightarrow\chi)$ and $M\vdash_{\overline{H(V)}} (\varphi\rightarrow(\chi\rightarrow\psi))$. With $Ax_2$ we obtain $M\vdash_{\overline{H(V)}} (((\varphi\rightarrow(\chi\rightarrow\psi))\rightarrow ((\varphi\rightarrow\chi)\rightarrow(\varphi\rightarrow\psi)))$. Two applications of modus ponens yield that $M\vdash_{\overline{H(V)}} (\varphi\rightarrow\psi)$, the desired result.

Finally, assume that $\psi$ is obtained from a formula $\chi$ by an application of GEN with generalized variable $X\in V$. Hence, $\psi$ is the formula $\forall X\chi$ and X is not free in $\varphi$. By induction hypothesis we know that $M\vdash_{\overline{H(V)}} (\varphi\rightarrow\chi)$. An application of GEN yields $M\vdash_{\overline{H(V)}} \forall X(\varphi\rightarrow\chi)$ (note that X was an element of V). Now we apply $Ax_5$ to derive $(\forall X(\varphi\rightarrow\chi)\rightarrow(\varphi\rightarrow\forall X\chi))$ (note that X is not free in $\varphi$) and modus ponens to obtain $M\vdash_{\overline{H(V)}} (\varphi\rightarrow\forall X\chi)$. ∎

From now on we will extensively use Theorem 3.2. The attentive reader will have noticed that the proof above used $Ax_1$, $Ax_2$ and $Ax_5$. It can indeed be shown that the use of the deduction theorem is a complete substitute for these axioms. From now on we will no longer be forced to construct concrete deductions, but may use the deduction theorem as a **meta-theorem** yielding derivability statements (meta-theorems) of the form $M\vdash_{\overline{H}} \varphi$. The prefix 'meta' is a hint that we do not reason *within* the calculus, but rather *about* the calculus.

### Tautology lemma 3.4

Let $\varphi$, $\psi$ and $\chi$ be formulas. Then, formulas (1)-(12) are tautologies:

| | |
|---|---|
| (1) | $(\varphi \to (\psi \to \psi))$ |
| (2) | $((\varphi \to \psi) \to ((\psi \to \chi) \to (\varphi \to \chi)))$ |
| (3) | $(\neg\neg\varphi \to \varphi)$ |
| (4) | $(\varphi \to \neg\neg\varphi)$ |
| (5) | $((\varphi \to \psi) \to (\neg\psi \to \neg\varphi))$ |
| (6) | $(\varphi \to (\neg\psi \to \neg(\varphi \to \psi)))$ |
| (7) | $(\neg\varphi \to (\varphi \to \psi))$ |
| (8) | $(\neg(\varphi \to \psi) \to \varphi)$ |
| (9) | $(\neg(\varphi \to \psi) \to \neg\psi)$ |
| (10) | $((\varphi \to \neg\varphi) \to \neg\varphi)$ |
| (11) | $((\neg\varphi \to \varphi) \to \varphi)$ |
| (12) | $(\neg(\varphi \to \varphi) \to \psi)$ |

*Proof* To obtain the following derivations, application of GEN is not needed. So Theorem 3.2 (DT) may be applied in an unlimited manner.

(1)   $\{\varphi, \psi\} \vdash_{\overline{H}} \psi$

   $\{\varphi\} \vdash_{\overline{H}} (\psi \to \psi)$                                        DT

   $\vdash_{\overline{H}} (\varphi \to (\psi \to \psi))$                                     DT

(2)   $\{(\varphi \to \psi), (\psi \to \chi), \varphi\} \vdash_{\overline{H}} \varphi$

   $\{(\varphi \to \psi), (\psi \to \chi), \varphi\} \vdash_{\overline{H}} (\varphi \to \psi)$

   $\{(\varphi \to \psi), (\psi \to \chi), \varphi\} \vdash_{\overline{H}} \psi$                        MP

   $\{(\varphi \to \psi), (\psi \to \chi), \varphi\} \vdash_{\overline{H}} (\psi \to \chi)$

   $\{(\varphi \to \psi), (\psi \to \chi), \varphi\} \vdash_{\overline{H}} \chi$                        MP

   $\{(\varphi \to \psi), (\psi \to \chi)\} \vdash_{\overline{H}} (\varphi \to \chi)$                     DT

   $\{(\varphi \to \psi)\} \vdash_{\overline{H}} ((\psi \to \chi) \to (\varphi \to \chi))$                  DT

   $\vdash_{\overline{H}} ((\varphi \to \psi) \to ((\psi \to \chi) \to (\varphi \to \chi)))$               DT

(3)   $\vdash_{\overline{H}} (\neg\neg\varphi \to (\neg\neg\neg\neg\varphi \to \neg\neg\varphi))$                 (Ax$_1$)

   $\{\neg\neg\varphi\} \vdash_{\overline{H}} (\neg\neg\neg\neg\varphi \to \neg\neg\varphi)$                  DT

   $\{\neg\neg\varphi\} \vdash_{\overline{H}} ((\neg\neg\neg\neg\varphi \to \neg\neg\varphi) \to (\neg\varphi \to \neg\neg\neg\varphi))$     (Ax$_3$)

   $\{\neg\neg\varphi\} \vdash_{\overline{H}} (\neg\varphi \to \neg\neg\neg\varphi)$                        MP

   $\{\neg\neg\varphi\} \vdash_{\overline{H}} ((\neg\varphi \to \neg\neg\neg\varphi) \to (\neg\neg\varphi \to \varphi))$     (Ax$_3$)

   $\{\neg\neg\varphi\} \vdash_{\overline{H}} (\neg\neg\varphi \to \varphi)$                              MP

$\{\neg\neg\varphi\} \vdash_{\mathrm{H}} \varphi$  ⠀⠀⠀⠀⠀⠀⠀⠀⠀DT

$\vdash_{\mathrm{H}} (\neg\neg\varphi\to\varphi)$  ⠀⠀⠀⠀⠀⠀⠀⠀DT

(4)⠀⠀$\vdash_{\mathrm{H}} (\neg\neg\neg\varphi\to\neg\varphi)$  ⠀⠀⠀⠀⠀⠀⠀(3)

$\vdash_{\mathrm{H}} ((\neg\neg\neg\varphi\to\neg\varphi)\to(\varphi\to\neg\neg\varphi))$  ⠀⠀$(\mathrm{Ax}_3)$

$\vdash_{\mathrm{H}} (\varphi\to\neg\neg\varphi)$  ⠀⠀⠀⠀⠀⠀MP

(5)⠀⠀$\{(\varphi\to\psi)\} \vdash_{\mathrm{H}} ((\neg\neg\varphi\to\neg\neg\psi)\to(\psi\neg\varphi\to\neg\varphi))$  ⠀$(\mathrm{Ax}_3)$

$\{(\varphi\to\psi)\} \vdash_{\mathrm{H}} (\neg\neg\varphi\to\varphi)$  ⠀⠀⠀(3)

$\{(\varphi\to\psi)\} \vdash_{\mathrm{H}} (\varphi\to\varphi)$

$\{(\varphi\to\psi)\} \vdash_{\mathrm{H}} (\neg\neg\varphi\to\psi)$  ⠀⠀⠀(2)+MP+MP

$\{(\varphi\to\psi)\} \vdash_{\mathrm{H}} (\psi\to\neg\neg\psi)$  ⠀⠀⠀(4)

$\{(\varphi\to\psi)\} \vdash_{\mathrm{H}} (\neg\neg\varphi\to\neg\neg\varphi)$  ⠀(2)+MP+MP

$\{(\varphi\to\psi)\} \vdash_{\mathrm{H}} (\neg\psi\to\neg\varphi)$  ⠀⠀⠀MP

$\vdash_{\mathrm{H}} ((\varphi\to\psi)\to(\neg\psi\to\neg\varphi))$  ⠀⠀DT

(6)⠀⠀$\{\varphi,(\varphi\to\psi)\} \vdash_{\mathrm{H}} \varphi$

$\{\varphi,(\varphi\to\psi)\} \vdash_{\mathrm{H}} (\varphi\to\varphi)$

$\{\varphi,(\varphi\to\varphi)\} \vdash_{\mathrm{H}} \psi$  ⠀⠀⠀MP

$\vdash_{\mathrm{H}} (\varphi\to((\varphi\to\psi)\to\psi))$  ⠀⠀DT

$\{\varphi\} \vdash_{\mathrm{H}} ((\varphi\to\varphi)\to\psi)$  ⠀⠀⠀DT

$\{\varphi\} \vdash_{\mathrm{H}} (((\varphi\to\psi)\to\varphi)\to(\neg\psi\to\neg(\varphi\to\psi)))$  ⠀(5)

$\{\varphi\} \vdash_{\mathrm{H}} (\neg\psi\to\neg(\varphi\to\psi))$  ⠀⠀MP

$\vdash_{\mathrm{H}} (\varphi\to(\neg\psi\to\neg(\varphi\to\psi)))$  ⠀DT

(7)⠀⠀$\vdash_{\mathrm{H}} (\neg\varphi\to(\neg\psi\to\neg\varphi))$  ⠀⠀⠀$(\mathrm{Ax}_1)$

$\{\neg\varphi\} \vdash_{\mathrm{H}} (\neg\psi\to\neg\varphi)$  ⠀⠀⠀DT

$\{\neg\varphi\} \vdash_{\mathrm{H}} ((\neg\psi\to\neg\varphi)\to(\varphi\to\psi))$  ⠀$(\mathrm{Ax}_3)$

$\{\neg\varphi\} \vdash_{\mathrm{H}} (\varphi\to\psi)$  ⠀⠀⠀⠀MP

$\vdash_{\mathrm{H}} (\neg\varphi\to(\varphi\to\psi))$  ⠀⠀⠀DT

(8)⠀⠀$\vdash_{\mathrm{H}} (\neg\varphi\to(\varphi\to\psi))$  ⠀⠀⠀(7)

$\vdash_{\mathrm{H}} (\neg(\varphi\to\psi)\to\neg\neg\varphi)$  ⠀⠀(5)+MP

$$\vdash_{\overline{H}} (\neg\neg\varphi \to \varphi) \qquad\qquad (3)$$

$$\vdash_{\overline{H}} (\neg(\varphi \to \psi) \to \varphi) \qquad\qquad (2)+MP+MP$$

(9) $\quad \vdash_{\overline{H}} (\psi \to (\varphi \to \varphi)) \qquad\qquad (Ax_1)$

$$\vdash_{\overline{H}} (\neg(\varphi \to \psi) \to \neg\varphi_2) \qquad\qquad (5)+MP$$

(10) $\quad \vdash_{\overline{H}} (\varphi \to (\neg\neg\varphi \to \neg(\varphi \to \neg\varphi))) \qquad\qquad (6)$

$$\{\varphi\} \vdash_{\overline{H}} (\neg\neg\varphi \to \neg(\varphi \to \neg\varphi)) \qquad\qquad DT$$

$$\{\varphi\} \vdash_{\overline{H}} \neg\neg\varphi \qquad\qquad (4)+DT$$

$$\{\varphi\} \vdash_{\overline{H}} \neg(\varphi \to \neg\varphi) \qquad\qquad MP$$

$$\vdash_{\overline{H}} ((\varphi \to \neg(\varphi \to \neg\varphi)) \to (\neg\neg(\varphi \to \neg\varphi) \to \neg\varphi)) \qquad\qquad (5)$$

$$\vdash_{\overline{H}} (\neg\neg(\varphi \to \neg\varphi) \to \neg\varphi) \qquad\qquad MP$$

$$\vdash_{\overline{H}} ((\varphi \to \neg\varphi) \to \neg\neg(\varphi \to \neg\varphi)) \qquad\qquad (4)$$

$$\vdash_{\overline{H}} ((\varphi \to \neg\varphi) \to \neg\varphi) \qquad\qquad (2)+MP+MP$$

(11) $\quad \vdash_{\overline{H}} ((\neg\varphi \to \neg\neg\varphi) \to \neg\neg\varphi) \qquad\qquad (10)$

Now use (3) and (4).

(12) $\quad \{\neg\psi\} \vdash_{\overline{H}} (\varphi \to \varphi) \qquad\qquad$ example

$$\vdash_{\overline{H}} ((\varphi \to \varphi) \to \neg\neg(\varphi \to \varphi)) \qquad\qquad (4)$$

$$\{\neg\psi\} \vdash_{\overline{H}} \neg\neg(\varphi \to \varphi) \qquad\qquad MP$$

$$\vdash_{\overline{H}} (\neg\psi \to \neg\neg(\varphi \to \varphi)) \qquad\qquad DT$$

$$\vdash_{\overline{H}} (\neg(\varphi \to \varphi) \to \psi) \qquad\qquad (Ax_3)+MP$$

■

So far we have derived tautologies, that is, we have made use only of $Ax_1$-$Ax_3$ and MP. Next we deal with quantifiers. Central to the following completeness proof of Hilbert's calculus H will be the possibility of renaming bound variables. Surprisingly, almost nothing else is needed for the completeness proof.

### Renaming Theorem 3.3

If Y does not occur free in $\varphi$ and if $\varphi\{X/Y\}$ is admissible, then $\vdash_{\overline{H}} (\forall X\varphi \to \forall Y\varphi\{X/Y\})$ and $\vdash_{\overline{H}} (\forall Y\varphi\{X/Y\} \to \forall X\varphi)$.

*Proof*    We successively obtain the following derivability statements (note that $H(\emptyset)$ is Hilbert's calculus without the generalization rule):

$$\{\forall X\varphi\} \vdash_{\overline{H(\varnothing)}} (\forall X\varphi \rightarrow \varphi\{X/Y\}) \qquad (Ax_4)$$

$$\{\forall X\varphi\} \vdash_{\overline{H(\varnothing)}} \forall X\varphi$$

$$\{\forall X\varphi\} \vdash_{\overline{H(\varnothing)}} \varphi\{X/Y\} \qquad\qquad MP$$

$$\{\forall X\varphi\} \vdash_{\overline{H(\{Y\})}} \forall Y\varphi\{X/Y\} \qquad\qquad GEN$$

$$\vdash_{\overline{H(\{Y\})}} (\forall X\varphi \rightarrow \forall Y\varphi\{X/Y\}) \qquad DT, \ Y \text{ is not free in } \forall X\varphi$$

$$\vdash_{\overline{H}} (\forall X\varphi \rightarrow \forall Y\varphi\{X/Y\}).$$

For the reverse direction we need the following results. For a term t, a formula $\varphi$, and variables X and Y:

(a)    If Y does not occur in t, then X does not occur in $t\{X/Y\}$ and $t\{X/Y\}\{Y/X\}$ is t.

(b)    If Y does not occur free in $\varphi$ and $\varphi\{X/Y\}$ is admissible, then

    (i)    X does not occur free in $\varphi\{X/Y\}$

    (ii)    $\varphi\{X/Y\}\{Y/X\}$ is admissible and

    (iii)    $\varphi\{X/Y\}\{Y/X\}$ is $\varphi$

Using these results we are able to complete the proof of the renaming theorem. We apply the already shown direction of the theorem on $\varphi\{X/Y\}$ instead of $\varphi$, X instead of Y and Y instead of X. The precondition for the application of this direction of the theorem as well as the preconditions of (a) and (b) above hold, therefore $\vdash_{\overline{H}} (\forall Y\varphi\{X/Y\} \rightarrow \forall X\varphi\{X/Y\}\{Y/X\})$, that is, $\vdash_{\overline{H}} (\forall Y\varphi\{X/Y\} \rightarrow \forall X\varphi)$.

So we are left with a proof of (a) and (b). Part (a) is clear. Part (b) is obtained as follows. We may assume that X and Y are different variables. Let us decompose $\varphi$ as a string into $a_1 X a_2 X....a_i X a_{i+1} X....a_k X a_{k+1}$ such that the indicated occurrences of X are all free occurrences of X in $\varphi$. These occurrences of X are thus not within the scope of a quantifier QX. Since $\varphi\{X/Y\}$ is admissible we furthermore know that these occurrences of X are not within the scope of a quantifier QY. Now let us consider $\varphi\{X/Y\}$, which is the string $a_1 Y a_2 Y....a_i Y a_{i+1} Y....a_k Y a_{k+1}$. The indicated occurrences of Y are *free occurrences* since the considered occurrences of X were not within the scope of a quantifier QX. In fact, they represent *all* free occurrences of Y in $\varphi\{X/Y\}$ as Y did not occur free in $\varphi$. Finally, the considered occurrences of Y are not within a quantifier QX, as was stated above. Hence $\varphi\{X/Y\}\{Y/X\}$ is admissible and $\varphi\{X/Y\}\{Y/X\}=\varphi$. ∎

**Exercise 3.2**   Give derivations (from the empty set) for $(\forall X\varphi \rightarrow \exists X\varphi)$ and $(\exists X\forall Y\varphi \rightarrow \forall Y\exists X\varphi)$ in H. Is it possible to give derivations if we change the implication direction?

**Exercise 3.3** (Equivalence theorem) Let $\psi_1$ be a subformula of $\varphi_1$. Let $\varphi_2$ be the formula obtained from $\varphi_1$ by replacing some occurrences of $\psi_1$ by $\psi_2$. Let $Y_1,\ldots,Y_k$ be all free variables of $\psi_1$ and $\psi_2$. Show that

$$\vdash_{\overline{H}} \forall Y_1\ldots\forall Y_k((\psi_1 \leftrightarrow \psi_2) \rightarrow (\varphi_1 \leftrightarrow \varphi_2)).$$

## 3.2.4   Consistency and completeness of formula sets

**Definition 3.5:**   Let M be a set of $\Sigma$-formulas. M is called $\Sigma$-**consistent** iff there does not exist a $\Sigma$-formula $\varphi$ such that $M\vdash_{\overline{H}} \varphi$ and $M\vdash_{\overline{H}} \neg\varphi$. M is called $\Sigma$-**complete** iff for every closed $\Sigma$-formula $\varphi$, $M\vdash_{\overline{H}} \varphi$ or $M\vdash_{\overline{H}} \neg\varphi$ (we shall see below that the restriction to closed formulas is natural and necessary). If $\Sigma$ is clear from the context, we simply talk of consistent and complete sets instead of $\Sigma$-consistent and $\Sigma$-complete sets.

We speak of completeness of sets of formulas although there is also the notion of completeness of calculi. We adopt this terminology because it is standard in the literature. It should not be difficult for the reader to distinguish these two concepts of completeness.

### Lemma 3.5

Let M be a set of $\Sigma$-formulas and $\varphi$ a $\Sigma$-formula.

(1)   M is $\Sigma$-consistent $\Leftrightarrow$

there does not exist a $\Sigma$-formula $\psi$ such that $M\vdash_{\overline{H}} \neg(\psi \rightarrow \psi) \Leftrightarrow$

there exists a $\Sigma$-formula $\psi$ that cannot be derived from M.

(2)   If M has a model, then M is $\Sigma$-consistent.

(3)   $M\vdash_{\overline{H}} \varphi \Leftrightarrow M\vdash_{\overline{H}} \forall(\varphi) \Leftrightarrow M\cup\{\neg\forall(\varphi)\}$ is not $\Sigma$-consistent.

*Proof* (1) Let M be $\Sigma$-consistent. Suppose there is a $\Sigma$-formula $\psi$ with $M\vdash_{\overline{H}}\neg(\psi\rightarrow\psi)$. By the Lemma 3.4 we obtain that $M\vdash_{\overline{H}}\psi$ and $M\vdash_{\overline{H}}\neg\psi$, a contradiction.

Suppose there does not exist a $\Sigma$-formula $\psi$ such that $M\vdash_{\overline{H}}\neg(\psi\rightarrow\psi)$. Let $\psi_0$ be an arbitrary $\Sigma$-formula. $\neg(\psi_0\rightarrow\psi_0)$ is then a $\Sigma$-formula not derivable from M.

Finally, suppose there exists a $\Sigma$-formula $\psi$ that cannot be derived from M. If M were not $\Sigma$-consistent, there would exist a $\Sigma$-formula $\varphi$ such that $M\vdash_{\overline{H}}\varphi$ and $M\vdash_{\overline{H}}\neg\varphi$. Using formula (7) from Lemma 3.4 we obtain $M\vdash_{\overline{H}}\psi$, a contradiction.

(2) Let $A$ be a model of M. If M is not $\Sigma$-consistent, there exists a $\Sigma$-formula $\varphi$ with $M\vdash_{\overline{H}}\varphi$ and $M\vdash_{\overline{H}}\neg\varphi$. Since H is already shown to be sound, it follows that $A\models\varphi$ and $A\models\neg\varphi$, a contradiction.

(3) The equivalence of statements $M\vdash_{\overline{H}}\varphi$ and $M\vdash_{\overline{H}}\forall(\varphi)$ is trivial because of the presence of $Ax_4$ and GEN. Suppose that $M\vdash_{\overline{H}}\varphi$. Then $M\cup\{\neg\forall(\varphi)\}\vdash_{\overline{H}}\varphi$, hence $M\cup\{\neg\forall(\varphi)\}\vdash_{\overline{H}}\forall(\varphi)$. Since also $M\cup\{\neg\forall(\varphi)\}\vdash_{\overline{H}}\neg\forall(\varphi)$, we obtain that $M\cup\{\neg\forall(\varphi)\}$ is not $\Sigma$-consistent. Conversely, suppose that $M\cup\{\neg\forall(\varphi)\}$ is not $\Sigma$-consistent. From (1) we know that there exists a $\Sigma$-formula $\psi$ such that $M\cup\{\neg\forall(\varphi)\}\vdash_{\overline{H}}\neg(\psi\rightarrow\psi)$. From point 11 of Lemma 3.4 it follows by MP $M\cup\{\neg\forall(\varphi)\}\vdash_{\overline{H}}\forall(\varphi)$. By Theorem 3.2 (it is applicable because $\neg\forall(\varphi)$ is closed) we obtain $M\vdash_{\overline{H}}(\neg\forall(\varphi)\rightarrow\forall(\varphi))$. From formula (11) of Lemma 3.4 we obtain that $M\vdash_{\overline{H}}((\neg\forall(\varphi)\rightarrow\forall(\varphi))\rightarrow\forall(\varphi))$. Finally, modus ponens implies $M\vdash_{\overline{H}}\forall(\varphi)$.    ∎

Note that statement (3) of Lemma 3.5 requires the universal closure in $M\cup\{\neg\forall(\varphi)\}$. In general, inconsistency of $M\cup\{\neg\varphi\}$ does not imply $M\vdash_{\overline{H}}\varphi$, as the example $M=\{\neg\forall X\neg p(X)\}$ and $\varphi=p(X)$ shows. (As an exercise, show that $M\vdash_{\overline{H}}\varphi$ does not hold.)

### Extension Theorem 3.4

Every $\Sigma$-consistent set M of formulas can be extended to a $\Sigma$-consistent and $\Sigma$-complete formula set M'.

*Proof* Let M be $\Sigma$-consistent. We enumerate all closed $\Sigma$-formulas $\varphi_0,\varphi_1,\varphi_2,\ldots$ (remember that signatures were defined as denumerable sets; if we admit arbitrary cardinalities we are forced to use transfinite enumeration and induction). Then we define an increasing sequence of sets of $\Sigma$-formulas

$M_0 \subseteq M_1 \subseteq M_2, \ldots$ as follows:

$$M_0 = M$$

$$M_{i+1} = M_i, \text{ if } M_i \vdash_{\overline{H}} \varphi_i, \text{ and } M_{i+1} = M_i \cup \{\neg \varphi_i\} \text{ otherwise}$$

By induction on i we show that every set $M_i$ is $\Sigma$-consistent. $M_0$ is $\Sigma$-consistent by definition and assumption on M, $M_{i+1}$ by induction hypothesis and part (3) of Lemma 3.5. Then $M' = \bigcup_{i \geq 0} M_i$ is the desired set of formulas. $M \subseteq M'$ is clear. M' is also $\Sigma$-consistent: suppose $M' \vdash_{\overline{H}} \neg(\psi \rightarrow \psi)$, for some $\Sigma$-formula $\psi$. Since only a finite number of  elements of M' occur in a derivation of $\neg(\psi \rightarrow \psi)$, there exists a number k with $M_k \vdash_{\overline{H}} \neg(\psi \rightarrow \psi)$. This is a contradiction to the consistency of $M_k$. Finally, M' is complete. Let $\varphi$ be an arbitrary closed formula. $\varphi$ occurs in the above enumeration as $\varphi_i$. If $M_i \vdash_{\overline{H}} \varphi_i$, then $M' \vdash_{\overline{H}} \varphi_i$. Else by definition $\varphi_i \in M_{i+1}$, hence $\varphi_i \in M'$ and $M' \vdash_{\overline{H}} \varphi_i$.     ∎

Let M be $\{\neg \forall X \neg p(X), \neg \forall X p(X)\}$. M is consistent, but there exists no consistent extension M' of M such that $M \vdash_{\overline{H}} \psi$ or $M \vdash_{\overline{H}} \neg \psi$, for *all formulas* $\psi$. Otherwise, either $M' \vdash_{\overline{H}} p(X)$ or $M' \vdash_{\overline{H}} \neg p(X)$, hence either $M' \vdash_{\overline{H}} \forall X p(X)$ or $M' \vdash_{\overline{H}} \forall X \neg p(X)$ (by generalization). In both cases M' would be inconsistent. This explains why we restricted consideration to closed formulas in the definition of completeness of an axiom system.

### Lemma 3.6

Let M be a set of formulas, $\varphi$ and $\psi$ formulas.

(1)   $M \vdash_{\overline{H}} \neg(\varphi \rightarrow \psi) \Leftrightarrow M \vdash_{\overline{H}} \varphi$ and $M \vdash_{\overline{H}} \neg \psi$.

(2)   If M is consistent and complete and $\varphi$ is closed, then

$$M \vdash_{\overline{H}} \neg \varphi \Leftrightarrow \text{ not } M \vdash_{\overline{H}} \varphi.$$

*Proof*   (1) follows from suitable formulas derived in Lemma 3.4, (2) is an immediate consequence of the definitions of consistency and completeness.     ∎

## 3.2.5   Witnesses and model construction

**Definition 3.6:**   Let M be a formula set. We say that M **contains witnesses** iff for each closed formula of the form $\forall X \varphi$ there exists a ground term $t_\varphi$ such that $M\vdash_{\overline{H}}(\neg\forall X\varphi\rightarrow\neg\varphi\{X/t_\varphi\})$. Informally stated, t is a concrete witness for the failure of the universal statement $\forall X\varphi$.

As an example let us consider the signature $\Sigma$ with a constant 0, 1-ary function symbol succ, and 1-ary predicate symbol p. Let M be the formula set $\{\neg\forall Xp(X)\} \cup \{p(succ^n(0)) \mid n\geq0\}$. M is consistent. A model is, for example, the interpretation with the set of integers as domain, number 0 as interpretation of the constant 0, the successor function as interpretation of succ, and the set of natural numbers as interpretation of predicate symbol p. M does not contain witnesses, since there is no ground term $succ^k(0)$ (only such ground terms are possible) such that $M\vdash_{\overline{H}}(\neg\forall Xp(X)\rightarrow\neg p(succ^k(0)))$.

### Witness Lemma 3.7
Let $\Sigma$ be a signature and M be a $\Sigma$-consistent set of $\Sigma$-formulas. We extend $\Sigma$ by an infinite number of new constants $c_1,c_2,...$ and obtain a new signature $\Sigma'$. Then there exists a set M' of $\Sigma'$-formulas such that:

- $M \subseteq M'$
- M' is $\Sigma'$-consistent
- M' contains witnesses

*Proof* We enumerate all closed $\Sigma'$-formulas of the form $\forall X\varphi$:

$$\forall Y_1\psi_1, \forall Y_2\psi_2,...$$

Then we define the following sequence of natural numbers $j_1<j_2<...$ :

$j_1$ is the least number such that $c_{j_1}$ does not occur in $\psi_1$

$j_{n+1}$ is the least number such that $j_{n+1}>j_n$ and $c_{j_{n+1}}$ does

not occur in $\psi_1,...,\psi_{n+1}$

We define formulas $EG_n=(\neg\forall Y_n\psi_n\rightarrow\neg\psi_n\{Y_n/c_{j_n}\})$ and finally the formula set $M'= M\cup\{EG_n \mid n\in \mathbb{N}\}$. We want to prove that the set M' so constructed is the one we are looking for. M' is clearly a superset of M. Furthermore, M' contains witnesses by definition. It remains to show that M' is $\Sigma'$-consistent. We show by induction over n that $M\cup\{EG_1,...,EG_n\}$ is $\Sigma'$-consistent.

For n=0 this means that M is $\Sigma'$-consistent. What we already know is that M is $\Sigma$-consistent. This does not automatically imply that M is $\Sigma'$-consistent. Since $\Sigma'$ contains more logical axioms, it could a priori be possible that they lead, when combined with M, to a contradiction, whereas the subset of $\Sigma$-axioms does not. We shall show that this is impossible. We shall hereby make use of Lemma 3.8 which allows us to substitute the new constants by variables. In the proof we assume an infinite number of variables.

### Lemma 3.8

Let $(\varphi_1,...,\varphi_m)$ be a derivation in Hilbert's calculus H from a formula set F. Further, let c be a constant symbol not occurring in any formula of F, and Z be a variable not occurring in any formula $\varphi_1,...,\varphi_n$. Define $t\{c/Z\}$ and $\varphi\{c/Z\}$ to be the term or formula which is obtained from t and $\varphi$ respectively by replacing every occurrence of c by Z. Then $(\varphi_1\{c/Z\},...,\varphi_m\{c/Z\})$ is a derivation in H from F.

Using Lemma 3.8 we are now able to carry out the inductive proof.

*Induction base: M is $\Sigma'$ consistent*    Suppose M is not $\Sigma'$consistent. Then there exists a $\Sigma$-formula $\varphi$ (here we are not forced into the extended signature $\Sigma'$) and an $\Sigma'$-derivation $(\varphi_1,...,\varphi_m)$ of $\neg(\varphi\rightarrow\varphi)$ from M. We successively replace the new constants occurring in $\varphi_1,...,\varphi_m$ by appropriate new variables according to Lemma 3.8 and obtain a $\Sigma'$-derivation $(\psi_1,...,\psi_m)$ of $\neg(\varphi\rightarrow\varphi)$ from M which does not contain any of the new constant symbols of $\Sigma'$. This means that we have a $\Sigma$-derivation of $\neg(\varphi\rightarrow\varphi)$ from M. It follows that M is not $\Sigma$-consistent, a contradiction.

*Induction step*    Assume that $M\cup\{EG_1,...,EG_n\}$ is already shown to be $\Sigma'$-consistent. We shall show that $M\cup\{EG_1,...,EG_n,EG_{n+1}\}$ is also $\Sigma'$-consistent. Assume the contrary. Then $M\cup\{EG_1,...,EG_n,\neg\neg EG_{n+1}\}$ is not $\Sigma'$-consistent, either. Thus,

$$M\cup\{EG_1,...,EG_n\}\vdash_{\overline{H}}\neg EG_{n+1}, \text{ that is}$$
$$M\cup\{EG_1,...,EG_n\}\vdash_{\overline{H}}\neg(\neg\forall Y_{n+1}\psi_{n+1}\rightarrow\neg\psi_{n+1}\{Y_{n+1}/c_{j_{n+1}}\})$$

The following formulas are tautologies:

$$(\neg(\neg\forall Y_{n+1}\psi_{n+1}\rightarrow\neg\psi_{n+1}\{Y_{n+1}/c_{j_{n+1}}\}))\rightarrow\neg\forall Y_{n+1}\psi_{n+1})$$
$$(\neg(\neg\forall Y_{n+1}\psi_{n+1}\rightarrow\neg\psi_{n+1}\{Y_{n+1}/c_{j_{n+1}}\}))\rightarrow\psi_{n+1}\{Y_{n+1}/c_{j_{n+1}}\})$$

Thus, we obtain:

(1)    $M\cup\{EG_1,...,EG_n\}\vdash_{\overline{H}}\neg\forall Y_{n+1}\psi_{n+1}$

(2)    $M\cup\{EG_1,...,EG_n\}\vdash_{\overline{H}}\psi_{n+1}\{Y_{n+1}/c_{j_{n+1}}\}$.

Let $(\varphi_1,...,\varphi_m)$ be a $\Sigma'$-derivation of $\psi_{n+1}\{Y_{n+1}/c_{j_{n+1}}\}$ from $M\cup\{EG_1,...,EG_n\}$. $c_{j_{n+1}}$ occurs neither in $EG_1,...,EG_n$ nor in $\psi_{n+1}$. Let Z be a variable not occurring in any formula $\varphi_1,\ldots,\varphi_m$. By Lemma 3.8 we know that $(\varphi_1\{c_{j_{n+1}}/Z\},\ldots,\varphi_m\{c_{j_{n+1}}/Z\})$ is also a $\Sigma'$-derivation from $M\cup\{EG_1,\ldots,EG_n\}$. Moreover, $\varphi_m\{c_{j_{n+1}}/Z\}=\psi_{n+1}\{Y_{n+1}/c_{j_{n+1}}\}\{c_{j_{n+1}}/Z\}=\psi_{n+1}\{Y_{n+1}/Z\}$, since $c_{j_{n+1}}$ does not occur in $\psi_{n+1}$. We have thus

$$M\cup\{EG_1,\ldots,EG_n\}\vdash_{\overline{H}}\psi_{n+1}\{Y_{n+1}/Z\}$$

and by generalization

$$M\cup\{EG_1,\ldots,EG_n\}\vdash_{\overline{H}}\forall Z\psi_{n+1}\{Y_{n+1}/Z\}.$$

By Theorem 3.3 we obtain

(3)    $$M\cup\{EG_1,\ldots,EG_n\}\vdash_{\overline{H}}\forall Y_{n+1}\psi_{n+1}$$

(1) and (3) contradict the consistency of $M\cup\{EG_1,\ldots,EG_n\}$.    ∎

---

**Exercise 3.4**    Prove Lemma 3.8. To do so, consider every $\varphi_i$ and distinguish whether it came into the given derivation as an axiom from $Ax_1$-$Ax_5$, as a formula from F, or via MP or GEN.

---

We already know that a formula set that has a model is consistent. Now we are able to show that the opposite is also true. This result is the basis for the subsequent completeness proof for Hilbert's calculus.

### Model Lemma 3.9

Let M be a $\Sigma$-consistent set of formulas. Then there exists a $\Sigma$-model $A$ of M. $A$ can be chosen in such a way that dom($A$) is denumerably infinite.

*Proof*    Let M be a $\Sigma$-consistent set of formulas. According to Lemma 3.7 we may choose an extension M' of M in an extended signature $\Sigma'$ which is $\Sigma'$-consistent and has witnesses. Then we apply Theorem 3.4 to obtain an extension M'' of M' in signature $\Sigma'$ such that M'' is $\Sigma'$-consistent and $\Sigma'$-complete. The property of possessing witnesses is immediately inherited from M' to M'' (see Definition 3.5). Now we define a Herbrand algebra $B$ in signature $\Sigma'$ as follows:

$$p_B = \{\ (t_1,\ldots,t_n)\ |\ t_1,\ldots,t_n \text{ are ground } \Sigma\text{'-terms with } M'' \vdash_{\overline{H}} p(t_1,\ldots,t_n)\ \}$$

for every n-ary predicate symbol p in $\Sigma$'. (Remember that dom($B$) and the interpretation of function symbols is fixed in a Herbrand algebra.) We show that for every *closed* $\Sigma$'-formula $\varphi$ the following statement is true:

$$B \models \varphi \Leftrightarrow M'' \vdash_{\overline{H}} \varphi$$

The proof is by induction over the number l of symbols from $\{\neg, \rightarrow, \forall\}$ occurring in $\varphi$. (Explain why structural induction on $\varphi$ is not appropriate here.) If $\varphi$ is a closed atomic formula $p(t_1,\ldots,t_n)$, our statement immediately follows from the definition of $p_B$ and the fact that the value of a ground term t in $B$ is t itself. If $\varphi$ is a closed formula of the form $\neg\psi$, we observe that $\psi$ is a closed formula with fewer logical symbols than $\varphi$. Hence, we may argue by induction as follows:

| | |
|---|---|
| $B \models \neg\psi$ | $\Leftrightarrow$ (because $\varphi$ is closed) |
| $B \not\models \psi$ | $\Leftrightarrow$ (by induction hypothesis) |
| not $M'' \vdash_{\overline{H}} \psi$ | $\Leftrightarrow$ (M'' is consistent and complete) |
| $M'' \vdash_{\overline{H}} \neg\psi$ | |

If $\varphi$ is a closed formula of the form $(\psi\rightarrow\chi)$ we argue as follows:

| | |
|---|---|
| $B \not\models (\psi\rightarrow\chi)$ | $\Leftrightarrow$ (since $(\psi\rightarrow\chi)$ is closed) |
| $B \models \psi$ and $B \not\models \chi$ | $\Leftrightarrow$ (by induction hypothesis) |
| $M'' \vdash_{\overline{H}} \psi$ and not $M'' \vdash_{\overline{H}} \chi$ | $\Leftrightarrow$ (M'' is consistent and complete) |
| $M'' \vdash_{\overline{H}} \psi$ and $M'' \vdash_{\overline{H}} \neg\chi$ | $\Leftrightarrow$ (see Lemma 3.6) |
| $M'' \vdash_{\overline{H}} \neg(\psi\rightarrow\chi)$ | $\Leftrightarrow$ (M'' is consistent and complete) |
| not $M'' \vdash_{\overline{H}} (\psi\rightarrow\chi)$ | |

Thus we obtain that $B \models (\psi\rightarrow\chi) \Leftrightarrow M'' \vdash_{\overline{H}} (\psi\rightarrow\chi)$.

If $\varphi$ is a closed formula of the form $\forall X\psi$, we first choose a ground term t such that $M'' \vdash_{\overline{H}} (\neg\forall X\psi \rightarrow \neg\psi\{X/t\})$ (remember that M'' possesses witnesses) and argue as follows:

$B \models \forall X\psi \Leftrightarrow$ ($B$ is a Herbrand algebra)

For every ground $\Sigma$'-term r, $B \models \psi\{X/r\}$ $\Leftrightarrow$

(by induction hypothesis, $\psi\{X/r\}$ contains fewer logical symbols than $\forall X\psi$)

For every ground $\Sigma$'-term r, $M'' \vdash_{\overline{H}} \psi\{X/r\}$ $\Leftrightarrow$

$M'' \vdash_{\overline{H}} \forall X\psi$.

The last equivalence is obtained as follows. First, $\vdash_{\overline{H}} (\forall X\psi \rightarrow \psi\{X/r\})$ for every ground term r. This gives us one direction. The reverse implication uses the chosen witness t from above. Assume that not $M'' \vdash_{\overline{H}} \forall X\psi$. As M'' is complete and consistent, we know that $M'' \vdash_{\overline{H}} \neg\forall X\psi$. This implies $M'' \vdash_{\overline{H}} \neg\psi\{X/t\}$. Thus, we obtain a ground $\Sigma'$-term r such that not $M'' \vdash_{\overline{H}} \psi\{X/r\}$.

Now we may complete the proof of Lemma 3.9. According to what we have just proved, we know that $B$ is a model of M''. (If φ is a formula from M'', then $\forall(\varphi)$ is a closed formula following from M'', so $B$ is a model of $\forall(\varphi)$, hence a model of φ.) In particular, $B$ is a model of the subset M of M''. Since M is a set of $\Sigma$-formulas, the reduct $A$ of $B$ to the signature $\Sigma$ is a model of M, too. It is the desired model. Its domain is denumerably infinite, as it is the set of all ground $\Sigma'$-terms.    ∎

### 3.2.6    The Gödel completeness theorem

**Completeness Theorem 3.5**

Let M be a formula set and φ a formula. Then $M \models \varphi$ implies $M \vdash_{\overline{H}} \varphi$.

*Proof*    Suppose that not $M \vdash_{\overline{H}} \varphi$. Then we know that $M \cup \{\neg\forall(\varphi)\}$ is consistent. By Lemma 3.9 we obtain a model $A$ of $M \cup \{\neg\forall(\varphi)\}$. Hence, $A$ is a model of M, but not a model of φ. This shows that $M \not\models \varphi$.    ∎

There are some important consequences from Gödel's theorem, some of them already proved in Chapter 2.

**Theorem 3.6 (Semantic characterization of consistency)**

For an axiom system M, M is consistent iff M has a model.

*Proof* Both directions have already been shown, one as a consequence of the soundness of Hilbert's calculus, the other as Lemma 3.9.    ∎

**Theorem 3.7 (Semantic characterization of completeness)**

For an axiom system M, M is complete iff, for any two models $A$ and $B$ of M, the same closed formulas are valid in $A$ as in $B$. (In the context of predicate logic with equality, which will be studied in Chapter 4, the latter property is called **elementary equivalence** of $A$ and $B$.)

*Proof*  This is left as an exercise for the reader.    ∎

### Compactness Theorem 3.8

A set of formulas M has a model iff every finite subset of M has a model.

*Proof*    Replace 'has a model' by 'is consistent' and note that in a derivation of a contradictory formula $\neg(\varphi \rightarrow \varphi)$ only a finite number of elements of M occur.    ∎

### Theorem (Löwenheim-Skolem) 3.9

Let M be a set of formulas. If M has a model, then it also has a model with a denumerably infinite domain.

*Proof*    Apply Lemma 3.9.    ∎

## 3.2.7  A Hilbert calculus for an alternative concept of logical conclusion based on state semantics ✂

In Section 2.4 we discussed an alternative concept of logical conclusion, not based on model semantics but on state semantics. $M \models_{alt} \varphi$ iff for every algebra $A$ and state sta over $A$: $A \models_{sta} \psi$, for every $\psi \in M$, implies $A \models_{sta} \varphi$. As was stated in Section 2.4, $\models_{alt}$ is a stronger concept than $\models$. Thus, we may expect that H is not an adequate calculus for $\models_{alt}$. Indeed it is the generalization rule that is not sound w.r.t. $\models_{alt}$, (remember that $\{p(X)\} \models \forall X p(X)$, but not $\{p(X)\} \models_{alt} \forall X p(X)$ ), whereas the remaining axioms and rules of H are sound. To obtain a sound and complete calculus for $\models_{alt}$ it is necessary to restrict the generalization rule. A solution, which is discussed in the literature (see Genesereth *et al.* (1987)) is the following calculus $H_{alt}$: for all formulas $\varphi$ and $\psi$, variables X and variable sequences $X_1 \ldots X_n$:

GEN(Ax$_1$-Ax$_5$) $\quad \dfrac{}{\forall X_1...\forall X_n \varphi} \quad$ for every instance $\varphi$ of Ax$_1$-Ax$_5$

DISTRIBUTION $\quad \dfrac{}{\forall X_1...\forall X_n(\forall X(\varphi \rightarrow \psi) \rightarrow (\forall X \varphi \rightarrow \forall X \psi))}$

restricted-GEN $\quad \dfrac{}{\forall X_1...\forall X_n(\varphi \rightarrow \forall X \varphi)} \quad$ if X is not free in $\varphi$

Modus Ponens (MP)

Derivability in H$_{alt}$ is denoted by $\vdash_{alt}$. Note that omission of the generalization rule GEN is compensated by the DISTRIBUTION and restricted-GEN axioms and the allowance for using arbitrary universally quantified versions of formulas from Ax$_1$-Ax$_5$. Soundness of H$_{alt}$ w.r.t. $\models_{alt}$ is obvious, since the new axioms are valid in every algebra and every state. Completeness of H$_{alt}$ w.r.t. $\models_{alt}$ is reduced to completeness of H w.r.t. $\models$ as follows. Assume that M $\models_{alt} \varphi$.

*Step 1*    Replace the free variables in M and $\varphi$ by new constants, thus obtaining a formula set M' and a formula $\varphi$' such that M' $\models \varphi$'. The compactness theorem implies that there is a finite subset $\{\varphi_1',...,\varphi_n'\}$ of M' such that $\{\varphi_1',...,\varphi_n'\} \models \varphi$'. Taking back the replacement of free variables by new constants we obtain that $\{\varphi_1,...,\varphi_n\} \models_{alt} \varphi$, for a finite subset $\{\varphi_1,...,\varphi_n\}$ of M.

*Step 2*    $\{\varphi_1,...,\varphi_n\} \models_{alt} \varphi$ is equivalent to $\models (\varphi_1 \rightarrow (\varphi_2 \rightarrow ... \rightarrow (\varphi_n \rightarrow \varphi)...))$. Thus we obtain $\vdash_H (\varphi_1 \rightarrow (\varphi_2 \rightarrow ... \rightarrow (\varphi_n \rightarrow \varphi)...))$.

*Step 3*    Now we show that H$_{alt}$ admits a restricted form of generalization, too: If M $\vdash_{alt} \varphi$ and X is not free in M, then M $\vdash_{alt} \forall X \varphi$. In particular, if $\vdash_{alt} \varphi$ then $\vdash_{alt} \forall X \varphi$. This statement can be easily shown by induction on the length of a derivation of $\varphi$ in H$_{alt}$ from M: if $\varphi$ is an element of M then X does not occur free in $\varphi$, hence we may use axiom $(\varphi \rightarrow \forall X \varphi)$ and MP to derive $\forall X \varphi$. If $\varphi$ is one of the axioms of H$_{alt}$ then $\forall X \varphi$ is an axiom, too. If $\varphi$ is obtained in H$_{alt}$ via MP from $(\psi \rightarrow \varphi)$ and $\psi$, then we may inductively conclude that M $\vdash_{alt} \forall X(\psi \rightarrow \varphi)$ and M $\vdash_{alt} \psi$, hence M $\vdash_{alt} (\forall X \psi \rightarrow \forall X \varphi)$, so M $\vdash_{alt} \forall X \varphi$.

*Step 4*    Next we show that $\vdash_H \varphi$ implies $\vdash_{alt} \varphi$. This is shown by induction on the length of a derivation of $\varphi$ in H. If $\varphi$ is an instance of Ax$_1$-Ax$_5$

then $\varphi$ is also an axiom of $H_{alt}$. If $\varphi$ is obtained in H via MP from $(\psi \to \varphi)$ and $\psi$, we may conclude by induction that $\vdash_{alt} (\psi \to \varphi)$ and $\vdash_{alt} \psi$, hence $\vdash_{alt} \varphi$ by MP. If $\varphi$ is obtained in H via GEN, say $\varphi$ is $\forall X\psi$, we may inductively conclude that $\vdash_{alt} \psi$, so (by step 3) also $\vdash_{alt} \forall X\psi$.

*Step 5*    Finally we come back to the second step. Using step 4, we obtain $\vdash_{alt} (\varphi_1 \to (\varphi_2 \to ... \to (\varphi_n \to \varphi)...))$. Then, n applications of MP lead to $M \vdash_{alt} \varphi$.

## 3.3    Sequent calculi and natural deduction

The Hilbert calculus H studied in Section 3.2 is most inconvenient in practical use. This is related to the dominant role of modus ponens. Let us take a closer look at the effects of the use of modus ponens.

First, thinking of the way a backward-chained automatic deduction system might try to derive a formula $\psi$, namely, to search through all possibilities to obtain $\psi$ from other formulas by application of a rule, and then to apply the same procedure to the formulas used to obtain $\psi$ until the search eventually arrives at axioms, the major drawback of modus ponens becomes obvious: in order to prove $\psi$, show $\varphi$ and $(\varphi \to \psi)$ for some formula $\varphi$. It thus introduces an infinite search space for possible lemmas, and complicates any attempt for a systematic top-down search. Such a top-down search, working backwards from the given formula to be derived to the axioms, seems to be no more goal directed than bottom-up search procedures which start from the given axioms and blindly derive more and more theorems until, eventually, the formula to be derived is obtained.

Second, the concrete derivation of even simple formulas may be a rather sophisticated and unnatural task, often running in a totally counter-intuitive way. The best example is the derivation of $(\varphi \to \varphi)$, even at the level of propositional logic. It was the introduction of the deduction theorem that gave us some more flexibility in the derivation process.

Forgetting at the moment the necessary conditions for application of the deduction theorem, we may look at the deduction theorem as two 'meta-rules' of the following form:

$$\frac{M \vdash_H (\varphi \to \psi)}{M \cup \{\varphi\} \vdash_H \psi} \qquad \frac{M \cup \{\varphi\} \vdash_H \psi}{M \vdash_H (\varphi \to \psi)}$$

An important difference between Hilbert's calculus and such meta-rules is that the latter allow change of premisses during a derivation, whereas deductions in calculus H refer to a fixed set of premisses M. This difference will again be found to be a major characteristic of the layout of the sequent calculi studied later.

Besides these critical comments, a positive remark on the role of modus ponens is in order, too. The use of a formula $\varphi$ with $M \vdash_{\overline{H}} \varphi$ and $M \vdash_{\overline{H}}$ ($\varphi \to \psi$) to obtain $M \vdash_{\overline{H}} \psi$ may well be a useful tool. Interpreting $\varphi$ as a lemma whose application leads to $\psi$, which is eventually already derived or whose derivation is postponed, is standard practice in mathematical deduction and supports the 'modular design' of mathematical proofs.

In the following we will present a straightforward introduction into a class of calculi called **sequent calculi**.

- First, we introduce the sequent calculus LK (with cut rule), show its correctness and prove its completeness, reducing it to the completeness result for Hilbert's calculus H.

- Then we mention the famous Gentzen cut-elimination theorem and Gentzen's sharpened *Hauptsatz*, and prove it for a special case.

- Finally, we discuss a modified sequent calculus G (without cut) and discuss its role in obtaining a model-theoretic proof for the cut-elimination result via a completeness proof for G.

A detailed discussion of the field may be found in Gallier (1986).

## 3.3.1    Logic of sequents

**Definition 3.7:**    A **sequent** over a signature $\Sigma$ is a pair $\Gamma \twoheadrightarrow \Delta$ consisting of finite sequences $\Gamma$ and $\Delta$ of $\Sigma$-formulas which are not both empty. (Do not confuse the sequent arrow $\twoheadrightarrow$ and the arrow $\to$ for logical implication.) $\Gamma$ is called the **antecedent** and $\Delta$ the **succedent** of $\Gamma \twoheadrightarrow \Delta$. For a $\Sigma$-algebra $A$ and a state sta over $A$, we say that $\Gamma \twoheadrightarrow \Delta$ is **valid in algebra $A$ in state** sta, iff the following is true:

$$A \models_{sta} \gamma, \text{ for all } \gamma \text{ in } \Gamma, \text{ implies } A \models_{sta} \delta, \text{ for some } \delta \text{ in } \Delta.$$

This property is denoted by $A \models_{sta} \Gamma \twoheadrightarrow \Delta$. We say that $A$ is a **model** of $\Gamma \twoheadrightarrow \Delta$ iff $A \models_{sta} \Gamma \twoheadrightarrow \Delta$ for all states sta over $A$. This property is denoted $A \models \Gamma \twoheadrightarrow \Delta$. We say that $\Gamma \twoheadrightarrow \Delta$ is **valid**, briefly denoted by $\models \Gamma \twoheadrightarrow \Delta$, iff $A \models \Gamma \twoheadrightarrow \Delta$, for every $\Sigma$-algebra $A$.

We may translate between formulas and sequents according to the following equivalences. Here, n and m are natural numbers $> 0$.

- $A \models \varphi_1,\ldots,\varphi_n \rightarrow \psi_1,\ldots,\psi_m$ iff $A \models \forall((\varphi_1 \wedge \ldots \wedge \varphi_n) \rightarrow (\psi_1 \vee \ldots \vee \psi_m))$

- $A \models \rightarrow \psi_1,\ldots,\psi_m$ iff $A \models \forall(\psi_1 \vee \ldots \vee \psi_m)$

- $A \models \varphi_1,\ldots,\varphi_n \rightarrow$ iff $A \models \forall(\neg(\varphi_1 \wedge \ldots \wedge \varphi_n))$

- $A \models \varphi$ iff $A \models \rightarrow \varphi$

- $\{\varphi_1,\ldots,\varphi_n\} \models \varphi$ iff $\models \forall(\varphi_1),\ldots,\forall(\varphi_n) \rightarrow \varphi$

### 3.3.2    The sequent calculus LK

Figures 3.1 and 3.2 contain the axioms and rules of sequent calculus LK. The formula in the conclusion of a rule that is introduced by that rule is called the **main formula** of that rule.

We will use LK only to derive sequents from the empty set of sequents. This is sufficient, since the notion of sequents already has incorporated an axiom set M into the antecedent part. Thus, we simply write 'a sequent is LK-derivable' instead of 'a sequent is LK-derivable from the empty set of sequents'.

Compared to H, which was axiom-dominated with rather unnatural axioms, LK is rule-dominated with rules for the different logical symbols which mirror the semantics of those symbols in a quite natural fashion.

### 3.3.3    Soundness of LK

*Theorem 3.10*

Every LK-derivable sequent is valid.

*Proof*    Let an algebra $A$ be given. We show that every axiom of LK is valid in $A$, and that validity in $A$ is preserved when going from the premisses of a rule of LK to its conclusion. The former proposition is obviously true, the latter one is proved for some cases, leaving the remaining ones as simple exercises for the reader.

**Axioms**
$$\varphi \rightarrow \varphi$$

**Structural rules**

(*weakening left*)

$$\frac{\Gamma \rightarrow \Delta}{\varphi, \Gamma \rightarrow \Delta}$$

(*weakening right*)

$$\frac{\Gamma \rightarrow \Delta}{\Gamma \rightarrow \Delta, \varphi}$$

(*contraction left*)

$$\frac{\varphi, \varphi, \Gamma \rightarrow \Delta}{\varphi, \Gamma \rightarrow \Delta}$$

(*contraction right*)

$$\frac{\Gamma \rightarrow \Delta, \varphi, \varphi}{\Gamma \rightarrow \Delta, \varphi}$$

(*exchange left*)

$$\frac{\Lambda, \varphi, \psi, \Gamma \rightarrow \Delta}{\Lambda, \psi, \varphi, \Gamma \rightarrow \Delta}$$

(*exchange right*)

$$\frac{\Gamma \rightarrow \Delta, \varphi, \psi, \Lambda}{\Gamma \rightarrow \Delta, \psi, \varphi, \Lambda}$$

(*cut rule*)

$$\frac{\Gamma \rightarrow \Delta, \varphi \qquad \varphi, \Sigma \rightarrow \Pi}{\Gamma, \Sigma \rightarrow \Delta, \Pi}$$

**Propositional rules**

(¬ *left*)

$$\frac{\Gamma \rightarrow \Delta, \varphi}{\neg \varphi, \Gamma \rightarrow \Delta}$$

(¬ *right*)

$$\frac{\varphi, \Gamma \rightarrow \Delta}{\Gamma \rightarrow \Delta, \neg \varphi}$$

(∧ *left*)

$$\frac{\varphi, \Gamma \rightarrow \Delta}{(\varphi \wedge \psi), \Gamma \rightarrow \Delta}$$

(∧ *left*)

$$\frac{\psi, \Gamma \rightarrow \Delta}{(\varphi \wedge \psi), \Gamma \rightarrow \Delta}$$

(∧ *right*)

$$\frac{\Gamma \rightarrow \Delta, \varphi \qquad \Gamma \rightarrow \Delta, \psi}{\Gamma \rightarrow \Delta, (\varphi \wedge \psi)}$$

**Figure 3.1**

$$(\vee \; left)$$

$$\frac{\varphi,\Gamma\to\Delta \qquad \psi,\Gamma\to\Delta}{(\varphi\vee\psi),\Gamma\to\Delta}$$

$$(\vee \; right)$$

$$\frac{\Gamma\to\Delta,\varphi}{\Gamma\to\Delta,(\varphi\vee\psi)}$$

$$(\vee \; right)$$

$$\frac{\Gamma\to\Delta,\psi}{\Gamma\to\Delta,(\varphi\vee\psi)}$$

$$(\to left)$$

$$\frac{\psi,\Gamma\to\Delta \qquad \Gamma\to\Delta,\varphi}{(\varphi\to\psi),\Gamma\to\Delta}$$

$$(\to right)$$

$$\frac{\varphi,\Gamma\to\Delta,\psi}{\Gamma\to\Delta,(\varphi\to\psi)}$$

**Quantifier rules**

$$(\forall \; left)$$

$$\frac{\varphi\{X/t\},\Gamma\to\Delta}{\forall X\varphi,\Gamma\to\Delta}$$

$$(\forall \; right)$$

$$\frac{\Gamma\to\Delta,\varphi\{X/Y\}}{\Gamma\to\Delta,\forall X\varphi}$$

$$(\exists \; left)$$

$$\frac{\varphi\{X/Y\},\Gamma\to\Delta}{\exists X\varphi,\Gamma\to\Delta}$$

$$(\exists \; right)$$

$$\frac{\Gamma\to\Delta,\varphi\{X/t\}}{\Gamma\to\Delta,\exists X\varphi}$$

It is required that $\varphi\{X/t\}$ and $\varphi\{X/Y\}$ are admissible, and Y does not occur free in $\Gamma\to\Delta,\forall X\varphi$ (in rule $\forall$ *right)* and $\exists X\varphi,\Gamma\to\Delta$ (in rule $\exists$ *left).*

**Figure 3.2**

Let us consider the cut rule. Assume that $A\models\Gamma\to\Delta,\varphi$ and $A\models\varphi,\Sigma\to\Pi$. We want to show that $A\models\Gamma,\Sigma\to\Delta,\Pi$. For this, let a state sta over $A$ be given. Assume that $A\models_{sta}\alpha$, for every formula $\alpha$ in $\Gamma,\Sigma$. Using $A\models\Gamma\to\Delta,\varphi$, hence $A\models_{sta}\Gamma\to\Delta,\varphi$, we may conclude that there exists some formula $\beta$ in $\Delta,\varphi$ such that $A\models_{sta}\beta$. If $\beta$ is a formula from $\Delta$, we are done: a formula $\beta$ in $\Delta,\Pi$ is found such that $A\models_{sta}\beta$. If $\beta$ is the formula $\varphi$ we may conclude from $A\models\varphi,\Sigma\to\Pi$ that there exists some formula $\gamma$ in $\Pi$, and therefore also in sequence $\Delta,\Pi$, such that $A \models_{sta}\gamma$.

Let us now consider the rule ($\rightarrow$ left). Assume that $A \models \psi,\Gamma\rightarrow\Delta$ and $A \models \Gamma\rightarrow\Delta,\varphi$. We want to show that $A \models (\varphi\rightarrow\psi),\Gamma\rightarrow\Delta$. For this, let a state sta over $A$ be given. Assume that $A \models_{sta}\alpha$, for every formula $\alpha$ in $(\varphi\rightarrow\psi),\Gamma$. In particular, $A \models_{sta}(\varphi\rightarrow\psi)$. If $A \models_{sta}\neg\varphi$ then we use $A \models\Gamma\rightarrow\Delta,\varphi$ to conclude that there exists a formula $\gamma$ in $\Delta,\varphi$ such that $A \models_{sta}\gamma$. $\gamma$ cannot be the formula $\varphi$, hence $\gamma$ is a formula in $\Delta$ such that $A_{sta} \models\gamma$. If $A \models_{sta}\psi$ we use $A \models\psi,\Gamma\rightarrow\Delta$ to obtain just the same conclusion, namely that there exists some formula $\gamma$ in $\Delta$ with $A \models_{sta}\gamma$.

Next, let us consider the rule ($\forall$ left). Assume that $\varphi\{X/t\}$ is admissible and $A \models\varphi\{X/t\},\Gamma\rightarrow\Delta$. We want to show that $A \models\forall X\varphi,\Gamma\rightarrow\Delta$. For this, let a state sta over $A$ be given. Assume that $A \models_{sta}\alpha$, for every formula $\alpha$ in $\forall X\varphi,\Gamma$. In particular, $A \models_{sta}\forall X\varphi$. Validity of $(\forall X\varphi\rightarrow\varphi\{X/t\})$ implies that $A \models_{sta}\alpha$, for every formula $\alpha$ in $\varphi\{X/t\},\Gamma$. Using $A \models \varphi\{X/t\},\Gamma\rightarrow\Delta$, we conclude that there exists some formula $\beta$ in $\Delta$ such that $A \models_{sta}\beta$.

Finally, we deal with rule ($\forall$ right). Assume that $\varphi\{X/Y\}$ is admissible, Y is not free in $\Gamma\rightarrow\Delta,\forall X\varphi$ and $A \models\Gamma\rightarrow\Delta,\varphi\{X/Y\}$. We want to show $A \models\Gamma\rightarrow\Delta,\forall X\varphi$. Let us, as a contrast to the earlier methods, show that the formula corresponding to the sequent $\Gamma\rightarrow\Delta,\forall X\varphi$ follows from the formula corresponding to $\Gamma\rightarrow\Delta,\varphi\{X/Y\}$. Assume that $\Gamma$ is $\gamma_1,...,\gamma_n$ and $\Delta$ is $\delta_1,...,\delta_m$, and finally assume that $\{Y,X_1,...,X_k\}$ covers the set of free variables of $\Gamma\rightarrow\Delta,\varphi\{X/Y\}$. Then the sequent $\Gamma\rightarrow\Delta,\varphi\{X/Y\}$ translates into the formula $\forall X_1...\forall X_k\forall Y(\neg\gamma_1\vee...\vee\neg\gamma_1\vee\delta_1\vee...\vee\delta_m\vee\varphi\{X/Y\})$. From this formula the formula $\forall X_1...\forall X_k(\neg\gamma_1\vee...\vee\neg\gamma_1\vee\delta_1\vee...\vee\delta_m\vee\forall Y\varphi\{X/Y\})$ follows, since (by assumption) Y does not occur free in $\gamma_1,...,\gamma_n$ and $\delta_1,...,\delta_m$. But $\forall Y\varphi\{X/Y\}$ is equivalent to $\forall X\varphi$ by Theorem 3.3, because Y does not occur free in $\forall X\varphi$ and $\varphi\{X/Y\}$ is admissible. So, the formula corresponding to the sequent $\Gamma\rightarrow\Delta,\forall X\varphi$ is valid in $A$.    ∎

---

**Exercise 3.6**    Complete the proof of Theorem 3.10.

**Exercise 3.7**    Show that the conditions in the rules of LK are necessary for the soundness of LK.

**Exercise 3.8**   Prove the validity of the formula $\exists X(\varphi \rightarrow \forall X\varphi)$ using LK.

**Exercise 3.9**   Using LK, check the validity of the following formulas:

$$((\exists Xp(X) \wedge q(a)) \rightarrow \forall Yp(f(Y)))$$

$$((\forall Xp(X) \wedge \exists Yq(Y)) \rightarrow (p(f(U)) \wedge \exists Zq(Z)))$$

---

### 3.3.4   Completeness of LK

**Deduction Theorem 3.11**

The following statements are equivalent:

(1)    $\vdash_{LK} \Gamma \rightarrow \Sigma, (\alpha \rightarrow \beta), \Lambda$

(2)    $\vdash_{LK} \alpha, \Gamma \rightarrow \Sigma, \Lambda, \beta$

*Proof*  Combination of rule ($\rightarrow$ right) and exchange rules shows that (2) implies (1). The converse implication may be shown by proving the following more general statement:

$$\vdash_{LK} \Gamma \rightarrow \Sigma_1, (\alpha_1 \rightarrow \beta_1), \Sigma_2, (\alpha_2 \rightarrow \beta_2), \ldots, \Sigma_n, (\alpha_n \rightarrow \beta_n), \Sigma_{n+1} \text{ implies}$$

$$\vdash_{LK} \alpha_1, \alpha_2, \ldots, \alpha_n, \Gamma \rightarrow \Sigma_1, \Sigma_2, \ldots, \Sigma_n, \Sigma_{n+1}, \beta_1, \beta_2, \ldots, \beta_n .$$

This statement is shown by induction on the height of an LK-derivation of the sequent $\Gamma \rightarrow \Sigma_1, (\alpha_1 \rightarrow \beta_1), \Sigma_2, (\alpha_2 \rightarrow \beta_2), \ldots, \Sigma_n, (\alpha_n \rightarrow \beta_n), \Sigma_{n+1}$.    ∎

---

**Exercise 3.10**   Carry out the inductive proof of the claim above. Which rule prevents that the simpler statement of the deduction theorem is directly proved by induction on the height of LK-derivations? Alternatively, show the deduction theorem directly by induction on the number of applications of the contraction rule with main formula $(\alpha \rightarrow \beta)$ together with a subinduction on the height of LK-derivations.

---

**Theorem 3.12 (Simulation of Hilbert's calculus H within LK)**

For every restricted predicate logic formula $\varphi$, if $\vdash_H \varphi$ then $\vdash_{LK} \rightarrow \varphi$.

*Proof*  We argue by induction on the length of a derivation of φ in Hilbert's calculus.

   *Case 1*  φ is the formula $(\alpha\rightarrow(\beta\rightarrow\alpha))$. Then, using the deduction theorem, we have to show that $\vdash_{\overline{LK}}\beta,\alpha\rightarrow\alpha$. This is easily obtained using (weakening left).

   *Case 2*  φ is the formula $((\alpha\rightarrow(\beta\rightarrow\gamma))\rightarrow((\alpha\rightarrow\beta)\rightarrow(\alpha\rightarrow\gamma)))$. Figure 3.3 shows that $\vdash_{\overline{LK}}\alpha,(\alpha\rightarrow\beta),(\alpha\rightarrow(\beta\rightarrow\gamma))\rightarrow\gamma$ (dashed lines indicate a number of applications of exchange rules).

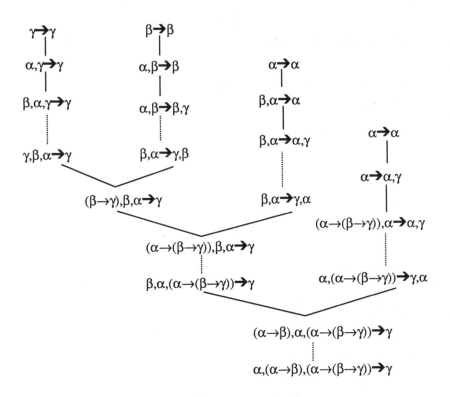

**Figure 3.3**

   *Case 3*  φ is the formula $((\neg\alpha\rightarrow\neg\beta)\rightarrow(\beta\rightarrow\alpha))$. We must give an LK-derivation tree for the sequent $\beta,(\neg\alpha\rightarrow\neg\beta)\rightarrow\alpha$. (See Figure 3.4.)

   *Case 4*  φ is the formula $(\forall X\psi\rightarrow\psi\{X/t\})$, with a term t such that $\psi\{X/t\}$ is admissible. Figure 3.5 contains an LK-derivation tree for the sequent $\forall X\psi\rightarrow\psi\{X/t\}$.

   *Case 5*  φ is the formula $(\forall X(\alpha\rightarrow\beta)\rightarrow(\alpha\rightarrow\forall X\beta))$ with a variable X which does not occur free in α. An LK-derivation tree for the sequent $\alpha,\forall X(\alpha\rightarrow\beta)\rightarrow\forall X\beta$ is given in Figure 3.6.

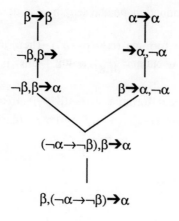

$$\beta \to \beta$$
$$|$$
$$\neg\beta,\beta\to$$
$$|$$
$$\neg\beta,\beta\to\alpha$$

$$\alpha \to \alpha$$
$$|$$
$$\to\alpha,\neg\alpha$$
$$|$$
$$\beta\to\alpha,\neg\alpha$$

$$(\neg\alpha\to\neg\beta),\beta\to\alpha$$
$$|$$
$$\beta,(\neg\alpha\to\neg\beta)\to\alpha$$

**Figure 3.4**

$$\psi\{X/t\}\to\psi\{X/t\}$$
$$|$$
$$\forall X\psi\to\psi\{X/t\}$$

**Figure 3.5**

Next we treat the cases that $\varphi$ is obtained via an application of a rule of H.

*Case 6*   $\varphi$ is obtained by an application of modus ponens from $\psi$ and $(\psi\to\varphi)$. By induction we may assume that the sequents $\to\psi$ and $\to(\psi\to\varphi)$ are LK-derivable. Then, by the deduction theorem, $\psi\to\varphi$ is LK-derivable, too. An application of the cut rule allows us to put together an LK-derivation of $\to\psi$ and an LK-derivation of $\psi\to\varphi$ to obtain an LK-derivation of $\to\varphi$.

*Case 7*   $\varphi$ is obtained by an application of generalization. So $\varphi$ is $\forall X\psi$, for some variable X and formula $\psi$. By induction hypothesis we may assume that the sequent $\to\psi$ is LK-derivable. Then an application of ($\forall$ right) extends an LK-derivation tree for $\to\psi$ to an LK-derivation tree for $\to\forall X\psi$.   ∎

### Corollary

For every restricted formula $\varphi$, $\models\varphi$ iff $\vdash_{LK}\to\varphi$.

*Proof*   $\models\varphi$ iff $\vdash_{H}\varphi$ iff $\vdash_{LK}\to\varphi$.   ∎

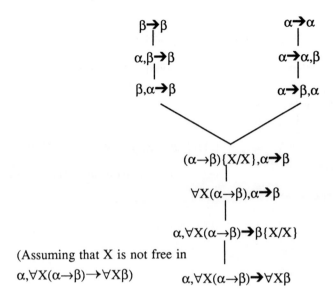

**Figure 3.6**

Theorem 3.12 was formulated for restricted formulas only (these are the only formulas that are treated in Hilbert's calculus H). Next, we extend the range of discussion to formulas with $\neg, \to, \wedge, \vee, \forall$ and $\exists$ as logical symbols. The equivalence symbol is left out since an adequate set of sequent rules is not as simple and natural as the rules for the other propositional connectives, and is therefore not included in LK.

**Definition 3.8:**  For a formula $\varphi$ with no occurrence of $\leftrightarrow$ we define a logically equivalent restricted formula $\varphi^*$ as follows:

$$p(t_1,\ldots,t_n)^* = p(t_1,\ldots,t_n)$$
$$(\neg\varphi)^* = \neg\varphi^*$$
$$(\varphi \to \psi)^* = (\varphi^* \to \psi^*)$$
$$(\varphi \vee \psi)^* = (\neg\varphi^* \to \psi^*)$$
$$(\varphi \wedge \psi)^* = \neg(\varphi^* \to \neg\psi^*)$$
$$(\forall X\varphi)^* = \forall X\varphi^*$$
$$(\exists X\varphi)^* = \neg\forall X\neg\varphi^*$$

It is easily shown by structural induction that $\varphi^*$ and $\varphi$ are equivalent.

### Lemma 3.10

For every formula $\varphi$ (without $\leftrightarrow$), $\vdash_{\overline{LK}} \varphi^* \rightarrow \varphi$ and $\vdash_{\overline{LK}} \varphi \rightarrow \varphi^*$.

*Proof*  By structural induction on $\varphi$. If $\varphi$ is $p(t_1,\ldots,t_n)$, $\varphi^* \rightarrow \varphi$ and $\varphi \rightarrow \varphi^*$ are obtained as axioms. Assume, by induction, that LK-derivation trees for the following sequents are given: $\varphi^* \rightarrow \varphi$, $\varphi \rightarrow \varphi^*$, $\psi^* \rightarrow \psi$, $\psi \rightarrow \psi^*$.

The following sequents have to be derived:

| | | |
|---|---|---|
| (1) | $\neg\varphi^* \rightarrow \neg\varphi$ | (see Figure 3.7) |
| (2) | $\neg\varphi \rightarrow \neg\varphi^*$ | (symmetrical situation as in (1)) |
| (3) | $(\varphi \rightarrow \psi) \rightarrow (\varphi^* \rightarrow \psi^*)$ | (see Figure 3.7) |
| (4) | $(\varphi^* \rightarrow \psi^*) \rightarrow (\varphi \rightarrow \psi)$ | (symmetrical situation as in (3)) |
| (5) | $(\varphi \vee \psi) \rightarrow (\neg\varphi^* \rightarrow \psi^*)$ | (see Figure 3.7) |
| (6) | $(\neg\varphi^* \rightarrow \psi^*) \rightarrow (\varphi \vee \psi)$ | (see Figure 3.7) |
| (7) | $(\varphi \wedge \psi) \rightarrow \neg(\varphi^* \rightarrow \neg\psi^*)$ | (see Figure 3.7) |
| (8) | $\neg(\varphi^* \rightarrow \neg\psi^*) \rightarrow (\varphi \wedge \psi)$ | (see Figure 3.8) |
| (9) | $\forall X \varphi \rightarrow \forall X \varphi^*$ | (see Figure 3.8) |
| (10) | $\forall X \varphi^* \rightarrow \forall X \varphi$ | (symmetrical situation as in (9)) |
| (11) | $\exists X \varphi \rightarrow \neg\forall X \neg\varphi^*$ | (see Figure 3.8) |
| (12) | $\neg\forall X \neg\varphi^* \rightarrow \exists X \varphi$ | (see Figure 3.8) |

(In Figures 3.7 and 3.8, dashed lines indicate applications of exchange and weakening rules.)                                                    ∎

### Corollary

For every sequent $\varphi_1,\ldots,\varphi_n \rightarrow \psi_1,\ldots,\psi_m$ (without $\leftrightarrow$),

$\vdash_{\overline{LK}} \varphi_1,\ldots,\varphi_n \rightarrow \psi_1,\ldots,\psi_m$ is equivalent to $\vdash_{\overline{LK}} \varphi_1^*,\ldots,\varphi_n^* \rightarrow \psi_1^*,\ldots,\psi_m^*$

*Proof*  Use the exchange rules, the cut rule and Lemma 3.10 to successively replace $\varphi_i$ and $\psi_j$ by $\varphi_i^*$ and $\psi_j^*$, and vice versa.                                                    ∎

### Corollary

For every formula $\varphi$ (without $\leftrightarrow$), $\models \varphi$ iff $\vdash_{\overline{LK}} \rightarrow \varphi$.

*Proof*  $\models \varphi \Leftrightarrow \models \varphi^* \Leftrightarrow \vdash_{\overline{LK}} \rightarrow \varphi^* \Leftrightarrow \vdash_{\overline{LK}} \rightarrow \varphi$.                                                    ∎

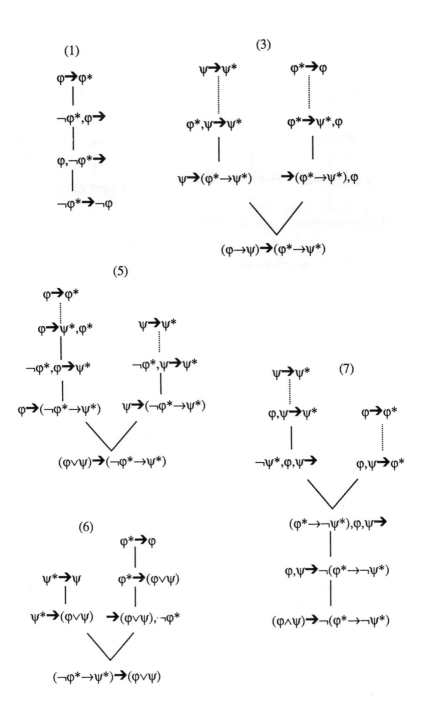

**Figure 3.7**

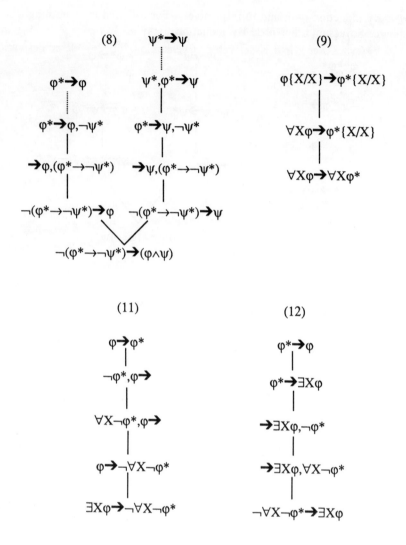

**Figure 3.8**

Finally, we show the completeness of LK for arbitrary sequents (so far, completeness is established only for sequents of the form $\rightarrow\varphi$).

### Lemma 3.11

For n>0, m>0 and formulas $\varphi_1,\ldots,\varphi_n,\psi_1,\ldots,\psi_m$ (without $\leftrightarrow$) the following sequents are LK-derivable:

$$\varphi_1,\ldots,\varphi_n\rightarrow(\varphi_1\wedge\ldots\wedge\varphi_n) \text{ and } (\psi_1\vee\ldots\vee\psi_m)\rightarrow\psi_1,\ldots,\psi_m$$

*Proof*  By induction on n and m respectively. For n=1 and m=1 nothing has to be shown. So assume that n>1. By induction hypothesis, we may assume that $\varphi_2,...,\varphi_n \rightarrow (\varphi_2 \wedge ... \wedge \varphi_n)$ is LK-derivable. (weakening left) yields a derivation of $\varphi_1,...,\varphi_n \rightarrow (\varphi_2 \wedge ... \wedge \varphi_n)$. (weakening left) and (exchange) yield a derivation of $\varphi_1,...,\varphi_n \rightarrow \varphi_1$. Then (∧ right) gives us an LK-derivation of $(\varphi_1 \wedge (\varphi_2 \wedge ... \wedge \varphi_n))$. A similar argument leads to an LK-derivation of $(\psi_1 \vee ... \vee \psi_m) \rightarrow (\psi_1 \vee ... \vee \psi_m)$.  ∎

### Completeness Theorem 3.13 for LK

For every sequent $\Gamma \rightarrow \Delta$, if $\Gamma \rightarrow \Delta$ is valid then $\Gamma \rightarrow \Delta$ is LK-derivable.

*Proof*  Let $\Gamma \rightarrow \Delta$ be the sequent $\varphi_1,...,\varphi_n \rightarrow \psi_1,...,\psi_m$. We treat the case that n>0 and m>0, leaving the (simpler) cases n=0 or m=0 to the reader. We successively conclude:

$\varphi_1,...,\varphi_n \rightarrow \psi_1,...,\psi_m$ is a valid sequent $\Rightarrow$

$((\varphi_1 \wedge ... \wedge \varphi_n) \rightarrow (\psi_1 \vee ... \vee \psi_m))$ is a valid formula $\Rightarrow$

$\rightarrow ((\varphi_1 \wedge ... \wedge \varphi_n) \rightarrow (\psi_1 \vee ... \vee \psi_m))$ is LK-derivable $\Rightarrow$

$(\varphi_1 \wedge ... \wedge \varphi_n) \rightarrow (\psi_1 \vee ... \vee \psi_m)$ is LK-derivable (deduction theorem) $\Rightarrow$

$\varphi_1,...,\varphi_n \rightarrow (\psi_1 \vee ... \vee \psi_m)$ is LK-derivable (Lemma 3.11 and cut rule) $\Rightarrow$

$\varphi_1,...,\varphi_n \rightarrow \psi_1,...,\psi_m$ is LK-derivable (Lemma 3.11 and cut rule).

∎

## 3.3.5   Cut elimination and Gentzen's sharpened *Hauptsatz* ✄

At the beginning of this section, we briefly discussed some drawbacks of modus ponens concerning automated theorem proving. The same applies to the cut rule of LK. In the LK-derivation of a sequent $\varphi_1,...,\varphi_n \rightarrow \psi_1,...,\psi_m$, not only subformulas of the formulas $\varphi_1,...,\varphi_n,\psi_1,...,\psi_m$ occur (by 'subformulas' we also mean in this Section formulas $\psi\{X/t\}$ for formulas $\forall X\psi$ and $\exists X\psi$), but eventually further formulas which are cut off by an application of the cut rule.

It is a central theorem in proof theory that the cut rule is, in fact, unnecessary for LK-derivations of special sequents, called pure sequents. This famous theorem of Gentzen, called **Gentzen's Hauptsatz**, is the basis of several rather deep proof theoretical and  model theoretical results, and serves as a basis for implementations of proof procedures. Its proof, a constructive method for the successive elimination of applications of the cut rule, is rather sophisticated and lengthy. Therefore, we present a proof only for a special case, namely for sequents whose antecedent consists of universally quantified formulas and whose succedent consists of existentially quantified formulas, a case which will play an important role in our later treatment of logic programming. For a proof of the general case we recommend Takeuti (1975), Smullyan (1968) or Börger (1985) as good expositions of this field.

**Definition 3.9:**    A sequent $\Gamma{\rightarrow}\Delta$ is called **pure** iff there does not exist a variable X such that X occurs both free and bound in $\Gamma{\rightarrow}\Delta$.

### Gentzen's Hauptsatz 3.14 (Cut elimination theorem)

Every LK-derivation tree for a pure sequent $\Gamma{\rightarrow}\Delta$ can be effectively transformed into a cut-free LK-derivation tree for $\Gamma{\rightarrow}\Delta$.

(As was shown in Kleene (1952) the pureness condition is necessary in Gentzen's *Hauptsatz*.) There is a sharpened version of Gentzen's *Hauptsatz* which states that the derivation of a sequent $\Gamma{\rightarrow}\Delta$ with formulas in prenex normal form can be done in two phases: first, all the propositional work is done, second, the quantifiers are treated.

### Theorem 3.13 (Gentzen's sharpened Hauptsatz)

Let $\Gamma{\rightarrow}\Delta$ be a pure, LK-derivable sequent consisting of formulas in prenex normal form. Then there exists an LK-derivation of the form indicated in Figure 3.9.

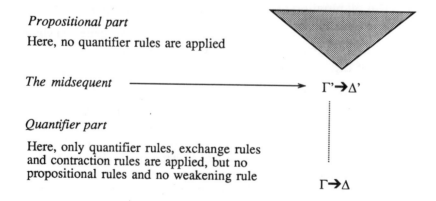

*Propositional part*
Here, no quantifier rules are applied

*The midsequent* ⟶ $\Gamma'{\rightarrow}\Delta'$

*Quantifier part*
Here, only quantifier rules, exchange rules and contraction rules are applied, but no propositional rules and no weakening rule

$\Gamma{\rightarrow}\Delta$

**Figure 3.9**

*Proof of Gentzen's sharpened Hauptsatz for a special case*    The case we will treat is for sequents $\forall(\varphi_1),\ldots,\forall(\varphi_n) \rightarrow \exists(\psi_1),\ldots,\exists(\psi_m)$ with quantifier-free formulas $\varphi_1,\ldots,\varphi_n,\psi_1,\ldots,\psi_m$. Let $X_1,\ldots,X_k$ be the variables occurring in $\varphi_1,\ldots,\varphi_n,\psi_1,\ldots,\psi_m$. We start with a valid sequent $\forall(\varphi_1),\ldots,\forall(\varphi_n) \rightarrow \exists(\psi_1),\ldots,\exists(\psi_m)$ as above. Validity of this sequent implies that the following set M of closed universal formulas does not possess a model:

$$\{\forall(\varphi_1),\ldots,\forall(\varphi_n),\forall(\neg\psi_1),\ldots,\forall(\neg\psi_m)\}$$

By Herbrand's theorem there is a finite set of ground instances of formulas from $\{\varphi_1,\ldots,\varphi_n,\neg\psi_1,\ldots,\neg\psi_m\}$ that does not possess a model. Let this set be

$$\{\varphi_{11},\ldots,\varphi_{1p_1}\}\cup\ldots\cup\{\varphi_{n1},\ldots,\varphi_{np_n}\}\cup$$
$$\{\neg\psi_{11},\ldots,\neg\psi_{1q_1}\}\cup\ldots\cup\{\neg\psi_{m1},\ldots,\neg\psi_{mq_m}\}$$

and assume that $\varphi_{ij}$ is a ground instance of $\varphi_i$, for $i=1,\ldots,n$ and $j=1,\ldots,p_i$, and $\psi_{ij}$ is a ground instance of $\psi_i$, for $i=1,\ldots,m$ and $j=1,\ldots,q_i$. Hence, the following sequent consisting of formulas without variables and quantifiers is valid, too:

$$\varphi_{11},\ldots,\varphi_{1p_1},\ldots,\varphi_{n1},\ldots,\varphi_{np_n} \rightarrow \psi_{11},\ldots,\psi_{1q_1},\ldots,\psi_{m1},\ldots,\psi_{mq_m} \qquad (3.1)$$

We will now construct a cut-free and quantifier-rule-free LK-derivation tree of this sequent; then we are done. Using (3.1) as the midsequent, we can derive $\forall(\varphi_1)\ldots,\forall(\varphi_n) \rightarrow \exists(\psi_1),\ldots,\exists(\psi_m)$ as indicated in Figure 3.10.

$$\varphi_{11},\ldots,\varphi_{1p_1},\ldots,\varphi_{n1},\ldots,\varphi_{np_n} \rightarrow \psi_{11},\ldots,\psi_{1q_1},\ldots,\psi_{m1},\ldots,\psi_{mq_m}$$

($\forall$ left)

(exchange left)          (Note that there are no problems with

($\exists$ right)          admissibility of substitutions since only

(exchange right)          ground terms are involved)

$$\forall(\varphi_1),\ldots,\forall(\varphi_1),\ldots,\forall(\varphi_n),\ldots,\forall(\varphi_n) \rightarrow \exists(\psi_1),\ldots,\exists(\psi_1),\ldots,\exists(\psi_m),\ldots,\exists(\psi_m)$$

(contraction left)

(exchange left)

(contraction right)

(exchange right)

$$\forall(\varphi_1),\ldots,\forall(\varphi_n) \rightarrow \exists(\psi_1),\ldots,\exists(\psi_m)$$

**Figure 3.10**

So let us finally derive (3.1) without cut and without quantifier rules. To show this, we introduce the following new axioms and propositional rules (see Figure 3.11).

Several statements are associated with these new rules:

(1)    Each of the new axioms is LK-derivable. Simply use the old axioms as well as exchange and weakening rules.

(2)    Each of the new rules can be simulated with the old propositional rules, exchange and contraction, but without cut. The only rules which deserve investigation are ( $\wedge$ left) and ( $\vee$ right). The others require the corresponding one of the old rules and exchanges. For LK-derivation trees see Figure 3.12.

(3)    Hence, validity is preserved when going from the premisses of one of these new rules to the conclusion.

(4)    The interesting new property of our new rules is that validity is also preserved in the opposite direction, that is, when going from the conclusion of one of these rules to the premisses. This was not true for the propositional rules of LK. The reason that this condition holds here is the modifications in the rules ($\wedge$ left) and ($\vee$ right). The simple proof is left to the reader.

(5)    Application of one of our new rules to a sequent increases the number of logical symbols $\neg, \vee, \wedge, \rightarrow, \forall, \exists$ occurring in the considered sequent by one.

(6)    Every valid sequent which consists of variable-free and quantifier-free formulas can be derived using only the new propositional rules above and the new axioms. This is almost trivially proved by induction on the number n of logical symbols $\neg, \vee, \wedge, \rightarrow, \forall, \exists$ occurring in the considered sequent. If n=0, then $\Gamma \rightarrow \Delta$ consists of ground atoms only. From Chapter 2 we know that $\Gamma \rightarrow \Delta$, written as a formula, is valid iff $\Gamma$ and $\Delta$ contain a common ground atom, i.e. iff $\Gamma \rightarrow \Delta$ is an axiom. If n>0, we use an appropriate rule to eliminate one of the logical symbols occurring in $\Gamma \rightarrow \Delta$. The premisses of the applied rule are again valid sequents with a smaller number of logical symbols, hence we can apply the induction hypothesis to conclude that they are derivable with the new rules and axioms.    ∎

### 3.3.6    The calculus G

Here, we present a slightly different calculus from LK, called G and intensively studied in Gallier (1986). There are three reasons for the introduction of G.

**Axioms**  Every sequent $\Gamma \rightarrow \Delta$ such that $\Gamma$ and $\Delta$ contain a common formula

**Propositional rules**

$$(\neg\ \text{left})\ \frac{\Gamma_1,\Gamma_2 \rightarrow \Delta,\varphi}{\Gamma_1,\neg\varphi,\Gamma_2 \rightarrow \Delta} \qquad\qquad (\neg\ \text{right})\frac{\varphi,\Gamma \rightarrow \Delta_1,\Delta_2}{\Gamma \rightarrow \Delta_1,\neg\varphi,\Delta_2}$$

$$(\wedge\ \text{left})\ \frac{\Gamma_1,\varphi,\psi,\Gamma_2 \rightarrow \Delta}{\Gamma_1,(\varphi\wedge\psi),\Gamma_2 \rightarrow \Delta} \qquad (\wedge\ \text{right})\frac{\Gamma \rightarrow \Delta_1,\varphi,\Delta_2 \quad \Gamma \rightarrow \Delta_1,\psi,\Delta_2}{\Gamma \rightarrow \Delta_1,(\varphi\wedge\psi),\Delta_2}$$

$$(\vee\ \text{left})\ \frac{\Gamma_1,\varphi,\Gamma_2 \rightarrow \Delta \quad \Gamma_1,\psi,\Gamma_2 \rightarrow \Delta}{\Gamma_1,(\varphi\vee\psi),\Gamma_2 \rightarrow \Delta} \qquad (\vee\ \text{right})\ \frac{\Gamma \rightarrow \Delta_1,\varphi,\psi,\Delta_2}{\Gamma \rightarrow \Delta_1,(\varphi\vee\psi),\Delta_2}$$

$$(\rightarrow\ \text{left})\ \frac{\Gamma_1,\psi,\Gamma_2 \rightarrow \Delta \quad \Gamma_1,\Gamma_2 \rightarrow \Delta,\varphi}{\Gamma_1,(\varphi\rightarrow\psi),\Gamma_2 \rightarrow \Delta} \qquad (\rightarrow\ \text{right})\frac{\varphi,\Gamma \rightarrow \Delta_1,\psi,\Delta_2}{\Gamma \rightarrow \Delta_1,(\varphi\rightarrow\psi),\Delta_2}$$

Figure 3.11

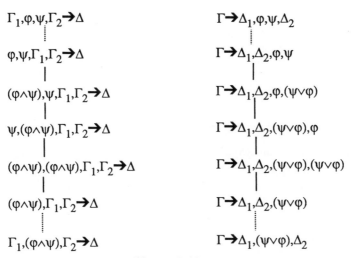

Figure 3.12

First, G is somehow easier to handle in practical applications. For example, the structural rules are incorporated into the logical rules. This saves lots of exchange steps.

Second, G does not contain a cut rule, therefore admits an alternative, namely semantic proof for the cut-elimination theorem of LK.

Third, restriction of G to propositional logic is the basis of Wang's algorithm, an algorithm that decides validity of propositional formulas. We will present a Prolog implementation of Wang's algorithm in Chapter 8.

Another motivation for the introduction of G can be found in the proof of the Sharpened *Hauptsatz* above. There, we use new propositional rules with the special property that they preserve validity in both directions. These new rules prevented us from using structural rules at all. Exchange rules are incorporated into the new rules by allowing arbitrary formulas (not only the leftmost or rightmost ones) to be processed. Weakening rules are incorporated into a broader concept of axioms. Contraction rules are incorporated into modified propositional rules. Following these lines, we define the calculus by the axioms and rules depicted in Figure 3.13.

### Theorem 3.15 (Soundness of G)

Validity of sequents is preserved when going from the premisses of a rule of G to its conclusion, and vice versa. In particular, if a sequent can be derived in G, it is valid.

*Proof*  This is left as an exercise for the reader.    ∎

### Theorem 3.16 (Simulation of G within LK-{cut})

Every axiom of G can be derived in LK-{cut}. Every rule of G can be simulated within LK-{cut}.

*Proof*  This is left as a simple exercise. In particular, we again obtain a proof of the soundness of G.    ∎

### Theorem 3.17 (Completeness of G)

Every valid, pure sequent can be derived in G.

*Proof*  See Gallier (1986).    ∎

### Corollary

For pure sequents, G and LK and LK-{cut} are equivalent. In particular, we obtain a semantical proof of Gentzen's cut-elimination theorem via a completeness proof for G.

**Axioms and propositional rules** as in Figure 3.11

**Quantifier rules:**

($\forall$ left) $\dfrac{\Gamma_1,\varphi\{X/t\},\forall X\varphi,\Gamma_2 \rightarrow \Delta}{\Gamma_1,\forall X\varphi,\Gamma_2 \rightarrow \Delta}$

if $\varphi\{X/t\}$ is defined

($\forall$ right) $\dfrac{\Gamma \rightarrow \Delta_1,\varphi\{X/Y\},\Delta_2}{\Gamma \rightarrow \Delta_1,\forall X\varphi,\Delta_2}$

if $\varphi\{X/Y\}$ is admissible and Y
is not free in $\Gamma \rightarrow \Delta_1,\forall X\varphi,\Delta_2$

($\exists$ left) $\dfrac{\Gamma_1,\varphi\{X/Y\},\Gamma_2 \rightarrow \Delta}{\Gamma_1,\exists X\varphi,\Gamma_2 \rightarrow \Delta}$

if $\varphi\{X/Y\}$ is admissible and Y
is not free in $\Gamma_1,\exists X\varphi,\Gamma_2 \rightarrow \Delta$

($\exists$ right) $\dfrac{\Gamma \rightarrow \Delta_1,\exists X\varphi,\varphi\{X/t\},\Delta_2}{\Gamma \rightarrow \Delta_1,\exists X\varphi,\Delta_2}$

if $\varphi\{X/t\}$ is defined

**Figure 3.13**

---

**Exercise 3.11**   Prove the soundness of G and its simulation within LK-{cut}

**Exercise 3.12**   Show the validity of the doctors-and-quacks example with LK and G.

**Exercise 3.13**   Using calculus LK or G prove the following: every barber shaves everyone who does not shave himself. No barber shaves someone who shaves himself. Then there exists no barber.

---

## 3.4    The resolution calculus RES

In Section 2.9 we introduced a first proof procedure for logical conclusion. Given a set $\{\varphi_1,\ldots,\varphi_n\}$ of closed formulas and a closed formula $\varphi$, property $\{\varphi_1,\ldots,\varphi_n\} \models \varphi$ was reduced to the statement that a certain set of closed universal formulas does not possess a model. If we transform the quantifier-free part of a universal formula into conjunctive normal form (that is, a conjunction of formulas which are disjunctions) and observe that a formula of the form $\forall(K_1\wedge\ldots\wedge K_n)$ is equivalent to $(\forall(K_1)\wedge\ldots\wedge\forall(K_n))$, we may finally turn our attention to the question of whether a set of universal formulas of the form

$\forall(L_1 \vee \ldots \vee L_m)$ with literals $L_1,\ldots,L_m$ does not possess a model. Such formulas are called formulas in **clausal form**.

In the following, we will represent such formulas in clausal form by finite sets of literals $\{L_1,\ldots,L_m\}$. This has the advantage that we abstract from multiple occurrences and the ordering of literals, a matter which is logically irrelevant.

### 3.4.1    The logic of clauses and the resolution calculus

**Definition 3.10 (Syntax and semantics of clauses):**    A **clause** C is a finite set of literals. A special clause is the **empty clause**, that is, the empty set of literals, denoted by □. A **ground clause** is a clause with no occurrences of variables. Given a clause $C=\{L_1,\ldots,L_m\}$ and a substitution $\sigma$, $C\sigma$ is the clause $\{L_1\sigma,\ldots,L_m\sigma\}$.

Concepts like **variants** and **renaming substitutions** carry over in a canonical way from formulas to clauses. Remember that it was not necessary to require admissibility of $\varphi\sigma$, for a universal formula $\varphi$, hence it is not necessary to introduce the idea of admissibility for $C\sigma$. For a set M of clauses let **variants(M)** be the set of all variants of clauses from M, and **ground-instances(M)** be the set of all ground clauses $C\sigma$, for C in M and a substitution $\sigma$.

An interpretation $A$ is called a **model** of clause $C=\{L_1,\ldots,L_m\}$ iff $m>0$ and $A$ is model of the formula $\forall(L_1 \vee \ldots \vee L_m)$. Thus, there is no model of the empty clause. A clause C **follows** from a set of clauses M iff every model of M is also a model of C.

*Theorem  3.18 (Herbrand's Theorem for clauses)*
A set M of clauses possesses a model iff every finite subset of ground-instances(M) possesses a model.

*Proof* Translate clauses into formulas and use Herbrand's theorem for formulas.                                                                                  ∎

Next we define a calculus allowing derivation of a clause from other clauses.

**Definition 3.11:**   The **resolution calculus** RES does not possess any axioms, its only rule is the following:

$$\frac{\{L_1,...,L_m\}\cup C_1 \qquad \{L_{m+1},...,L_{m+n}\}\cup C_2}{(C_1\cup C_2)\mu}$$

for all $m\geq 1$, $n\geq 1$, literals $L_1,...,L_m,L_{m+1},...,L_{m+n}$, clauses $C_1$ and $C_2$, and a most general unifier $\mu$ of $\{L_1,...,L_m,\sim L_{m+1},...,\sim L_{m+n}\}$. It is assumed that the clauses $\{L_1,...,L_m\}\cup C_1$ and $\{L_{m+1},...,L_{m+n}\}\cup C_2$ do not contain common variables. $(C_1\cup C_2)\mu$ is called a **resolvent** of the clauses $\{L_1,...,L_m\}\cup C_1$ and $\{L_{m+1},...,L_{m+n}\}\cup C_2$ **via substitution** $\mu$.

Let $\{L_1,...,L_m\}\mu$ be $\{L\}$ and thus $\{\not\!\!L_{m+1},...,\not\!\!L_{m+n}\}\mu$ be $\{\sim L\}$. In other words, the resolution calculus allows us to cut off a complementary pair $(L,\sim L)$, after application of a most general unifier $\mu$, and collect the remaining literals.

For technical reasons we will need some variants of the resolution calculus. The **extended resolution calculus** $RES_{ext}$ admits not only most general unifiers, but also arbitrary unifying substitutions $\sigma$ of the literal set $\{L_1,...,L_m,\sim L_{m+1},...,\sim L_{m+n}\}$ occurring in the rule of RES above, while the **propositional resolution calculus** $RES_{prop}$ admits only the empty substitution. Hence, the rule of $RES_{prop}$ is:

$$\frac{\{L\}\cup C_1 \qquad \{\sim L\}\cup C_2}{C_1\cup C_2}$$

$RES_{prop}$ does nothing but cut off a complementary pair. As above, we introduce the notion of a propositional resolvent and an extended resolvent via substitution $\sigma$.

The notion of a **deduction** from a set of clauses, or a **deduction tree** from a set of clauses in any one of the calculi above, is defined according to the general definitions from Section 3.1. At every non-leaf node of a deduction tree we indicate the substitution used to derive the clause at the considered node as a resolvent of its parent nodes. Thus, interior nodes of a deduction tree together with their parent nodes look like Figure 3.14:

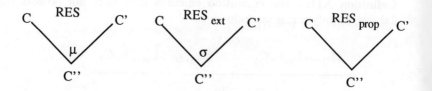

**Figure 3.14**

# EXAMPLE 3.1

We want to formalize and prove the following informal conclusion:

> If all students are citizens, then students' votes are citizens' votes.

We shall use 1-ary predicate symbols $s(X)$ and $cit(X)$ for 'X is student' an 'X is citizen', and a 2-ary predicate symbol $v(X,Y)$ for 'X is a vote of Y'. Th premise of the deduction above can be written as the closed formul $\varphi=\forall X(s(X)\to cit(X))$, the conclusion as the closed formul $\psi=\forall X(\exists Y(s(Y)\wedge v(X,Y))\to\exists Z(cit(Z)\wedge v(X,Z)))$. We want to show tha $\{\varphi\}\models\psi$, or equivalently that $\{\varphi,\neg\psi\}$ does not possess a model. We trans form $\varphi$ and $\neg\psi$ into clausal form as described in Chapter 2.

| | |
|---|---|
| original formula | $\forall X(s(X)\to cit(X))$ |
| prenex normal form | $\forall X(s(X)\to cit(X))$ |
| Skolem normal form | $\forall X(s(X)\to cit(X))$ |
| conjunctive NF | $\forall X(\neg s(X)\vee cit(X))$ |
| clausal form | $\{\neg s(X),cit(X)\}$ |

| | |
|---|---|
| original formula | $\neg\forall X(\exists Y(s(Y)\wedge v(X,Y))\to\exists Z(cit(Z)\wedge v(X,Z)))$ |
| prenex normal form | $\exists X\neg(\exists Y(s(Y)\wedge v(X,Y))\to\exists Z(cit(Z)\wedge v(X,Z)))$ |
| | $\exists X(\exists Y(s(Y)\wedge v(X,Y))\wedge\neg\exists Z(cit(Z)\wedge v(X,Z)))$ |
| | $\exists X(\exists Y(s(Y)\wedge v(X,Y))\wedge\forall Z\neg(cit(Z)\wedge v(X,Z)))$ |
| | $\exists X\exists Y((s(Y)\wedge v(X,Y))\wedge\forall Z\neg(cit(Z)\wedge v(X,Z)))$ |
| | $\exists X\exists Y\forall Z((s(Y)\wedge v(X,Y))\wedge\neg(cit(Z)\wedge v(X,Z)))$ |
| Skolem normal form | $\forall Z(s(b)\wedge v(a,b)\wedge\neg(cit(Z)\wedge v(a,Z)))$ |
| conjunctive NF | $\forall Z((s(b)\wedge v(a,b))\wedge(\neg cit(Z)\vee\neg v(a,Z)))$ |
| clausal form | $\{s(b)\}$     $\{v(a,b)\}$     $\{\neg cit(Z),\neg v(a,Z)\}$ |

A derivation tree in RES of the empty clause from the four clauses $\{\neg s(X),cit(X)\}$, $\{s(b)\}$, $\{v(a,b)\}$, $\{\neg cit(Z),\neg v(a,Z)\}$ and their variants looks like Figure 3.15.

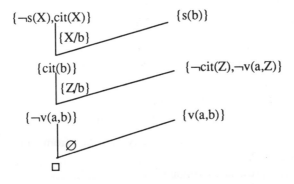

**Figure 3.15**

# EXAMPLE 3.2

*Doctors and quacks* We formalize the following conclusion:

Some patient likes every doctor.
No patient likes any quack.
Therefore no doctor is a quack.

We use predicate symbols p, doc and q of arity 1 (for 'is a patient', 'is a doctor' and 'is a quack') and a 2-ary relation symbol likes (with intended meaning of 'likes(X,Y)' as 'X likes Y'). We can now express the above statements by formulas:

$\exists X(p(X) \land \forall Y(doc(Y) \rightarrow likes(X,Y)))$
$\neg \exists X \exists Y(p(X) \land q(Y) \land likes(X,Y))$
$\forall X(doc(X) \rightarrow \neg q(X))$

The third formula follows from the first two. We show this by proving that the set consisting of the first two formulas and the negation of the third does not possess a model. First we transform the formulas into Skolem normal form. Some steps of this transformation are:

| | |
|---|---|
| original formula | $\exists X (p(X) \land \forall Y(doc(Y) \rightarrow likes(X,Y)))$ |
| prenex normal form | $\exists X \forall Y(p(X) \land (doc(Y) \rightarrow likes(X,Y)))$ |
| Skolem normal form | $\forall Y(p(c) \land (doc(Y) \rightarrow likes(c,Y)))$ |
| conjunctive normal form | $\forall Y (p(c) \land (\neg doc(Y) \lor likes(c,Y)))$ |
| clausal form | $\{p(c)\} \quad \{\neg doc(Y),likes(c,Y)\}$ |

| | |
|---|---|
| original formula | $\neg \exists X \exists Y(p(X) \land q(Y) \land likes(X,Y))$ |
| prenex normal form | $\forall X \forall Y \neg (p(X) \land q(Y) \land likes(X,Y))$ |
| Skolem normal form | $\forall X \forall Y \neg (p(X) \land q(Y) \land likes(X,Y))$ |

| | |
|---|---|
| conjunctive normal form | $\forall X\forall Y(\neg p(X)\vee\neg q(Y)\vee\neg likes(X,Y))$ |
| clausal form | $\{\neg p(X),\neg q(Y),\neg likes(X,Y)\}$ |
| | |
| original formula | $\neg\forall X(doc(X)\rightarrow\neg q(X))$ |
| prenex normal form | $\exists X\neg(doc(X)\rightarrow\neg q(X))$ |
| Skolem normal form | $\neg(doc(d)\rightarrow\neg q(d))$ |
| conjunctive normal form | $(doc(d)\wedge q(d))$ |
| clausal normal form | $\{doc(d)\}$      $\{q(d)\}$ |

One of the possibilities for deriving □ from the clauses $\{p(c)\}$, $\{\neg doc(Y),likes(c,Y)\}$,  $\{\neg p(X),\neg q(Y),\neg likes(X,Y)\}$,  $\{doc(d)\}$,  $\{q(d)\}$ and their variants is presented in Figure 3.16.

A comparison with the derivation presented in Figure 3.17 in the extended resolution calculus shows the main advantage of most general unifiers over arbitrary unifiers. In the latter derivation tree, the binding of variable Y took place at a stage where we could not really know which was the right term to be substituted for Y, whereas the former derivation tree postponed the binding of Y up to the stage where it had become necessary to bind Y, and also recognizable which was the right term to be substituted for Y.

---

**Exercise 3.14**   Transform the formula

$$\forall X\forall Y((\neg p(X)\vee\neg p(f(a))\vee q(Y))\wedge p(Y)\wedge\neg p(g(b,X)\wedge\neg q(b)))$$

into clausal form and prove by resolution that it does not possess a model.

**Exercise 3.15**   Formalize and prove by resolution the following informal deduction:  if the professor is happy if all his students like logic, then he is happy if he has no students.

**Exercise 3.16**   Find all resolvents of the following pairs of clauses.

(a)    $\{p(X,Y),p(Y,Z)\}$, $\{\neg p(a,f(a))\}$

(b)    $\{p(X,X),\neg q(X,f(X))\},\{q(X,Y),r(Y,Z)\}$

(c)    $\{p(X,Y),\neg p(X,X),q(X,f(X),Z)\},\{\neg q(f(X),X,Z),p(X,Z)\}$

(d)    $\{\neg p(X,Y),\neg p(f(a),g(U,b)),q(X,U)\},\{p(f(X),g(a,b)),\neg q(f(a),b),\neg q(a,b)\}$

(e)    $\{p(X,f(X),Z),p(U,W,W)\},\{\neg p(X,Y,Z),\neg p(Z,Z,Z)\}$

**Exercise 3.17**   Show by resolution that $(\forall X\exists Yp(X,Y) \wedge \exists X\forall Y\neg p(X,Y))$ does not possess a model.

**Exercise 3.18**   Express in predicate logic and prove by resolution the following result: if a binary relation r is symmetric, transitive and total (that is, $\forall X\exists Y\ r(X,Y)$), then it is also reflexive.

---

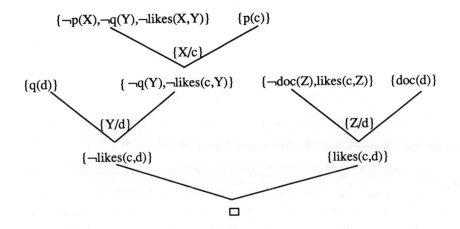

**Figure 3.16**

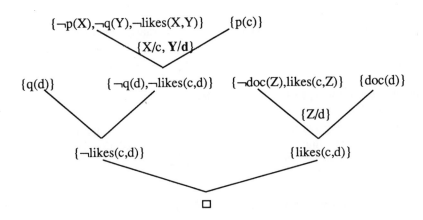

**Figure 3.17**

## 3.4.2    Soundness of resolution as a deduction calculus

### Theorem 3.19

Let M be a set of clauses. If a clause C can be derived from M in the extended resolution calculus $RES_{ext}$, then C follows from M. In particular, if C can be derived from M in the resolution calculus RES or the propositional resolution calculus $RES_{prop}$, then C follows from M.

*Proof*  We must show that application of the rule of $RES_{ext}$ preserves validity in an algebra $A$. So assume that $A$ is a model of clauses $\{L_1,...,L_m\}\cup C_1$ and $\{L_{m+1},...,L_{m+n}\}\cup C_2$. Then, $A$ is also a model of $(\{L_1,...,L_m\}\cup C_1)\sigma$ and $(\{L_{m+1},...,L_{m+n}\}\cup C_2)\sigma$, for arbitrary substitution $\sigma$ (remember that admissibility is not required for universal formulas). Now assume that $\{L_1,...,L_m\}\sigma=\{L\}$ and $\{L_{m+1},...,L_{m+n}\}\sigma=\{\sim L\}$, for a literal L. Then, $A$ is a model of $\{L\}\cup C_1\sigma$ and $\{\sim L\}\cup C_2\sigma$. To show that $A$ is a model of $C_1\sigma\cup C_2\sigma$, we translate back the clauses $\{L\}\cup C_1\sigma$, $\{\sim L\}\cup C_2\sigma$ and $C_1\sigma\cup C_2\sigma$ into formula notation. Let $C_1\sigma$ be $\{A_1,...,A_k\}$ and $C_2\sigma$ be $\{B_1,...,B_l\}$. Table 3.1 shows corresponding formulas (implicitly universally quantified). In any case we see that application of $RES_{ext}$ preserves validity in an interpretation $A$, since the following formulas are valid:

$((\varphi\rightarrow\psi)\rightarrow((\psi\rightarrow\chi)\rightarrow(\varphi\rightarrow\chi)))$          transitive conclusion

$(\varphi\rightarrow((\varphi\rightarrow\psi)\rightarrow\psi))$                    modus ponens

$(\varphi\rightarrow(\neg\varphi\rightarrow false))$                      ex falso quodlibet

for an arbitrary formula *false* without a model.          ■

The following corollary is formulated in a way that might look strange at first sight (concerning the role of variants). The formulation is chosen to fit well with the completeness theorem to be shown later.

### Corollary (Soundness of RES as a refutation calculus)

Let M be a set of clauses. If the empty clause can be derived in $RES_{ext}$ from variants(M), then M does not possess a model. In particular, the same holds if we replace $RES_{ext}$ by RES or $RES_{prop}$.

*Proof*    Assume M had a model $A$. Then $A$ is a model of variants(M), too. Hence, by the soundness of the resolution calculus, $A$ is a model of the empty clause, a contradiction.          ■

**Table 3.1**

| $\{L,A_1,\ldots,A_k\}$ | $\{\sim L,B_1,\ldots,B_l\}$ | $\{A_1,\ldots,A_k\}\cup\{B_1,\ldots,B_l\}$ |
|---|---|---|
| For $k>0$, $l>0$ | | |
| $L\vee A_1\vee\ldots\vee A_k$ | $\sim L\vee B_1\vee\ldots\vee B_l$ | $A_1\vee\ldots\vee A_k\vee B_1\vee\ldots\vee B_l$ |
| $\neg(A_1\vee\ldots\vee A_k)\to L$ | $L\to(B_1\vee\ldots\vee B_l)$ | $\neg(A_1\vee\ldots\vee A_k)\to(B_1\vee\ldots\vee B_l)$ |
| For $k>0$, $l=0$ | | |
| $L\vee A_1\vee\ldots\vee A_k$ | $\sim L$ | $A_1\vee\ldots\vee A_k$ |
| $\sim L\to(A_1\vee\ldots\vee A_k)$ | | |
| For $k=0$, $l>0$ | | |
| $L$ | $\sim L\vee B_1\vee\ldots\vee B_l$ | $B_1\vee\ldots\vee B_l$ |
| | $L\to(B_1\vee\ldots\vee B_l)$ | |
| For $k=0$, $l=0$ | | |
| $L$ | $\sim L$ | empty clause |

### 3.4.3    Completeness of resolution as a refutation calculus

*Theorem    3.20 (Completeness of resolution as a refutation calculus for ground clauses)*

Let M be a set of ground clauses. If M does not possess a model, then the empty clause can be derived from M in $\text{RES}_{\text{prop}}$.

*Proof*    Because of Theorem 3.8 we may assume that M is finite. Furthermore, we may assume that □ is not a member of M. We shall prove the following statement:

For every sequence of ground clauses $(C_1,\ldots,C_n)$: if $\{C_1,\ldots,C_n\}$ does not possess a model then the empty clause can be derived in $\text{RES}_{\text{prop}}$ from $\{C_1,\ldots,C_n\}$.

We prove this by induction on the number

$k(C_1,\ldots,C_n)$ = number of literals in $C_1$ +...+ number of literals in $C_n$ - n.

*Induction base*    Let $k(C_1,...,C_n)=0$. This means that each clause $C_i$ contains exactly one literal $L_i$ (i=1,...,n). If the clause $\{C_1,...,C_n\}$ does not possess a model, then the formula $(L_1\wedge...\wedge L_n)$ does not possess a model, either. In Chapter 2 we showed that in this case $\{L_1,...,L_n\}$ contains a complementary pair $L_i=\sim L_j$. Then, of course, $\square$ can be derived from $C_i=\{\sim L_j\}$ and $C_j=\{L_j\}$ by a single application of $RES_{prop}$.

*Induction step*    Let $k(C_1,C_2,...,C_n)>0$. This means that there is a clause with at least two literals. We assume without loss of generality that $C_1$ is such a clause. Let L be a literal in $C_1$, D the clause $\{L\}$ and E the non-empty clause $C_1-\{L\}$. Apparently $k(D,C_2,...,C_n)<k(C_1,C_2,...,C_n)$ and $k(E,C_2,...,C_n)=k(C_1,C_2,...,C_n)-1<k(C_1,C_2,...,C_n)$. Furthermore, neither $\{D,C_2,...,C_n\}$ nor $\{E,C_2,...,C_n\}$ possesses a model (such a model would also be a model of $\{C_1,C_2,...,C_n\}$ because $C_1$ has to be read as a disjunction).

It follows by induction hypothesis that the empty clause can be derived in $RES_{prop}$ from the clause set $\{\{L\},C_2,...,C_n\}$ as well as from the clause set $\{E,C_2,...,C_n\}$. Now we take a derivation tree of $\square$ from $\{E,C_2,...,C_n\}$. It is a binary tree with ground clauses as nodes; no substitutions are used to derive a successor clause from given clauses. The root of this tree is labelled with $\square$, the leaves are labelled with clauses from $\{E,C_2,...,C_n\}$. Occurrences of E in leaves disturb us, so we put back the literal L at all places where it was deleted (E becomes again $C_1$), and make use of the following observation. If C'' is derived from ground clauses C and C' by a single application of the rule of $RES_{prop}$, then either $C''\cup\{L\}$ or C'' can be derived from $C\cup\{L\}$ and C' by a single application of the rule of $RES_{prop}$ (note that here is the point where we use the fact that clauses are *sets*, rather than *lists* of literals). This point is illustrated in Figure 3.18. This fact allows us to insert L into the considered derivation tree of $\square$ from clause set $\{E,C_2,...,C_n\}$ from the leaves to the root. We obtain thus a derivation tree whose leaves are clauses from $\{C_1,C_2,...,C_n\}$; its root is either $\square$ or $\{L\}$. In the first case we are through. In the second we observe that deducibility in $RES_{prop}$ of clause $\{L\}$ from $\{C_1,C_2,...,C_n\}$ and of $\square$ from $\{\{L\},C_2,...,C_n\}$ implies derivability of $\square$ from $\{C_1,C_2,...,C_n\}$ (compositionality of deductions).    ∎

For the generalization of Theorem 3.20 to the case of arbitrary clauses we need the following lemma.

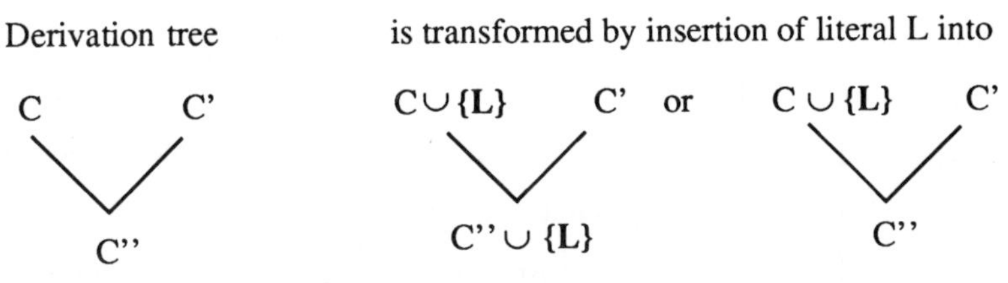

Derivation tree
C
C'
C''
is transformed by insertion of literal L into
C∪{L}
C'
C''∪{L}
or
C ∪ {L}
C'
C''

**Figure 3.18**

***Lifting Lemma 3.12***

Let $C_1$ and $C_2$ be clauses without common variables. Further, let $\sigma_1$ and $\sigma_2$ be substitutions such that $C_1\sigma_1$ and $C\sigma_2$ have no common variables. Finally, let C be obtained from $C_1\sigma_1$ and $C_2\sigma_2$ by a single application of the rule of $RES_{ext}$ via substitution $\sigma$. Then there exists a clause C' and substitutions $\mu$ and $\tau$ such that C' is obtained from $C_1$ and $C_2$ by a single application of the rule of RES via $\mu$, and $C'\tau=C$. (See Figure 3.19.)

C1
C2
σ1
σ2
C1σ1
C2σ2
RESext
σ
C
is 'lifted' to
RES
μ
C'
τ

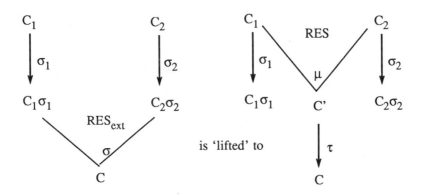

**Figure 3.19**

*Proof* Let $C_1=\{L_1,\ldots,L_m\}\cup Rest_1$ and $C_2=\{L_{m+1},\ldots,L_{m+n}\}\cup Rest_2$ (with m>0, n>0) be clauses such that $\sigma$ is a unifier of $\{L_1\sigma_1,\ldots,L_m\sigma_1,\sim L_{m+1}\sigma_2,\ldots,\sim L_{m+n}\sigma_2\}$ and $C=Rest_1\sigma_1\sigma\cup Rest_2\sigma_2\sigma$. We may assume that $\sigma_1$ contains only pairs X/t with variables X occurring in $C_1$, and $\sigma_2$ contains only pairs X/t with variables X occurring in $C_2$. Since $C_1\sigma_1$ and $C_2\sigma_2$ do not contain common variables, we may define $(\sigma_1\cup\sigma_2)$ and note that $(\sigma_1\cup\sigma_2)\sigma$ is a unifier of $\{L_1,...,L_m\}\cup\{{\sim}L_{m+1},...,{\sim}L_{m+n}\}$. Let $\mu$ be a most general unifier of $\{L_1,...,L_m\}\cup\{{\sim}L_{m+1},...,{\sim}L_{m+n}\}$ and $\tau$ be a substitution such that $(\sigma_1\cup\sigma_2)\sigma=\mu\tau$. Then, the clause $C'=\text{Rest}_1\mu \cup \text{Rest}_2\mu$ is a resolvent of $C_1$ and $C_2$ via $\mu$. Now we calculate:

$$
\begin{aligned}
&C'\tau \\
&=(\text{Rest}_1\mu \cup \text{Rest}_2\mu)\tau \\
&=\text{Rest}_1\mu\tau \cup \text{Rest}_2\mu\tau \\
&=\text{Rest}_1(\sigma_1\cup\sigma_2)\sigma \cup \text{Rest}_2(\sigma_1\cup\sigma_2)\sigma \\
&=\text{Rest}_1\sigma_1\sigma \cup \text{Rest}_2\sigma_2\sigma \\
&=C \qquad\qquad\qquad\qquad\qquad\qquad\qquad\qquad\qquad\blacksquare
\end{aligned}
$$

As an example consider the clauses $C_1=\{p(X,Y),p(f(a),Z)\}$ and $C_2=\{\neg p(f(X'),g(Y')),\neg q(X',Y')\}$ and the derivation tree in Figure 3.20.

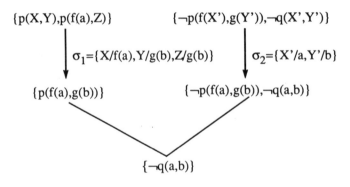

**Figure 3.20**

Lifting yields the deduction tree in Figure 3.21, with most general unifier $\mu=\{X/f(a),Y/g(Y'), Z/g(Y'),X'/a\}$.

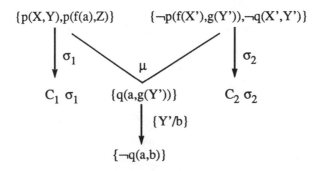

**Figure 3.21**

*Theorem 3.21 (Completeness of resolution as refutation calculus)*

If M is a set of clauses without a model, then the empty clause is derivable in RES from variants(M).

*Proof*    Assume that M does not possess a model. By Herbrand's Theorem for clauses, there exists a finite subset of ground-instances(M) without a model. By the completeness of the resolution calculus for ground clauses, we obtain a derivation of the empty clause from that finite subset of ground-instances(M). Now we lift, working from the leaves to the root, a derivation tree of the empty clause from ground-instances(M) according to the lifting lemma. In order to meet the condition of variable disjointness required in the lifting lemma, we use suitably chosen variants of clauses from M, that are mapped by a substitution to the used clauses from ground-instances(M). The root of the lifted tree is a clause which is transformed into the empty clause when a suitable substitution is applied; this is only possible if the root is already □. So we have obtained a refutation from M.

Figures 3.22 and 3.23 illustrates the lifting of a tree for a typical situation. Here, $C_1, C_2, C_3, C_4, C_5$ are clauses from M, $C_1\sigma_1, C_2\sigma_2, C_3\sigma_3, C_4\sigma_4, C_5\sigma_5$ are ground clauses, and C,D,E are the ground-clauses used at interior nodes. Note that the occurring resolvents are obtained in $RES_{prop}$, that is, with the empty substitution. Since it is allowed to use any element of variants(M) in the derivation, we may assume that any two of the clauses $C_1, C_2, C_3, C_4, C_5$ have disjoint variable sets (this may require a renaming $C_i\rho$ for some of the $C_i$ and an adaption of $\sigma_i$ into $(\rho^{-1}\sigma_i)$, where $\rho$ is a renaming substitution for $C_i$). Note that the clauses C' and D' that are obtained after two applications of lifting do not possess common variables (since no two of $C_1, C_2, C_3, C_4$ did). C and D do not possess common variables since they are ground clauses. Thus, the lifting lemma is applicable again to C', C, D' and D. After four applications we obtain $C_{final}\tau_{final} = $ □, hence $C_{final} = $ □. Hence, the final lifted tree is a derivation tree in RES of □ from variants(M).    ■

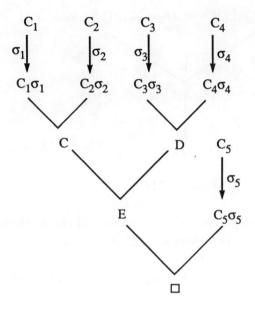

After two applications of lifting

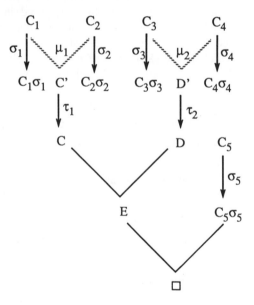

**Figure 3.22**

After third application of lifting

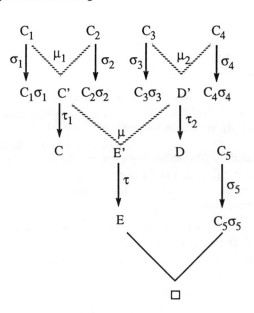

After fourth application of lifting

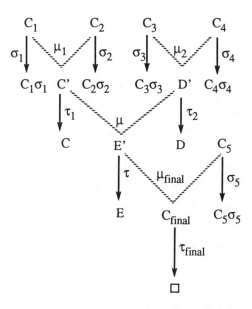

**Figure 3.23**

**Remarks**

(1)    The following example shows that it is necessary to allow variants of clauses in order to achieve completeness: the set of clauses $\{\{p(X)\},\{\neg p(f(X))\}\}$ does not possess a model. $p(X)$ and $p(f(X))$ are, however, not unifiable; only if it is allowable to use a variant $\{p(Y)\}$ of $\{p(X)\}$, can the empty clause be derived from $\{p(Y)\}$ and $\{\neg p(f(X))\}$.

(2)    Resolution calculus is sound as a deduction calculus, that is, w.r.t. logical conclusion. A consequence was that it is also sound as a refutation calculus, that is, w.r.t. the property of a set of clauses to possess no model. The completeness result above states that it is also a complete refutation calculus. The following example shows that it is *not complete as a deductive calculus*. Let M consist of the single clause $\{p(a)\}$. Obviously the clause $\{p(a),p(b)\}$ follows from M, but it is not derivable from M by resolution.

(3)    Resolution is tailored to the representation of formulas as clauses, that is, as sets of literals. An adaptation of RES to clausal *formulas* (lists of literals) with the rule to cut off a complementary pair from two clausal formulas does not lead to a complete calculus, as the following example shows. The set M consisting of the two clausal formulas $(p \wedge p)$ and $(\neg p \wedge \neg p)$, which obviously does not possess a model, does not allow to derive the empty clause, since the only obtainable formulas from M are $(p \wedge p)$, $(\neg p \wedge \neg p)$, $(p \wedge \neg p)$ and $(\neg p \wedge p)$.

## 3.5    Analytic tableaux

Analytic tableaux are a method of implementing Gentzen-like calculi in a user-friendly form, often used today as the basis for mechanical theorem provers. Unlike the calculi presented so far which are tailored to *validity*, analytic tableaux provide a method for showing **unsatisfiability**. This concentration on unsatisfiability is, of course, not a restriction in the applicability of the tableaux method. Recall that a formula $\varphi$ is valid iff its negation is unsatisfiable. Likewise, a formula $\varphi$ follows from a finite set $\{\varphi_1,...,\varphi_n\}$ of formulas iff the implication $(\forall(\varphi_1) \wedge ... \wedge \forall(\varphi_n)) \to \varphi$ is valid iff $(\forall(\varphi_1) \wedge ... \wedge \forall(\varphi_n) \wedge \neg \varphi)$ is unsatisfiable. Hence, $\varphi$ follows from a set of closed formulas $\{\varphi_1,...,\varphi_n\}$ iff $(\varphi_1 \wedge ... \wedge \varphi_n \wedge \neg \varphi)$ is unsatisfiable.

Our approach is as follows: given a formula $\varphi$ whose unsatisfiablility has to be proven, we start with $\varphi$ and, by contradiction, assume it was satisfiable. Using **decomposition rules** we break down $\varphi$ into formulas with fewer logical symbols which, together with $\varphi$, were also satisfiable if $\varphi$ was satisfiable. As an example:

If $\forall X \varphi$ is satisfiable then $\{\forall X \varphi, \varphi\{X/t\}\}$ is satisfiable, too,

for every term t such that $\varphi\{X/t\}$ is admissible.

(Note that here, if $\forall X \varphi$ has a model, then $\{\forall X \varphi, \varphi\{X/t\}\}$ has a model (just the same as $\forall X \varphi$)). As a propositional example take the following:

If $(\varphi \rightarrow \psi)$ is satisfiable then either $\{(\varphi \rightarrow \psi), \neg\varphi\}$ or $\{(\varphi \rightarrow \psi), \psi\}$ is satisfiable.

(Note that in this case it is not true that, if $(\varphi \rightarrow \psi)$ has a model, then either $\{(\varphi \rightarrow \psi), \neg\varphi\}$ or $\{(\varphi \rightarrow \psi), \psi\}$ must have a model, as the example $(p(X) \rightarrow p(X))$ shows). A further example:

If $\neg(\varphi \rightarrow \psi)$ is satisfiable then $\{\neg(\varphi \rightarrow \psi), \varphi, \neg\psi\}$ is satisfiable.

Finally an example concerning existential quantification:

If $\exists X \varphi$ is satisfiable then $\{\exists X \varphi, \varphi\{X/Y\}\}$, with a fresh variable Y, is satisfiable, too.

(Again this statement is false with respect to validity in a model.)

# EXAMPLE 3.3

Before giving an exact definition of the decomposition rules and the way they are used to define analytic tableaux let us look at a simple example. Assume that we want to show the validity of the formula

$$((p \rightarrow (q \rightarrow r)) \rightarrow ((p \rightarrow q) \rightarrow (p \rightarrow r)))$$

with atomic formulas p, q and r. Thus we should try to show unsatisfiability of the negated formula. Assume that the formula

$$\neg((p \rightarrow (q \rightarrow r)) \rightarrow ((p \rightarrow q) \rightarrow (p \rightarrow r)))$$

is satisfiable. Then also the set consisting of formulas

$$\neg((p \rightarrow (q \rightarrow r)) \rightarrow ((p \rightarrow q) \rightarrow (p \rightarrow r))),$$

$$(p \rightarrow (q \rightarrow r)) \text{ and } \neg((p \rightarrow q) \rightarrow (p \rightarrow r))$$

would be satisfiable  (see stage (2) and (3) in Figure 3.24). At stage (7) we use that satisfiability of  $(p \rightarrow (q \rightarrow r))$ implies satisfiablity of either $\neg p$ or else $(q \rightarrow r)$. Looking at the whole tree, the assumption that $\neg((p \rightarrow (q \rightarrow r)) \rightarrow ((p \rightarrow q) \rightarrow (p \rightarrow r)))$ was satisfiable implies that there exists a path in the tree such that the set of formulas occurring along that path forms a satisfiable set, too. But as can be seen, each such path contains a complementary pair. Such paths are called **closed paths** and are indicated by a star at their leaves.

If a path is closed, satisfiability is impossible along this path. So, if all paths are closed, unsatisfiability of the formula considered is shown (supposing that the method of analytic tableaux is complete). In our example, we established that the original formula $\neg((p \rightarrow (q \rightarrow r)) \rightarrow ((p \rightarrow q) \rightarrow (p \rightarrow r)))$ is unsatisfiable, hence $((p \rightarrow (q \rightarrow r)) \rightarrow ((p \rightarrow q) \rightarrow (p \rightarrow r)))$  is valid.

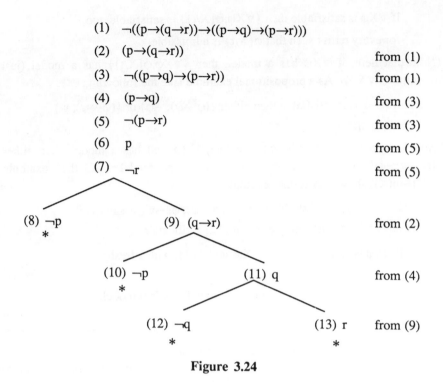

**Figure 3.24**

**Definition 3.12    Analytic tableaux** are finite trees labelled with formulas. They are inductively defined as follows:

- Every non-branching tree with arbitrary formulas $\varphi_1,...,\varphi_n$ at its nodes is an analytic tableau:

$$\varphi_1$$
$$\cdot$$
$$\cdot$$
$$\cdot$$
$$\varphi_n$$

- Assume that T is an analytic tableau. Fix an arbitrary leaf L and a formula $\varphi$ that occurs along the path from the root of T to leaf L. Depending on the form of $\varphi$, leaf L may be equipped with successor nodes as shown in Figure 3.25. The resulting extended tree is also an analytic tableau.

A path from the root of an analytic tableaux to some leaf is called a **closed path** iff among the formulas along the considered path a complementary pair occurs. Otherwise it is called an **open path**. An analytic tableau is called **closed** iff it contains only closed paths from its root to its leaves.

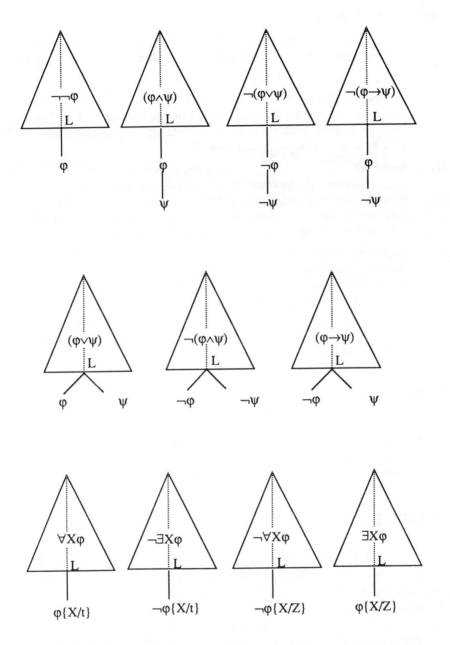

With an arbitrary term t such that φ{X/t} is admissible

With an arbitrary variable Z that does not occur free in a formula between the root and leaf L

**Figure 3.25**

**Remarks and facts**

(1)     *Preservation of satisfiability* If we generate an analytic tableau T by starting with a non-branching tree labelled with formulas $\varphi_1,...,\varphi_n$ such that $\{\varphi_1,...,\varphi_n\}$ is satisfiable, then there is at least one leaf L of T such that the set of formulas occurring between the root of T and L is satisfiable, too. The proof is by an induction on the definition of analytic tableaux and uses nothing other than well-known simple facts about validity in an interpretation and state.

(2)     Some authors use a new constant c instead of a new variable Z in the decomposition rule for quantifiers. This does not affect the results to be proved soon. We prefer to use a new variable since this allows us to stay within the given signature, and simplifies the later simulation of cut-free Gentzen derivations within the calculus of analytic tableaux.

(3)     Implementations of the tableau calculus usually allow us to attach the formulas resulting from a decomposition step *simultaneously at all leaves of open paths*. Then it is only necessary to choose the same formula twice in case of formulas $\forall X\varphi$ and $\neg\exists X\varphi$, to be able to provide for several terms t as instances for the variable X. (See also Example 3.4.) We do not worry about such implementation details and concentrate only on the logical aspects of analytic tableaux.

## EXAMPLE 3.4

Consider the unsatisfiable set $\{\neg p(c),\ p(f(f(c))),\ \forall X(p(X)\vee \neg p(f(X)))\}$. Figure 3.26 presents a closed analytic tableau starting with the considered formula at its first 3 nodes.

     Note that formula (3) is processed twice.

## EXAMPLE 3.5

Finally let us consider the example of doctors and quacks from the section on resolution. We want to show that formula $\forall X(doc(X)\rightarrow\neg q(X))$ follows from $\exists X(p(X)\wedge\forall Y(doc(Y)\rightarrow likes(X,Y)))$     and     $\neg\exists X\exists Y(p(X)\wedge q(Y)\wedge likes(X,Y))$. To do so, we construct a closed analytic tableau with the negation of the first formula and the remaining two formulas at its first three nodes. Note that for convenience we freely use a conjunction with more than two operands and we use a modified rule for such conjunctions. It should be clear how to transform the tableau in Figure 3.27 into a standard one.

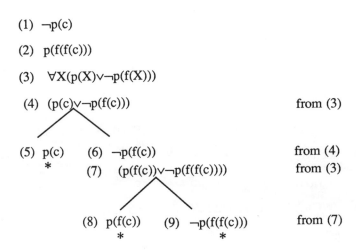

(1)  ¬p(c)

(2)  p(f(f(c)))

(3)  ∀X(p(X)∨¬p(f(X)))

(4)  (p(c)∨¬p(f(c)))                                    from (3)

(5) p(c)     (6) ¬p(f(c))                               from (4)
     *       (7)  (p(f(c))∨¬p(f(f(c))))                 from (3)

          (8) p(f(c))     (9) ¬p(f(f(c)))              from (7)
              *                 *

**Figure 3.26**

Having illustrated the working of the tableaux method with some examples, we shall finally turn our attention to some theoretical statements describing its purpose in general. In particular, these results will show that the tableau method is indeed nothing more than an appropriate reformulation of Gentzen's calculus. Besides its practical advantages, the tableaux method also exhibits nice theoretical properties. In fact, Smullyan (1968) is a logic book based totally on tableaux.

### Theorem 3.22 (Soundness of the tableaux method)

If a closed analytic tableau T can be generated by starting with a non-branching tree labelled with formulas $\varphi_1,...,\varphi_n$, then $\{\varphi_1,...,\varphi_n\}$ is unsatisfiable.

*Proof* Assume that $\{\varphi_1,...,\varphi_n\}$ is a satisfiable set. According to the 'preservation of satisfiability' property, there would exist at least one path of T from the root to a leaf such that the set M of formulas occurring along that path is satisfiable. This is a contradiction, since M would be a superset of $\{\varphi_1,...,\varphi_n\}$, hence unsatisfiable, too.                    ■

### Theorem 3.23 (Completeness of the tableaux method)

If a formula set $\{\varphi_1,...,\varphi_n\}$ is unsatisfiable, then a closed analytic tableaux can be generated by starting with a non-branching tree labelled with formulas $\varphi_1,...,\varphi_n$.

(1)   ∃X(p(X)∧∀Y(doc(Y)→likes(X,Y)))

(2)   ¬∃X∃Y(p(X)∧q(Y)∧likes(X,Y))

(3)   ¬∀X(doc(X)→¬q(X))

(4)   ¬(doc(A)→¬q(A)), with a fresh variable A          from (3)

(5)   doc(A)                                              from (4)

(6)   ¬¬q(A)                                              from (4)

(7)   (p(C)∧∀Y(doc(Y)→likes(C,Y))), with a fresh variable C    from (1)

(8)   p(C)                                                from (7)

(9)   ∀Y(doc(Y)→likes(C,Y))                               from (7)

(10)  (doc(A)→likes(C,A))                                 from (9)

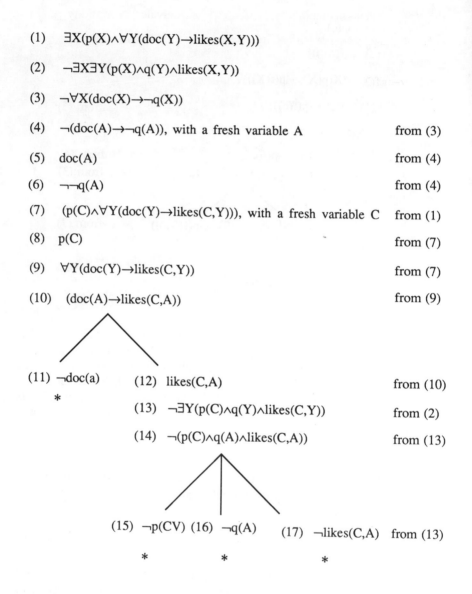

(11) ¬doc(a)      (12)  likes(C,A)                        from (10)
     *
                  (13)   ¬∃Y(p(C)∧q(Y)∧likes(C,Y))        from (2)

                  (14)   ¬(p(C)∧q(A)∧likes(C,A))          from (13)

          (15)  ¬p(CV) (16)  ¬q(A)      (17)  ¬likes(C,A)  from (13)

             *           *                *

**Figure 3.27**

*Proof* ✂    Unsatisfiability of $\{\varphi_1,...,\varphi_n\}$ is equivalent to validity of the sequent $\varphi_1,...,\varphi_n \rightarrow$. For simplicity, we assume that $\varphi_1,...,\varphi_n \rightarrow$ is a pure sequent. (The reader may try to remove this assumption.) By Gentzen's theorem, there exists a cut-free derivation of $\varphi_1,...,\varphi_n \rightarrow$. Now it suffices to show the following.

### Simulation lemma 3.13

If $\Gamma \rightarrow \Delta$, with $\Gamma = \varphi_1,...,\varphi_n$ and $\Delta = \psi_1,...,\psi_m$ is derivable in LK-{cut}, then there exists a closed analytic tableau of the form:

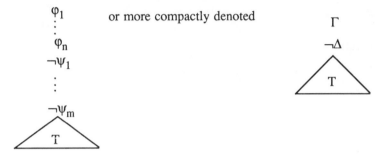

*Proof of the simulation lemma*    We argue by induction on the height of a deduction tree for $\Gamma \rightarrow \Delta$ in LK-{cut}. If $\Gamma \rightarrow \Delta$ is an axiom we are done. If $\Gamma \rightarrow \Delta$ is obtained by application of one of the structural rules of LK-{cut}, the claim of our lemma is obviously proved. Let us finally check the remaining rules of LK-{cut}. Table 3.2 illustrates the situation for some of the rules. It contains the analytic tableaux corresponding to the premisses of the applied LK-rule (that are assumed to exist by induction hypothesis), as well as the analytic tableau corresponding to the conclusion of the considered LK-rule. Note that every propositional and quantifier rule of LK is mirrored by a corresponding decomposition rule for tableaux. Also, the critical variable conditions in the quantifier rules of LK mirror those for tableaux.     ■

---

**Exercise 3.17**  Treat the remaining rules in the proof above.

---

Implementations of the tableaux method may even be equipped with a limited ability to *recognize satisfiability* of formulas. Contrary to its ability to recognize unsatisfiability always, satisfiability can be recognized only in certain cases, because of the general undecidability result for validity and satisfiability. We again recommend Smullyan (1968) as a profound exposition of such questions.

**Table 3.2**

| LK-rule | Tableaux corresponding to the premisses (assumed to exist by induction) | Tableau corresponding to the conclusion (exists by definition of tableaux) |
|---|---|---|

**(¬ left)**

Premiss:
Γ
¬Δ
¬φ
◺ T ◹

Conclusion:
¬φ
Γ
¬Δ
◺ T ◹

**(∧ left)**

Premiss:
φ
Γ
¬Δ
◺ T ◹

Conclusion:
(φ∧ψ)
Γ
¬Δ
φ
ψ
◺ T ◹

**(→ left)**

Premisses:
ψ                    Γ
Γ                    ¬Δ
¬Δ                   ¬φ
◺ T ◹              ◺ S ◹

Conclusion:
(φ→ψ)
Γ
¬Δ
      ¬φ          ψ
    ◺ S ◹      ◺ T ◹

**(∀ left)**

Premiss:
φ{X/t}
Γ
¬Δ
◺ T ◹          with φ{X/t} admissible

Conclusion:
∀Xφ
Γ
¬Δ
φ{X/t}
◺ T ◹

**(∃ left)**

Premiss:
φ{X/Z}
Γ
¬Δ
◺ T ◹

Conclusion:
∃Xφ
Γ
¬Δ
φ{X/Z}
◺ T ◹

with Z not free in Γ, Δ and φ such that φ{X/Z} is admissible

# Chapter 4
# Predicate Logic with Equality

---

---

Given a $\Sigma$-interpretation $A$, there is a fundamental mathematical relation on dom($A$) called the **equality** on dom($A$), written $=_A$ and defined as the 2-ary relation $\{(a,a) \mid a \in \text{dom}(A)\}$. Owing to the importance of equality in mathematics and its omnipresence in applications, we study in this chapter extensions of pure predicate logic by equality. Two fundamental ways to deal with equality have been established in the literature.

The first one takes a fixed 2-ary, infixedly written relation symbol $=$, called the equality symbol, and interprets it in any interpretation $A$ with domain A as the equality on A. This leads to **predicate logic with equality**. This is the method we adopt in this book.

The second one treats $=$ as part of the set of relation symbols and hence admits a priori arbitrary interpretations for it. Being interested only in equality-like interpretations for $=$, we therefore restrict our attention to interpretations of $=$ which exhibit basic properties of equality. These properties are formulated as a set Eq of formulas of pure predicate logic (with relation symbol $=$). The models of Eq will be just those interpretations interpreting $=$ as a congruence relation.

Though with respect to their models these two approaches to equality are different (the second variant admits more models than the first one), they will be shown to be equivalent from a logical point of view. This equivalence will supply us at once with a sound and complete Hilbert calculus for predicate logic with equality, so it will not be necessary to repeat all the constructions and proofs used for pure predicate logic. Finally, corresponding extensions of the sequent calculus LK and the resolution calculus RES are introduced.

## 4.1    Syntax and semantics of predicate logic with equality

- *Vocabulary*

  We introduce a new fixed symbol = and call it the **equation symbol**.

- *Signatures, Σ-interpretations and states, Σ-terms*

  The concepts are carried over from predicate logic without changes.

- *Σ-formulas*

  A new type of Σ-formula is introduced, namely **equations** $t_1=t_2$, consisting of Σ-terms $t_1$ and $t_2$. Hence, = is treated like a 2-ary, infixedly written relation symbol.

- *Validity*

  Given a Σ-interpretation $A$, a state sta over $A$ and an equation $t_1=t_2$, we define:

$$A \models_{\text{sta}} t_1=t_2 \text{ iff } \text{val}_{A,\text{sta}}(t_1)=\text{val}_{A,\text{sta}}(t_2)$$

- *Models, semantic conclusion, satisfiability and further semantic concepts*

  As before.

## 4.2    Reduction of predicate logic with equality to predicate logic

**Definition 4.1:**    For a signature Σ, let Eq(Σ) be the set consisting of the following formulas:

$$X=X$$
$$X=Y \rightarrow Y=X$$
$$X=Y \rightarrow (Y=Z \rightarrow X=Z)$$
$$(X_1=Y_1 \rightarrow (X_2=Y_2 \ldots \rightarrow (X_n=Y_n \rightarrow f(X_1,\ldots,X_n)=f(Y_1,\ldots,Y_n))\ldots))$$
$$(X_1=Y_1 \rightarrow (X_2=Y_2 \ldots \rightarrow (X_n=Y_n \rightarrow (p(X_1,\ldots,X_n) \rightarrow p(Y_1,\ldots,Y_n)))\ldots))$$

for all variables $X,Y,Z,X_1,\ldots,X_n,Y_1,\ldots,Y_n$, all n-ary function symbols f and n-ary relation symbols p in Σ. Eq(Σ) is called the set of **equality axioms** in signature Σ.

**Fact**   Every $\Sigma$-interpretation is a model of $Eq(\Sigma)$.

*Proof*   $Eq(\Sigma)$ expresses well-known properties of equality, namely reflexivity, symmetry and transitivity (usually collected under the notion of an equivalence relation), as well as substitutivity (equal objects may be interchangeably used).                                                                                 ∎

To reduce predicate logic with equality to pure predicate logic, we introduce, given a signature $\Sigma$, an extended signature $\Sigma_=$, which contains = as a further, 2-ary, infixedly written predicate symbol. Thus, from now on, we may consider a $\Sigma$-formula $\varphi$ of predicate logic with equality also as a $\Sigma_=$-formula of pure predicate logic. Given a $\Sigma$-interpretation $A$, we consider the expansion $(A, =_{\text{dom}(A)})$ which interprets symbols from $\Sigma$ the same way $A$ does, and in addition the equality predicate by the equality relation $=_{\text{dom}(A)}$ on $\text{dom}(A)$.

**Fact**   For every $\Sigma$-formula $\varphi$ and $\Sigma$-interpretation $A$, $A \models \varphi$ iff $(A, =_{\text{dom}(A)}) \models \varphi$. Note that the latter notion refers to predicate logic over signature $\Sigma_=$, whereas the former notion refers to predicate logic with equality over signature $\Sigma$.                                                                       ∎

For every $\Sigma$-interpretation $A$, $(A, =_{\text{dom}(A)})$ is a model of $Eq(\Sigma)$, but the converse does not hold. As an example, consider an interpretation $A$ with $\text{dom}(A)=\{1,2\}$ and the relation $\{(1,1),(2,2),(1,2),(2,1)\}$ as interpretation of = in $A$. This example shows that $Eq(\Sigma)$ admits more models than only those that interpret = as the equality relation on the domain of the considered interpretation.

> **Definition 4.2:**   Let $A$ be a $\Sigma$-algebra and $\sim$ a 2-ary relation on $\text{dom}(A)$. We say that $\sim$ is a **congruence relation** on $A$ iff it is an equivalence relation and the following conditions are fulfilled:
>
> - $a_1 \sim b_1, \ldots, a_n \sim b_n$ implies $f(a_1, \ldots, a_n) \sim f(b_1, \ldots, b_n)$, for all $a_1, \ldots, a_n$, $b_1, \ldots, b_n \in \text{dom}(A)$ and every n-ary function symbol f in $\Sigma$.
> - $a_1 \sim b_1, \ldots, a_n \sim b_n$ and $(a_1, \ldots, a_n) \in p_A$ imply that $(b_1, \ldots, b_n) \in p_A$.
>
> For a $\Sigma$-algebra $A$ and a 2-ary relation $\sim$ on $\text{dom}(A)$, $(A, \sim)$ is the expansion of $A$ to the $\Sigma_=$-algebra that interprets predicate symbol = by relation $\sim$.

**Fact**    For a $\Sigma$-algebra $A$ and an arbitrary 2-ary relation $\sim$ on dom($A$), $(A,\sim)$ is a model of Eq($\Sigma$) iff $\sim$ is a congruence relation on $A$.

**Definition 4.3:**    Let $A$ be a $\Sigma$-algebra and $\sim$ a congruence relation on $A$. For $x \in$ dom($A$), $[x]_\sim = \{y \in$ dom($A$)| $y \sim x\}$ is the equivalence class of x w.r.t. $\sim$. Let dom($A$)$_{/\sim}$ be the set of all equivalence classes of elements of dom($A$). Finally, $A_{/\sim}$ is defined as the following $\Sigma$-algebra $B$:

- dom($B$)=dom($A$)$_{/\sim}$
- $f_B([a_1]_\sim,\ldots,[a_n]_\sim)=[f_A(a_1,\ldots,a_n)]_\sim$
- $p_B=\{([a_1]_\sim,\ldots,[a_n]_\sim) \mid (a_1,\ldots,a_n) \in p_A\}$

for all $a_1,\ldots,a_n \in$ dom($A$), every n-ary function symbol f in $\Sigma$, and every n-ary predicate symbol p in $\Sigma$.

The assumption that $\sim$ is a congruence relation guarantees that $f_B$ and $p_B$ are uniquely defined, that is, the possibility of choosing other representatives for the equivalence classes $[a_1]_\sim$ ,...,$[a_n]_\sim$ than just $a_1,\ldots,a_n$ does not affect the definitions of $f_B$ and $p_B$. $A_{/\sim}$ is called the **quotient interpretation of $A$ modulo congruence relation $\sim$.**

**Definition 4.4:**    Let $A$ and $B$ be $\Sigma$-interpretations. A function $h$:dom($A$)$\rightarrow$ dom($B$) is called a **homomorphism** from $A$ to $B$ iff for every n-ary function symbol f, predicate symbol p from $\Sigma$ and all $a_1,\ldots,a_n \in$ dom($A$) the following conditions are fulfilled:

- $h(f_A(a_1,\ldots,a_n))=f_B(h(a_1),\ldots,h(a_n))$
- $(a_1,\ldots,a_n) \in p_A \Leftrightarrow (h(a_1),\ldots,h(a_n)) \in p_B$

That h is a homomorphism from $A$ to $B$ is denoted by $h$:$A \rightarrow B$. If in addition h is surjective, it is called a homomorphism from $A$ onto $B$. If h is even bijective it is called an **isomorphism** from $A$ onto $B$. $A$ and $B$ are called **isomorphic**, written $A \cong B$, iff there exists an isomorphism from $A$ onto $B$.

**Fact**    A surjective homomorphism h from $A$ onto $B$ is an isomorphism iff h is a homomorphism from $(A,=_{\text{dom}(A)})$ onto $(B,=_{\text{dom}(B)})$.

This is clear since h is a homomorphism from $(A,=_{dom(A)})$ onto $(B,=_{dom(B)})$ iff h is a homomorphism from $A$ onto $B$ and, for all $x,y \in dom(A)$, $x=_{dom(A)}y$ iff $h(x)=_{dom(B)}h(y)$.

**Definition 4.5:** Let $A$ be a $\Sigma$-algebra and $\sim$ a congruence relation on $A$. The function $hom:dom(A) \to dom(A)/_\sim$ defined by $hom(x)=[x]_\sim$, for every $a \in dom(A)$, is called the **canonical homomorphism** from $A$ onto $A/_\sim$.

**Fact** The function hom defined above is indeed a surjective homomorphism from $A$ onto $A/_\sim$. It is even a surjective homomorphism from $(A,\sim)$ onto $(A/_\sim,=)$, if $=$ is the equality relation on $dom(A/_\sim)$.

*Proof* The first claim follows immediately from the definition of hom and the way function and predicate symbols are interpreted in $A/_\sim$. For the second claim, observe that $x \sim y$ iff $[x]_\sim=[y]_\sim$ iff $hom(x)=hom(y)$. ∎

*Lemma 4.1*

Let $A$ and $B$ be $\Sigma$-interpretations, h be a homomorphism from $A$ onto $B$, and sta be a state over $A$. (Hence h∘sta is a state over $B$.) Then:

(1)  $val_{B,h \circ sta}(t)=h(val_{A,sta}(t))$

(2)  $B \models_{h \circ sta} \varphi$ iff $A \models_{sta} \varphi$

(3)  $B \models \varphi$ iff $A \models \varphi$

for every $\Sigma$-term t and every $\Sigma$-formula $\varphi$ without equality. But they are not true for formulas with equality. If h is an isomorphism from $A$ onto $B$ then the claims above are true even for formulas $\varphi$ with equality.

*Proof* The first claim is shown by structural induction on term t. If t is a variable $X$ then $val_{B,h \circ sta}(t)=h \circ sta(X)=h(val_{A,sta}(X))=h(val_{A,sta}(t))$. If t is a composed term, say t is $f(t_1,...,t_n)$, then

$$val_{B,h \circ sta}(t)$$
$$=val_{B,h \circ sta}(f(t_1,...,t_n))$$
$$=f_B(val_{B,h \circ sta}(t_1),...,val_{B,h \circ sta}(t_n))$$
$$= \text{(by induction hypothesis) } f_B(h(val_{A,sta}(t_1)),...,h(val_{A,sta}(t_n)))$$

$$=h(f_A(val_{A,sta}(t_1),...,val_{A,sta}(t_n)))$$
$$=h(val_{A,sta}(f(t_1,...,t_n)))$$
$$=h(val_{A,sta}(t))$$

The second claim is shown by structural induction on formula $\varphi$. If $\varphi$ is an atomic formula $p(t_1,...,t_n)$ then

$B \models_{h \circ sta} \varphi \Leftrightarrow$

$B \models_{h \circ sta} p(t_1,...,t_n) \Leftrightarrow$

$(val_{B,h \circ sta}(t_1),...,val_{B,h \circ sta}(t_n)) \in p_B \Leftrightarrow$ ( just proved)

$(h(val_{A,sta}(t_1)),...,h(val_{A,sta}(t_n))) \in p_B \Leftrightarrow$ (h is a homomorphism)

$(val_{A,sta}(t_1),...,val_{A,sta}(t_n)) \in p_A \Leftrightarrow$

$A \models_{sta} p(t_1,...,t_n) \Leftrightarrow$

$A \models_{sta} \varphi$

If $\varphi$ is of the form $\neg \psi$ or $(\psi \rightarrow \chi)$ then a trivial application of induction proves the claim. If $\varphi$ is of the form $\forall X \psi$ then we argue as follows:

$B \models_{h \circ sta} \varphi \Leftrightarrow$

$B \models_{h \circ sta} \forall X \psi \Leftrightarrow$

$B \models_{(h \circ sta)(X/b)} \psi$, for all $b \in dom(B) \Leftrightarrow$ (surjectivity of h)

$B \models_{(h \circ sta)(X/h(a))} \psi$, for all $a \in dom(A) \Leftrightarrow$

(note that $(h \circ sta)(X/h(a))=h \circ (sta(X/a))$)

$B \models_{(h \circ sta(X/a))} \psi$, for all $a \in dom(A) \Leftrightarrow$ (induction hypothesis)

$A \models_{sta(X/a)} \psi$, for all $a \in dom(A) \Leftrightarrow$

$A \models_{sta} \forall X \psi \Leftrightarrow$

$A \models_{sta} \varphi$

The third claim is shown as follows:

$B \models \varphi \Leftrightarrow$

$B \models_{sta'} \varphi$ for every state sta' over $B \Leftrightarrow$

(using the surjectivity of h, write sta' as h∘sta, for a state sta over $A$)

$B \models_{h \circ sta} \varphi$, for every state sta over $A \Leftrightarrow$ ( just shown)

$A \models_{\text{sta}} \varphi$, for every state sta over $A$ $\Leftrightarrow$

$A \models \varphi$.

That these statements are not true, in general, for formulas with equality, can be seen from the following simple example. In the empty signature consider $A$ with dom($A$)={1,2} and $B$ with dom($B$)={3} and the surjective homomorphism h with h(1)=3 and h(2)=3. Then, $B \models \forall X \forall Y \ X=Y$, but $A \not\models \forall X \forall Y$ $X=Y$.

Finally assume that h is an isomorphism from $A$ onto $B$. Hence, $h$ is also a homomorphism from $(A,=_{\text{dom}(A)})$ onto $(B,=_{\text{dom}(B)})$, and we may argue for a formula $\varphi$ in predicate logic with equality:

$B \models \varphi \Leftrightarrow$ (via reduction to predicate logic)

$(B,=_{\text{dom}(B)}) \models \varphi \Leftrightarrow$ (h is a homomorphism from $(A,=_{\text{dom}(A)})$ onto

$(B,=_{\text{dom}(B)})$ and $\varphi$ is interpreted in $(B,=_{\text{dom}(B)})$ and $(A,=_{\text{dom}(A)})$ as a

formula *without equality*, hence we may use point (3) from Lemma 4.1)

$(A,=_{\text{dom}(A)}) \models \varphi \Leftrightarrow$

$A \models \varphi$ ■


### Theorem 4.1

Let $A$ be a $\Sigma$-algebra and $\sim$ a congruence relation on dom($A$). Then $(A,\sim) \models \varphi$ iff $A_{/\sim} \models \varphi$, for every formula $\varphi$ in predicate logic with equality. (The latter notion of validity refers to predicate logic with equality, the former to predicate logic with = as a predicate symbol.)

*Proof* We have already shown that hom is a surjective homomorphism from $(A,\sim)$ onto $(A_{/\sim},=)$ if = denotes the equality relation on dom($A_{/\sim}$). ■


### Corollary

Let M be a set of $\Sigma$-formulas and $\varphi$ be a $\Sigma$-formula in predicate logic with equality. Then the following statements are equivalent:

(1)  Every model $A$ of M is also a model of $\varphi$, in the sense of predicate logic with equality.

(2)  Every model $(A,\sim)$ of M$\cup$Eq($\Sigma$) is also a model of $\varphi$, in the sense of predicate logic (without equality).

*Proof*   (2) implies (1), since for a model $A$ of M, $(A,=_{dom(A)})$ is a model of $M \cup Eq(\Sigma)$. Conversely, (1) implies (2): assume that $(A,\sim)$ is a model of $M \cup Eq(\Sigma)$ (in logic without equality). Then, $\sim$ is a congruence relation on $A$. It follows that $A_{/\sim}$ is a model of M (in logic with equality), hence by (1), $A_{/\sim}$ is a model of $\varphi$, so $(A,\sim)$ is also a model of $\varphi$.    ∎

### Compactness Theorem 4.2 for predicate logic with equality

Let M be a set of $\Sigma$-formulas in predicate logic with equality. Then M has a model iff every finite subset of M has a model. Likewise, M is satisfiable iff every finite subset of M is satisfiable.

*Proof*   If every finite subset of M has a model (in predicate logic with equality) then every finite subset of $M \cup Eq(\Sigma)$ has a model (in predicate logic without equality). By Theorem 2.8 (compactness theorem for predicate logic without equality) $M \cup Eq(\Sigma)$ has a model $(A,\sim)$. Then, $A_{/\sim}$ is the desired model of M (in logic with equality). The same argument proves the claim for satisfiability.    ∎

The following theorem carries over the statements of the Löwenheim Skolem theorem from logic without equality to logic with equality, and sharpens some of them.

### Theorem 4.3 (Löwenheim Skolem theorems)

Let M be a set of $\Sigma$-formulas in predicate logic with equality. If M has a model then M has also a model whose domain is either finite or denumerably infinite. If M has a model with an infinite domain then it has also a model with a denumerably infinite domain. If, for every natural number k, M has a model whose domain contains at least k elements, then M has also a model with infinite domain.

*Proof*   If M has a model (in predicate logic with equality) then $M \cup Eq(\Sigma)$ has a model $(A,\sim)$ (in predicate logic without equality) with a denumerably infinite domain. Then $A_{/\sim}$ is a model of M (in predicate logic with equality). Since the domain of $A_{/\sim}$ consists of equivalence classes of elements of the domain of $A$ w.r.t. the congruence relation $\sim$, it follows that $dom(A_{/\sim})$ is either finite or denumerably infinite.

If M has a model (in predicate logic with equality) with an infinite domain, then $M \cup Inf$ has a model $(A,\sim)$ (in predicate logic with equality), with the formula set $Inf=\{\varphi_m | m>0\}$ where

$$\varphi_m \text{ is the formula } \exists X_1 \exists X_2 ... \exists X_m \bigwedge_{i<j} \neg X_i = X$$

(Obviously, this formula expresses that the domain of a model includes at least m different elements.) By the claim just proved we conclude that M∪Inf has a model $A$ with a domain which is either finite or denumerably infinite. Since $A$ is a model of Inf, its domain cannot be finite.

Finally, assume that M has models with domains of arbitrarily large finite cardinality. Then, every finite subset of M∪Inf has a model. By Theorem 4.2 we conclude that M∪Inf has a model, too. Such a model is a model of M and has an infinite domain.                                             ∎

### Corollary (Non-characterizability of finiteness)

Given a signature $\Sigma$, there exists no set of $\Sigma$-formulas whose models are just those $\Sigma$-interpretations with finite domain.

**Definition 4.6:**  Let M be a set of $\Sigma$-formulas in predicate logic with equality. M is called **complete** iff for every closed $\Sigma$-formula $\varphi$ in predicate logic with equality, M $\models \varphi$ or M $\models \neg\varphi$. For a $\Sigma$-interpretation $A$, we define the **theory of algebra $A$**, briefly denoted Th($A$), to be the set of all closed formulas in predicate logic with equality that are valid in $A$. Two $\Sigma$-interpretations $A$ and $B$ are called **elementary equivalent**, briefly denoted $A \equiv B$, iff the same closed formulas are valid in $A$ as in $B$.

### Lemma 4.2

(1)     Th($A$) is a complete set of formulas.

(2)     $A \equiv B$ iff Th($A$)=Th($B$).

(3)     If $A \cong B$ then $A \equiv B$.

(4)     M is complete iff every two models of M are elementary equivalent.

*Proof*   Only the last claim requires a proof. Assume that M is complete, $A$ and $B$ are models of M, and $\varphi$ is a closed formula. If M $\models \varphi$ then $\varphi$ is valid in $A$ as well as in $B$. If M $\models \neg\varphi$ then $\varphi$ is neither valid in $A$, nor in $B$. Conversely, assume that every two models of M are elementary equivalent. Let a closed formula $\varphi$ be given. If M does not possess any models, we know that M $\models \varphi$. So let us assume that M has a model $A$. If $A \models \varphi$, then $B \models \varphi$ for every interpretation $B$ that is elementary equivalent to $A$, hence $B \models \varphi$ for every model $B$ of M, hence M $\models \varphi$. If $A \models \neg\varphi$ we obtain similarly that M $\models \neg\varphi$.     ∎

## 4.3    Proof theory for predicate logic with equality

We extend the calculi discussed in Chapter 3 by mechanisms that adequately capture the concept of equality.

### 4.3.1    Hilbert's calculus for predicate logic with equality

**Definition 4.7:**    For a signature $\Sigma$, let $H(\Sigma,=)$ be the extension of Hilbert's calculus $H(\Sigma)$ by the formulas from $Eq(\Sigma)$ as further axioms. Usually we write $H(=)$ instead of $H(\Sigma,=)$.

***Theorem 4.4***

For an arbitrary set M of $\Sigma$-formulas and a $\Sigma$-formula $\varphi$ in predicate logic with equality, $M \models \varphi$ (in predicate logic with equality) is equivalent to $M \vdash_{H(=)} \varphi$. Hence, $H(\Sigma,=)$ is a sound and complete calculus for logical conclusion in predicate logic with equality.

*Proof*    $M \models \varphi$ (in predicate logic with equality) $\Leftrightarrow$

$M \cup Eq(\Sigma) \models \varphi$ (in predicate logic without equality) $\Leftrightarrow$

$M \cup Eq(\Sigma) \vdash_{H} \varphi \Leftrightarrow$

$M \vdash_{H(=)} \varphi$    ∎

### 4.3.2    Sequent calculus for predicate logic with equality

Syntax and semantics of the logic of sequents is extended in an obvious way to sequents with formulas from predicate logic with equality.

**Definition 4.8:**    For a signature $\Sigma$ extend the sequent calculus $LK(\Sigma)$ introduced in Section 3.3 by the following rules to a calculus $LK(\Sigma,=)$:

$$\frac{}{\rightarrow t=t} \qquad \frac{}{r=s \rightarrow s=r} \qquad \frac{}{r=s, s=t \rightarrow r=t}$$

$$\frac{}{r_1=t_1, r_2=t_2, \ldots, r_n=t_n \rightarrow f(r_1, r_2, \ldots, r_n)=f(t_1, t_2, \ldots, t_n)} \qquad \text{For n-ary f}$$

$$\frac{}{r_1=t_1, r_2=t_2, \ldots, r_n=t_n, p(r_1, r_2, \ldots, r_n) \rightarrow p(t_1, t_2, \ldots, t_n)} \qquad \text{For n-ary p}$$

### Theorem 4.5

For a sequent $\Pi \rightarrow \Gamma$ in the logic of sequents with equality, $\Pi \rightarrow \Gamma$ is valid iff $\Pi \rightarrow \Gamma$ can be derived in $LK(\Sigma,=)$.

*Proof*    Soundness of the extended sequent calculus is clear. Looking back at the way the completeness of the sequent calculus without equality was shown in Section 3.3, we see what must be checked here. First note that the deduction theorem is also true for $LK(\Sigma,=)$. In fact, there are no more cases to be treated other than for sequent logic without equality. Next, show that, for every $\Sigma$-formula $\varphi$ in predicate logic with equality, if $\varphi$ can be derived in $H(\Sigma,=)$ then the sequent $\rightarrow \varphi$ can be derived in the extended sequent calculus. Only formulas in $Eq(\Sigma)$ must be checked. An application of the rule that corresponds to a formula of $Eq(\Sigma)$ and a finite number of applications of ($\rightarrow$ right) and structural rules proves the claim.    ■

### 4.3.3    Resolution calculus with equality and paramodulation

Syntax and semantics of the logic of clauses is extended in an obvious way to clauses with literals from predicate logic with equality.

Looking back at the constructions in Section 3.4 and the required preparations from Chapter 2, in particular those concerning Herbrand interpretations, the reader will recognize the following difficulty. Two of our theorems that were used to establish the completeness of the resolution calculus are not true for formulas which involve built-in equality. The first one is the theorem on satisfiability of ground clauses not containing complementary literals. Consider the following conjunction of literals: $p(a) \wedge a=b \wedge \neg p(b)$. Though there is no complementary pair among the involved literals, the whole formula does not possess a model (in logic with equality). The second theorem is Herbrand's theorem. Consider set $M = Eq \cup \{a=b\}$. Obviously, $M$ does not possess a Herbrand model with = interpreted as equality, since such a model would interpret $a$ and $b$ differently. Nevertheless, it possesses a model (interpret = by a congruence relation ~ with a~b).

But nobody forces us to adapt the prior proofs to the situation of logic with equality. Rather, we may reduce the whole problem to predicate logic without equality, again interpreting = as a normal 2-ary relation symbol. The result is as follows.

### Theorem 4.6

Let $M$ be a set of clauses (with literals that may contain equality). Consider Eq-clauses, the clause representation of Eq:

$\{X=X\}$
$\{\neg X=Y, Y=X\}$

$$\{\neg X=Y,\neg Y=Z,X=Z\}$$
$$\{\neg X_1=Y_1,...,\neg X_n=Y_n,f(X_1,...,X_n)=f(Y_1,...,Y_n)\}$$
$$\{\neg X_1=Y_1,...,\neg X_n=Y_n,\neg p(X_1,...,X_n),p(Y_1,...,Y_n)\}$$

Then the following statements are equivalent:

(a)   M does not possess a model (in clause logic with equality).

(b)   The empty clause can be derived in RES from $M \cup Eq$-clauses.

*Proof* M does not possess a model (in clause logic with equality) iff $M \cup Eq$-clauses does not possess a model (in clause logic without equality).     ∎

For practical applications, the use of set Eq-clauses as a sound and complete axiomatization of equality cannot be recommended. The reason is that the use of Eq-clauses leads to an enormous increase in the number of possibilities for constructing resolvents. Therefore, more efficient substitutes for the use of Eq-clauses have been proposed, the most prominent being paramodulation.

**Definition 4.9:**   We replace resolution by two rules, the binary resolution and the factoring rule. Then we add a paramodulation rule and some axioms. These are depicted in Figure 4.1.

Note   how   the   paramodulant   $\{(a\mu)(r\mu)(b\mu)\} \cup (C_1 \cup C_2)\mu$   is constructed:

> Choose a subterm t of a literal atb and an mgu $\mu$ of t and l, apply $\mu$ and replace subterm $l\mu$ of $(a\mu)(l\mu)(b\mu)$ by $r\mu$. Alternatively, we can replace the right-hand side of an equation by the corresponding left-hand side.

### Theorem 4.7

Let M be a set of clauses in clause logic with equality. Then, M does not possess a model (in logic with equality) iff the empty clause can be derived from M together with the extra clauses above via binary resolution, factoring and paramodulation.

For a proof the reader may consult Loveland (1978).

---

**Exercise 4.1**   Consider the group theory example treated in Section 2.4. Show using paramodulation that existence of a right-neutral element and of a right inverse for each group element follows from associativity, existence of a left-neutral element and existence of left inverses.

---

*Binary resolution*

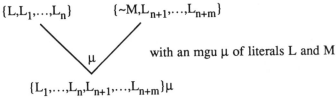

$$\{L,L_1,\ldots,L_n\} \qquad \{\sim M,L_{n+1},\ldots,L_{n+m}\}$$

with an mgu $\mu$ of literals L and M

$$\{L_1,\ldots,L_n,L_{n+1},\ldots,L_{n+m}\}\mu$$

*Factoring rule*    $\{L_1,\ldots,L_n,L_{n+1},\ldots,L_{n+m}\}$

$\mu$

if $n>1$ and $\mu$ is an mgu of $\{L_1,\ldots,L_n\}$
($n=2$ would already be sufficient)

$$\{L_1,\ldots,L_n,L_{n+1},\ldots,L_{n+m}\}\mu$$

*Paramodulation rule* (string atb denotes a literal with subterm t)

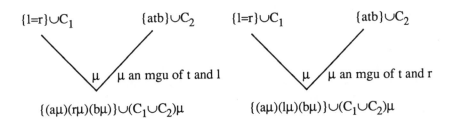

$$\{l=r\}\cup C_1 \qquad \{atb\}\cup C_2 \qquad \{l=r\}\cup C_1 \qquad \{atb\}\cup C_2$$

$\mu$  $\mu$ an mgu of t and l          $\mu$  $\mu$ an mgu of t and r

$$\{(a\mu)(r\mu)(b\mu)\}\cup(C_1\cup C_2)\mu \qquad \{(a\mu)(l\mu)(b\mu)\}\cup(C_1\cup C_2)\mu$$

*Extra axioms*        $\{t=t\}$ for all terms t

$$\{f(t_1,\ldots,t_n)=f(t_1,\ldots,t_n)\}$$

for all terms $t_1,\ldots,t_n$ and n-ary function symbols f

**Figure 4.1**

# Chapter 5
# Basic Concepts from Model Theory

This chapter introduces a couple of basic concepts and facts from model theory that will be required in later chapters. Briefly stated, model theory is the combination of logic and algebra. Properties of interpretations that occur in algebra are investigated using logical tools, and conversely, logical problems are attacked by using algebraic methods. The field is quite extended. We will concentrate on a few basic concepts that will be required later in program verification. For a thorough introduction to model theory the reader may refer to Chang and Keisler (1973).

## 5.1  Substructures and elementary substructures

**Definition 5.1:**  Let $A$ and $B$ be $\Sigma$-interpretations. $A$ is called a **subinterpretation** (or **subalgebra**) of $B$, briefly denoted $A \subseteq B$, iff

- $\mathrm{dom}(A) \subseteq \mathrm{dom}(B)$
- $f_A(a_1,\ldots,a_n) = f_B(a_1,\ldots,a_n)$
- $(a_1,\ldots,a_n) \in p_A \Leftrightarrow (a_1,\ldots,a_n) \in p_B$

for all $a_1,\ldots,a_n \in \mathrm{dom}(A)$, every n-ary function symbol f and every n-ary predicate symbol p from $\Sigma$.

As an example, the algebra of natural numbers with addition, multiplication and linear order is a subalgebra of the algebra of integers with corresponding functions and relations.

### Theorem 5.1

Let $A$ be a subinterpretation of $B$ and sta be a state over $A$ (hence also a state over $B$). Then

- $val_{A,sta}(t)=val_{B,sta}(t)$, for every term t

- $A \models_{sta} \varphi \Leftrightarrow B \models_{sta} \varphi$, for every quantifier-free formula $\varphi$

- $B \models \varphi \Rightarrow A \models \varphi$, for every universal formula $\varphi$.

*Proof*  The first and the second claims, to be proved by structural induction on t and $\varphi$, are immediately obtained from Definition 5.1. The third claim follows from the second one and the inclusion $dom(A) \subseteq dom(B)$.    ■

**Definition 5.2:**  Let $A$ and $B$ be $\Sigma$-interpretations. $A$ is called an **elementary subinterpretation** (or **elementary subalgebra**) of $B$, briefly denoted $A \prec B$, iff $A \subseteq B$ and for all formulas $\varphi$ and states sta over $A$, $A \models_{sta} \varphi$ iff $B \models_{sta} \varphi$. In this case, $B$ is called an **elementary extension** of $A$.

If $A \prec B$ then not only the same closed formulas are valid in $A$ and $B$ (thus $A \equiv B$), but also formulas 'with parameters from $dom(A)$'. This can be made precise as follows.

### Lemma 5.1

Let $A$ and $B$ be $\Sigma$-interpretations such that $A \subseteq B$. Extend $\Sigma$ by a new constant a, for every $a \in dom(A)$. Let $(A,a)_{a \in dom(A)}$ and $(B,a)_{a \in dom(A)}$ be the expansions of $A$ and $B$ that interpret constant a by element a, for every $a \in dom(A)$. Then $A \prec B$ iff $(A,a)_{a \in dom(A)} \equiv (B,a)_{a \in dom(A)}$.

*Proof*  Follows immediately from the definitions.    ■

As was stated above, $A \prec B$ implies $A \subseteq B$ and $A \equiv B$. The converse direction is not true. As an example, consider $B$=the linear order of natural numbers and $A$=the linear order of positive natural numbers. Obviously, $A \subseteq B$. Furthermore, $A$ is isomorphic to $B$, hence $A \equiv B$. But $A$ is not an elementary subalgebra of $B$ since, for example, element 1 of the domain of $A$ as interpretation of variable X satisfies formula $\exists Z\ Z{<}X$ in $A$, but not in $B$.

Further examples for elementary subalgebras, in particular positive ones, are postponed until Section 5.2.

The following results and definitions require (unfortunately) some elementary knowledge about cardinal numbers.

### Löwenheim-Skolem Theorems 5.2 (upwards and downwards)

(Remember here that signatures, as we introduced them, are always at most denumerable sets. If arbitrary sets are admitted as signatures, as required for some applications in model theory, the present theorem needs some modifications and additional assumptions.)

(1)    (*downwards*)  For every interpretation $A$ with infinite domain and every subset X of dom($A$) there exists an elementary substructure $B$ of $A$ such that $X{\subseteq}$dom($B$) and the cardinality of dom($B$) is not greater than the maximum of $\omega$ (=the cardinality of the set of natural numbers) and the cardinality of X.

(2)    (*upwards*)   For every interpretation $A$ with infinite domain and every infinite cardinal number $\kappa$ there is an elementary extension $B$ whose domain is of cardinality $\kappa$.

For a proof consult Chang and Keisler (1973).


## 5.2    Categoricity ✂

An important question is how precisely a given interpretation $A$ can be characterized by an axiom system Ax. Obviously, Ax should be chosen in such a way that $A$ is a model of Ax. Is it possible that conversely $A$ is the only model of Ax? Of course not. Every interpretation $B$ that is isomorphic to $A$ is also a model of Ax.

Hence, our question should be posed more moderately as follows: can Ax be chosen in such a way that the only models of Ax are the interpretations $B$ that are isomorphic to $A$? If the domain of $A$ is infinite, the theorem of Löwenheim-Skolem (upwards) immediately shows that this cannot be achieved. (For finite domain it is indeed possible: see Exercise 5.1.) We can readily construct an elementary extension $B$ of $A$ such that the cardinality of dom($B$) is greater than the cardinality of dom($A$). Since $B$ is elementary equivalent to $A$, $B$ is a model of Ax, too. But $B$ cannot be isomorphic to $A$ as there is no bijective function from dom($B$) onto dom($A$).

The final question which now makes sense is: is there an axiom system Ax such that the models $B$ of Ax with a domain of the same cardinality as the domain of $A$ are just the interpretations $B$ that are isomorphic to $A$? This leads to the following notion.

**Definition 5.3:** For a given cardinal number κ, a formula set Ax (in predicate logic with equality) is called **κ-categorical**, iff every two models $A$ and $B$ with domains of cardinality κ are isomorphic.

---

**Exercise 5.1** Show that every term-generated algebra **A** with finite domain may be categorically characterized by an axiom system Ax, in the sense that every model of Ax is isomorphic to **A**.

---

The following theorem of Los is an often used (and easily applicable) sufficient criterion for the completeness of an axiom system.

### Theorem 5.3 (Los)

Assume that a formula set Ax has only models with infinite domain and is κ-categorical, for an infinite cardinal number κ. Then Ax is complete.

*Proof*   We must show that every two models of Ax are elementary equivalent. So let models $A$ and $B$ of Ax be given. By assumption, the domains of $A$ and $B$ are infinite. With the Löwenheim-Skolem theorem (upwards if the the cardinality of dom($A$) is <κ, and downwards if the the cardinality of dom($A$) is >κ) we choose an interpretation $A'$ with a domain of cardinality κ that is either an elementary extension or elementary substructure of $A$. In any case, $A'$ is elementary equivalent to $A$. The same is done for $B$, leading to an elementary equivalent interpretation $B'$ with a domain of cardinality κ. Since $A'$ and $B'$ are models of Ax with domains of cardinality κ, κ-categoricity of Ax implies that $A'$ and $B'$ are isomorphic. Hence, $A'$ and $B'$ are elementary equivalent. Altogether, we obtain $A \equiv A' \equiv B' \equiv B$.     ∎

## EXAMPLE 5.1

The axiom systems for *dense linear orderings without least and greatest element* is the most prominent example of an ω-categorical axiom system. Its axioms are (using a 2-ary predicate symbol < that is infixedly denoted):

| | |
|---|---|
| $\forall X \neg X < X$ | irreflexivity |
| $\forall X \forall Y \forall Z((X<Y \wedge Y<Z) \rightarrow X<Z)$ | transitivity |
| $\forall X \forall Y(X<Y \vee X=Y \vee Y<X)$ | linearity |
| $\forall X \forall Y(X<Y \rightarrow \exists Z(X<Z \wedge Z<Y))$ | density |
| $\forall X \exists Y \ X<Y$ | no greatest element |
| $\forall X \exists Y \ Y<X$ | no least element |

To show that this axiom system is $\omega$-categorical, consider two models $(A,<_A)$ and $(B,<_B)$ with domains that are denumerably infinite. We enumerate A and B:

$$A=\{a_n \mid n\in \mathbb{N}\} \text{ and } B=\{b_n \mid n\in \mathbb{N}\}$$

An isomorphism h from $(A,<_A)$ onto $(B,<_B)$ is constructed in stages as the union of an ascending sequence of partial functions $h_k$ from A into B, with $k\in \mathbb{N}$. The functions $h_k$ are defined in such a way that:

(1)     $\{a_n \mid n<k\}$ is contained in the domain of $h_k$.

(2)     $\{b_n \mid n<k\}$ is contained in the range of $h_k$.

(3)     For all x,y in the domain of $h_k$, $x<_A y$ iff $h_k(x)<_B h_k(y)$.

(4)     $h_k$ is injective.

Hence, with increasing k, $h_k$ captures more and more elements of A and B in its domain and range. Every $h_k$ is injective and partially fulfils the compatibility condition of an isomorphism from $(A,<_A)$ onto $(B,<_B)$, at least for all x,y where it is already defined. As a consequence of (1)-(4), if we define h as the union of all $h_k$, we obtain an isomorphism from $(A,<_A)$ onto $(B,<_B)$.

The definition of $h_k$ proceeds as a so-called **back-and-forth proce-dure**. We start with $h_0=\varnothing$, the empty function. Obviously, (1)-(4) are fulfilled. Having defined $h_k$ with (1)-(4), we define $h_{k+1}$ as follows.

(Forth)   If $a_{k+1}$ is already contained in the domain of $h_k$, define an inter-mediate function $f_k$ to be $h_k$. If $a_{k+1}$ is not contained in the domain of $h_k$ choose an arbitrary element $b_l$ that is not in the range of $h_k$ in such a way that, for eve-ry x in the domain of $h_k$, the relative position of $a_{k+1}$ to element x in the order-ing $<_A$ is the same as the relative position of $b_l$ to element $h_k(x)$ in the ordering $<_B$. Note that such a choice is always possible for $b_l$ since $<_B$ is a dense linear ordering without least and greatest element, and (3) was as-sumed for $h_k$. Then define $f_k$ to be $h_k\cup\{(a_{k+1},b_l)\}$. Note that conditions (3) and (4) hold for $f_k$.

(Back)   If $b_{k+1}$ is already contained in the range of $f_k$, define $h_{k+1}$ to be $f_k$. If $a_{k+1}$ is not contained in the range of $f_k$, choose an arbitrary element $a_m$ that is not in the domain of $f_k$ in such a way that, for every x in the domain of $f_k$, the relative position of $a_m$ to element x in the ordering $<_A$ is the same as the relative position of $b_{k+1}$ to element $f_k(x)$ in the ordering $<_B$. Note that such a choice is always possible for $a_m$, since $<_A$ is a dense linear ordering without least and greatest element, and (3) was shown for $f_k$. Then define $h_{k+1}$ to be $f_k \cup \{(a_m, b_{k+1})\}$. Note that conditions (1)-(4) are fulfilled for $h_{k+1}$.

The axiom system treated here is complete by Los's theorem as it possesses only infinite models (by definition) and is $\omega$-categorical.

# EXAMPLE 5.2

A later application will require the fact that the *theory of a free term-algebra* is decidable. Let us first introduce the required concepts. We start with a finite signature $\Sigma$ that consists only of function symbols, contains at least one constant and one proper function symbol. Then we consider the only Herbrand interpretation in signature $\Sigma$ and call it $T(\Sigma)$. Remember how it was defined. Its domain consists of all ground $\Sigma$-terms. By assumption on $\Sigma$, this domain is non-empty, even infinite. Now consider the axiom system FREE($\Sigma$) consisting of the following formulas:

$$f(X_1,...,X_n)=f(Y_1,...,Y_n)\rightarrow(X_1=Y_1\wedge...\wedge X_n=Y_n)$$

$$\neg f(X_1,...,X_n)=g(Y_1,...,Y_m)$$

$$\neg X=t$$

for every n-ary function symbol f, m-ary function symbol g different from f, variable X and term t that is not a variable and contains X. It is clear that $T(\Sigma)$ is a model of FREE($\Sigma$), and all models of FREE($\Sigma$) have infinite domain.

We shall show that FREE($\Sigma$) is $\omega_1$-categorical ($\omega_1$ is the next cardinal number beyond $\omega$. The next theorem will show that FREE($\Sigma$) is not $\omega$-categorical). We require a minimal amount of knowledge about ordinal numbers and transfinite induction. To show $\omega_1$-categoricity, let us consider two models $A$ and $B$ of FREE($\Sigma$) with domains of cardinality $\omega_1$. Take transfinite enumerations $A=\{a_\xi \mid \xi$ an ordinal number $<\omega_1\}$ and $B=\{b_\xi \mid \xi$ an ordinal number $<\omega_1\}$. As in Example 5.1, we construct an isomorphism h from $A$ onto $B$ in states as the union of the ascending transfinite sequence $(h_\xi)_{\xi<\omega_1}$. We must take care that in every stage $\xi<\omega_1$ of the construction the following conditions are fulfilled:

(1)     $\{a_\eta \mid \eta<\xi\}$ is contained in the domain of $h_\xi$.

The domain of $h_\xi$ is countable and closed under all functions of $A$.

(2)     $\{b_\eta \mid \eta<\xi\}$ is contained in the range of $h_\xi$.

The range of $h_\xi$ is countable and closed under all functions of $B$.

(3)     $h_\xi$ is an isomorphism from the substructure of $A$ whose domain is the domain of $h_\xi$ onto the substructure of $B$ whose domain is the range of $h_\xi$. Equivalently stated, $h_\xi$ is injective and for every term $t(X_1,...,X_m)$ and elements $x_1,...,x_m$ of the domain of $h_\xi$:

$$h_\xi(t_A(x_1,...,x_m))=t_B(h_\xi(x_1),...,h_\xi(x_m))$$

(The difference in invariants (1)-(3) of the construction to those in Example 5.1 is caused by the fact that we are now dealing with a structure with *functions*, whereas Example 5.1 dealt with purely *relational* structures. Under the presence of functions we are no longer as free to construct an isomorphism h from $A$ onto $B$: having defined h(a)=b, h(x) is fixed for every element x that can be generated from a by the constants and functions of $A$!)

The construction is again a back-and-forth procedure. We start with $h_0=\{(t_A,t_B)|\ t$ is a ground term$\}$. Since different ground terms evaluate to different values in a model of FREE($\Sigma$), conditions (1)-(3) are fulfilled. Next, consider a limit ordinal number $\lambda<\omega_1$ (that is, an ordinal that is neither 0 nor the successor of an ordinal). Assume that $h_\xi$ has already been defined such that (1)-(3) hold for every $\xi<\lambda$. Then define $h_\lambda$ as the union of all $h_\xi$, with $\xi<\lambda$. Conditions (1)-(3) carry over in a trivial manner. Finally, consider a successor ordinal $\xi+1$. Assume that $h_\xi$ has already been defined such that (1)-(3) hold.

*(Forth)*     We define an intermediate function $f_\xi$. If $a_\xi$ is already an element of the domain of $h_\xi$, define $f_\xi$ to be $h_\xi$. Otherwise choose an element $b_\rho$ with $\rho<\omega_1$ such that $b_\rho$ is not an element of the range of $h_\xi$. Then define $f_\xi$ on all elements that can be generated by the functions of $A$ from the elements of the domain of $h_\xi$ as follows:

$$f_\xi(t_A(a_\xi,x_1,...,x_m))=t_B(b_\rho,h_\xi(a_1),...,h_\xi(a_m))$$

for every term $t(X,X_1,...,X_m)$ with at most the variables $X,X_1,...,X_m$, and all $x_1,...,x_m$ from the domain of $h_\xi$.

It has to be shown that $f_\xi$ is uniquely defined. For this purpose, consider terms $t(X,X_1,...,X_m)$ and $r(X,X_1,...,X_m)$ and assume that $t_A(a_\xi,x_1,...,x_m)=r_A(a_\xi,x_1,...,x_m)$. Since $A$ is a model of FREE($\Sigma$), we may reduce terms t and r by successively cutting off a common leading function symbol, until terms $t'(X,X_1,...,X_m)$ and $r'(X,X_1,...,X_m)$ are obtained such that $t'_A(a_\xi,x_1,...,x_m)=r'_A(a_\xi,x_1,...,x_m)$ and one of them is a variable. (Note that we cannot end with terms with different leading function symbols.)

If neither $t'(X,X_1,\ldots,X_m)$ nor $r'(X,X_1,\ldots,X_m)$ contains X, say $t'(X,X_1,\ldots,X_m)=t''(X_1,\ldots,X_m)$ and $r'(X,X_1,\ldots,X_m)=r''(X_1,\ldots,X_m)$, we argue as follows. We know that $t''_A(x_1,\ldots,x_m)= r''_A(x_1,\ldots,x_m)$. Hence by assumption (3) on $h_\xi$ we may conclude that

$$t''_B(h_\xi(x_1),\ldots,h_\xi(x_m))$$
$$=h_\xi(t''_A(x_1,\ldots,x_m))$$
$$=h_\xi(r''_A(x_1,\ldots,x_m))$$
$$=r''_B(h_\xi(x_1),\ldots,h_\xi(x_m))$$

and also

$$f_\xi(t_A(a_\xi,x_1,\ldots,x_m))$$
$$=t_B(b_\rho,h_\xi(a_1),\ldots,h_\xi(a_m))$$
$$=r_B(b_\rho,h_\xi(a_1),\ldots,h_\xi(a_m))$$
$$=f_\xi(r_A(a_\xi,x_1,\ldots,x_m))$$

As $a_\xi$ was not an element of the domain of $h_\xi$, we know that both $t'(X,X_1,\ldots,X_m)$ and $r'(X,X_1,\ldots,X_m)$ must contain X. So assume that one of $t'(X,X_1,\ldots,X_m)$ and $r'(X,X_1,\ldots,X_m)$ coincides with variable X. Because of the axiom $\neg X = t$ both $t'(X,X_1,\ldots,X_m)$ and $r'(X,X_1,\ldots,X_m)$ must coincide with X. So, again we obtain that $t_B(b_\rho,h_\xi(a_1),\ldots,h_\xi(a_m))=r_B(b_\rho,h_\xi(a_1),\ldots,h_\xi(a_m))$. So the uniqueness of the definition of $f_\xi$ is established.

$\{a_\eta \mid \eta<\xi+1\}$ is contained in the domain of $f_\xi$, the domain of $f_\xi$ is countable (since $\xi<\omega_1$ and the set of terms is countable) and closed under all functions of $A$. The range of $f_\xi$ is also countable and closed under all functions of $A$. $f_\xi$ is an extension of $h_\xi$. All these statements follow in a trivial manner from the definition of $f_\xi$. $f_\xi$ is injective. This is shown with just the same argument that proved the uniqueness of the definition of $f_\xi$, with the roles of $A$ and $B$ interchanged. Finally, condition (3) is fulfilled for $f_\xi$. This follows again immediately from the definition of $f_\xi$.

(Back) The same procedure as in (Forth), with the roles of domains and ranges interchanged, enables us to press $b_\xi$ into the range of an extension $h_{\xi+1}$ of $f_\xi$ such that properties (1)-(3) are true for $h_{\xi+1}$. ∎

---

**Exercise 5.2** (quantifier elimination)     Consider again the axiom system T for dense linear ordering without least and greatest element. For every formula $\varphi(X_1,...,X_n)$ with $n>0$ and free variables among $X_1,...,X_n$ construct a quantifier-free formula $\psi(X_1,...,X_n)$ such that $T \models (\varphi(X_1,...,X_n) \leftrightarrow \psi(X_1,...,X_n))$. (Hint: Using induction on the number of quantifiers, prenex normal form and disjunctive normal form, it suffices to consider formulas $\varphi(X_1,...,X_n)$ of the form $\exists X(A_1 \wedge ... \wedge A_n)$, where $A_1,...,A_n$ are atomic formulas of the form $Y=Z$ or $Y<Z$.)

**Exercise 5.3**     Show that the result of Exercise 5.2 gives us another proof of the completeness of formula set T. (Hint: There is a little difficulty with the requirement that the number n of variables above is $>0$. For a closed formula $\varphi$ consider the equivalent formula $(\varphi \wedge X_1=X_1)$. Then examine what a quantifier-free formula with the only variable $X_1$ looks like.)

**Exercise 5.4**     Using the quantifier elimination result show that for any two models $(A,<_A)$ and $(B,<_B)$ of T, if $(A,<_A)$ is a subalgebra of $(B,<_B)$, then $(A,<_A)$ is an elementary subalgebra of $(B,<_B)$. As an example, the dense order of rational numbers is an elementary subalgebra of the dense order of real numbers.

---

For those interpretations which occur frequently in computer science, namely those with an infinite domain and the property that every element in its domain can be named by a ground term, a negative result w.r.t. categoricity is shown next.

### Theorem 5.4

Let $A$ be a $\Sigma$-interpretation with an infinite domain and the property that for every $x \in \text{dom}(A)$ there exists a ground $\Sigma$-term t such that $x=\text{val}_A(t)$. (Such interpretations are called **term generated**.) Then $\text{Th}(A)$ is not $\omega$-categorical. As a consequence, there is no axiom system Ax with the property that the models of Ax with countably infinite domain are just the $\Sigma$-interpretations that are isomorphic to $A$.

*Proof*     We construct a $\Sigma$-interpretation $B$ with a denumerably infinite domain such that $B$ is elementary equivalent to $A$, but not isomorphic to $A$. Then the theorem is proved. To construct $B$, we extend $\Sigma$ by a new constant c and consider the following axiom system in signature $\Sigma \cup \{c\}$:

Th($A$)$\cup\{\neg$c=t| t is a ground $\Sigma$-term$\}$

We show that every finite subset of Th($A$)$\cup\{$c=t| t is a ground $\Sigma$-term$\}$ has a $\Sigma\cup\{$c$\}$-model. For such a finite subset Ax$\cup\{\neg$c=$t_1$,...,$\neg$c=$t_m\}$, with Ax$\subseteq$Th($A$), we obtain a $\Sigma\cup\{$c$\}$-model as an expansion of $A$ which interprets constant c by an arbitrary element x of dom($A$)-$\{$val$_A(t_1)$,...,val$_A(t_m))$. (Such a choice for x is always possible since dom($A$) is infinite.) By the compactness theorem we obtain a model $C$ of Th($A$)$\cup\{\neg$c=t| t is ground $\Sigma$-term$\}$. Let x be the interpretation of constant c in $C$ and $B$ be the reduct of $C$ to signature $\Sigma$. (Note that x$\in$dom($B$) as the construction of reducts does not affect domains.) Since also $B$ is a model of Th($A$), we obtain that $B$ is elementary equivalent to $A$.

We also show that $B$ is not isomorphic to $A$. Assume, by contradiction, that h is an isomorphism from $A$ onto $B$. Take y$\in$dom($A$) such that h(y)=x. Since $A$ is term generated, we may write y as val$_A$(t), for a ground $\Sigma$-term t. Thus we obtain val$_C$(c)=x=h(y)=h(val$_A$(t))= val$_B$(t)=val$_C$(t). This implies that $C$ is a model of c=t, a contradiction.                     ∎

The reader should keep in mind that there exist non-standard models of arithmetic, that is, algebras that are elementary equivalent to **Nat**= ($\mathbb{N}$,0,succ,+,*) but are not term generated. Such models will be extensively used in Chapter 16.

---

**Exercise 5.5**   Show that every non-standard model of arithmetic is an elementary extension of **Nat**. More generally, show that $A \prec B$ provided that $A\subseteq B$, $A\equiv B$ and $A$ is term generated.

---

## 5.3   Definability

**Definition 5.4:**   Let $A$ be a $\Sigma$-interpretation, $B$ be an $\Omega$-interpretation, r be an n-ary relation on dom($A$), f be an n-ary function on dom($A$), and S be a set of states over $A$. We say that:

- r is **definable** in $A$ iff there is a $\Sigma$-formula $\varphi_r(X_1,...,X_n)$ such that for all $a_1,...,a_n\in$ dom($A$),

    $(a_1,...,a_n)\in$ r iff $A \models \varphi_r(a_1,...,a_n)$.

- f is **definable** in $A$ iff there is a $\Sigma$-formula $\varphi_f(X_1,...,X_n,Y)$ such that for all $a,a_1,...,a_n \in dom(A)$,

  $f(a_1,...,a_n)=a$ iff $A \models \varphi_f(a_1,...,a_n,a)$.

  Thus, definability of f means definability of its graph.

- $S$ is **definable** in $A$ (or **expressible** in $A$) iff there is a $\Sigma$-formula $\varphi$ such that for all states sta over A, sta $\in$ S iff $A \models_{sta} \varphi$.

- $B$ is **definable within** $A$ iff $dom(B) \subseteq dom(A)$ and the following relations are definable in $A$: $dom(B)$, $f_B$ and $r_B$, for all function symbols f and all relation symbols r in $\Omega$.

**Fact** Let $B$ be a subalgebra of $A$. Then $B$ is definable within $A$ iff $dom(B)$ is definable in $A$. In this case we say that $B$ is a **definable subalgebra** of $A$.

*Proof* Let $\varphi(X)$ be a formula that defines $dom(B)$ in $A$. For n-ary relation symbol r, $r_B$ can be defined in $A$ by formula $\varphi(X_1) \wedge ... \wedge \varphi(X_n) \wedge r(X_1,...,X_n)$. A similar construction applies to functions. ∎

## EXAMPLE 5.3

Consider algebras $N=(\mathbb{N},0,succ)$, $Nat=(\mathbb{N},0,succ,+,*)$ and $Int=(Z,0,succ, pred,+,*)$ with set of integers Z, successor function succ, addition and multiplication, and predecessor function pred.

(1)    The $<$-relation on $\mathbb{N}$ is definable in *Nat*, since

   a$<$b iff $Nat \models \exists Z\ b=a+succ(Z)$.

(2)    The function DIV (with a DIV $0=0$) is definable in *Nat*, since for all $a,b,c \in \mathbb{N}$,

   a DIV b=c iff $Nat \models (b=0 \wedge c=0) \vee (\neg b=0 \wedge \exists R(R<b \wedge a=b*c+R))$.

(3)    *Nat* is a definable subalgebra of *Int*. A definition of $\mathbb{N}$ in *Int* is obtained using the famous number theoretic result that every natural number is the sum of four square numbers. Hence, for all $a \in Z$ we know that

   $a \in \mathbb{N}$ iff $Int \models \exists A\ \exists B\ \exists C\ \exists D\ a=(A*A+(B*B+(C*C+D*D)))$.

(4)    The only subsets of $\mathbb{N}$ that are definable in $N$ are the finite subsets and their complements. This is shown by constructing for every formula $\varphi(X_1,...,X_n)$ in the signature of $N$ a quantifier-free formula $\psi(X_1,...,X_n)$ in the signature of $N$ such that

$$N \models \forall X_1 \ldots \forall X_n (\varphi(X_1,\ldots,X_n) \leftrightarrow \psi(X_1,\ldots,X_n))$$

This construction, by induction on formula $\varphi(X_1,\ldots,X_n)$, is left to the reader as an exercise. Now it is easily recognized that a quantifier-free formula $\psi(X)$ defines either a finite set or the complement of a finite set.

(5)    $(\mathbb{N},0,\text{succ},\text{pred},+,*,\text{DIV},\text{MOD},<)$ is definable within $(Z,0,1,+,*)$.

---

Given an algebra $B$ that is definable within algebra $A$, we may talk about $B$ within $A$. This is done as follows.

### Definability Lemma 5.2

Let $B$ be definable within $A$ as in the definition above. For every $\Sigma'$-formula $\psi(X_1,\ldots,X_n)$ we may construct a $\Sigma$-formula $[\psi]^*(X_1,\ldots,X_n)$ such that for all $b_1,\ldots,b_n \in \text{dom}(B)$ the following statements are equivalent:

(1)    $B \models \psi(b_1,\ldots,b_n)$

(2)    $A \models [\psi]^*(b_1,\ldots,b_n)$

*Proof* We may assume that the atomic formulas and equations occurring in $\psi$ are of the following normal form: $p(X_1,\ldots,X_n)$, $X=Y$, $Y=c$, $Y=f(X_1,\ldots,X_n)$ with different variables $Y, X, X_1,\ldots,X_n$. (It is an easy exercise to transform a given formula into an equivalent one in the required normal form.) Now we define $[\psi]^*$ as follows:

- $[p(X_1,\ldots,X_n)]^*$ is $\varphi_p(X_1,\ldots,X_n)$
- $[X=Y]^*$ is $X=Y$
- $[Y=c]^*$ is $\varphi_c(Y)$
- $[Y=f(X_1,\ldots,X_n)]^*$ is $\varphi_f(X_1,\ldots,X_n,Y)$
- $[\neg\psi]^*$ is $\neg[\psi]^*$
- $[(\psi \text{ op } \chi)]^*$ is $([\psi]^* \text{ op } [\chi]^*)$, for every op$\in \{\wedge,\vee,\to,\leftrightarrow\}$
- $[\forall X \psi]^*$ is $\forall X(\varphi(X) \to [\psi]^*)$
- $[\exists X \psi]^*$ is $\exists X(\varphi(X) \wedge [\psi]^*)$    ∎

**Definition 5.5:** Let $B$ be a definable subalgebra of $A$ and $\varphi(X)$ a formula that defines $\text{dom}(B)$ in algebra $A$. So, $B$ is definable within $A$. Given a formula $\psi$, the formula $\psi^*$ constructed in the proof of Lemma 5.2 is denoted for this special case of a subalgebra $B$ definable within another one $A$ by $[\psi]^{\varphi(X)}$, and is called the **relativization of** $\psi$ **to** $\varphi(X)$.

### Theorem 5.5

Let $\Sigma$ be a finite signature. Let $A$ be a term generated $\Sigma$-algebra and $B$ be a $\Sigma$-algebra that is elementary equivalent to $A$, but not term generated. Then the subset G of dom($B$) consisting of all values of ground terms t in $B$ (the so-called **term generated part** of dom($B$)) is not definable in $B$.

*Proof*  Since $A$ is term-generated it is a model of the following axiom of structural induction SInd($\varphi(X)$), for every formula $\varphi(X)$ with exactly one free variable:

$$\left( \bigwedge_{f \text{ n-ary}} \forall X_1...\forall X_n((\varphi(X_1)\wedge...\wedge\varphi(X_n))\rightarrow\varphi(f(X_1,...,X_n))) \rightarrow \forall X\varphi(X) \right)$$

The formula SInd($\varphi(X)$) obviously says: 'If property $\varphi(X)$ is invariant under application of all available functions in the underlying signature (including the constant functions), then $\varphi(X)$ is true for arbitrary elements X of the domain of a considered algebra.' This is just what structural induction means.

Now assume that the term-generated part G of dom($B$) is definable in $B$ by a formula $\varphi(X)$. Since $A$ is a model of SInd($\varphi(X)$), $B$ is a model of SInd($\varphi(X)$), too. Obviously, G is invariant under application of all available functions of $B$. Hence, $B$ is a model of $\forall X\varphi(X)$. This contradicts the assumption that $B$ is not term generated. ∎

Theorem 5.5 will be applied to a non-term-generated algebra $B$ that is elementary equivalent to the algebra $N=(\mathbb{N},0,\text{succ})$. Then it shows that the set of all 'standard elements of dom($B$)', that is, the set of all elements of the form $\text{succ}_B(\text{succ}_B(...\text{succ}_B(0_B)...))$, is not definable in $B$. Formula SInd($\varphi(X)$) reads as follows for this special case:

$$(\varphi(0)\wedge\forall X(\varphi(X)\rightarrow\varphi(\text{succ}(X))))\rightarrow\forall X\varphi(X)$$

the well-known induction principle for natural numbers.

## 5.4    Conservative extensions

**Definition 5.6:**    Let $\Sigma$ and $\Sigma'$ be signatures such that $\Sigma'$ is an extension of $\Sigma$. Let Ax be a set of $\Sigma$-formulas and Ax' be a set of $\Sigma'$-formulas such that $Ax \subseteq Ax'$. We say that Ax' is a **conservative extension** of Ax iff for all closed $\Sigma$-formulas $\varphi$, $Ax \models \varphi$ iff $Ax' \models \varphi$.

How can we show that an extension Ax' is a conservative extension of a set Ax? Of course, a direct application of Definition 5.6 may be problematic, since it requires consideration of all $\Sigma$-formulas. The following gives a model-theoretic criterion for conservativity.

**Theorem 5.6**

Let $\Sigma, \Sigma', Ax$ and Ax be given as above such that $Ax \subseteq Ax'$. Then, (1),(2) and (3) below are equivalent:

(1)    Ax' is a conservative extension of Ax.

(2)    For every $\Sigma$-algebra $A$ which is a model of Ax there exists a $\Sigma$-algebra $B$ and an expansion $B_{exp}$ of $B$ to a $\Sigma'$-algebra such that $B$ is elementary equivalent to $A$ and $B_{exp}$ is a model of Ax':

$$A \equiv B \lhd B_{exp} \models Ax'$$

(3)    For every $\Sigma$-algebra $A$ which is a model of Ax there exists a $\Sigma$-algebra $B$ and an expansion $B_{exp}$ of $B$ to a $\Sigma'$-algebra such that $B$ is elementary extension of $A$ and $B_{exp}$ is a model of Ax':

$$A \prec B \lhd B_{exp} \models Ax'$$

A sufficient (but not necessary), purely algebraic criterion for conservativity is

(4)    For every $\Sigma$-algebra $A$ which is a model of Ax there exists an expansion $A_{exp}$ of $A$ to a $\Sigma'$-algebra such that $A'$ is a model of Ax':

$$A \lhd A_{exp} \models Ax'$$

*Proof* Since $A \prec B$ implies $A \equiv B$, (3) implies (2). Now assume (2). In order to show (1) let a closed $\Sigma$-formula $\varphi$ be given. Obviously, $Ax \models \varphi$ implies $Ax' \models \varphi$, since $Ax \subseteq Ax'$. Conversely, assume that $Ax' \models \varphi$. Consider a $\Sigma$-algebra $A$ which is a model of Ax. Using (2) we may choose a $\Sigma$-algebra $B$ and an expansion $B_{exp}$ of $B$ to a $\Sigma'$-algebra such that $A \equiv B \lhd B_{exp} \models Ax'$. So, $B_{exp}$ is a model of $\varphi$. Since $\varphi$ is a $\Sigma$-formula we obtain that $B$ is a model of $\varphi$, therefore so is $A$ (as it is elementary equivalent to $B$). This shows that $Ax \models \varphi$.

Finally, assume (1). We must prove (3). Let a $\Sigma$-algebra $A$ be given which is a model of Ax. Consider the expansion $(A,a)_{a \in A}$ of $A$ by element a as a new constant, for every $a \in A$, and the set $\text{Th}((A,a)_{a \in A}) \cup \text{Ax}'$ of closed formulas in the signature $\Sigma$ extended by these new constants. We want to show that this set of formulas has a model. By the compactness theorem it suffices to show that for every finite subset $E=\{\varphi_1(a_1,\ldots,a_m),\ldots,\varphi_n(a_1,\ldots,a_m)\}$ of $\text{Th}((A,a)_{a \in A})$ with new constants $a_1,\ldots,a_m$, $E \cup \text{Ax}'$ has a model. Assume, by contradiction, that $E \cup \text{Ax}'$ does not possess a model. Thus

$$\text{Ax}' \models \neg(\varphi_1(a_1,\ldots,a_m) \wedge \ldots \wedge \varphi_n(a_1,\ldots,a_m))$$

Since $a_1,\ldots,a_m$ do not occur in Ax', we conclude that

$$\text{Ax}' \models \forall X_1 \ldots \forall X_n \neg(\varphi_1(X_1,\ldots,X_m) \wedge \ldots \wedge \varphi_n(X_1,\ldots,X_m))$$

with new variables $X_1,\ldots,X_n$. Since $\forall X_1 \ldots \forall X_n \neg(\varphi_1(X_1,\ldots,X_m) \wedge \ldots \wedge \varphi_n(X_1,\ldots,X_m))$ is a closed $\Sigma$-formula, we obtain from assumption (1) that

$$\text{Ax} \models \forall X_1 \ldots \forall X_n \neg(\varphi_1(X_1,\ldots,X_m) \wedge \ldots \wedge \varphi_n(X_1,\ldots,X_m))$$

As $A$ is a model of Ax, we conclude that

$$A \models \forall X_1 \ldots \forall X_n \neg(\varphi_1(X_1,\ldots,X_m) \wedge \ldots \wedge \varphi_n(X_1,\ldots,X_m))$$

Instantiating $X_1,\ldots,X_n$ with $a_1,\ldots,a_m$, respectively, we obtain that

$$(A,a)_{a \in A} \models \neg(\varphi_1(a_1,\ldots,a_m) \wedge \ldots \wedge \varphi_n(a_1,\ldots,a_m))$$

This contradicts the fact that E was a subset of $\text{Th}((A,a)_{a \in A})$. Thus, (1),(2) and (3) are shown to be equivalent.

Apparently, (4) implies (2). We provide an example which shows that (4) is indeed weaker than (2). Consider the signature $\Sigma$ with a constant 0 and a 1-ary function symbol succ. Let Ax be the empty set of axioms. Extend $\Sigma$ by a new 1-ary predicate symbol p to a signature $\Sigma'$. Let Ax' consist of the following three formulas:

$$p(0), (p(X) \rightarrow p(\text{succ}(X))), \exists X \neg p(X)$$

Ax' is a conservative extension of Ax. This can be shown by proving condition (2) above: let $A=(A,a,f)$ be a $\Sigma$-algebra. Using the compactness theorem we can choose a $\Sigma$-algebra $B=(B,b,g)$ such that $B \equiv A$ and $\{g^n(b)|n \in \mathbb{N}\}$ is a proper subset of B (consider the set of formulas $\{\neg c=\text{succ}^n(0)|n \in \mathbb{N}\}$, with new constant c). Expand $B$ to a $\Sigma'$-algebra $B_{\exp}= (B,b,g,P)$ with the following set $P=\{g^n(b)|n \in \mathbb{N}\}$. It is clear that $B_{\exp}$ is a model of Ax'. Thus (2) is shown.

Condition (4) on the other hand is not true, since the $\Sigma$-algebra $(\mathbb{N},0,\text{succ})$ with successor function succ cannot be expanded to a model of Ax'. This example shows that it is indeed sometimes necessary to introduce an elementary equivalent algebra before expanding.

# Chapter 6
# Many-sorted Logic

Many applications in mathematics and computer science deal with domains of discourse containing objects of different types. Here are some examples: linear algebra deals with vector spaces consisting of two domains of discourse, scalars and vectors. Geometry deals with points, lines, planes, and so on. Programming languages contain lots of different data types (numbers, arrays, records,...).

Usage of more than one domain of discourse may always be replaced by the consideration of one universal domain U containing all the different objects, and distinguishing the types of objects by a 1-ary relation for each occurring sort. In the geometry example, we could consider the domain U consisting of all points, lines and planes. Using sort names 'point', 'line' and 'plane', a 'many-sorted formula' like $\forall X$:point $\varphi$ may be simulated by the formula $\forall X(\text{isPoint}(X) \rightarrow \varphi)$, where isPoint is a new 1-ary predicate symbol. Similarly, $\exists X$:line $\varphi$ may be simulated by $\exists X(\text{isLine}(X)\ \varphi)$.

Nevertheless, is seems to be more satisfactory to use a logic which allows us to deal with such structured areas in a more direct way. Besides an improvement in readability, there are also efficiency reasons in favour of many-sorted logic. For example, a theorem-prover processing formula $\exists X$:line $\varphi$ has to search through the space of lines in order to find an appropriate instantiation for X, whereas it searches through the larger space of points, lines and planes to satisfy $(\text{isLine}(X)\ \varphi)$ when processing $\exists X(\text{isLine}(X)\ \varphi)$.

We will introduce the necessary generalizations of the concepts introduced so far for predicate logic leading to many-sorted logic. All these generalizations are quite simply obtained in a canonical manner. Concerning the generalization of the theorems proved for predicate logic we recommend that the reader should work through Chapters 1-5 and convince him/herself that, down to the very last detail, there are no problems in generalizing theorems and proofs to the many-sorted case. Actually, we could have dealt from the beginning with many-sorted logic. This has not been done, since the consequence would have been a considerable notational burden in definitions, theorems and proofs, despite the fact that, conceptually, the matter would not have become more difficult.

## 6.1    Syntax

**Definition 6.1:**  A **many-sorted signature** is a pair $(S, \Sigma)$ such that $S$ is a finite non-empty set of so-called **sort symbols**, and $\Sigma$ is a denumerable set of strings $f:s_1s_2\ldots s_n \to s$ or $p:s_1s_2\ldots s_n$, with $s_1, s_2, \ldots, s_n, s \in S$.

If $f:s_1s_2\ldots s_n \to s$ is an element of $\Sigma$, we call it a **function symbol of type** $s_1s_2\ldots s_n \to s$. If $n=0$, we call $f$ a **constant of type** $s$. If $p:s_1s_2\ldots s_n$ is an element of $\Sigma$, we call $p$ a **relation symbol of type** $s_1s_2\ldots s_n$. Often, we simply write $f \in \Sigma$ to express that there are sorts $s_1, s_2, \ldots, s_n, s$ such that $f:s_1s_2\ldots s_n \to s$ is an element of $\Sigma$, and $p \in \Sigma$ likewise, and say that $f$ ($p$) is a function (or, respectively relation) symbol from $\Sigma$.

In concrete examples, a signature $(S, \Sigma)$ is usually defined by listing the function and relation symbols together with their types. As an example:

0:nat
succ:nat→nat
less:nat nat
∅:set
insert:set nat→set
element:nat set

denotes a many-sorted signature with sorts nat and set, constants 0 of type nat and ∅ of type set, a function symbol succ of type nat→nat, a predicate symbol of type nat nat, and so on.

**Definition 6.2:**  Let $V=(V_s)_{s \in S}$ be an S-indexed family of disjoint sets, called a (S-indexed) **family of variables**. An element $X$ of $V_s$ is called a **variable of type** s. Furthermore, let $(S,\Sigma)$ be a many-sorted signature such that no function and predicate symbol from $\Sigma$ occurs among the variables in V. Inductively, we define $\Sigma$-**terms of type s** (over the family V of variables) as follows:

*   Every variable of sort s is a $\Sigma$-term of type s.

*   If $f:s_1s_2...s_n \rightarrow s$ is a function symbol in $\Sigma$ and $t_i$ is a $\Sigma$-term of type $s_i$, for i=1,...,n, then $f(t_1,...,t_n)$ is a $\Sigma$-term of type s.

We denote by $T_{\Sigma,s}(V)$, for $s \in S$, the set of all $\Sigma$-terms of type s (over the family V of variables). A term of type s without variables is called a ground term of sort s. $T_{\Sigma,s}$ is the set of all ground terms in $T_{\Sigma,s}(V)$.

**Definition 6.3:**   Let $(S,\Sigma)$ be a many-sorted signature and $V=(V_s)_{s \in S}$ be a family of variables. An **atomic** $(S,\Sigma)$-**formula** over V is a string of the form $p(t_1,...,t_n)$ with $p:s_1...s_n$ in $\Sigma$ and a $\Sigma$-term $t_i$ of type $s_i$, for i=1,...,n. In the case of predicate logic with equality, strings $t_1=t_2$ with terms $t_1,t_2$ of the same sort are called $(S,\Sigma)$-**equations** over V. $(S,\Sigma)$-**formulas** over V are inductively defined as follows:

*   Every atomic $(S,\Sigma)$-formula over V is an $(S,\Sigma)$-formula over V.

*   Every $(S,\Sigma)$-equation over V is an $(S,\Sigma)$-formula over V.

*   If $\varphi,\psi$ are $(S,\Sigma)$-formulas over V and X is a variable, then $(S,\Sigma)$-formulas over V are also: $\neg\varphi$, $(\varphi \rightarrow \psi)$, $(\varphi \wedge \psi)$, $(\varphi \vee \psi)$, $(\varphi \leftrightarrow \psi)$, $\forall X \varphi$ and $\exists X \varphi$.

$F_\Sigma(V)$ denotes the set of all $(S,\Sigma)$-formulas over V.

**Free** and **bound occurrence** of variables in formulas as well as the notion of a closed formula are introduced as in Chapter 2. A **substitution** $\sigma$ is a finite set $\{X_1/t_1,...,X_n/t_n\}$ consisting of pairs $X_i/t_i$ with a variable $X_i$ and a term $t_i$ of the same type, such that the usual properties are fulfilled:

*   $X_i \neq t_i$, for i=1,...,n

*   $X_1,...,X_n$ are different variables

**Application of a substitution** to a term or formula, and **admissibility** of the application of a substitution to a formula are defined as before.

## 6.2    Semantics

**Definition 6.4:**    Let $(S,\Sigma)$ be a many-sorted signature and $V=(V_s)_{s \in S}$ a family of variables. An $(S,\Sigma)$-**interpretation** $A$ consists of:

- a non-empty set $A_s$ (the **domain of sort** $s$), for every $s \in S$;
- a function $f_A:A_{s_1} \times A_{s_2} \times ... \times A_{s_n} \rightarrow A_s$, for every function symbol $f:s_1 s_2 ... s_n \rightarrow s$ (for a constant f of type s, $f_A$ is an element of $A_s$);
- a relation $p_A \subseteq A_{s_1} \times A_{s_2} \times ... \times A_{s_n}$, for every relation symbol $p:s_1 s_2 ... s_n$ (for n=0, $p_A$ is one of the truth values true or false).

The notion of an **expansion** and **reduct** is generalized in such a way that we now also allow new sorts and corresponding domains in an expansion.

A **state** sta over $A$ is a family $(sta_s)_{s \in S}$ of functions $sta_s:V_s \rightarrow A_s$. For a state sta over $A$, variable X of type $s_0$, and $d_0 \in A_{s_0}$, let $sta(X/d_0)$ be the state $sta'=(sta'_s)_{s \in S}$ over $A$ defined as follows:

- $sta'_s=sta_s$, for all $s \neq s_0$
- $sta'_{s_0}(X)=d_0$
- $sta'_{s_0}(Y)=sta_{s_0}(Y)$, for all $Y \neq X$

**Definition 6.5:**    Let $A$ be an $(S,\Sigma)$-interpretation as in Definition 6.4, and $sta=(sta_s)_{s \in S}$ be a state over $A$. Given an $(S,\Sigma)$-term t of type s, we define $val_{A,sta}(t)$ as an element of $A_s$ inductively as follows:

- $val_{A,sta}(X)=sta_s(X)$, for every variable X of type s.
- $val_{A,sta}(f(t_1,...,t_n))=f_A(val_{A,sta}(t_1),...,val_{A,sta}(t_n))$, if f is a function symbol in $\Sigma$ of type $s_1 s_2 ... s_n \rightarrow s$, and $t_i$ is a $(S,\Sigma)$-term t of type $s_i$, for i=1,...,n.

**Definition 6.6:**    Let $A$ be an $(S,\Sigma)$-interpretation as in Definition 6.4, and $sta=(sta_s)_{s \in S}$ be a state over $A$. Given an $(S,\Sigma)$-formula $\varphi$, we define **validity** of $\varphi$ in interpretation $A$ and state sta, briefly denoted $A \models_{sta} \varphi$, inductively as usual:

- $A \vDash_{sta} p(t_1,...,t_n)$ iff $((val_{A,sta}(t_1),...,val_{A,sta}(t_n)) \in p_A$

  for a predicate symbol p of type $s_1...s_n$ and a term $t_i$ of type $s_i$, for i=1,...,n.

- $A \vDash_{sta} t_1 = t_2$ iff $val_{A,sta}(t_1) = val_{A,sta}(t_2)$

- $A \vDash_{sta} \neg\varphi$ iff $A \nvDash_{sta}\varphi$

- $A \vDash_{sta}(\varphi \rightarrow \psi)$ iff $A \nvDash_{sta}\varphi$ or $A \vDash_{sta}\psi$

- $A \vDash_{sta}(\varphi \wedge \psi)$ iff $A \vDash_{sta}\varphi$ and $A \vDash_{sta}\psi$

- $A \vDash_{sta}(\varphi \vee \psi)$ iff $A \vDash_{sta}\varphi$ or $A \vDash_{sta}\psi$

- $A \vDash_{sta}(\varphi \leftrightarrow \psi)$ iff

  $A \vDash_{sta}\varphi$ and $A \vDash_{sta}\psi$, or $A \nvDash_{sta}\varphi$ and $A \nvDash_{sta}\psi$

- $A \vDash_{sta}\forall X\varphi$ iff $A \vDash_{sta(X/a)}\varphi$, for all $a \in A_s$

- $A \vDash_{sta}\exists X\varphi$ iff $A \vDash_{sta(X/a)}\varphi$, for at least one $a \in A_s$

We say that formula $\varphi$ is **valid in** $A$, or $A$ is a **model** of $\varphi$, and denote this as before by $A \vDash \varphi$, iff $A \vDash_{sta}\varphi$, for every state sta over $A$. We say that formula $\varphi$ **follows** from a set of formulas M, denoted $M \vDash \varphi$, iff every model of M is also a model of $\varphi$.


## 6.3    Proof theory

**Definition 6.7 (Hilbert calculus):**    We adapt Hilbert's calculus presented in Section 3.2 to many-sorted logic. The necessary steps are obvious:

- Axioms from $(Ax_1)$-$(Ax_3)$ and $(Ax_5)$ are carried over without changes.
- Axioms $\forall X\varphi \rightarrow \varphi\{X/t\}$ from $(Ax_4)$ require that variable X and term t are of the same type.
- In the case of predicate logic with equality, for every equation r=t occurring in one of the axioms of set Eq, r and t must be terms of the same type.
- Modus ponens and generalization are carried over without changes.

A sound and complete calculus is obtained. The reader is encouraged to work out the necessary proofs. In a similar manner, the sequent calculus LK and the resolution calculus RES may be generalized to the many-sorted case.

## 6.4    Homomorphisms and isomorphisms

**Definition 6.8:**    Let $A$ and $B$ be $(S,\Sigma)$-interpretations with domains $A_s$ and $B_s$, for every sort $s \in S$. A **homomorphism** h from $A$ into $B$ is a family $(h_s)_{s \in S}$ consisting of functions $h_s : A_s \to B_s$, for $s \in S$, with the following properties:

- $h_s(f_A(x_1,\dots,x_n)) = f_B(h_{s_1}(x_1),\dots,h_{s_n}(x_n))$
- $(x_1,\dots,x_n) \in p_A \Leftrightarrow (h_{s_1}(x_1),\dots,h_{s_n}(x_n)) \in p_B$

for every function f symbol in $\Sigma$ of type $s_1 s_2 \dots s_n \to s$, predicate symbol p in $\Sigma$ of type $s_1 s_2 \dots s_n$ and $x_1 \in A_{s_1},\dots,x_n \in A_{s_n}$. h is called an **isomorphism** from A onto B iff all of the functions $h_s$ are bijective.

The reader may also generalize the model-theoretic notions and results from Chapter 5 to the many-sorted case. For the remaining chapters we present concepts and results for the 1-sorted case thus concentrating on the actual problem without the additional burden of many-sortedness, and leave it to the reader as a simple exercise to generalize to the many-sorted case.

## 6.5    A collection of frequently used algebras

In later chapters we will frequently use a couple of basic algebras. These are introduced next. For convenience, we use the same notation for a function symbol and its interpretation in an algebra.

(1)    **Boolean**=({true,false},true,false,not,and,or,if) in signature

true: $\to$ boole
false: $\to$ boole
not:boole $\to$ boole
and:boole boole $\to$ boole
or:boole boole $\to$ boole

if:boole boole boole → boole

Functions 'not', 'and', 'or' are logical negation, conjunction, and disjunction respectively. 'if' is a conditional defined by if(true,C,D)=C and if(false,C,D)=D.

(2)    $N$=($\mathbb{N}$,0,succ) in signature

0: → nat

succ: nat → nat

with successor function succ. This algebra is the prototype of a term-generated algebra.

(3)    $Nat$=($\mathbb{N}$,0,succ,+,*) in signature

0: →nat

succ: nat → nat

+: nat nat → nat

*: nat nat → nat

with successor function succ. addition + and multiplication *. This algebra is the prototype of an arithmetical algebra, a concept that is of importance in Hoare logic.

(4)    ***Cardinal***, the expansion of the 2-sorted algebra (***Boolean,Nat***) by

(a)    functions pred, -, MOD and DIV

(b)    functions equal, unequal, less, less_or_equal, greater, greater_or_equal, $if_{nat}$

defined as follows:

x-y is x minus y for x≥y, and x-y=0 for x<y

pred(x)=x-1 for x>0, and pred(0)=0

x DIV y is least number t such that (t+1)*y>x if y>0

x DIV 0 = 0

x MOD y is x- (x DIV y) if y>0

x MOD 0 = 0

equal(x,y)=true if x=y, and false otherwise

less(x,y)=true if x<y, and false otherwise

less_or_equal(x,y)=true if x≤y, and false otherwise

greater(x,y)=true if x>y, and false otherwise

greater_or_equal(x,y)=true if x≥y, and false otherwise

$if_{nat}$(true,x,y)=x and $if_{nat}$(false,x,y)=y

Sometimes it is convenient to write

$x<y$ instead of less(x,y)=true

$x \leq y$ instead of less_or_equal(x,y)=true

$x>y$ instead of greater(x,y)=true

$x \geq y$ instead of greater_or_equal(x,y)=true

$x \neq y$ instead of unequal(x,y)=true

(5)    *I*=(Z,0,succ,pred) in signature

0: → nat

succ: nat → nat

pred: nat → nat

with the set of integers Z as domain, successor function succ and predecessor function pred.

(6)    *Int*=(Z,0,succ,pred,+,*) in signature

0: → nat

succ: nat → nat

pred: nat → nat

+: nat nat → nat

*: nat nat → nat

with successor function succ, predecessor function pred, addition + and multiplication * on integers.

(7)    *Integer*, the expansion of the 2-sorted algebra (*Boolean,Int*) by corresponding function as in the definition of *Cardinal*.

(8)    *Set(Data)* is defined for a 1-sorted algebra *Data* as the expansion (*Data*,$A_{set}$,∅,insert) of *Data* in a signature with additional sort 'set' and function symbols

∅: → set

insert: set data → set

$A_{set}$ is the set of all finite subsets of dom(*Data*), ∅ is interpreted by the empty set, 'insert' by the function insert(set,x)=set∪{x}.

(9)    *Stack(Data)* is defined for a 1-sorted algebra *Data* with a distinguished    element    ⊥    of    dom(*Data*)    as    the    expansion

(*Data*,A$_{stack}$,emptystack,push,top,pop) of *Data* in a signature with additional sort 'stack' and function symbols

> emptystack: → stack
>
> push: stack data → stack
>
> top: stack → data
>
> pop: stack → stack

A$_{stack}$ is the set of all finite strings of elements from dom(*Data*), 'emptystack' is interpreted as the empty string $\varepsilon$, 'push' as the function push(w,x)=wx, 'top' as the function top($\varepsilon$)=⊥ and top(wx)=x, and 'pop' as the function defined by pop($\varepsilon$)=$\varepsilon$ and pop(wx)=w.

(10)   *Queue(Data)* is defined for a 1-sorted algebra *Data* with a distinguished element   ⊥   of   dom(*Data*)   as   the   expansion (*Data*,A$_{queue}$,emptyqueue,enqueue, first,rest) of *Data* in a signature with additional sort 'queue' and function symbols

> emptyqueue: → queue
>
> enqueue: queue data → queue
>
> first: queue → data
>
> rest: queue → queue

A$_{queue}$ is again the set of all finite strings of elements from dom(*Data*), 'emptyqueue' is interpreted as the empty string $\varepsilon$, 'enqueue' as the function enqueue(w,x)=wx, 'first' as the function first($\varepsilon$)=⊥ and first(w)=the leftmost data element in string w, for non-empty string w, and 'rest' as the function defined as rest($\varepsilon$)=$\varepsilon$ and rest(w)=w without its leftmost element, for non-empty string w.

Hence, *Queue(Data)* differs from *Stack(Data)* in that it generated strings from the right (as *Stack(Data)* does), but accesses them from the left (whereas *Stack(Data)* accesses them from the right).

(11)   *Array* [0..N] *of Data* is defined for a 1-sorted algebra *Data* with a distinguished element ⊥ of dom(*Data*) as the following expansion of (*Data,Boolean*) in a signature with additional sorts 'index' and 'array' and function symbols as follows:

> if$_{data}$: boole data data → data
>
> if$_{array}$: boole array array → array
>
> 0: → index

up: index $\rightarrow$ index
down: index $\rightarrow$ index
less: index index $\rightarrow$ boole
$equal_{index}$: index index $\rightarrow$ boole
initarray: $\rightarrow$ array
update: array index data $\rightarrow$ array
read: array index $\rightarrow$ data

$A_{index}$ is the set of numbers $\{0,...,N\}$, $A_{array}$ is the set of all vectors $(a_1,...,a_N)$ with elements from dom(**Data**), the functions above are interpreted as follows:

$if_{data}(true,x,y)=x$,
$if_{data}(false,x,y)=y$
$if_{array}(true,A,B)=A$,
$if_{array}(false,A,B)=B$
$up(i)=i+1$, for $i<N$, and $up(N)=N$
$down(i)=i-1$, for $i>0$, and $down(0)=0$
$less(i,j)=true$ if $i<j$, $less(i,j)=false$ otherwise
$equal_{index}(i,j)=true$ for $i=j$, $equal_{index}(i,j)=false$ otherwise
initarray is the vector $(\bot,\bot,...,\bot)$
update(Arr,i,x) results from vector Arr by replacing the i-th element of Arr by x
read(Arr,i) is the i-th element of vector Arr

When using **Array** [0..N] **of Data** in imperative programming, we adopt the following usual notation:

i+1 instead of term up(i)
i-1 instead of term down(i)
Arr[i] instead of term read(Arr,i)
Arr[i]:=x instead of assignment Arr:=update(Arr,i,x)
i<j instead if less(i,j)=true

---

**Exercise 6.1** Generalize the definitions of **Set(Data)**, **Stack(Data)**, **Queue(Data)** and **Array** [0..N] **of Data** to the case of a many-sorted algebra **Data**.

---

(12)    Now let ***Data*** be a many-sorted algebra in signature $(S, \Sigma)$ with domains $D_s$, for $s \in S$. Let $s_1, \ldots, s_n$ be sorts from S. For $i = 1, \ldots n$, let distinguished elements $\perp_i \in D_{s_i}$ be given. Then ***Record*** $sel_1:s_1, \ldots, sel_n:s_n$ *of* ***Data*** is defined as the following expansion of ***Data*** in a signature with additional sort 'rec' and function symbols

initrecord: $\rightarrow$ rec

$sel_i$: rec $\rightarrow s_i$, for $i = 1, \ldots, n$

$assign_i$: rec $s_i \rightarrow$ rec, for $i = 1, \ldots, n$

$A_{rec}$ is $D_{s_1} \times D_{s_2} \times \ldots \times D_{s_n}$, the cartesian product, and the functions above are interpreted as follows:

initrecord is the vector $(\perp_1, \perp_2, \ldots, \perp_n)$

$sel_i(d_1, d_2, \ldots, d_n) = d_i$, for $i = 1, \ldots, n$

$assign_i((d_1, d_2, \ldots, d_n), d) = (d_1, d_2, \ldots, d_{i-1}, d, d_{i+1}, \ldots, d_n)$, for $i = 1, \ldots, n$

As usual in programming, we write

$R.sel_i$ instead of term $sel_i(R)$

$R.sel_i := d$ instead of assignment $R := assign_i(R, d)$

## 6.6    Order-sorted logic

There is a slight generalization of many-sorted logic, called **order-sorted logic**. The idea is to allow some sort s to be a subsort of another sort t, indicated in the denotation of a signature by s<t. Then, from a syntactical point of view, terms of type s may be used in every context where a term of type t is expected. From a semantical point of view we require, in an order-sorted algebra with sorts s<t, the domain of sort s to be a subset of the domain of sort t. Using order-sorted logic sometimes exhibits advantages in that it saves duplication of functions and predicates (using sort nat as a subsort of sort int, we require only a single order relation less of type int). A further advantage is that the efficiency of automated theorem-proving may be improved by using order-sorted logic. Since, from a logical point of view, the field does not lead to deeper insights and since we do not apply it later, we do not enlarge upon it here. More information may be found in Smolka *et al.*(1989).

## Historical remarks and recommended reading for Part I

The history of logic is closely related to the history of philosophy and goes back to ancient Greece, the most prominent logician of that time being Aristotle (see Russell (1946)). However, logic did not develop rapidly until after the middle of the last century. Predicate logic as it is known today was introduced by Frege (1879), the usual notations being given by Peano 1889. The completeness of predicate logic was first shown in 1930 by Gödel.

At the beginning of the twentieth century, attempts were made to describe mathematics completely by means of formal systems, thus leading to a mechanization of mathematics; this task is known by the name *Hilbert's Programme*. In his famous work, Gödel (1931) proved that this task was totally unrealistic, because for every sufficiently rich formal system a valid assertion could be constructed which could not be derived in the formal system. Another fundamental negative experience concerning Hilbert's programme was the famous undecidability result of Turing and Church, Church (1936). For the history of logic, the reader should consult Russell (1946) or van Heijenoort (1967).

There are many text books on mathematical logic, for example Ebbinghaus *et al.* (1978), Gallier (1986), Kleene (1967), Shoenfield (1967) and Mendelson (1964). In particular, Gallier (1986) is a modern and quite exhaustive presentation of sequent calculi, while most other books adopt Hilbert-like calculi. A comparison of various approaches to completeness of predicate calculi can be found in Smullyan (1968).

Automated theorem-proving is a research line in logic primarily interested in automatizing formal reasoning. The systems are usually based on resolution, first introduced by Robinson (1965). In Siekmann and Wrightson (1983a and 1983b) the reader can find many classical papers from this field as well as an overview of the beginnings of automated deduction.

Books covering automated reasoning, including many refinements of resolution and paramodulation, are Chang and Lee (1973), Loveland (1978), Robinson (1979) and Wos *et al.* (1984). An interesting extension of resolution is theory resolution introduced in Stickel (1985). Siekmann (1984) is a good overview of the unification field.

# Part II

# LOGIC PROGRAMMING
# AND
# PROLOG

So far we have seen how logic can be used to represent declarative knowledge about several domains of discourse, as well as to extract further knowledge from such a representation by proving theorems.

One way of proving theorems is to use resolution. By restricting the set of admitted formulas to so-called **Horn formulas** we arrive at a sublogic of predicate logic which allows a more efficient implementation of theorem proving. Even better, theorem proving can be turned into computing by *procedurally interpreting* formulas.

It is this new operational semantics of formula sets (now called **logic programs**) that allows us to abandon the static view of logic and bridge the gap to programming. The central idea of this approach is expressed by the famous equation due to Kowalski, *algorithm = logic + control*, logic standing for the static (declarative) information about the problem to be solved, and control standing for the solution strategy. The ideal case would be that control is fully processed by a machine, and the 'programmer' only describes the problem to be solved.

The field of logic programming is dominated by the programming language Prolog. Therefore we introduce the basic concepts of this language and show how it can be used for some practical problems.

# Part II

# LOGIC PROGRAMMING

# AND

# PROLOG

# Chapter 7
# Horn Logic

## 7.1    The logic of Horn formulas

Let us take a look back at Section 3.4 where the resolution calculus RES was studied. The objects of interest in resolution theory were clauses, that is, sets $\{L_1,...,L_k\}$ of literals. Literals may be positive or negative. Let us partition a clause into the subset of positive literals $\{A_1,...,A_n\}$ and the subset of negative literals $\{\neg B_1,...,\neg B_m\}$: $\{L_1,...,L_k\}=\{A_1,...,A_n\}\cup\{\neg B_1,...,\neg B_m\}$. Horn logic restricts attention to clauses with *at most one positive literal*, i.e. requires $n\le1$. This leads to clauses with one of the following forms:

$\{A,\neg B_1,...,\neg B_m\}$, with $m>0$    called **rule clause**

$\{A\}$    called **fact clause**

$\{\neg B_1,...,\neg B_m\}$, with $m>0$    called **goal clause**

the empty set of of literals    called **empty goal**

In Horn logic we take back the representation of formulas (of clausal form) as finite *sets* of literals, that is, clauses, and work again with formulas where there is an ordering of literals. The reason for this is that it fits better into the procedural interpretation of logical programs to be discussed later. Let us first look at how the Horn clauses above translate back into formula form.

| Horn clause (m>0) | Corresponding formula(s) | Called |
|---|---|---|
| $\{A, \neg B_1, \ldots, \neg B_m\}$ | $\forall(A \lor \neg B_1 \lor \ldots \lor \neg B_m)$ or equivalently $\forall((B_1 \land \ldots \land B_m) \rightarrow A)$ | rule |
| $\{A\}$ | $\forall(A)$ | fact |
| $\{\neg B_1, \ldots, \neg B_m\}$ | $\forall(\neg B_1 \lor \ldots \lor \neg B_m)$ or equivalently $\neg\exists(B_1 \land \ldots \land B_m)$ | goal |
| empty clause | false, an unsatisfiable formula | empty goal |

There are several alternative notations for these three types of formula in use.

| Formula (m>0) | Notation used in logic programming | Prolog notation |
|---|---|---|
| $\forall((B_1 \land \ldots \land B_m) \rightarrow A)$ | $A \leftarrow B_1, \ldots, B_m$ | $A :\text{-} B_1, \ldots, B_m.$ |
| $\forall(A)$ | $A \leftarrow$ | $A.$ |
| $\neg\exists(B_1 \land \ldots \land B_m)$ | $\leftarrow B_1, \ldots, B_m$ | $?\text{-} B_1, \ldots, B_m.$ |

These alternative notations mirror the procedural interpretation of formulas in logic programming. The reversed implication in a rule $A \leftarrow B_1, \ldots, B_m$ indicates that the corresponding implication $\forall((B_1 \land \ldots \land B_m) \rightarrow A)$ may be used in a backward manner: to show A, show all the formulas $B_1, \ldots, B_m$. The same for facts: the fact $A \leftarrow$ is readily shown (by reducing it to the task 'to do nothing'). A corresponding interpretation of a goal $\leftarrow B_1, \ldots, B_m$ requires a closer look at the way clauses were used. Remember that the main purpose of a set of clauses was to show that is does not possess a model. Let P be a set of rules and facts, and $\leftarrow B_1, \ldots, B_m$ be a goal, with m>0. To show that P together with $\leftarrow B_1, \ldots, B_m$ does not possess a model means that P together with $\neg\exists(B_1 \land \ldots \land B_m)$ does not possess a model, that is, $\exists(B_1 \land \ldots \land B_m)$ follows from P. So seen, we may interpret a goal $\leftarrow B_1, \ldots, B_m$ in the context of a set P of rules and facts as the task to satisfy $B_1, \ldots, B_m$ successively by reducing these formulas to facts of P using the rules of P.

To prepare for Prolog programming in the following sections, we will adopt the Prolog notation with a minor modification: in the current text we will omit the closing '.' required in Prolog notation, since this notation would disrupt the flow of normal English sentences.

**Definition 7.1:**   A **Horn formula** is a string of one of the following forms, with literals $A, B_1, \ldots, B_m$:

- $A:-B_1, \ldots, B_m$ , for m>0, called a **rule**
- $A$ , called a **fact**
- $?-B_1, \ldots, B_m$ , for m>0, called a **goal**
- $\square$ , called the **empty goal**

In a canonical way we carry over to Horn formulas the notion of substitution application, variant and renaming.

Rules and facts are collected under the name **program formulas**. A finite set P of program formulas is called a **logic program**.

An algebra $A$ is called a **model** of rule $A:-B_1, \ldots, B_m$ or fact A or non-empty goal $?-B_1, \ldots, B_m$ iff it is a model of $\forall((B_1 \wedge \ldots \wedge B_m) \rightarrow A)$ or $\forall(A)$ or $\neg \exists(B_1 \wedge \ldots \wedge B_m)$, respectively. The empty goal does not possess a model. $A$ is called a model of a set of Horn formulas iff it is a model of every element of that set. A formula (here we admit predicate logic formula) $\varphi$ *follows* from a logic program P iff every model of P is also a model of $\varphi$.

**Fact**     Let P be a logic program and $?-B_1, \ldots, B_m$ be a non-empty goal. Then the following statements are equivalent:

(1)     $P \cup \{?-B_1, \ldots, B_m\}$ does not possess a model.

(2)     The formula $\exists(B_1 \wedge \ldots \wedge B_m)$ follows from P.

For universal formulas like Horn formulas, the notion of a Herbrand model plays a central role. In logic programming a special representation of Herbrand algebras is usual, which greatly simplifies the notation of Herbrand algebras. This representation is introduced next.

**Definition 7.2:**    Let $\Sigma$ be a signature with at least one constant symbol. For a Herbrand interpretation $A$ over signature $\Sigma$ we define $I(A)$ to be the following set of ground atoms:

$$\{r(t_1,...,t_n) \mid r \in \Sigma, r \text{ n-ary}, t_1,...,t_n \text{ ground } \Sigma\text{-terms and } (t_1,...,t_n) \in r_A\}$$

$A$ can be recovered from $I(A)$ as follows. For an arbitrary set $I$ of ground terms define $Alg(I)$ to be the Herbrand algebra interpreting every n-ary predicate symbol r in $\Sigma$ by

$$r_{Alg(I)} = \{(t_1,...,t_n) \mid t_1,...,t_n \text{ are ground } \Sigma\text{-terms and } r(t_1,...,t_n) \in I\}$$

Then, $Alg(I(A))=A$ and $I(Alg(I))=I$. Thus, we can use $A$ equally entitled with $I(A)$ without loss of information. We will also call sets $I$ of ground clauses Herbrand algebras. On the basis of this correspondence we say that $I$ is a model of a formula $\varphi$ or formula set M, written $I \models \varphi$ and $I \models M$, iff $I(A)$ is a model of $\varphi$ or M.

For a Herbrand algebra $I$ and a *closed* formula $\varphi$, let us express what it means when $I(A)$ is a model of $\varphi$ directly in terms of $I$:

$$I \models r(t_1,...,t_n) \Leftrightarrow r(t_1,...,t_n) \in I$$

$$I \models \neg\varphi \Leftrightarrow \text{not } I \models \varphi$$

$$I \models (\varphi \wedge \psi) \Leftrightarrow I \models \varphi \text{ and } I \models \psi$$

$$I \models (\varphi \vee \psi) \Leftrightarrow I \models \varphi \text{ or } I \models \psi$$

$$I \models (\varphi \rightarrow \psi) \Leftrightarrow \text{If } I \models \varphi \text{ then } I \models \psi$$

$$I \models (\varphi \leftrightarrow \psi) \Leftrightarrow I \models \varphi \text{ iff } I \models \psi$$

$$I \models \forall X\varphi \Leftrightarrow \text{For every ground term t, } I \models \varphi\{X/t\}$$

$$I \models \exists X\varphi \Leftrightarrow \text{There is a ground term t such that } I \models \varphi\{X/t\}$$

Herbrand algebras, in their representation as sets of ground atoms, may be partially ordered according to set inclusion. Note that $I \subseteq J$ does not affect the domain of the Herbrand algebras associated with $I$ and $J$, since domains are always fixed as the set of all ground terms. $I \subseteq J$ expresses that the interpretation of a predicate p in $I$ is a subset of the corresponding interpretation of p in $J$. There are two extreme cases, namely a Herbrand algebra whose predicates do not hold for any n-tuples (this Herbrand algebra corresponds to $I=\varnothing$), and a Herbrand algebra whose predicates hold for all n-tuples, which corresponds to $I=$the set of all ground atoms. The former is the least element in the partial ordering of Herbrand algebras, the latter is the greatest element.

## 7.2    Declarative semantics of logic programs

Now we investigate the Herbrand models of a logic program P. The main point is that among these Herbrand models there always exists a least one, that is, one that is contained in every other Herbrand model of P. This peculiarity of logic programs, which is not true for arbitrary sets of clauses, forms the basis for the completeness of a calculus called **SLD-resolution**, which will be introduced soon.

> ### Theorem 7.1
>
> Let P be a logic program. Among the Herbrand algebras $I$ (in the representation as sets of ground atoms) that are models of P, there is a **least Herbrand model** $M_P$. It is defined by
>
> $$M_P = \{r(t_1,...,t_n) \mid t_1,...,t_n \text{ ground terms, } r(t_1,...,t_n) \text{ follows from P}\}$$

*Proof*  Let $M_P$ be defined as above. We show that $M_P$ is model of every formula in P. Let $A \leftarrow B_1,...,B_m$ be a program formula in P (facts are subsumed under this case for m=0). Let $X_1,...,X_n$ be the variables occurring in $A \leftarrow B_1,...,B_m$. We conclude:

$M_P \models A \leftarrow B_1,...,B_m$

$\Leftrightarrow$

$M_P \models \forall X_1...\forall X_n((B_1 \wedge ... \wedge B_m) \rightarrow A) \Leftrightarrow$

For all ground terms $t_1,...,t_n$, $M_P \models ((B_1 \wedge ... \wedge B_m) \rightarrow A)\{X_1/t_1,...,X_n/t_n\}$

$\Leftrightarrow$

For all ground terms $t_1,...,t_n$:

$M_P \models B_i\{X_1/t_1,...,X_n/t_n\}$, for all i=1,...,n $\Rightarrow$ $M_P \models A\{X_1/t_1,...,X_n/t_n\}$

$\Leftrightarrow$

For all ground terms $t_1,...,t_n$:

$B_i\{X_1/t_1,...,X_n/t_n\} \in M_P$, for all i=1,...,n $\Rightarrow$ $A\{X_1/t_1,...,X_n/t_n\} \in M_P$

$\Leftrightarrow$

For all ground terms $t_1,...,t_n$:

$P \models B_i\{X_1/t_1,...,X_n/t_n\}$, for all i=1,...,n $\Rightarrow$ $P \models A\{X_1/t_1,...,X_n/t_n\}$

The last statement is true, since $P \models ((B_1 \wedge ... \wedge B_m) \rightarrow A)\{X_1/t_1,...,X_n/t_n\}$. Thus we know that $M_P$ is a model of P. Next we show that $M_P$ is the least

Herbrand interpretation that is a model of P. Consider an arbitrary Herbrand model $I$ of P. Since $I$ is also a model of all formulas that follow from P, we conclude that $I$ is a model of all formulas of $M_P$. Since formulas in $M_P$ are ground atoms, it follows that $M_P \subseteq I$.                                                                                   ∎

Proving that the least Herbrand model of a logic program P coincides with a given Herbrand interpretation $I$ can be regarded as a **correctness statement** about program P w.r.t. to the specification $I$. How do we show, for a concrete program P and Herbrand interpretation $I$, that indeed $M_P = I$ ?

(1)    $M_P \subseteq I$ is usually shown by proving that $I$ is a model of P. Informally stated, when developing P, we must take care that only valid program formulas w.r.t. $I$ are used.

(2)    $I \subseteq M_P$ requires that we show that every ground atom in $I$ follows from P. Usually, induction over a problem-dependent parameter solves the problem.

## EXAMPLE 7.1    Paths in a directed graph

Assume that a finite, directed graph is given by a set EDGE of facts of the form edge(a,b), with constants a and b (nodes of the graph) and a 2-ary predicate edge modelling the edges. As an example, consider the following directed graph with set EDGE of facts:

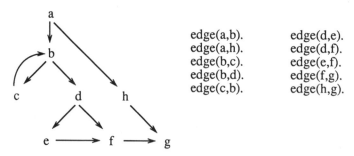

edge(a,b).          edge(d,e).
edge(a,h).          edge(d,f).
edge(b,c).          edge(e,f).
edge(b,d).          edge(f,g).
edge(c,b).          edge(h,g).

Now consider the following logic program PATH which defines the 2-ary predicate path:

(1)    path(X,X).

(2)    path(X,Y):-edge(X,Z),path(Z,Y).

We want to show that the least Herbrand model of logic program EDGE∪PATH is the following Herbrand interpretation $I$:

EDGE∪{path(u,v)| there exists a path from u to v}

Let us first show that *I* is a model of P: obviously, *I* is a model of the facts in EDGE, since any such fact is an element of *I*. *I* is a model of program formula (1), since path(u,u) is an element of *I*, for every node u. *I* is a model of program formula (2), since the following implications are true, for all nodes u,v and w:

> edge(u,v) and path(v,w) are elements of *I* $\Rightarrow$
> there is an edge from u to v and a path from v to w $\Rightarrow$
> there is a path from u to w $\Rightarrow$
> path(u,w) is an element of *I*

Let us now verify that $I \subseteq M_P$. It suffices to show that for any two nodes u and w such that there exists a path from u to w, the ground atom path(u,w) follows from EDGE∪PATH. This is best shown by induction on the length l of a path from u to w. If l=0, then u=w and program formula (1) entails that path(u,w). If l>0 then there exists an intermediate node v with an edge from u to v and a path of length l-1 from v to w. By induction we know that the ground atom path(v,w) follows from EDGE∪PATH; furthermore, edge(u,v) follows from EDGE∪PATH. Together with program formula (2) we obtain that path(u,w) follows from EDGE∪PATH.

---

The verification above is typical for the verification of logic programs. We will address the verification discussion again in more detail in Chapter 9.

### Theorem 7.2

Let P be a logic program. We define a sequence of Herbrand interpretations $I_0(P) \subseteq I_1(P) \subseteq ... \subseteq I_n(P) \subseteq ...$ and its limit $I_\infty(P)$ as follows:

- $I_0(P) = \varnothing$
- $I_{n+1}(P) = I_n(P) \cup \{A \mid$ there is a ground instance $A \leftarrow A_1,...,A_m$ of a program formula in P such that $A_1,...,A_m \in I_n(P)\}$
- $I_\infty(P)$ is the union of all $I_n(P)$

Then $I_\infty(P) = M_P$.

Note the intuitive understanding of the sequence above as a deductive calculus. Given a logic program P, we are interested in the set of all ground atoms which follow from P. This set is computed by a sort of information-enriching, monotonic construction as follows. We start with the empty set. Obviously, $I_1(P)$ is the set of all ground instances of facts in P. To compute $I_2(P)$ we look at all ground instances of rules in P whose premisses are already contained in

$I_1(P)$. Then we collect the conclusions of all such ground instances of rules into $I_2(P)$. Briefly stated, $I_2(P)$ consists of all ground atoms that can be derived from P by at most one application of a rule. More generally, $I_n(P)$ consists of all ground atoms which can be derived from P with a maximal nesting n-1 of rule applications.

*Proof of the theorem*   We show by induction over n that $I_n(P) \subseteq M_P$. The claim is trivial for n=0. If $I_n(P) \subseteq M_P$ is proven, then $I_{n+1}(P) \subseteq M_P$ can be shown as follows: let $A \in I_{n+1}(P)$ be given. Then there is a ground instance $A \leftarrow A_1,...,A_m$ of a program formula in P such that $A_1,...,A_m \in I_n(P)$. Then $A_1,...,A_m$ are, by induction hypothesis, members of $M_P$, that is, $M_P$ is a model of $A_1,...,A_m$. Since $M_P$ is a model of P, it is also a model of every ground instance of a program formula in P, in particular a model of $A \leftarrow A_1,...,A_m$. Hence, $M_P$ is a model of A, so $A \in M_P$. We have thus shown that $I_\infty(P) \subseteq M_P$. Now we show that $I_\infty(P)$ is a model of P. Let $A \leftarrow B_1,...,B_m$ be a program formula in P with variables $X_1,...,X_n$. We must show that for all ground terms $t_1,...,t_n$, $I_\infty(P) \models ((B_1 \wedge ... \wedge B_m) \rightarrow A)\{X_1/t_1,...,X_n/t_n\}$. So assume that $I_\infty(P) \models (B_1 \wedge ... \wedge B_m)\{X_1/t_1,...,X_n/t_n\}$. This implies that $B_i\{X_1/t_1,...,X_n/t_n\} \in I_\infty(P)$, for all i=1,...,m. Let N be big enough so that $B_i\{X_1/t_1,...,X_n/t_n\} \in I_N(P)$, for all i=1,...,m. Then $A\{X_1/t_1,...,X_n/t_n\} \in I_{N+1}(P)$, by the definition of $I_{N+1}(P)$, therefore $A\{X_1/t_1,...,X_n/t_n\} \in I_\infty(P)$. Hence $I_\infty(P) \models A\{X_1/t_1,...,X_n/t_n\}$. Since $M_P$ is the least Herbrand model of P, it follows that $M_P \subseteq I_\infty(P)$.     ∎

### Theorem 7.3

Let $\exists X_1 ... \exists X_k (B_1 \wedge ... \wedge B_m)$ be a closed existential formula. Then the following statements are equivalent:

(1)    $P \models \exists X_1 ... \exists X_k (B_1 \wedge ... \wedge B_m)$.

(2)    $P \models (B_1 \wedge ... \wedge B_m)\{X_1/t_1,...,X_k/t_k\}$, for ground terms $t_1,...,t_k$.

(3)    $M_P$ is a model of $\exists X_1 ... \exists X_k (B_1 \wedge ... \wedge B_m)$.

(4)    $M_P \models (B_1 \wedge ... \wedge B_m)\{X_1/t_1,...,X_k/t_k\}$, for ground terms $t_1,...,t_k$.

In general, it is not true that (1) and (3) are equivalent for universal formulas.

*Proof*     (2) implies (1), since formulas $(\varphi\sigma \to \exists(\varphi))$ are always valid, for a quantifier-free formula $\varphi$ and arbitrary substitution $\sigma$. (1) implies (3) since $M_P$ is a model of P. (3) implies (4) by the definition of semantics and the fact that the domain of $M_P$ is the set of ground terms.

Finally we show that (4) implies (2). Assume $M_P$ is a model of $(B_1 \wedge ... \wedge B_m)$ $\{X_1/t_1,...,X_k/t_k\}$, for ground terms $t_1,...,t_k$. Then $B_1\{X_1/t_1,...,X_k/t_k\},..., B_m\{X_1/t_1,...,X_k/t_k\}$ are elements of the Herbrand interpretation $M_P$. Since $M_P$ is the least Herbrand model of P it follows that $B_1\{X_1/t_1,...,X_k/t_k\},...,B_m\{X_1/t_1,...,X_k/t_k\}$ are also elements of every Herbrand model $I$ of P. Hence, every Herbrand model $I$ of P is a model of the formula $(B_1 \wedge ... \wedge B_m)\{X_1/t_1,...,X_k/t_k\}$. Thus $P \models (B_1 \wedge ... \wedge B_m)\{X_1/t_1,...,X_k/t_k\}$.

In the next example we shall see that (1) and (3) are not equivalent for universal formulas.     ■

Note the importance of Theorem 7.3. The standard situation in logic programming is to have a logic program P and a goal ?- $B_1,...,B_m$, and to ask whether $P \cup \{?- B_1,...,B_m\}$ does not possess a model. As we already know, this is equivalent to asking whether the formula $\exists X_1 ... \exists X_k (B_1 \wedge ... \wedge B_m)$ follows from P. If indeed $\exists X_1 ... \exists X_k (B_1 \wedge ... \wedge B_m)$ follows from P, this can always be witnessed by concrete instances for the variables $X_1,...,X_k$ in the form of ground terms $t_1,...,t_k$. Moreover, whether $\exists X_1 ... \exists X_k (B_1 \wedge ... \wedge B_m)$ follows from P and, if this is the case, what are the corresponding witnesses $t_1,...,t_k$, can be obtained by looking at $M_P$. Thus, $M_P$ is really the *meaning of logic program* P.

## EXAMPLE 7.2

As an example of the way existential questions can be concretely answered, consider the following logic program P in a signature with a constant 0, 1-ary function symbol succ, and 3-ary predicate symbol add.

add(X,0,X).
add(X,succ(Y),succ(Z)):-add(X,Y,Z).

---

**Exercise 7.1**     Show that $M_P = \{ \text{add}(\text{succ}^n(0),\text{succ}^m(0),\text{succ}^{n+m}(0)) \mid n,m \in \mathbb{N} \}$.

Having determined $M_P$ we may predict how existential questions posed to P should be answered.

(Question 1)          ?-add(succ$^3$(0),succ$^8$(0),Z),

that is, the question whether the existential formula $\exists Z$ add(succ$^3$(0), succ$^8$(0),Z) follows from P. Our knowledge about $M_P$ allows us to give the answer:

yes, namely with Z=succ$^{11}$(0)

Here we used that $M_P$ is a model of add(succ$^3$(0),succ$^8$(0),Z){Z/succ$^{11}$(0)}.

(Question 2)          ?-add(X,succ$^8$(0),Z),

that is, the question whether the existential formula $\exists X \exists Z$ add(X,succ$^8$(0),Z) follows from P. Following the lines of (Question 1) we could correctly answer:

yes, namely with X=0 and Z=succ$^8$(0),
but also with X=succ(0) and Z=succ$^9$(0),
but also with X=succ$^2$(0) and Z=succ$^{10}$(0), etc.

To show that the existential formula $\exists X \exists Z$ add(succ$^3$(0),succ$^8$(0),Z) follows from P by instantiating X and Z by appropriate **ground witnesses** (as was done above) is not the best way to witness the considered logical conclusion. A better way, which would be much more informative, is to show the considered logical conclusion by a **parametrized witness** Z=succ$^8$(X). Indeed, the universal formula $\forall X$ add(X,succ$^8$(0),Z){Z/succ$^8$(X)} follows from P. (Note that for universal formulas $\varphi$, the property that $\varphi$ follows from P is stronger than validity of $\varphi$ in $M_P$.) In some sense, {Z/succ$^8$(X)} is the *most general way to witness* the existential question $\exists X \exists Z$ add(X,succ$^8$(0),Z).

(Question 3)          ?-add(succ$^3$(0),Y,Z),

that is, the question whether the existential formula $\exists Y \exists Z$ add(succ$^3$(0),Y,Z) follows from P. As for (Question 2) we may answer

yes, namely with Y=0 and Z=succ$^3$(0),
but also with Y=succ(0) and Z=succ$^4$(0),
but also with Y=succ$^2$(0) and Z=succ$^5$(0), ...

But note here that the parametrized witness Z=succ$^3$(Y) would not be a correct way to show that $\exists Y \exists Z$ add(succ$^3$(0),Y,Z) follows from P. The explanation is simple: It is not true that the universal formula $\forall Y$add(succ$^3$(0),Y,Z){Z/succ$^3$(Y)} follows from P, although the formula $\forall Y$add(succ$^3$(0),Y,Z){Z/succ$^3$(Y)} is valid in $M_P$ (see Exercise 7.2).

**Exercise 7.2**    Construct a model of the logic program P above (not a Herbrand model!) with the integers as domain, which is not model of the universal formula $\forall Y$ add($succ^3(0)$,Y,Z)$\{Z/succ^3(Y)\}$.

Thus, we again must warn the reader to be very careful when answering a question and to use only the information present in a logic program, and nothing else.

(Question 4)            ?-add(X,Y,$succ^3(0)$),

that is, the question whether the existential formula $\exists X \exists Y$ add(X,Y,$succ^3(0)$) follows from P. Formally stated, this is the task of computing 'inverses X,Y of value $succ^3(0)$' for the function addition encoded by predicate add. Possible witnesses are:

X=0 and Y=$succ^3(0)$, but also
X=succ(0) and Y=$succ^2(0)$, but also
X=$succ^2(0)$ and Y=succ(0), but also
X=$succ^3(0)$ and Y=0

There are no parametrized, more general, witnesses. The parametrized witnesses discussed in this example lead to Definition 7.3.

**Definition 7.3:**    Let G=?-$B_1$,...,$B_m$ be a goal and P a logic program. For a substitution $\sigma$, let $\sigma|_G$ be the restriction of substitution $\sigma$ to the variables occurring in G, that is, $\sigma|_G=\{X/t \mid X/t \in \sigma$ and X occurs in G$\}$. A substitution $\sigma=\{X_1/t_1,...,X_n/t_n\}$ is called a **correct answer substitution** for $P \cup \{G\}$ iff $X_1,...,X_n$ occur in G and $P \models \forall((B_1 \wedge ... \wedge B_m)\sigma)$.

Again note that the statement $P \models \forall((B_1 \wedge ... \wedge B_m)\sigma)$ is not equivalent to $M_P \models (\forall(B_1 \wedge ... \wedge B_m)\sigma)$, unless $(B_1 \wedge ... \wedge B_m)\sigma$ is variable free.

The next theorem sheds some light on the concept of correct answer substitutions. It tells us that correct answer substitutions are just that sort of answer that we would like to obtain when asking existential questions.

### Theorem 7.4

Let P be a logic program. Then the following statements are true:

(1)  Let G be the goal $?\text{-}B_1,...,B_m$ whose variables are $Y_1,...,Y_k,X_1,...,X_n$. If $\forall Y_1...\forall Y_k \exists X_1...\exists X_n(B_1 \wedge...\wedge B_m)$ follows from P, then there is a correct answer substitution $\sigma$ for $P \cup \{G\}$ of the form $\{X_1/t_1,...,X_n/t_n\}$, with terms $t_1,...,t_n$ that contain only variables from $\{Y_1,...,Y_k\}$.

(2)  Conversely, let $G = ?\text{-}B_1,...,B_m$ be a goal and $\sigma = \{X_1/t_1,...,X_n/t_n\}$ a correct answer substitution for $P \cup \{G\}$, with terms $t_1,...,t_n$ containing variables $Y_1,...,Y_k$. Assume that $Y_1,...,Y_k,X_1,...,X_n$ are different variables. Then the formula $\forall Y_1...\forall Y_k \exists X_1...\exists X_n$ $(B_1 \wedge...\wedge B_m)$ follows from P.

*Proof*  (2) is obvious from the definition of correct answer substitutions and the validity of a suitably chosen formula. For statement (1), let $c_1,...,c_k$ be new constants. In the subsequent proof we make use of a new signature which contains all the symbols of the old one plus new constant symbols $c_1,...,c_k$. We conclude:

$$P \models \forall Y_1...\forall Y_k \exists X_1...\exists X_n(B_1 \wedge...\wedge B_m) \Leftrightarrow$$

$$P \models \exists X_1...\exists X_n((B_1 \wedge...\wedge B_m)\{Y_1/c_1,...,Y_k/c_k\}) \Leftrightarrow \text{(see Theorem 7.3)}$$

There are ground terms $s_1,...,s_n$ in the extended signature such that

$$P \models (B_1 \wedge...\wedge B_m)\{Y_1/c_1,...,Y_k/c_k\}\{X_1/s_1,...,X_n/s_n\}$$

If we take $t_1,...,t_n$ as the terms in the original signature that are obtained from $s_1,...,s_n$ by replacing all occurrences of $c_i$ by $Y_i$ $(i=1,...,k)$, then we have:

$$P \models \forall Y_1...\forall Y_k((B_1 \wedge...\wedge B_m)\{X_1/t_1,...,X_n/t_n\}). \qquad \blacksquare$$

The question arises of how correct answer substitutions can be computed. In the subsequent sections we shall see that we do not need the full power of resolution calculus.

## 7.3    SLD-resolution

**Definition 7.4:**  Let G be the goal $?\text{-}A_1,...,A_m,...,A_k$ and C be the program formula $A:\text{-}B_1,...,B_q$. (C may also be a fact when $q=0$.) Assume that G and C do not contain common variables. Finally, let $\mu$ be an mgu of $A_m$ and A. Then, the following goal G' is called an **SLD-resolvent** of G and C via $\mu$:

$$?\text{-}A_1\mu,\ldots,A_{m\text{-}1}\mu,B_1\mu,\ldots,B_q\mu,A_{m+1}\mu,\ldots,A_k\mu.$$

That G' is an SLD-resolvent of G and C via $\mu$ is indicated as follows:

(SLD-resolution stands for Selective Linear resolution for Definitive clauses). If instead of a most general unifier $\mu$ of $A_m$ and A we use an arbitrary unifier $\sigma$ of $A_m$ and A, we call G' an **extended SLD-resolvent**. (Extended SLD-resolution will be used only in the proof of the completeness of SLD-resolution.)

Thus, SLD-resolution can be considered as a calculus with a single 2-ary rule allowing us to deduce from variable-disjoint goal $?\text{-}A_1,\ldots,A_m,\ldots,A_k$ and program formula $A:\text{-}B_1,\ldots,B_q$ goal $?\text{-}A_1\mu,\ldots,A_{m\text{-}1}\mu,B_1\mu,\ldots,B_q\mu,A_{m+1}\mu,\ldots,A_k\mu$, if $\mu$ is a mgu of $A_m$ and A.

In this chapter we do not refer to the definitions of the derivation and deduction tree laid down in Section 3.1 for arbitrary calculi, but define corresponding concepts separately. The reasons are that we need a somewhat more informative notion of derivation, and that derivations of infinite length will be admitted, too.

**Definition 7.5:**   An **SLD-derivation** is a finite or infinite tree of the form indicated in Figure 7.1, consisting of goals $G_0,G_1,\ldots,G_n,\ldots$, program formulas $C_0,C_1,\ldots,C_n,\ldots$, and substitutions $\mu_0,\mu_1,\ldots,\mu_n,\ldots$, such that, for all n, $G_{n+1}$ is an SLD-resolvent of $G_n$ and $C_n$ via $\mu_n$, and the following **variable disjointness condition** holds:

Variables occurring in $C_{n+1}$ may not occur in any of the formulas $G_0,G_1,\ldots,G_n,C_0,C_1,\ldots,C_n$ and substitutions $\mu_0,\mu_1,\ldots,\mu_n$.

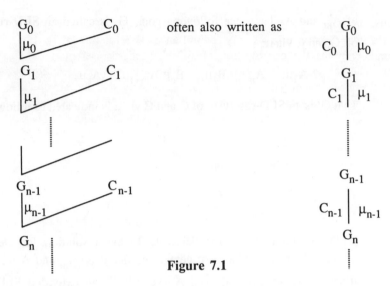

**Figure 7.1**

In the case of a finite SLD-derivation, say up to goal $G_n$, we call n the length of the considered SLD-derivation. Sometimes, we talk of an SLD-derivation with substitutions $\mu_0, \mu_1, \ldots, \mu_n$.

Corresponding notions are introduced for extended SLD-resolution.

**Definition 7.6:** Let P be a logic program and G be a goal. An **SLD-refutation** of $P \cup \{G\}$ is a finite SLD-derivation as in the definition above, say up to goal $G_n$, such that the root goal $G_n$ is the empty goal, and the program formulas $C_0, C_1, \ldots, C_{n-1}$ are variants of program formulas from P.

Note that allowing variants of program formulas from P enables us always to realize the variable disjointness condition required in the definition of an SLD-derivation.

**Definition 7.7:** Let P be a logic program. The **success set** of P is the set of all ground atoms A such that $P \cup \{?\text{-}A\}$ has an SLD-refutation.

**Definition 7.8:** Let P be a logic program and G be a goal. A substitution $\mu$ is called a **computed answer substitution** for $P \cup \{G\}$ iff there exists an SLD-refutation of some length n with a list of most general unifiers $\mu_0, \mu_1, \ldots, \mu_{n-1}$ such that $\mu = (\mu_0 \mu_1 \ldots \mu_{n-1})|_G$.

Thus, the way we obtain computed answer substitutions for a logic program P and a goal G is as follows. Try to find an SLD-refutation of $P \cup \{G\}$. If one is found, compute the composition of all substitutions used to build this refutation. The resulting substitution will contain lots of pairs X/t, for variables X that do not occur in the initial goal G. (Remember that, because of the variable disjointness condition, lots of auxiliary variables will be used in an SLD derivation.) These are of no interest w.r.t. G, so we will discard them.

In the following sections we will deal with the question of soundness and completeness of SLD-resolution. Investigations will be done on three levels of increasing sophistication.

(1)    The level of yes/no-answers to goals ?-A, for ground atoms A. There is the semantic notion of the least Herbrand model of a logic program P, and the proof theoretic notion of the *success set* of a logic program P.

(2)    The level of yes/no-answers to arbitrary goals. There is the semantic notion that an existential formula $\exists(B_1 \wedge ... \wedge B_m)$ follows from a logic program P, and the proof theoretic notion that there exists an SLD-refutation of $P \cup \{?\text{-}B_1,...,B_m\}$.

(3)    The level of answer substitutions. There is the semantic notion of correct answer substitution for $P \cup \{G\}$, and the proof theoretic notion of a computed answer substitution for $P \cup \{G\}$.

Semantical and proof theoretical notions will prove to be equivalent (as far as this makes sense) on each of the three levels.

## 7.4    Soundness of SLD-resolution

We begin by treating the most sophisticated third level, and carry over the obtained soundness result to the other levels as corollaries.

### Theorem 7.5

Let P be a logic program and G be the goal $?\text{-}A_1,...,A_k$. Then every computed answer substitution $\mu$ for $P \cup \{G\}$ is a correct answer substitution for $P \cup \{G\}$. Furthermore, if $\mu$ is a computed answer substitution for $P \cup \{G\}$ and $\tau$ is an arbitrary substitution, then $(\mu\tau)|_G$ is also a correct answer substitution for $P \cup \{G\}$.

*Proof* Let $\mu$ be a computed answer substitution for $P \cup \{G\}$. Choose an SLD-refutation as in Figure 7.2 with $\mu = (\mu_0\mu_1...\mu_{n-1})|_G$. We show that $\forall((A_1 \wedge ... \wedge A_k)(\mu_0\mu_1...\mu_{n-1}))$ follows from P by induction on $n > 0$.

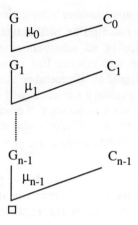

**Figure 7.2**

*Induction base*  Assume that n=1. Then, G has the form ?-$A_1$ and $C_0$ is a fact A such that $A_1\mu_0 = A\mu_0$. Since $C_0$ is a variant of a formula in P, we may successively conclude that  $P \models A$,  $P \models A\mu_0$,  $P \models A_1\mu_0$  and  $P \models \forall(A_1\mu_0)$. The last is the statement to be shown.

*Induction step*  Assume that n>1. We consider the first step in the SLD-refutation above. Assume that $C_0$ is a program formula $A\text{:-}B_1,...,B_q$ with q>0 (q=0 is even simpler to treat). Furthermore, let $A_m\mu_0 = A\mu_0$ (for some m≤k) and $G_1$ be the goal:

$?\text{-}(A_1,...,A_{m-1},B_1,...,B_q, A_{m+1},...,A_k)\mu_0.$

Since we have an SLD-refutation of $P\cup\{G_1\}$ of length n-1, we may use induction hypothesis to conclude:

$P \models \forall((A_1\wedge...\wedge A_{m-1}\wedge B_1\wedge...\wedge B_q\wedge A_{m+1}\wedge...\wedge A_k)\mu_0\mu_1...\mu_{n-1})$

Furthermore, since $A\text{:-}B_1,...,B_q$ is a variant of a program formula from P, we know that $((B_1\wedge...\wedge B_q)\rightarrow A)$ follows from P. Thus:

$P \models ((B_1\wedge...\wedge B_q)\rightarrow A)\mu_0\mu_1...\mu_{n-1}$, hence (as $A_m\mu_0 = A\mu_0$)

$P \models ((B_1\wedge...\wedge B_q)\rightarrow A_m)\mu_0\mu_1...\mu_{n-1}$

We conclude that

$P \models \forall((A_1\wedge...\wedge A_{m-1}\wedge A_m\wedge A_{m+1}\wedge...\wedge A_k)\mu_0\mu_1...\mu_{n-1})$

Since $\mu=\mu_0\mu_1...\mu_{n-1}|_G$, we may conclude that $P \models \forall((A_1\wedge...\wedge A_k)\mu)$, that is, $\mu$ is a correct answer substitution for $P\cup\{G\}$.

Let $\tau$ be a further substitution. We have already shown that $\forall((A_1\wedge...\wedge A_k)\mu)$ follows from P, therefore $(A_1\wedge...\wedge A_k)(\mu\tau)$ and $\forall((A_1\wedge...\wedge A_k)(\mu\tau))$ also follow from P. So, $P \models \forall((A_1\wedge...\wedge A_k)(\mu\tau)|_G)$. This means that $(\sigma\tau)|_G$ is a correct answer substitution for $P\cup\{G\}$.   ■

### Corollary

Let P be a logic program and G be the goal $?\text{-}A_1,...,A_k$. If there is an SLD-refutation of $P\cup\{G\}$, then $P\cup\{G\}$ does not possess a model, hence $P \models \exists(A_1\wedge...\wedge A_k)$.

*Proof*  Let an SLD-refutation of $P\cup\{G\}$ be given, and let $\mu$ be the computed answer substitution for $P\cup\{G\}$. From Theorem 7.5 we know that $\mu$ is also a correct answer substitution, that is, $P \models \forall((A_1\wedge...\wedge A_k)\mu)$. Since $(\forall((A_1\wedge...\wedge A_k)\mu)\rightarrow\exists(A_1\wedge... \wedge A_k))$ is a valid formula, $\exists(A_1\wedge...\wedge A_k)$ also follows from P.   ■

### Corollary

The success set of a logic program P is a subset of $M_P$.

*Proof*  Let ground atom A be an element of the success set of P. By the corollary above, the existential closure of A, in this case A, follows from P. So, A is a member of $M_P$.   ■

## 7.5    Completeness of SLD-resolution

The discussion of completeness of SLD-resolution starts with the simplest level of consideration, the level of ground atoms. First, we adapt the lifting lemma proved in Section 3.4 to the peculiarities of SLD-resolution. Since we require a stronger form of lifting (because of our interest in computed answer substitutions, and not only in yes/no answers) the premisses of the following lifting lemma are rather strong. Nevertheless, they will be fulfilled in every application.

### Lifting lemma 7.1 for SLD-resolution

Given a program formula C and a goal G without common variables, a substitution $\theta$ and an extended SLD-resolvent G' of $G\theta$ and C via a substitution $\sigma$ such that $\theta$ does not act on C, then there is an SLD-resolvent G'' of G and C via a substitution $\mu$ and a substitution $\tau$ such that G''$\tau$=G' and ($\theta\sigma$)=($\mu\tau$). (See Figure 7.3.)

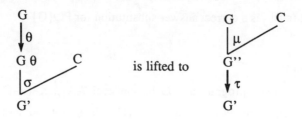

**Figure 7.3**

*Proof*    Let G be ?-$A_1,...,A_{m-1},A_m,A_{m+1},...,A_k$ and C be $A$:-$B_1,...,B_q$. Assume that $A_m\theta$ is the literal resolved with the head A of C, that is, $A_m\theta\sigma=A\sigma$, and G' is the goal ?-$(A_1,...,A_{m-1},B_1,...,B_q,A_{m+1},...,A_k)\theta\sigma$. Since $\theta$ does not act on C, we may conclude that $A\theta=A$. Hence, $A_m(\theta\sigma)=A\sigma=A\theta\sigma$, that is, $A_m$ and A are unifiable with unifier ($\theta\sigma$). Let $\mu$ be an mgu of $A_m$ and A, and $\tau$ be a substitution with ($\theta\sigma$)=($\mu\tau$). Define G'' to be the goal

$$?-(A_1,...,A_{m-1},B_1,...,B_q, A_{m+1},..., A_k)\mu.$$

Then, G'' is an SLD-resolvent of G and C via $\mu$, and G''$\tau$ is the goal

$$?-(A_1,...,A_{m-1},B_1,..., B_q,A_{m+1},...,A_k)(\mu\tau)$$
$$= ?-(A_1,...,A_{m-1},B_1,...,B_q,A_{m+1},...,A_k)(\theta\sigma)$$
$$= G'.$$    ∎

### *Corollary*

Let a logic program P, a goal G, a substitution $\theta$ and an extended SLD-refutation of $P\cup\{G\theta\}$ be given, in which substitutions $\sigma_0,\sigma_1,...,\sigma_{n-1}$ and program formulas $C_0,C_1,...,C_{n-1}$ are successively applied (see Figure 7.1). Assume that $G,C_0,C_1,...,C_{n-1}$ do not contain common variables and $\theta$ does not act on $C_0,C_1,...,C_{n-1}$. Then there exists an SLD-refutation of $P\cup\{G\}$ with substitutions $\mu_0,\mu_1,...,\mu_{n-1}$. Furthermore, there is a substitution $\tau$ such that $\theta\sigma_0\sigma_1...\sigma_{n-1}=\mu_0\mu_1...\mu_{n-1}\tau$. (See Figure 7.4.)

is lifted to

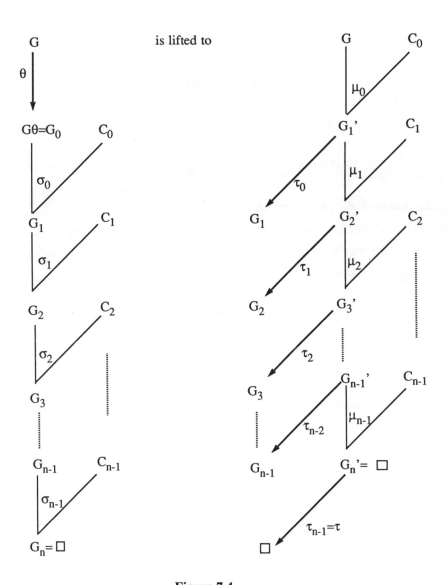

**Figure 7.4**

*Proof*  We argue by induction on n≥1. For n=1 the claim of our corollary is an immediate consequence of the lifting lemma. Assume that n>1, and the corollary has already been shown for n-1. Consider an SLD-refutation of length n as in the statement of the corollary. We lift the first step. This is possible since we know that the given substitution $\theta$ does not act on $C_0$. We obtain substitutions $\mu_0$ and $\tau_0$ and a goal $G_1'$ such that $G_1'$ is an SLD-resolvent of G and $C_0$ via $\mu_0$, $G_1'\tau_0=G_1$ and $(\theta\sigma_0)=(\mu_0\tau_0)$. Thus we have achieved the first level of the lifted derivation in Figure 7.4. Now we consider the initial substitution $\tau_0$ mapping $G_1'$ into $G_1$, and the SLD-derivation starting at $G_1$. If we knew that $\tau_0$ did not act on $C_1,...,C_{n-1}$ we could lift by induction hypothesis the SLD-derivation below $G_1$, thus obtaining a lifted SLD-refutation with substitutions $\mu_1,...,\mu_{n-1}$ and a substitution $\tau$ with $\tau_0\sigma_1...\sigma_{n-1}=\mu_1...\mu_{n-1}\tau$. Then we would be done since

$$\theta\sigma_0\sigma_1...\sigma_{n-1}=\mu_0\tau_0\sigma_1...\sigma_{n-1}=\mu_0\mu_1...\mu_{n-1}\tau$$

So we are left with the task of showing that $\tau_0$ does not act on $C_1,...,C_{n-1}$. Assume that some pair $X/t$ occurs in $\tau_0$. Then, some pair $X/r$ occurs in $\mu_0\tau_0$, thus in $\theta\sigma_0$. This implies that some pair $X/s$ occurs in $\theta$ or $\sigma_0$. In both cases we may conclude that X does not occur in $C_1,...,C_{n-1}$: in the former case because it was assumed that $\theta$ does not act on $C_1,...,C_{n-1}$, in the latter case because the variables of $C_1,...,C_{n-1}$ do not occur in $\sigma_0$, owing to the variable disjointness condition for an SLD-derivation.    ∎

### Lemma 7.2

Let SLD-refutations of $P\cup\{?\text{-}B_i\}$, for $i=1,...,n$, be given such that none of the substitutions used in these refutations acts on $B_1,...,B_n$. Then we obtain an SLD-refutation of $P\cup\{?\text{-}B_1,...,B_n\}$ by composing the n given SLD-refutations as in Figure 7.5 .

### Lemma 7.3 (Renaming of program formulas)

In a finite SLD-derivation it is always possible to rename the variables occurring in the used program formulas and substitutions in such a way that the resulting tree again forms an SLD-derivation, possesses the same goals (these are not affected by the renaming), and has the property that the renamed program formulas do not contain variables from a given finite set E of variables.

*Proof*    It suffices to consider a single SLD-step as in Figure 7.6. There, $\rho$ is a renaming substitution of the form $\{X_1/Y_1,...,X_n/Y_n\}$ of program formula C with fresh variables $Y_1,...,Y_n$ that do not occur in goal $G'$ and $\sigma$, nor in the whole SLD-derivation above G (including G), nor in the given finite set E. Note that $\rho^{-1}$ in $\rho^{-1}\sigma$ takes back the renaming done in $C\rho$, but does not act on G according to the choice of $Y_1,...,Y_n$. For the same reason, the variable disjointness condition for an SLD-derivation is again fulfilled.    ∎

?−$B_1$,...,$B_m$.

  SLD refutation of P ∪{ ?-$B_1$}
  +  $B_2$,...,$B_m$ added to all intermediate goals

?-$B_2$,...,$B_m$. (unchanged, as the substitutions applied above
       do not act on $B_2$,...,$B_m$)

  SLD-refutation of P ∪{ ?-$B_2$}
  +  $B_3$,...,$B_m$ added to all intermediate goals

?-$B_3$,...,$B_m$. (unchanged, as the substitutions applied above
       do not act on $B_3$,...,$B_m$)

 &bull;
 &bull;
 &bull;

?-$B_m$.

  SLD refutation of P ∪{ ?-$B_m$.}

 □

**Figure 7.5**

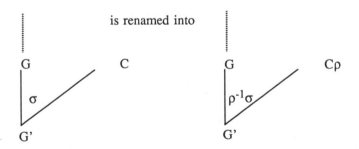

**Figure 7.6**

Now we are able to consider completeness of SLD-resolution and treat
first the case of goals ?-A, where A is a ground atomic formula.

***Theorem 7.6 (Completeness of SLD-resolution for ground atoms)***
The success set of a logic program P coincides with its least Herbrand model $M_P$. Thus, ground atom A follows from P iff there is an SLD-refutation of $P \cup \{?\text{-}A\}$.

*Proof*  It has already been shown that the success set of P is a subset of $M_P$. To show the converse inclusion we refer to the approximative computation of $M_P$ by the sequence of sets $I_n(P)$, for $n \in \mathbb{N}$. We show by induction over n that $I_n(P)$ is subset of the success set of P. The claim is trivial for n=0. So assume it is already shown for a natural number n. Let A be an element of $I_{n+1}(P)$. If A belongs to $I_n(P)$, then there is nothing to show. If, on the other hand, there is a ground instance $B\sigma\text{:-}B_1\sigma,...,B_m\sigma$ of a program formula $B\text{:-}B_1,...,B_m$ from P such that $A=B\sigma$ and $B_1\sigma,...,B_m\sigma \in I_n(P)$, then we may continue as follows: by induction hypothesis, $B_1\sigma,...,B_m\sigma$ are elements of the success set of P. So there is an SLD-refutation of $P \cup \{?\text{-}B_1\sigma\}$, for i=1,...,m. Since $B_1\sigma,...,B_m\sigma$ do not contain variables, the preconditions of the renaming lemma are fulfilled. Hence, we may compose these m refutations into a single SLD-refutation of $P \cup \{?\text{-}B_1\sigma,...,B_m\sigma\}$. Since $?\text{-}B_1\sigma,...,B_m\sigma$ is an extended SLD-resolvent of goal ?-A and rule $B\text{:-}B_1,...,B_m$ via substitution $\sigma$, we obtain the extended SLD-refutation of $P \cup \{?\text{-}A\}$ depicted in Figure 7.7 with the empty substitution as initial substitution $\theta$.

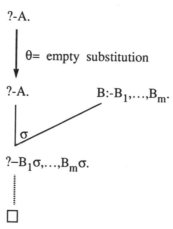

**Figure 7.7**

Using the lifting lemma (renaming is not necessary because $\theta$ is the empty substitution, hence the precondition of the lifting lemma concerning $\theta$ is trivially fulfilled) we obtain an SLD-refutation of $P \cup \{?\text{-}A.\}$.  ∎

Next we treat arbitrary goals, but we do not consider answer substitutions yet.

### Theorem 7.7 (Completeness of SLD-resolution for arbitrary goals)

Let P be a logic program and G be the goal $?\text{-}A_1,...,A_k$. If $P \models \exists(A_1 \wedge ... \wedge A_k)$ then there exists an SLD-refutation of $P \cup \{G\}$.

*Proof*   Assume that $P \models \exists(A_1 \wedge ... \wedge A_k)$. Our result on definite answers implies that there is a substitution $\theta$ such that $(A_1 \wedge ... \wedge A_k)\theta$ does not contain variables and $P \models (A_1 \wedge ... \wedge A_k)\theta$. So, all $A_i\theta$, for $i=1,...,k$, are elements of $M_P$. Therefore there are SLD-refutations of $P \cup \{?\text{-}A_i\theta\}$, for every $i=1,...,k$. The substitutions used in these refutations do not act on $A_i\theta$, for $i=1,...,k$, since these are ground atoms. Using the compositionality lemma we may compose these refutations to an SLD-refutation of $P \cup \{?\text{-}A_1\theta,...,A_k\theta\}$, that is, to an SLD-refutation of $P \cup \{G\theta.\}$. By suitable renamings we may assume that the program formulas used in the considered SLD-refutation of of $P \cup \{?\text{-}A_1\theta,...,A_k\theta\}$ fulfil the required precondition of the lifting lemma, that is, that $\theta$ does not act on them. Using the lifting lemma (here we have a situation with a non-empty initial substitution $\theta$) we finally obtain an SLD-refutation of $P \cup \{G\}$.                                                                  ∎

Finally we turn our attention to answer substitutions. First we treat a special case.

### Lemma 7.4

Let P be a logic program and A be an atom. Further, assume that $P \models \forall(A)$ (equivalently stated, the empty substitution is a correct answer substitution for $P \cup \{?\text{-}A\}$). Then there is an SLD-refutation of $P \cup \{?\text{-}A\}$ with a sequence of substitutions $\sigma_0, \sigma_1, ..., \sigma_{n-1}$ that do not act on A (as a consequence, the empty substitution is a computed answer substitution for $P \cup \{?\text{-}A\}$). The variables X occurring in pairs X/t in $\sigma_0\sigma_1...\sigma_{n-1}$ may even be chosen in such a way that they do not belong to a given finite set E of variables.

*Proof*   Let A contain only variables $X_1,...,X_n$. Let $c_1,...,c_n$ be new constants and $\theta$ be the substitution $\{X_1/c_1,...,X_n/c_n\}$. Then $P \models A\theta$. By Theorem 7.7 there is an SLD-refutation of $P \cup \{?\text{-}A\theta\}$. In this SLD-refutation we may use variants of program formulas of P in such a way that no pair X/t, with

$X \in \{X_1,...,X_n\} \cup E$, occurs in any substitution of the refutation (recall that $A\theta$ contains no variables). In this refutation we replace all occurrences of $c_i$ by $X_i$, for $i=1,...,n$, and obtain an SLD-refutation of $P \cup \{?\text{-}A\}$ such that no used substitution contains a pair $X/t$ with $X \in \{X_1,...,X_n\} \cup E$. Therefore the answer substitution computed by this refutation is $\varnothing$.          ∎

### Theorem 7.8 (Completeness of SLD resolution w.r.t. answer substitutions)

Let P be a logic program and G be the goal $?\text{-}A_1,...,A_k$. Let $\theta$ be a correct answer substitution for $P \cup \{?\text{-}G\}$. Then there is a computed answer substitution $\mu$ for $P \cup \{?\text{-}G\}$ and a substitution $\tau$ such that $\theta = (\mu\tau)|_G$.

*Proof* Let $\theta$ be a correct answer substitution for $P \cup \{?\text{-}G\}$. Hence, $\forall((A_1 \wedge ... \wedge A_k)\theta)$ follows from P, therefore also $\forall(A_i\theta)$, for $i=1,...,k$. By Lemma 7.4 (with the set of variables of G and $G\theta$ as the finite set E) there exists, for each $i=1,...,k$, an SLD-refutation of $P \cup \{?\text{-}A_i\theta\}$ such that the used substitutions do not act on G and $G\theta$. Hence, these substitutions do not act on $A_i\theta$, for $i=1,...,k$. As a consequence, we may compose these refutations to a single refutation of $P \cup \{?\text{-}A_1\theta,...,A_k\theta\}$ with a sequence $\sigma_0,\sigma_1,...,\sigma_{n-1}$ of substitutions that do not act on G and $G\theta$. By the lifting lemma (applicable after suitable renaming of the used program formulas in the considered refutation) there is an SLD-refutation of $P \cup \{?\text{-}G\}$ with substitutions $\mu_0,\mu_1,...,\mu_{n-1}$, and a substitution $\tau$ such that $\theta\sigma_0\sigma_1...\sigma_{n-1} = \mu_0\mu_1...\mu_{n-1}\tau$.

Now we define $\mu = (\mu_0\mu_1...\mu_{n-1})|_G$. Then, $\mu$ is a computed answer substitution for $P \cup \{?\text{-}G\}$. We restrict both sides of the equation $\theta\sigma_0\sigma_1...\sigma_{n-1} = \mu_0\mu_1...\mu_{n-1}\tau$ to G. On the right-hand side we obtain $(\mu_0\mu_1...\mu_{n-1}\tau)|_G$, which is obviously the same as $(\mu\tau)|_G$.

On the left-hand side we obtain $\theta$. This can be shown in the following way: Let X/t be an element of $\theta$. X occurs in G, since $\theta$ is a correct answer substitution for $P \cup \{?\text{-}G\}$. Next we show that $t\sigma_0\sigma_1...\sigma_{n-1} = t$. If Y is a variable in t, then Y occurs in $G\theta$, therefore substitution $\sigma_0\sigma_1...\sigma_{n-1}$ does not act on Y. This means that pairs X/t occurring in $\theta$ are not deleted when restricted to the variables of G. Further, $\sigma_0\sigma_1...\sigma_{n-1}$ contains no pair X/t with a variable X occurring in G, so $\sigma_0\sigma_1...\sigma_{n-1}$ completely vanishes when we restrict the equation to variables occurring in G.                                    ■

## Remark

Although it is often claimed in the literature, it is not always possible to express a correct answer substitution $\theta$ as $\mu\tau$, where $\mu$ is a computed answer substitution and $\tau$ another substitution. To see this, let us consider the following example.

In a signature with constant symbols a and b, a 2-ary function symbol f and a 1-ary predicate symbol p, we consider the logic program $P=\{p(f(X,Y))\}$ and the goal ?-p(Z). A correct answer substitution for $P\cup\{?-p(Z)\}$ is $\theta=(Z/f(a,b)\}$, since p(f(a,b)) follows from P. All variants $\mu$ of the mgu $\{Z/f(X,Y)\}$ of f(X,Y) and Z are computed answer substitutions. Each such $\mu$ must have the form $\{U/f(V,W)\}$, where U,V,W are different variables. So, in f(V,W) not only Z occurs, but at least one further variable Z'. Now suppose that we may write $\theta$ as $(\mu\tau)$. Then $\tau$ must contain a pair Z'/t (where t is a or b, dependent on whether V or W is Z'). This pair must also occur in $(\mu\tau)$. But $\theta$ contains no pair Z'/s. This is a contradiction.

# Chapter 8
# Steps Towards Programming in Logic

We present a couple of logic programs from several fields of application. The reader may direct his/her special attention to the problem of determining the least Herbrand model of these programs, thereby establishing that they are indeed correct with respect to the intended Herbrand interpretation $I$.

## 8.1    Preparing for the verification of logic programs

We have already discussed in Section 7.2 the standard method of how to show a claim of the form $M_P = I$, for a logic program P and intended Herbrand interpretation $I$ (see Example 7.1).

One inclusion, $M_P \subseteq I$, may be shown by proving that $I$ is a model of P.

The reversed inclusion is usually shown by induction on a suitably chosen application dependent parameter associated with ground atoms from $I$.

In the following examples, it is not necessary (or even desirable) to show

$M_P \subseteq I$. Sometimes, it suffices to know which ground atoms *from a distinguished set of ground atoms Restr* are in $M_P$. So let us discuss the verification of statements of the form $M_P \cap \text{Restr} \subseteq I$.

### Theorem 8.1

Assume that the following two conditions 'Inheritance of the restriction' and 'Relativized validity of P in *I* ' below are fulfilled, for a logic program P, a set of ground atoms Restr and a Herbrand interpretation *I*. Then $M_P \cap \text{Restr} \subseteq I$.

- *Inheritance of the restriction*
  For every ground instance $B\sigma$:-$B_1\sigma,\ldots,B_m\sigma$ of a program formula in P, $B_1\sigma,\ldots,B_m\sigma$ can be arranged into an ordering $A_1,\ldots,A_m$ such that:
  (a)     $B\sigma \in$ Restr implies that $A_1 \in$ Restr.
  (b)     For all k=2,...,m, $A_1,\ldots,A_{k-1} \in I$ implies $A_k \in$ Restr.

- *Relativized validity of P with respect to I*
  *I* is a model of all ground instances $B\sigma$:-$B_1\sigma,\ldots,B_m\sigma$ of program formulas in P with *ground atoms $B\sigma,B_1\sigma,\ldots,B_m\sigma$ taken from set Restr*. If this is the case we say that *I* is a **model** of P **relativized to** Restr.

That we only require relativized validity of P w.r.t. *I* is clear, since the claim $M_P \cap \text{Restr} \subseteq I$ requires P to work well only for ground atoms from Restr. Indeed, we cannot expect that *I* is a model of P, since this even implies $M_P \subseteq I$. The sense of the inheritance property is to guarantee that the literals of the body of a rule instance $B\sigma$:-$B_1\sigma,\ldots,B_m\sigma$ with $B\sigma \in$ Restr may be satisfied in a suitable order in such a way that, whenever a ground atom $A_k$ is treated, it is an element of Restr.

*Proof*     We show that $M_P \cap \text{Restr} \subseteq I$ by showing that $I_n(P) \cap \text{Restr} \subseteq I$, for every number n. (Here we refer to the approximate characterization of $M_P$ introduced in Section 7.2.) The latter claim is proved by induction on n. Consider a ground atom in $I_{n+1}(P) \cap \text{Restr}$, that is, a ground atom $B\sigma$ such that $B\sigma$:-$B_1\sigma,\ldots,B_m\sigma$ is a ground instance of a program formula in P and $B_1\sigma,\ldots,B_m\sigma \in I_n(P)$. Arrange $B_1\sigma,\ldots,B_m\sigma$ into an ordering $A_1,\ldots,A_m$ such that properties (a) and (b) of  the inheritance condition are fulfilled. Since $B\sigma \in$ Restr, we obtain from (a) that $A_1 \in$ Restr, hence $A_1 \in I_n(P) \cap$ Restr. By induction hypothesis we conclude that $A_1 \in I$. From property (b) we obtain that $A_2 \in$ Restr, hence $A_2 \in I_n(P) \cap$ Restr. Again by induction hypothesis we obtain that $A_2 \in I$. Proceeding this way we finally obtain that $A_1, \ldots, A_m$ are elements of $I_n(P) \cap$ Restr, hence of $I \cap$ Restr. Thus, $B_1 \sigma, \ldots, B_m \sigma$ are elements of $I \cap$ Restr. Application of the relativized validity of P w.r.t. $I$ gives us that $B\sigma \in I$, hence $A \in I$. ∎

Examples showing the application of this theorem can be found in the following subsections.

## 8.2    List processing

Lots of applications are best treated using data type 'list'. So we will start with a definition of lists and some simple logic programs processing lists.

We consider a 2-ary function symbol, written in infix notation [L|Rest], and a constant []. Using these two function symbols, together with further ones from a signature $\Sigma$ and variables, we may form **lists over** $\Sigma$ as the following terms:

- [] is a list over $\Sigma$, called the **empty list**.

- [F|Rest] is a list over $\Sigma$, whenever Rest is a list over $\Sigma$ and F is either a $\Sigma$-term or a list over $\Sigma$. Look at [X|Rest] as a list with **first element** F and **tail** Rest. Note that F may itself be a list, whereas Rest must be a list.

Lists without variables are called **ground lists**. Often, a graphical representation helps us to imagine particular lists. We represent a list [F|Rest] as in Figure 8.1.

F          Rest

**Figure 8.1**

To decrease the burden resulting from the multitude of brackets in nested list expressions, we introduce the following standard abbreviations:

- $[F_1, F_2, \ldots, F_n | Rest]$ is $[F_1 | [F_2 | [\ldots | [F_n | Rest] \ldots ]]]$
- $[F_1, F_2, \ldots, F_n]$ is $[F_1, F_2, \ldots, F_n | []]$

Here we assume that n>0, Rest is a list, and $F_1, F_2, \ldots, F_n$ are $\Sigma$-terms or lists. Graphically these lists look like Figure 8.2. Here the notion of lists finds its natural explanation. Let us treat some examples (see Figures 8.3 and 8.4).

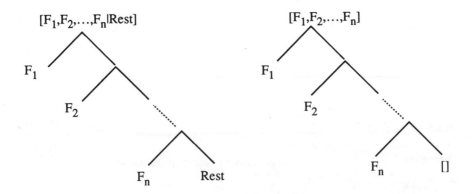

**Figure 8.2**

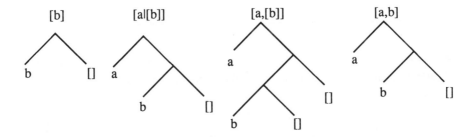

**Figure 8.3**

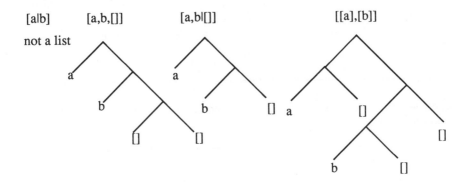

**Figure 8.4**

**Exercise 8.1**    Show that every ground list is of the form [] or $[F_1,F_2,...,F_n]$, for $n>0$ and lists or terms $F_1,F_2,...,F_n$.

# EXAMPLE 8.1

A frequently used function in list processing is concatenation •:

$$[X_1,...,X_n]•[Y_1,...,Y_m]=[X_1,...,X_n,Y_1,...,Y_m]$$

Consider the following logic program APPEND:

    append([],M,M).
    append([X|L],M,[X|N]):-append(L,M,N).

**Exercise 8.2**    Let Restr be the set of all ground atoms of the form $append(l_1,l_2,l)$ with lists $l$, $l_1$ and $l_2$. Let $I$ be the Herbrand interpretation consisting of all ground atoms $append(l_1,l_2,l)$ with lists $l_1$ and $l_2$ and $l=l_1•l_2$. Show that $M_{APPEND} \cap Restr=I$. What is the reason for the intersection with Restr? Note that the inheritance condition of Theorem 8.1 expresses the true statement that L is a list, provided that [X|L] is a list.

Let us treat some goals with corresponding computed answer substitutions. Consider the goal ?-append([a,b],[a,Y],Z). A most general correct answer substitution is {Z/[a,b,a,Y]}. This is also the computed answer substitution.

**Exercise 8.3**    Show how the correct answer substitution above is obtained as a computed answer substitution.

Next we treat the goal ?-append (X,Y,[a,b,c]). Whereas the former goal intends to compute the concatenation Z of two inputs [a,b] and [a,Y], the latter intends to compute the inverse images of output [a,b,c] with respect to concatenation. Now there are four correct answer substitutions, {X/[],Y/[a,b,c]}, {X/[a],Y/[b,c]}, {X/[a,b],Y/[c]} and {X/[a,b,c],Y/[]}.

---

**Exercise 8.4**  Show how the correct answer substitutions above are obtained as computed answer substitutions.

---

# EXAMPLE 8.2

The following logic program defines element test on lists:

```
element(X,[X|L]).
element(X,[Y|L]):-element(X,L).
```

# EXAMPLE 8.3

The following logic program defines predicate mirror with the following intended meaning:

mirror(l,m) iff list m results from list l by reversing the order of elements

The program uses the append predicate introduced and defined above:

```
mirror([],[]).
mirror([X|L],M):-mirror(L,N),append(N,[X],M).
```

# EXAMPLE 8.4

In the presence of a predicate unequal defining inequality on the domain of elements that are used within lists, we may introduce two predicates delete and delete1 with the following meanings:

delete(l,x,m) iff list m results from list l by deleting every occurrence of x

delete1(l,x,m) iff x occurs in list l and list m results from l by removing the leftmost occurrence of x

```
delete([],X,[]).
delete([X|L],X,M):-delete(L,X,M).
delete([Y|L],X,[Y|M]):-unequal(Y,X),delete(L,X,M).
delete1([X|L],X,L).
delete1([Y|L],X,[Y|M]):-unequal(Y,X),delete1(L,X,M).
```

**Exercise 8.5**   Verify the programs for delete and delete1 in Example 8.4. What are the intended least Herbrand models?

## 8.3   Arithmetics

In Section 7.2 natural numbers were introduced via a constant 0 and 1-ary function symbol succ. There, we also presented a first logic program defining addition:

>     add(X,0,X).
>     add(X,succ(Y),succ(Z)):-add(X,Y,Z).

For further use, let us introduce some more logic programs defining basic functions and relations on natural numbers:

>     minus(X,0,X).
>     minus(0,succ(Y),0).
>     minus(succ(X),succ(Y),Z):-minus(X,Y,Z)

obviously defines non-negative subtraction;

>     mult(X,0,0).
>     mult(X,succ(Y),Z):-mult(X,Y,U),add(U,X,Z).

defines multiplication;

>     less(0,succ(Y)).
>     less(succ(X),succ(Y)):-less(X,Y).

defines the linear order on natural numbers;

>     unequal(X,Y):-less(X,Y).
>     unequal(X,Y):-less(Y,X).

defines inequality.

Let us now answer the more theoretical question as to which functions on natural numbers can be 'computed by logic programs'. As we will show, these are just the computable functions well known from recursive function theory. Before we state precisely what 'computable by a logic program' means and then prove the claim, we must consider some aspects of recursive function theory.

Recursive function theory deals with partial functions $f: \mathbb{N}^n \to \mathbb{N}$, with $n \geq 0$. For reasons of uniformity, $n=0$ is admitted leading to 0-ary functions which are nothing more than constants.

(1)    The following total functions are called **basic functions**:

(a)    the **constant** 0

(b)    the **successor function** succ

(c)    the **projection function** $p_{n,i}$ with $p_{n,i}(x_1,...,x_n)=x_i$, for $1\leq i\leq n$

(2)    The partial function $f:\mathbb{N}^n\rightarrow\mathbb{N}$ is said to be obtained by **composition** from the partial functions $g:\mathbb{N}^m\rightarrow\mathbb{N}$, $h_1:\mathbb{N}^n\rightarrow\mathbb{N},...,h_m:\mathbb{N}^n\rightarrow\mathbb{N}$ iff:

$$f(x_1,...,x_n)=g(h_1(x_1,...,x_n),...,h_m(x_1,...,x_n))$$

(with the usual understanding that $f(x_1,...,x_n)$ is undefined iff any one of the subterms of $g(h_1(x_1,...,x_n),...,h_m(x_1,...,x_n))$ is undefined).

(3)    The partial function $f:\mathbb{N}^{n+1}\rightarrow\mathbb{N}$ is said to be obtained by **primitive recursion** from the partial functions $g:\mathbb{N}^n\rightarrow\mathbb{N}$ and $h:\mathbb{N}^{n+2}\rightarrow\mathbb{N}$ iff:

$$f(x_1,...,x_n,0)=g(x_1,...,x_n)$$
$$f(x_1,...,x_n,y+1)=h(x_1,...,x_n,f(x_1,...,x_n),y)$$

(with the usual understanding that $f(x_1,...,x_n,0)$ is undefined iff $g(x_1,...,x_n)$ is undefined, and $f(x_1,...,x_n,y+1)$ is undefined iff $f(x_1,...,x_n)$ is undefined or $h(x_1,...,x_n,f(x_1,...,x_n),y)$ is undefined).

(4)    A function f is called **primitive recursive** iff it can be obtained from the basic functions by finitely applying composition and primitive recursion. Note that every primitive recursive function is total.

As is well known from recursive function theory, the set of all primitive recursive functions does not include every function which we would like to call computable in a naive sense. What is needed is a further closure operation, called minimization.

(5)    A partial function $f:\mathbb{N}^n\rightarrow\mathbb{N}$ is said to be obtained from the partial function $g:\mathbb{N}^{n+1}\rightarrow\mathbb{N}$ by **minimization** iff, $f(x_1,...,x_n)$ is the least number y such that

(a)    $g(x_1,...,x_n,z)$ is defined, for all $z\leq y$

(b)    $g(x_1,...,x_n,z)\neq 0$, for all $z<y$

(c)    $g(x_1,...,x_n,y)=0$

Of course, $f(x_1,...,x_n)$ is undefined iff there is no number y. In the case of a total function g, the definition of $f(x_1,...,x_n)$ simply reads as follows:

$$f(x_1,\ldots,x_n)=\text{the least number } y \text{ such that } g(x_1,\ldots,x_n,y)=0$$

We denote the function f defined above by $\mu y.g(x_1,\ldots,x_n,z)$.

A partial function f is called **computable** (or **partial recursive**) iff it can be obtained from the basic functions by iterated application of composition, primitive recursion and minimization.

It is a non-trivial theorem from recursive function theory that one application of minimization suffices to obtain every computable function. As a consequence, every total computable function f can be obtained from the basic functions by iterated application of composition, primitive recursion and minimization in such a way that all intermediate functions constructed in the process of obtaining f are total. This fact will be used later.

### Theorem 8.2

For every partial recursive function $f: \mathbb{N}^n \to \mathbb{N}$ we construct a logic program Prog(f) in a signature containing (among others) a constant 0, a 1-ary function symbol succ and an n+1-ary relation symbol $p_f$, such that for all $x_1,\ldots,x_n,y \in \mathbb{N}$ the following statements are equivalent:

(1)     $f(x_1,\ldots,x_n)=y$

(2)     $\text{Prog}(f) \models p_f(\text{succ}^{x_1}(0),\ldots,\text{succ}^{x_n}(0),\text{succ}^{y}(0))$

In this case we say that Prog(f) computes f via relation symbol $p_f$.

*Proof*   Prog(f) is constructed inductively over the definition of partial recursive functions:

(1)     Prog(0) is the logic program consisting of fact $p_0(0)$.

(2)     Prog(succ) is the logic program consisting of fact $p_{succ}(X,\text{succ}(X))$.

(3)     If f is the projection function $p_{n,i}$, then we take as Prog(f) the logic program consisting of fact $p_f(X_1,\ldots,X_n,X_i)$.

(4)     Let f be defined by $f(x_1,\ldots,x_n)=g(h_1(x_1,\ldots,x_n),\ldots,h_m(x_1,\ldots,x_n))$. Assume, by induction, that we have already constructed logic programs $\text{Prog}(g),\text{Prog}(h_1),\ldots,\text{Prog}(h_m)$ which compute $g,h_1,\ldots,h_m$ via predicate symbols $p_g,p_{h_1},\ldots,p_{h_m}$, respectively. We may assume that these programs do not contain common function symbols and predicates, with the exception of 0 and succ. Then we take as Prog(f) the union of programs $\text{Prog}(g),\text{Prog}(h_1),\ldots,\text{Prog}(h_m)$ together with the following additional program formulas:

$$p_f(X_1,...,X_n,Y):-$$
$$p_{h_1}(X_1,...,X_n,U_1),...,p_{h_m}(X_1,...,X_n,U_m),p_g(U_1,...,U_m,Y).$$

(5)    Let f be defined by

$$f(x_1,...,x_n,0)=g(x_1,...,x_n)$$
$$f(x_1,...,x_n,y+1)=h(x_1,...,x_n,f(x_1,...,x_n,y),y)$$

By induction, we may assume that we have already constructed logic programs Prog(g) and Prog(h) which compute g and h via $p_g$ and $p_h$, respectively. Again we may assume that the two programs contain at most 0 and succ as common symbols. Then we take as Prog(f) the union of Prog(g) and Prog(h) together with the following additional program formulas:

$$p_f(X_1,...,X_n,0,Z):-p_g(X_1,...,X_n,Z).$$
$$p_f(X_1,...,X_n,succ(Y),Z):-p_f(X_1,...,X_n,Y,U),p_h(X_1,...,X_n,U,Y,Z).$$

So far, all quite trivial. The only interesting case is definition by minimization.

(6)    Let f be defined by $f(x_1,...,x_n)= \mu y.g(x_1,...,x_n,y)=0$. By induction hypothesis we may assume that a logic program Prog(g) computing g via $p_g$ has already been constructed. Besides $p_f$ we use a further new n+1-ary relation symbol noZero with the following intended meaning:

noZero$(x_1,...,x_n,b)$ iff for all $y<b$, $g(x_1,...,x_n,y)$ is defined and $>0$

Then we may take as Prog(f) the program Prog(g) together with the following additional program formulas:

$$p_f(X_1,...,X_n,Y):-p_g(X_1,...,X_n,Y,0),noZero(X_1,...,X_n,Y).$$
$$noZero(X_1,...,X_n,0).$$
$$noZero(X_1,...,X_n,succ(Z)):-$$
$$p_g(X_1,...,X_n,Z,succ(V)),noZero(X_1,...,X_n,Z).$$

## 8.4  Knapsack problem

Assume we have a knapsack of capacity k litres and want to fill it with a choice among objects of volumes $v_1,...,v_n$, in such a way that the knapsack is completely filled. Such a choice does not exist in general. Hence, we might ask how to decide whether such a choice is possible, given k and $v_1,...,v_n$, and in this case, how to obtain such a choice.

We introduce a 2-ary predicate packable(list,capacity) with the following intended meaning:

packable(list,cap) iff list is a finite list $[v_1,...,v_n]$ of numbers and cap is a number such that there is a sequence $b_1,...,b_n$ of numbers from $\{0,1\}$ with $b_1 \cdot v_1+...+b_n \cdot v_n=cap$

Of course, the sequence $b_1,...,b_n$ encodes the choice among the packable objects; $b_i=0$ means that the i-th object with volume $v_i$ is not packed, whereas $b_i=1$ means that it is packed.

Now consider the following logic program which defines predicate packable recursively on the length of the list of packable objects:

(1)     packable([],0).

(2)     packable([V|Rest],Cap):-less(Cap,V),packable(Rest,Cap).

(3)     packable([V|Rest],Cap):-less(V,Cap),packable(Rest,Cap).

(4)     packable([V|Rest],V):-packable(Rest,V).

(5)     packable([V|Rest],Cap):-

          less(V,Cap),minus(Cap,V,Capnew),packable(Rest,Capnew).

(6)     packable([V|Rest],V):-packable(Rest,0).

Fact (1) expresses that with the empty list of available objects a knapsack of capacity 0 can be packed. Since the heads of rules (2)-(6) deal with non-empty lists of packable objects, it is expressed that only the knapsack of capacity 0 can be packed using an empty list of objects. Rule (2) says that an object of volume V exceeding capacity Cap is not used in packing a knapsack of capacity Cap. Rules (3), (4) and (5), (6), applicable in the case that the volume V of the first object does not exceed the capacity Cap of our knapsack, express the two possibilities of packing our knapsack: either omit to pack the first object into the knapsack and fill it with the remaining objects (rules (3) and (4)), or pack the first object into the knapsack and fill the remaining knapsack of rest capacity Cap-V with the remaining objects (rules (5) and (6)).

Our program decides whether a suitable choice among the packable objects completely fills the given knapsack, but does not pass back such a choice where it exists. The program can be easily improved to pass back such a choice. This is done by maintaining as a third argument a list $[b_1,...,b_n]$. We thus deal with a predicate packable1 with the following intended meaning:

packable1($[v_1,...,v_n]$,cap,$[b_1,...,b_n]$) iff $b_1 \cdot v_1+...+b_n \cdot v_n=cap$

**Exercise 8.6**     Write and verify a logic program that defines predicate packable1.

## 8.5    Wang's algorithm

Let us implement a theorem-prover for predicate logic formulas without variables and quantifiers and $\neg$, $\vee$, $\wedge$ as the only propositional connectives. As we know from Chapter 3, such a quantifier free and variable free formula $\varphi$ is valid iff the sequent $\rightarrow \varphi$ is valid. Now, for the validity of sequents $\varphi_1,...,\varphi_n \rightarrow$ $\psi_1,...,\psi_m$ consisting of quantifier free and variable free formulas, a complete calculus was developed in the proof of the special case of the sharpened *Hauptsatz* of Gentzen (Theorem 3.13). The rules of this calculus, a subcalculus of calculus G, had the useful property that they preserved validity in both directions. Interpreted in this sense they lead to the following true statements (all occurring formulas are quantifier free and variable free):

(1)     $\varphi_1,...,\varphi_n \rightarrow \psi_1,...,\psi_m$ is valid if $\{\varphi_1,...,\varphi_n\}$ and $\{\psi_1,...,\psi_m\}$ contain a common element. In the case of atomic formulas $\varphi_1,...,\varphi_n,\psi_1,...,\psi_m$, the converse is also true: $\varphi_1,...,\varphi_n \rightarrow \psi_1,...,\psi_m$ is valid iff $\{\varphi_1,...,\varphi_n\}$ and $\{\psi_1,...,\psi_m\}$ contain a common element.

(2)     $\varphi_1,...,\neg\varphi,...,\varphi_n \rightarrow \psi_1,...,\psi_m$ is valid iff $\varphi_1,...,\varphi_n \rightarrow \varphi,\psi_1,...,\psi_m$ is valid.

(3)     $\varphi_1,...,\varphi_n \rightarrow \psi_1,...,\neg\psi,...,\psi_m$ is valid iff $\psi,\varphi_1,...,\varphi_n \rightarrow \psi_1,...,\psi_m$ is valid.

(4)     $\varphi_1,...,(\varphi\wedge\psi),...,\varphi_n \rightarrow \psi_1,...,\psi_m$ is valid iff $\varphi,\psi,\varphi_1,...,\varphi_n \rightarrow \psi_1,...,\psi_m$ is valid.

(5)     $\varphi_1,...,\varphi_n \rightarrow \psi_1,...,(\varphi\vee\psi),...,\psi_m$ is valid iff $\varphi_1,...,\varphi_n \rightarrow \varphi,\psi,\psi_1,...,\psi_m$ is valid.

(6)     $\varphi_1,...,\varphi_n \rightarrow \psi_1,...,(\varphi\wedge\psi),...,\psi_m$ is valid iff both $\varphi_1,...,\varphi_n \rightarrow \varphi,\psi_1,...,\psi_m$ and $\varphi_1,...,\varphi_n \rightarrow \psi_1,...,\psi_m$ are valid.

(7)     $\varphi_1,...,(\varphi\vee\psi),...,\varphi_n \rightarrow \psi_1,...,\psi_m$ is valid iff both $\varphi,\varphi_1,...,\varphi_n \rightarrow \psi_1,...,\psi_m$ and $\psi,\varphi_1,...,\varphi_n \rightarrow \psi_1,...,\psi_m$ are valid.

(2)-(7) enable the algorithm successively to simplify composed formulas within the considered sequents. Validity thereby is preserved. Having arrived at atoms as the only constituents of our sequents, rule (1) finishes the

game correctly. The usual exponential complexity of algorithms deciding validity or satisfiablity of formulas of propositional logic is mirrored in the exponential generation of successively more sequents due to rules (6) and (7).
Nevertheless, with some luck, rule (1) may enable us to finish the game before the involved formulas are cut down to the atomic level. Now it is a simple
task to realize Wang's algorithm as a logic program. For simplicity we represent negation, conjunction and disjunction by function symbols 'neg', 'and',
'or' of arity 1,2,2 respectively. Thus, $(A \wedge B)$ reads somewhat unusually as
and(A,B). For a sequent $\varphi_1,...,\varphi_n \rightarrow \psi_1,...,\psi_m$, antecedent and succedent are
separately represented as lists ante=$[\varphi_1,...,\varphi_n]$ and suc=$[\psi_1,...,\psi_m]$. We
make use of predicates 'unequal' for formulas (exercise), 'element' and
'delete' for lists. Here is the program:

(0)    valid(Formula):-wang([],[Formula]).

(1)    wang(Ante,Suc):-intersects(Ante,Suc).

intersects([X|L],M):-element(X,M).

intersects([X|L],M):-intersects(L,M).

(2)    wang(Ante,Suc):-
               element(not(A),Ante),
               delete(Ante,not(A),Antenew),
               wang(Antenew,[A|Suc]).

(3)    wang(Ante,Suc):-
               element(not(A),Suc),
               delete(Suc,not(A),Sucnew),
               wang([A|Ante],Sucnew).

(4)    wang(Ante,Suc):-
               element(and(A,B),Ante),
               delete(Ante,and(A,B),Antenew),
               wang([A,B|Antenew],Suc).

(5)    wang(Ante,Suc):-
               element(or(A,B),Suc),
               delete(Suc,or(A,B),Sucnew),
               wang(Ante,[A,B|Sucnew]).

(6)    wang(Ante,Suc):-
               element(and(A,B),Suc),
               delete(Suc,and(A,B),Sucnew),
               wang(Ante,[A|Sucnew]),
               wang(Ante,[B|Sucnew]).

(7)    wang(Ante,Suc):-
               element(or(A,B),Ante),
               delete(Ante,or(A,B),Antenew),
               wang([A|Antenew],Suc),
               wang([B|Antenew],Suc).

## 8.6    Slowsort

Let a logic program LESS be given that defines a linear order less(X,Y). The purpose of the following programs is to sort a given input list L consisting of elements from the considered linear order. Inequality is defined in terms of less, so we may use predicate delete1 for lists as discussed in Section 8.2.

```
sort(L,M):-permute(L,M),sorted(M).

permute([],[]).
permute([X|L],M):-
            element(X,M),
            delete1(M,X,Mnew),
            permute(L,Mnew).

sorted([]).
sorted([X]).
sorted([X,X|M]):-sorted([X|M]).
sorted([X,Y|M]):-less(X,Y),sorted([Y|M]).
```

## 8.7    Mergesort

An input list L consisting of elements of a partially ordered set defined by a set of ground facts LESS is sorted as follows: divide L into two sublists L1 and L2 whose lengths differ by at most 1 (unless L is the empty list or consists of exactly one element), then *recursively* apply the mergesort algorithm to L1 and L2 to obtain sorted lists M1 and M2 which are permutations of L1 and L2, and finally merge together M1 and M2 to obtain a sorted list M:

```
mergesort([],[]).
mergesort([X],[X]).
mergesort([X,Y|L],M):-
            divide([X,Y|L],L1,L2),
            mergesort(L1,M1),
            mergesort(L2,M2),
            merge(M1,M2,M).

divide([],[],[]).
divide([X],[X],[]).
divide([X,Y|L],[X|U],[Y|V]):-divide(L,U,V).

merge([],M2,M2).
merge(M1,[],M1).
merge([X|M1],[X|M2],[X|M]):-merge(M1,[X|M2],M).
merge([X|M1],[Y|M2],[X|M]):-less(X,Y),merge(M1,[Y|M2],M).
merge([X|M1],[Y|M2],[Y|M]):-less(Y,X),merge([X|M1],M2,M).
```

Let us treat the correctness of the mergesort program MERGE above. We start with the definition of predicate merge. Let MERGE be the logic program that consists of the last five program formulas above. Can we easily say what $M_{MERGE}$ is? Since the process of merging two not necessarily ordered lists together may be rather intricate, the best we can say is:

Ground atom merge(l,m,n) is an element of $M_{MERGE}$ iff n results from l and m by application of the algorithm described by program MERGE.

This is rather unsatisfactory since it says nothing more than 'MERGE does what it does'. What we would like to have is a more abstract description of the meaning of program MERGE. Such a description may be obtained iff we restrict discussion to sorted input lists l and m. Let us define

Restr$_1$={merge(l,m,n) | l and m are sorted lists}.

Then we claim that $M_P \cap$Restr$_1$=$I_1$ with the Herbrand interpretation $I_1$ consisting of the following ground facts:

*   All ground facts less(a,b) from set LESS.

*   All ground facts merge(l,m,n) where l,m and n are ordered lists and n is a permutation of the concatenation l•m.

$I_1 \subseteq M_{MERGE} \cap$Restr$_1$ is easily shown by induction on the sum of the lengths of lists l and m in a ground atom merge(l,m,n). $M_{MERGE} \cap$Restr$_1 \subseteq I_1$ is obtained by application of Theorem 8.1. Both conditions from the theorem are readily checked. The inheritance condition reduces to the fact that sortedness of [X|L] implies sortedness of L. Relativized correctness with respect to set Restr$_1$, for example of the fifth program formula of MERGE, reduces to the observation that [Y|M] is sorted and and a permutation of [X|M1]•[Y|M2], provided that [X|M1] and [Y|M2] are sorted, less(Y,X) holds, M is sorted and a permutation of [X|M1]•M2.

Next we deal with program DIVIDE defining predicate divide.

---

**Exercise 8.7**   Let Restr$_2$ be the set of all ground atoms divide(l,m,n) with list l. Show that $M_{DIVIDE} \cap$Restr$_2$=$I_2$ with the Herbrand interpretation $I_2$ consisting of all ground facts

divide([a$_1$,b$_1$,...,a$_n$,b$_n$],[a$_1$,...,a$_n$],[b$_1$,...,b$_n$])

divide([a$_1$,b$_1$,...,a$_n$,b$_n$,a$_{n+1}$],[a$_1$,...,a$_n$,a$_{n+1}$],[b$_1$,...,b$_n$]).

---

Finally, we show that $M_P \cap \text{Restr}=I$ for the complete program P and set Restr consisting of all ground atoms

- mergesort(l,m) with list l
- divide(l,m,n) with list l
- merge(l,m,n) with sorted lists l and m

and the Herbrand interpretation $I$ consisting of all ground atoms from $I_1 \cup I_2$ and ground atoms mergesort(l,m,n) such that l, m and n are lists and n is a sorted permutation of l•m. We concentrate on the claim $M_P \cap \text{Restr} \subseteq I$ and apply Theorem 8.1. Relativized correctness of the program formulas with respect to Restr is easily shown. Let us see what the inheritance condition expresses. Consider a ground instance

> mergesort([x,y|l],m):-
>                 divide([x,y|l],l1,l2),mergesort(l1,m1),
>                 mergesort(l2,m2),merge(m1,m2,m).

The ordering of ground atoms of the rule body that we use is divide([x,y|l],l1,l2), mergesort(l1,m1), mergesort(l2,m2), merge(m1,m2,m). The following implications are to be checked:

- mergesort([x,y|l],m)∈ Restr implies divide([x,y|l],l1,l2)∈ Restr  (trivial)
- divide([x,y|l],l1,l2)∈ $I$ implies mergesort(l1,m1)∈ Restr  (trivial)
- divide([x,y|l],l1,l2)∈ $I$ and   mergesort(l1,m1)∈ $I$ imply mergesort(l1,m1) ∈ Restr  (trivial)
- divide([x,y|l],l1,l2)∈ $I$,   mergesort(l1,m1)∈ $I$   and   mergesort(l1,m1)∈ $I$ imply merge(m1,m2,m)∈ Restr (this is the only non-trivial part shown as follows: mergesort(l1,m1)∈ $I$ and mergesort(l1,m1)∈ $I$ imply that m1 and m2 are sorted lists, thus mergesort(l1,m1)∈ Restr)

## 8.8  Quicksort

An input list L consisting of elements of a partially ordered set (D,<) is sorted as follows: select an arbitrary (pivot) element P from list L, then split L into the three sublists L1, L2 and L3 which consist of all list elements X which are less than P, equal to P and greater than P respectively (unless L is empty), recursively apply the quicksort algorithm to L1 and L3 to obtain sorted lists M1 and M3 which are permutations of L1 and L3, respectively, and finally concatenate M1, L2 and M3 to obtain a final sorted list M. Thus we may implement mergesort as a logic program as follows:

```
quicksort([],[]).
quicksort([P],[P]).
quicksort([P,X|L],M):-
          split(P,[P,X|L],[],[],[],L1,L2,L3),
          quicksort(L1,M1),
          quicksort(L3,M3),
          append(M1,L2,Laux),
          append(Laux,M3,M).

split(P,[],A,B,C,A,B,C).
split(P,[X|L],A,B,C,U,V,W):-
          less(X,P),
          split(P,L,[X|A],B,C,U,V,W).
split(P,[P|L],A,B,C,U,V,W):-
          split(P,L,A,[X|B],C,U,V,W).
split(P,[X|L],A,B,C,U,V,W):-
          less(P,X),
          split(P,L,A,B,[X|C],U,V,W).
```

Note what happens when quicksort arrives at a unit list [D]. Split computes L1=[], L2=[D], L3=[]. Then it halts with M=[]•[D]•[]=[D].

---

**Exercise 8.8**    Verify the above quicksort program along the lines presented in the Section 8.7.

**Exercise 8.9**    A further sorting algorithm, called selection sort, sorts a list L as follows: select the least element in L and put it at the first place in the output list, then put the second least element at the second place, and so on. Define selection sort by a logic program.

---

## 8.9    Path searching in a directed graph

In Section 7.2 a logic program was discussed that, given a directed graph via a set EDGE of ground facts edge(a,b), defined a predicate path(X,Y) with the declarative semantics 'there exists a path from node X to node Y'.

The disadvantage of this predicate is that in a graph with cycles infinite SLD-derivations are possible. To avoid this, we may introduce a predicate restrictedPath(X,Y,L) with the intended meaning 'there exists a path from X to Y with interior nodes not taken from list L' as follows:

```
restrictedPath(X,X,L).
restrictedPath(X,Y,L):-
            edge(X,Z),
            new(Z,L),
            restrictedPath(Z,Y,[X|L]).

new(Z,[]).
new(Z,[X|L]):-
            unequal(Z,X),
            new(Z,L).
```

+ a definition of inequality

Defining path(X,Y):-restrictedPath(X,Y,[]) then keeps track of the nodes and avoids using a node twice along the considered actual path. With this new definition, SLD-derivations starting with goals ?-path(x,y) are always finite.

# Chapter 9
# Verification of Logic Programs

9.1        Validity of formulas in least Herbrand models

9.2        Verification of asserted predicates

Do logic programs require verification? Verification of programs always means verification against a specification of the desired behaviour of the program, which preferably should be expressed on a more abstract level than the program itself.

In the earlier days of logic programming some authors claimed that 'Logic programs are their own specifications, and thus are automatically correct and need not be verified'. Such statements are driven by the opinion that logic programs are the ultimate level of abstraction, behind which the level of informal requirement specifications begins.

Already, our first experience with logic programming and verification in Chapter 7 should have shown that such a view of logic programming is rather unrealistic. It is often possible to describe the behaviour of a logic program in a more abstract and perspicuous way by using higher languages like full predicate logic, or an even stronger logic.

Even within Horn logic itself there are lots of levels of abstraction. Think of the abstract, but inefficient version of sorting, in comparison with the two proposed efficient implementations by mergesort and quicksort. Or think of the first program searching for a path in a directed path which is not even effective, in the sense that it may loop within a cycle of the given graph. (This idea leads to the verification of a logic program against another one.) Taking into account the impure procedural predicates that will be discussed in Chapter 10 (leading to Prolog), the necessity for abstract specifications and verification will become still more urgent.

To date, the verification of logic programs is still a matter in evolution. We will therefore restrict discussion to those aspects of verification of logic programs, whose *raisons d'être* are certainly not controversial, and which have already been mentioned in prior chapters.

In contrast to the verification methods that were discussed in Chapter 8, which referred to infinite mathematical objects like Herbrand interpretation, emphasis is put here on **finitary methods** as a formal basis for semi-automated systems for the verification of logic programs.

## 9.1     Validity of formulas in least Herbrand models

The sort of problem we will treat is the following:

- Show $M_P \models \varphi$, for a logic program P and a predicate logic formula $\varphi$.

As an example, consider the joint logic program P from Sections 8.6 and 8.7 defining predicates sort and mergesort. We might wish to prove that

$$M_P \models \forall L \forall M(\text{mergesort(L,M)} \leftrightarrow \text{sort(L,M)})$$

thus claiming the equivalence of predicates mergesort and sort.

We might also consider the joint logic program Q from Sections 7.2 and 8.9 that defines predicates path and restrictedPath. Then we might wish to show that

$$M_Q \models \forall X \forall Y(\text{path(X,Y)} \leftrightarrow \text{restrictedPath(X,Y,[])})$$

thus showing that restrictedPath 'correctly implements' predicate path.

As was said in the introduction to this chapter, we are interested in finitary methods to show $M_P \models \varphi$, that is, methods that do not refer to the infinite object $M_P$, but manipulate only the finite objects P and $\varphi$. A first attempt (that will prove to be unsuccessful) would be to replace claim $M_P \models \varphi$ by claim $P \vdash \varphi$, where $\vdash$ denotes derivability in any sound and complete calculus for predicate logic. This causes several difficulties that will be described next.

### 9.1.1     Validity in the least Herbrand models of logic program P versus derivability from P

What is the relation between claims $M_P \models \varphi$ and $P \vdash \varphi$?

**Fact**   $P \vdash \varphi$ implies $M_P \models \varphi$.

*Proof*   Clear since $M_P$ is a model of logic program P.                    ■

Can we expect the converse to be true?

**Fact**   If $\varphi$ is a ground atom then $M_P \models \varphi$ implies $P \vdash \varphi$.

*Proof*   Clear from the definition of $M_P$.                    ■

### Corollary

If $\varphi$ is a variable free formula without quantifiers containing only conjunction and disjunction as propositional connectives, then $M_P \models \varphi$ implies $P \vdash \varphi$.

*Proof*  By structural induction on $\varphi$.                                    ∎

Note that our claim becomes incorrect if we allow universal quantifiers or negation or implication. As an example, consider logic program $P=\{p(a),$ $p(b), q(a)\}$. Then, $M_P=P$. Hence, $M_P$ is a model of $\neg q(b)$ and also a model of $\forall X\, p(X)$. Of course, neither one of these formulas can be derived from P.

Concerning negation, this is because the definition of $M_P$ interprets logic program P under the so-called **closed world assumption**:

The ground atoms true in $M_P$ are those that follow from P, and *only* those.

The standard way to incorporate such 'negative information' is to use the **completion** of logic program P instead of P. We shall define and investigate this concept in the next subsection.

Concerning universal quantification, the described weakness of P as a set of first-order formulas results from the fact that the definition of $M_P$ interprets P under the **term-generatedness assumption**:

The domain of $M_P$ is the set of all ground terms.

Again, this is a sort of negative information not present within P as a set of first-order axioms. In the example above, we could have captured the concept of term-generatedness in a simple manner by the **domain-closure axiom**:

$\forall X(X=a \vee X=b)$

More generally, if there is only a finite number of ground terms in the considered signature, say $t_1,...,t_n$, then we can always express the concept of term-generatedness by the axiom $\forall X(X=t_1 \vee ... \vee X=t_n)$. Of course, usually there is an infinite number of ground terms. Then we cannot formulate a domain-closure axiom as above. Let us consider an example. Let P consist of program formulas $add(X,0,X)$ and $add(X,succ(Y),succ(Z)):-add(X,Y,Z)$ that defines addition on natural numbers. Then, $M_P$ is a model of formula

$\forall X \forall Y \forall Z(add(X,Y,Z) \rightarrow add(Y,X,Z))$.

As was shown in Chapter 2, the considered formula does not follow from P. What may we offer now as a counterpart to the domain-closure axiom from above that allows the derivation of our formula from P? Induction will help us. Several sorts of induction, structural and well-founded induction, will be introduced in Subsection 9.1.3.

## 9.1.2    Closed-world assumption and the completion of a logic program

Consider a logic program P consisting of a single rule $p(X):-q(X)$ defining predicate p together with rules that define predicate q. The considered rule for p expresses that every X satisfying q also satisfies p. The closed-world assumption supplements this by saying that there are no further X satisfying p. This could be easily grasped by turning the implication in $p(X):-q(X)$ into an equivalence: $p(X) \leftrightarrow q(X)$. This equivalence will be called the completion of the considered definition of predicate p.

In the general case, there may be several program formulas that define predicate p. Furthermore, there may be rule heads of the form $p(t)$ with a composed term t instead of a simple variable. Finally, the bodies of the rules that define p may contain variables that do not occur in the corresponding head. This makes it necessary to define the completion of a predicate p more carefully.

> **Definition 9.1:**    Let P be a logic program. The **completion** of P, denoted by compl(P), is constructed in the following steps as a set of predicate logic formulas with equality.
>
> $\quad$ *Step 1* $\quad$ Every program formula $p(t_1,\ldots,t_n):-B_1,\ldots,B_m$ of P is replaced by the following equivalent formula of predicate logic with equality (using fresh variables $X_1,\ldots,X_n$):
>
> $\quad p(X_1,\ldots,X_n) \leftarrow (X_1 = t_1 \wedge \ldots \wedge X_n = t_n \wedge B_1,\ldots,B_m)$
>
> $\quad$ *Step 2* $\quad$ Now we collect all program formulas that define the same predicate and make explicit quantifiers: for a predicate symbol p, let
>
> $\quad p(X_1,\ldots,X_n):-B_{11},\ldots,B_{1m(1)}$
>
> $\qquad \vdots$
>
> $\quad p(X_1,\ldots,X_n):-B_{k1},\ldots,B_{km(k)}$
>
> be all the program formulas defining predicate p that are obtained after step 1. Of course, we may assume that the same set $X_1,\ldots,X_n$ of fresh formulas is used in all these program formulas.
>
> $\quad$ *Step 3* $\quad$ Make quantifiers explicit. For a formula $p(X_1,\ldots,X_n):-B_{i1},\ldots,B_{im(i)}$ let $Y_{i1},\ldots,Y_{il(i)}$ be the variables other than $X_1,\ldots,X_n$, that occur in the program formulas obtained after step 2 that define p. Note that the considered program formulas read as predicate logic formulas as follows:

$$\forall X_1...\forall X_n \forall Y_{i1}...\forall Y_{il(i)}(p(X_1,...,X_n) \leftarrow (B_{i1} \wedge ... \wedge B_{im(i)}))$$

*Step 4*   Move quantifiers into the definition of p. From Chapter 2, the formula from step 3 may be equivalently rewritten into the formula

$$\forall X_1...\forall X_n(p(X_1,...,X_n) \leftarrow \exists Y_{i1}...\exists Y_{il(i)}(B_{i1} \wedge ... \wedge B_{im(i)}))$$

*Step 5*   A set of rules and facts is interpreted conjunctively. So consider the single formula that is the conjunction of all the formulas from step 4, or equivalently the formula

$$\forall X_1...\forall X_n(p(X_1,...,X_n) \leftarrow$$

$$(\exists Y_{11}...\exists Y_{1l(1)}(B_{11} \wedge ... \wedge B_{1m(1)}) \vee ... \vee$$

$$\exists Y_{k1}...\exists Y_{kl(k)}(B_{k1} \wedge ... \wedge B_{km(k)})))$$

(For k=0, the right-hand side of this equivalence is a non-satisfiable formula false.)

*Step 6*   Replace $\leftarrow$ by $\leftrightarrow$ in the formula from step 5 to obtain

$$\forall X_1...\forall X_n(p(X_1,...,X_n) \leftrightarrow$$

$$(\exists Y_{11}...\exists Y_{1l(1)}(B_{11} \wedge ... \wedge B_{1m(1)}) \vee ... \vee$$

$$\exists Y_{k1}...\exists Y_{kl(k)}(B_{k1} \wedge ... \wedge B_{km(k)})))$$

*Step 7*   Axiomatize equality. Let Eq be the following set of equality axioms: for every n-ary predicate p and n-ary function symbol f:

$X=X$

$X=Y \rightarrow Y=X$

$X=Y \rightarrow (Y=Z \rightarrow X=Z)$

$(X_1=Y_1 \wedge ... \wedge X_n=Y_n) \rightarrow f(X_1,...,X_n)=f(Y_1,...,Y_n)$

$(X_1=Y_1 \wedge ... \wedge X_n=Y_n) \rightarrow (p(X_1,...,X_n) \rightarrow p(Y_1,...,Y_n))$

So far, we have collected the usual equality axioms. Eq also contains the following axioms expressing that we are dealing with a free term algebra: for any two different function symbols f of arity n and g of arity m and every term t(X) which contains X and is not a variable:

$\neg f(X_1,...,X_n)=g(Y_1,...,Y_m)$

$f(X_1,...,X_n)=f(Y_1,...,Y_n) \rightarrow (X_1=Y_1 \wedge ... \wedge X_n=Y_n)$

$\neg X(t)=t$

The formulas obtained after step 6, together with the equality axioms from step 7, form the **completion** compl(P) of program P.

As an example, let us complete the program formulas defining predicates mergesort from the logic program from Section 8.7. The original definition of mergesort was:

    mergesort([],[]).
    mergesort([X],[X]).
    mergesort([X,Y|L],M):-
            divide([X,Y|L],L1,L2),
            mergesort(L1,M1),
            mergesort(L2,M2),
            merge(M1,M2,M).

*After step 1:*

    mergesort(U,V)←(U=[]∧V=[])
    mergesort(U,V)←(U=[X]∧V=[X])
    mergesort(U,V)←(U=[X,Y|L]∧
                    V=M∧divide([X,Y|L],L1,L2)∧
                    mergesort(L1,M1)∧
                    mergesort(L2,M2)∧
                    merge(M1,M2,M))

*After step 5:*

    mergesort(U,V)←(U=[]∧V=[])
    mergesort(U,V)←∃X(U=[X]∧V=[X])
    mergesort(U,V)←∃X ∃Y ∃L ∃M ∃L1 ∃L2 ∃M1 ∃M2
                    ( U=[X,Y|L]∧
                      V=M∧
                      divide([X,Y|L],L1,L2)∧
                      mergesort(L1,M1)∧
                      mergesort(L2,M2)∧
                      merge(M1,M2,M))

*After step 6:*

    mergesort(U,V)↔(U=[]∧V=[])∨∃X(U=[X]∧V=[X])∨
                    ∃X ∃Y ∃L ∃M ∃L1 ∃L2 ∃M1 ∃M2
                    (U=[X,Y|L]∧
                     V=M∧
                     divide([X,Y|L],L1,L2)∧
                     mergesort(L1,M1)∧
                     mergesort(L2,M2)∧
                     merge(M1,M2,M))

The main point about compl(P) is that is does not allow us to derive any more positive information than P did. It may well allow negative information (negated formulas) to be derived, whereas P does not allow any negative information to be derived.

### Theorem 9.1

(1)    P follows from compl(P).

(2)    $M_P$ is a model of compl(P).

(3)    For every ground atom A (without equality), A follows from P iff A follows from compl(P).

*Proof*

(1)    Using the direction $\leftarrow$ of the formulas from step 6, suitable equality axioms and standard equivalences of predicate logic, we may derive $p(t_1,\ldots,t_n){:-}B_1,\ldots,B_m$ from

$$\forall X_1\ldots\forall X_n(p(X_1,\ldots,X_n)\leftrightarrow$$

$$(\exists Y_{11}\ldots\exists Y_{1l(1)}(B_{11}\wedge\ldots\wedge B_{1m(1)})\vee\ldots\vee$$

$$\exists Y_{k1}\ldots\exists Y_{kl(k)}(B_{k1}\wedge\ldots\wedge B_{km(k)})))$$

So, P follows from compl(P).

(2)    Obviously, $M_P$ is a model of Eq. The direction $\leftarrow$ of the formulas from step 2 is obvious. The reverse implication is shown as follows. Consider the formula constructed in step 2 for a predicate p. Assume that $s_1,\ldots,s_n$ are ground terms such that $p(s_1,\ldots,s_n)$ is valid in $M_P$. Hence, $p(s_1,\ldots,s_n)$ is an element on $I_{n+1}(P)$, for some number n. This means that there is a ground instance $p(t_1,\ldots,t_n)\sigma{:-}B_1\sigma,\ldots,B_m\sigma$ of a program formula $p(t_1,\ldots,t_n){:-}B_1,\ldots,B_m$ of P such that $p(t_1,\ldots,t_n)\sigma=p(s_1,\ldots,s_n)$ and $B_1\sigma,\ldots,B_m\sigma\in I_n(P)$. Thus, $M_P$ is a model of $B_1\sigma,\ldots,B_m\sigma$. Let $Y_1,\ldots,Y_l$ be the variables occurring in $p(t_1,\ldots,t_n){:-}B_1,\ldots,B_m$. Let $X_1,\ldots,X_n$ be the fresh variables chosen in step 2. $M_P$ is also a model of $s_1=t_1\sigma\wedge\ldots\wedge s_n=t_n\sigma$. Hence, $M_P$ is a model of $(s_1=t_1\wedge\ldots\wedge s_n=t_n\wedge B_1\wedge\ldots\wedge B_m)\sigma$. As we know from predicate logic, $M_P$ is a model of $\exists Y_1\ldots\exists Y_l(X_1=t_1\wedge\ldots\wedge X_n=t_n\wedge B_1\wedge\ldots\wedge B_m)$. This proves the desired direction $\rightarrow$ of the formula constructed in step 2.

(3)    Finally, the same ground atoms A without equality follow from P as from compl(P). One direction is obtained from (1), the other follows from part (2).    ■

## 9.1.3    Structural and well-founded induction

**Definition 9.2:**    Let $\Sigma$ be a finite signature whose function symbols are $f_1,...,f_n$. For an n-ary function symbol f in $\Sigma$ and a $\Sigma$-formula $\varphi(X)$ with a free variable X let closed($\varphi$,f) be the following formula:

$$\forall X_1...\forall X_n((\varphi(X_1)\wedge...\wedge\varphi(X_n))\rightarrow\varphi(f(X_1,...,X_n)))$$

In particular, for a constant c, closed($\varphi$,c) is the formula $\varphi$(c). The variables $X_1,...,X_n$ have to be chosen in such a way that all occurring substitutions are admissible. The **structural induction formula** for $\varphi$ is the following formula SInd($\varphi$):

$$((closed(\varphi,f_1)\wedge...\wedge closed(\varphi,f_n))\rightarrow\forall X\varphi(X))$$

By SInd we denote the set of all formulas SInd($\varphi$) as above.

**Fact**    Every Herbrand interpretation over signature $\Sigma$ is a model of SInd.

*Proof*    Clear, since Herbrand interpretations are term generated.    ∎
     Structural induction is not always applicable in concrete examples. Proof of a property E(L), for all lists L, by structural induction requires showing E([]) and concluding E([X|L]) from E(L). Sometimes, it is more convenient to use induction on the length of list L, that is, to show E(L) under the assumption that E(L') has been shown for all lists L' of shorter length than L. This sort of induction is called **well-founded induction** and formalized as follows.

**Definition 9.3:**    Let $\Sigma$ be a signature with a distinguished 2-ary relation symbol <, written in infix notation. For a $\Sigma$-formula $\varphi(X)$ with a free variable X define the **well-founded induction formula** $WInd_<(\varphi)$ of $\varphi(X)$ with respect to < to be the formula

$$(\forall X(\forall Y(Y<X \rightarrow \varphi(Y))\rightarrow\varphi(X))\rightarrow\forall X\varphi(X))$$

'In order to prove $\forall X\varphi(X)$ it suffices to prove $\varphi(X)$ for arbitrary X under the assumption that $\varphi(Y)$ is already shown for every Y<X'.By $WInd_<$ we denote the set of all formulas $WInd_<(\varphi)$ as above.

**Fact**    Formula $WInd_<$ is true in every interpretation that interprets < by a well-founded relation.

*Proof*    Let < be interpreted by a well-founded relation $<_A$ in a given interpretation $A$. Assume that $\text{WInd}(\Sigma,<,\varphi)$ is not valid in $A$ in a given state sta. This means that formula $\forall X(\forall Y(Y<X\rightarrow\varphi\{X/Y\})\rightarrow\varphi)$ is valid in $A$ in state sta, but $\forall X\varphi$ is not. Hence there is an element $a_0$ of $A$ that does not satisfy $\varphi(X)$. Instantiating $a_0$ for X in $\forall X(\forall Y(Y<X\rightarrow\varphi\{X/Y\})\rightarrow\varphi)$, we obtain an element $a_1<_A a_0$ of $A$ that does not satisfy $\varphi(X)$, either. Proceeding this way we obtain an infinite decreasing sequence, contradicting the well-foundedness of $<_A$.    ∎

As an example, induction on the length of lists is the same as well-founded induction w.r.t. the relation < defined by the following logic program Ext:

[]<[Y|M]
[X|L]<[Y|M]:-L<M.

Obviously, the interpretation of < in an arbitrary Herbrand model of Ext is the relation 'shorter' between lists, a well-founded relation.


### 9.1.4    Conservative extensions of logic programs

A common experience in inductive proofs is that auxiliary predicates may be required to express suitable 'induction invariants'. We will provide for the possibility of introducing such auxiliary predicates with the idea of a conservative extension. 'Conservativity', of course, refers to the need to preserve the semantics of the old predicates.


**Definition 9.4:**
(1)    Let Q be a logic program and $\Omega$ be a set of predicates. We say that Q is **conservative with respect to** $\Omega$ iff Q does not define predicates from $\Omega$. Thus, the facts and rules of Q may use predicates from $\Omega$, but are not allowed to define them.
(2)    Let P and Q be logic programs over the same set of function symbols. We say that $P\cup Q$ is a **conservative extension** of P iff Q is conservative with respect to the set of predicates that occur in P.


*Lemma 9.1*

Assume that P and Q are logic programs over the same set of function symbols such that $P\cup Q$ is a conservative extension of P. Denote by H(P) the set of all ground atoms $p(t_1,...,t_n)$ with predicates p occurring in P. Then $M_{P\cup Q}\cap H(P)=M_P$, that is, $P\cup Q$ does not allow the derivation of more ground atoms from the signature of P than P does. Extending P by Q is thus conservative in the sense that it does not destroy the semantics of P.

*Proof*   $M_P\subseteq M_{P\cup Q}\cap H(P)$ is trivial, so we are left with the reversed inclusion. This is shown by using the approximations $I_n(P\cup Q)$ to $M_{P\cup Q}$. Assume that $I_n(P\cup Q)\cap H(P)\subseteq M_P$ has already been shown. Consider some $A\in I_{n+1}(P\cup Q)\cap H(P)$. If $A\in I_n(P\cup Q)\cap H(P)$ we are done. Otherwise, there is a ground instance $B\sigma\text{:-}B_1\sigma,...,B_m\sigma$ of a program formula of $P\cup Q$ with $A=B\sigma$ and $B_1\sigma,...,B_m\sigma\in I_n(P\cup Q)$. Since $A=B\sigma\in H(P)$ and Q is conservative with respect to the set of predicates of P, we know that $B\sigma\text{:-}B_1\sigma,...,B_m\sigma$ must be a ground instance of a program formula of P. Hence $B_1\sigma,...,B_m\sigma\in I_n(P\cup Q)\cap H(P)$, and inductively we may conclude that $B_1\sigma,...,B_m\sigma\in M_P$. This implies that $A=B\sigma\in M_P$, too.                                          ∎

### Theorem 9.2

Assume that P and Q are logic programs over the same set of function symbols such that $P\cup Q$ is a conservative extension of P. Then for all formulas φ which contain only the predicates from P the following is true:

$$M_P\models\varphi\ \text{iff}\ M_{P\cup Q}\models\varphi$$

*Proof*   For ground atoms φ this is the statement of Lemma 9.1. Now the theorem is easily shown by induction on the number of logical symbols in φ, using (for the case of quantifiers) the fact that the domains of $M_P$ and $M_{P\cup Q}$ coincide and consist of the set of all ground terms.                                          ∎

Hence, in order to show claim $M_P\models\varphi$ we may always freely introduce auxiliary predicates as long as they are defined within a conservative extension of P.


## 9.1.5    Verification strategy for validity in $M_P$

- Choose a suitable conservative extension $P\cup Q$ of P.

- If $WInd_<$ is used, show that predicate symbol $<$ is interpreted in every Herbrand model of $P\cup Q$ by a well-founded relation.

- Show that $\text{compl}(P \cup Q) \cup \text{SInd} \cup \text{WInd}_< \vdash \varphi$.

To show well-foundedness of $<$ is usually not as a hard problem as one might expect. The proof theory described here is sound with respect to the original claim $M_P \models \varphi$. It is not complete but, in general, seems to be sufficient for a wide range of practical problems.

## 9.2    Verification of asserted predicates✀

A common situation is that we want to show validity in $M_P$ of a universally quantified implication, for example, in Section 9.1 where we treated the claim $M_P \models \forall L \forall M(\text{mergesort}(L,M) \rightarrow \text{sort}(L,M))$. Similarly, if we want to show that $M_P \cap \text{Restr} \subseteq M_Q$, where P, Q and R are logic programs and Restr is a special restriction set, namely the set of all ground atoms of $M_R$ with predicate symbol p (see Chapter 8 for problems of this type), we may proceed as follows. Rename all predicates q occurring in Q into q', and all predicates r in R into r''. Then show that $M_{P \cup Q' \cup R''}$ is a model of the formula

$$\forall X_1 \ldots \forall X_n((p'(X_1,\ldots,X_n) \wedge p(X_1,\ldots,X_n)) \rightarrow p''(X_1,\ldots,X_n))$$

Formulas of this type are investigated in Bossi and Cocco (1989) under the name 'module specifications'. There, they are written in a suggested form resembling formulas in Hoare's logic (see Chapters 14-18), namely as

$$\{p'(X_1,\ldots,X_n)\}p(X_1,\ldots,X_n))\{p''(X_1,\ldots,X_n)\}$$

Here, $p'(X_1,\ldots,X_n)$ is read as a **precondition** that must be satisfied before processing predicate p, and $p''(X_1,\ldots,X_n)$ as a **postcondition** that holds after satisfaction of p. More generally, we will consider formulas $\{\varphi\}p(X_1,\ldots,X_n)\{\psi\}$, called **asserted predicates**.

The advantage of dealing with asserted predicates is that in proving a claim of the form $M_P \models \{\varphi\}p(X_1,\ldots,X_n)\{\psi\}$ we need not start by applying the verification strategy described in Subsection 9.1.5 (a task that may require considerable effort), but may decompose the problem into smaller subproblems of the same type by a special verification strategy for asserted predicates.

**Definition 9.5:**    An **asserted predicate** is a string of the form $\{\varphi\}p(X_1,\ldots,X_n)\{\psi\}$, where p is a predicate and $\varphi$ and $\psi$ are arbitrary formulas. $\{\varphi\}p(X_1,\ldots,X_n)\{\psi\}$ is taken as an abbreviation for the formula $((\varphi \wedge p(X_1,\ldots,X_n)) \rightarrow \psi)$.

**Definition 9.6: (Verification strategy for asserted predicates)**     Let
P be a logic program and $\{\varphi\}p(X_1,...,X_n)\{\psi\}$ be an asserted predicate.
In order to show the validity of asserted predicate $\{\varphi\}p(X_1,...,X_n)\{\psi\}$
in $M_P$ we may proceed as follows:

*Step 0: Variable condition*     Let the free variables of $\varphi$ and $\psi$
other than $X_1,...,X_n$ be $Y_1,...,Y_l$. Check that $Y_1,...,Y_l$ do not occur in
logic program P. (Variables $X_1,...,X_n$ are allowed to occur free in $\varphi$ and
$\psi$.) Also check that there is no quantification in $\varphi$ and $\psi$ over a variable
that occurs in P. This may require the renaming of some variables.

*Step 1: Consistency with the facts*     Show that for every fact
$p(t_1,...,t_n)$ of P

$$M_P \models ( \varphi\{X_1/t_1,...,X_n/t_n\} \rightarrow \psi\{X_1/t_1,...,X_n/t_n\} )$$

*Step 2: Intermediate assertions for the rules*     For every rule
$p(t_1,...,t_n):-p_1(t_{11},...,t_{1n(1)}),...,p_k(t_{k1},...,t_{kn(k)})$ of P, equip its body
with 'intermediate assertions' $\chi_0,...,\chi_k$ as follows:

$\{\chi_0\}$
  $p_1(t_{11},...,t_{1n(1)})$
$\{\chi_1\}$
  $p_2(t_{21},...,t_{2n(2)})$
$\{\chi_2\}$

  .
  .
  .

  $p_{k-1}(t_{k-11},...,t_{k-1n(k-1)})$
$\{\chi_{k-1}\}$
  $p_k(t_{k1},...,t_{kn(k)})$
$\{\chi_k\}$

with $\chi_0=\varphi\{X_1/t_1,...,X_n/t_n\}$ and $\chi_k=\psi\{X_1/t_1,...,X_n/t_n\}$.

*Step 3: Match of intermediate assertions with pre- and postcon-
dition*     Given an asserted rule body as in step 2 and an $i=1,...,n$ with
$p_i=p$, show

$$M_P \models ((\chi_0\wedge...\wedge\chi_{i-1})\rightarrow\varphi\{X_1/t_{i1},...,X_n/t_{in(i)}\})$$     'match pre'

$$M_P \models ((\psi\{X_1/t_{i1},...,X_n/t_{in(i)}\}\wedge\chi_0\wedge...\wedge\chi_{i-1})\rightarrow \chi_i)$$     'match post'

*Step 4: Subverification*    Given an asserted rule body as in step 2 and an i=1,...,k with $p_i \neq p$, show that

$$M_P \models \{(\chi_0 \wedge \ldots \wedge \chi_{i-1})\}\, p_i(t_{i1},\ldots,t_{in(i)})\, \{\chi_i\}$$

Here, we may apply the same strategy again or the one described in Subsection 9.1.5.

## Remarks

(1)     The idea behind this verification strategy can be described informally as follows. We want to show the validity in $M_P$ of asserted predicate $\{\varphi\}p(X_1,\ldots,X_n)\{\psi\}$. We separately investigate the two possibilities of satisfying $p(X_1,\ldots,X_n)$, namely via unification with a fact $p(t_1,\ldots,t_n)$ or with the head of a rule $p(t_1,\ldots,t_n):\text{-}p_1(t_{11},\ldots,t_{1n(1)}),\ldots,p_k(t_{k1},\ldots,t_{kn(k)})$. The first case obviously leads to a check of the 'consistency' condition of step 1. To treat the case of a rule, we start with precondition $q_0 = \varphi\{X_1/t_1,\ldots,X_n/t_n\}$ and work through the rule body collecting more and more information, obtained by satisfaction of the literals of the rule's body. This leads to intermediate formulas $q_1,\ldots,q_{k-1}$.

Let us see what happens when we arrive at literal $p_i(t_{i1},\ldots,t_{in(i)})$. The information accumulated so far is $(\chi_0 \wedge \ldots \wedge \chi_{i-1})$. If $p_i$ is different from our asserted predicate $p$, we are left with a subverification of asserted predicate $\{(\chi_0 \wedge \ldots \wedge \chi_{i-1})\}p_i(t_{i1},\ldots,t_{in(i)})\{\chi_i\}$. The interesting case is that $p_i$ coincides with $p$ (recursive call of $p$). Inductively we may assume that

$$\{\varphi\{X_1/t_{i1},\ldots,X_n/t_{in(i)}\}\}p(t_{i1},\ldots,t_{in(i)})\{\psi\{X_1/t_{i1},\ldots,X_n/t_{in(i)}\}\}$$

has already been verified. (As the proof below will show, the appropriate sort of 'induction' will be induction on the length of a proof for literal $p(t_{i1},\ldots,t_{in(i)})$.) Then we are left with the task of showing that the information $(\chi_0 \wedge \ldots \wedge \chi_{i-1})$ that is present at the considered stage is stronger than the required precondition $\varphi\{X_1/t_{i1},\ldots,X_n/t_{in(i)}\}$ above, and that the information obtained after satisfaction of $p(t_{i1},\ldots,t_{in(i)})$, namely $(\psi\{X_1/t_{i1},\ldots,X_n/t_{in(i)}\} \wedge \chi_0 \wedge \ldots \wedge \chi_{i-1})$, is stronger than the desired information $\chi_i$. This is expressed by the properties 'match pre' and 'match post'.

(2)     What is the 'variable condition' in the strategy above good for? First, it guarantees that all substitutions that are applied in steps 1-3 are in-

deed admissible. Let us see what happens if we violate it. Let logic program P consist of facts $q(a,b)$ and $p(Y_1)$, with constants a and b.

*Example 1*: $M_P$ is not a model of asserted predicate

$$\{q(Y_1,X_1)\}p(X_1)\{q(X_1,X_1)\},$$

though condition 'consistency from step 1, namely $(q(Y_1,Y_1)\rightarrow q(Y_1,Y_1))$, is valid. Since there are no rules in P, steps 2-4 are empty. The failure of our strategy is due to the fact that variable $Y_1$ occurred in program P.

*Example 2*: $M_P$ is not a model of asserted predicate

$$\{\exists X_1 q(X_1,Y_1)\}p(X_1)\{\exists Y_1 q(Y_1,Y_1)\},$$

though condition 'consistency', namely $(\exists X_1 q(Y_1,Y_1)\rightarrow\exists X_1 q(Y_1,Y_1))$ is valid. The failure of our strategy is due to the fact that quantified variable $X_1$ occurs in program P. This leads to a substitution $\{Y_1/X_1\}$ whose application to formula $\exists X_1 q(X_1,Y_1)$ is not admissible.

(3)     The verification strategy described above may be generalized by allowing us to equip *several predicates simultaneously* with pre- and post-conditions. This is useful if we deal with predicates defined by mutual recursion.


**Theorem 9.3**

If the verification strategy is successfully applied to logic program P and asserted predicate $\{\varphi\}p(X_1,...,X_n)\{\psi\}$, then $\{\varphi\}p(X_1,...,X_n)\{\psi\}$ is valid in $M_P$.

*Proof*   Let the free variables of $\varphi$ and $\psi$ other than $X_1,...,X_n$ be $Y_1,...,Y_l$. Assume that $Y_1,...,Y_l$ do not occur in logic program P and that there is no quantification in $\varphi$ and $\psi$ over a variable that occurs in P. The former condition implies that, for all terms $r_1,...,r_n$ and every substitution $\rho$ that acts only on variables of P (thus does not act on free variables $Y_1,...,Y_l$ of $\varphi$ and $\psi$) we obtain:

(1)     $\varphi\{X_1/r_1,...,X_n/r_n\}\rho=\varphi\{X_1/r_1\rho,...,X_n/r_n\rho\}$

(2)     $\psi\{X_1/r_1,...,X_n/r_n\}\rho=\psi\{X_1/r_1\rho,...,X_n/r_n\rho\}$

The latter condition guarantees that all applications of substitutions that occur in the proof below are indeed admissible.

It has to be shown that formula $(p(X_1,\ldots,X_n) \to (\varphi \to \psi))$ is valid in $M_P$. Interpreting the implicit universal quantification over variables $X_1,\ldots,X_n$, this is equivalent to: for all ground substitutions $\sigma = \{X_1/s_1,\ldots,X_n/s_n\}$, $M_P \models (p(s_1,\ldots,s_n) \to (\varphi\sigma \to \psi\sigma))$. Note that this reformulation treats variables $Y_1,\ldots,Y_l$ correctly. Since $M_P \models p(s_1,\ldots,s_n)$ means that there is a number $m>0$ with $p(s_1,\ldots,s_n) \in I_m(P)$, we may reformulate our claim as follows:

For all $m>0$ and all ground substitutions $\sigma = \{X_1/s_1,\ldots,X_n/s_n\}$,

$$\text{if } p(s_1,\ldots,s_n) \in I_m(P) \text{ then } M_P \models (\varphi\sigma \to \psi\sigma)). \tag{9.1}$$

Now, (9.1) is shown by induction on m. Let $\sigma = \{X_1/s_1,\ldots,X_n/s_n\}$ be a ground substitution with $p(s_1,\ldots,s_n) \in I_m(P)$. We distinguish two cases, depending on whether $p(s_1,\ldots,s_n)$ has come into $I_m(P)$ as a ground instance of a fact of P, or via a rule of P.

*Case 1*    There is a fact $p(t_1,\ldots,t_n)$ of P and a ground substitution $\rho$ acting only on variables of this fact such that $p(t_1,\ldots,t_n)\rho$ coincides with $p(s_1,\ldots,s_n)$. Thus $t_i\rho = s_i$, for $i=1,\ldots,n$. Using the condition 'consistency' of step 1 of the described strategy, we know that

$$M_P \models (\varphi\{X_1/t_1,\ldots,X_n/t_n\} \to \psi\{X_1/t_1,\ldots,X_n/t_n\})$$

By further instantiating variables via substitution $\rho$ we obtain

$$M_P \models (\varphi\{X_1/t_1,\ldots,X_n/t_n\}\rho \to \psi\{X_1/t_1,\ldots,X_n/t_n\}\rho)$$

Using (1) and (2) we obtain

$$M_P \models (\varphi\{X_1/t_1\rho,\ldots,X_n/t_n\rho\} \to \psi\{X_1/t_1\rho,\ldots,X_n/t_n\rho\})$$

Since $t_i\rho = s_i$ we conclude that $M_P \models (\varphi\sigma \to \psi\sigma))$. Thus (9.1) is true for case 1.

*Case 2*    There is a rule $p(t_1,\ldots,t_n) \text{:-} p_1(t_{11},\ldots,t_{1n(1)}),\ldots,p_k(t_{k1},\ldots,t_{kn(k)})$ of P and a ground substitution $\rho$ acting only on variables of this rule such that $p(t_1,\ldots,t_n)\rho$ coincides with $p(s_1,\ldots,s_n)$ and $p_1(t_{11},\ldots,t_{1n(1)})\rho,\ldots,$ $p_k(t_{k1},\ldots,t_{kn(k)})\rho \in I_{m-1}(P)$ (in particular, m-1 is still >0). Thus $t_i\rho = s_i$, for $i=1,\ldots,n$. Assume that the body of the considered rule has been equipped with intermediate assertions as in step 2 of the strategy above. Inductively on $i=0,\ldots,k$ we show that $M_P \models (\varphi\sigma \to \chi_i\rho)$. We start with i=0. Since $\chi_0\rho = \varphi\{X_1/t_1,\ldots,X_n/t_n\}\rho = \varphi\sigma$, nothing remains to be shown. Assume that we already have shown

(3)     $M_P \models (\varphi\sigma\rightarrow\chi_0\rho),..., M_P \models (\varphi\sigma\rightarrow\chi_{i-1}\rho)$

Now we take a look at literal $p_i(t_{i1},...,t_{in(i)})$.

*Subcase 2a* $p_i\neq p$. Then condition 'subverification' tells us that

(4)     $M_P \models \{(\chi_0\wedge...\wedge\chi_{i-1})\}p_i(t_{i1},...,t_{in(i)})\{\chi_i\}$

This implies that

(5)     $M_P \models ((p_i(t_{i1},...,t_{in(i)})\rho\wedge\chi_0\rho\wedge...\wedge\chi_{i-1}\rho)\rightarrow\chi_i\rho)$

Since $p_i(t_{i1},...,t_{in(i)})\rho\in I_{m-1}(P)$ we conclude that $M_P \models (\varphi\sigma\rightarrow\chi_i\rho)$.

*Subcase 2b* $p_i=p$. Conditions 'match pre' and 'match post' tell us that

(6)     $M_P \models ((\chi_0\wedge...\wedge\chi_{i-1})\rightarrow\varphi\{X_1/t_{i1},...,X_n/t_{in(i)}\})$

(7)     $M_P \models ((\psi\{X_1/t_{i1},...,X_n/t_{in(i)}\}\wedge\chi_0\wedge...\wedge\chi_{i-1})\rightarrow\chi_i)$

Since $p(t_{i1},...,t_{in(i)})\rho\in I_{m-1}(P)$ we may inductively apply (9.1) with $\tau = \{X_1/t_{i1}\rho,...,X_n/t_{in(i)}\rho\}$ to obtain:

(8)     $M_P \models (\varphi\tau\rightarrow\psi\tau)$

Successively we conclude that

$$M_P \models (\varphi\sigma\rightarrow(\chi_0\rho\wedge...\wedge\chi_{i-1}\rho)) \qquad \text{use (3)}$$

$$M_P \models (\varphi\sigma\rightarrow\varphi\{X_1/t_{i1},...,X_n/t_{in(i)}\}\rho) \qquad \text{use (6)}$$

$$M_P \models (\varphi\sigma\rightarrow\varphi\{X_1/t_{i1}\rho,...,X_n/t_{in(i)}\rho\}) \qquad \text{use (1)}$$

$$M_P \models (\varphi\sigma\rightarrow\varphi\tau) \qquad \text{see definition of } \tau$$

$$M_P \models (\varphi\sigma\rightarrow\psi\tau) \qquad \text{use (8)}$$

$$M_P \models (\varphi\sigma\rightarrow\psi\{X_1/t_{i1}\rho,...,X_n/t_{in(i)}\rho\}) \qquad \text{see definition of } \tau$$

$$M_P \models (\varphi\sigma\rightarrow\psi\{X_1/t_{i1},...,X_n/t_{in(i)}\}\rho) \qquad \text{use (2)}$$

$$M_P \models (\varphi\sigma\rightarrow\chi_i\rho) \qquad \text{use (7) and (3)}$$

Having thus shown $M_P \models (\varphi\sigma\rightarrow\chi_i\rho)$, for $i=0,...,k$, we obtain $M_P \models (\varphi\sigma\rightarrow\chi_k\rho)$. By definition of $\chi_k$ this means that $M_P \models (\varphi\sigma\rightarrow\psi\{X_1/t_1,...,X_n/t_n\}\rho)$, that is, by (2), $M_P \models (\varphi\sigma\rightarrow\psi\{X_1/t_1\rho,...,X_n/t_n\rho\})$, therefore $M_P \models (\varphi\sigma\rightarrow\psi\{X_1/s_1,...,X_n/s_n\})$, so that $M_P \models (\varphi\sigma\rightarrow\psi\sigma))$. Thus (9.1) is shown for case 2.  ∎

## EXAMPLE 9.1 ✂—————————————————————————

Let P be the joint logic program from Sections 8.6 and 8.7 defining predicates sort and mergesort, together with a logic program LESS defining a linear order less. We show validity in $M_P$ of asserted predicate

> {true}mergesortsort(L,M){sort(L,M)}

It says that predicate mergesort indeed implements sorting. The proof is uniform in the set LESS: all we use is the transitivity of predicate less. Let us collect the program formulas of P:

(1)    sort(L,M):-permute(L,M),sorted(M).

(2)    permute([],[]).
       permute([X|L],M):-
              element(X,M),delete1(X,M,N),permute(L,N).

(3)    element(X,[X|L]).
       element(X,[Y|M]):-element(X,M).

(4)    delete1([X|L],X,L).
       delete1([Y|L],X,[Y|M]):-unequal(X,Y),delete1(L,X,M).

(5)    unequal(X,Y):-less(X,Y).
       unequal(X,Y):-less(Y,X).

(6)    sorted([]).
       sorted([X]).
       sorted([X,X|L]):-sorted([X|L]).
       sorted([X,Y|L]):-less(X,Y),sorted([Y|L]).

(7)    mergesort([],[]).
       mergesort([X],[X]).
       mergesort([X,Y|L],M):-
              divide([X,Y|L],L1,L2),
              mergesort(L1,M1),
              mergesort(L2,M2),
              merge(M1,M2,M).

(8)    divide([],[],[]).
       divide([X],[X],[]).
       divide([X,Y|L],[X|M],[Y|N]):-divide(L,M,N).

(9)    merge(L,[],L).
       merge([],M,M).
       merge([X|L],[Y|M],[X|N]):-less(X,Y),merge(L,[Y|M],N).

merge([X|L],[Y|M],[Y|N]):-less(Y,X),merge([X|L],M,N).
merge([X|L],[X|M],[X,X,N]):-merge(L,M,N).

For the verification we use three auxiliary predicates append(L,M,N), split(L,M,N) and L<M that are defined as follows:

(10)    append([],L,L).
        append([X|L],M,[X|N]):-append(L,M,N).

(11)    split(L,M,N):-append(M,N,K),permute(L,K).

(12)    []<[Y|M].
        [X|L]<[Y|M]:-L<M.

Thus, append defines concatenation, split(L,M,N) says that M,N is a partition of list L into two parts, not necessarily of almost equal length as required for predicate divide(L,M,N). As will be seen later, the proof that mergesort indeed sorts does not require that the division of list L via predicate divide is done in such a balanced manner. Even division of L into [],L or L,[] would not destroy the 'relative correctness' of the mergesort predicate. Balancing only allows us to obtain the usual $O(nlog(n))$ running time for the mergesort algorithm. L<M means that list L is shorter than list M. Obviously, < is interpreted in every Herbrand model of its definition by a well-founded relation. Thus we may use well-founded induction with respect to <, nothing more than induction on the length of a list. Note that the rules from (1)-(12) form a conservative extension of (1)-(9).

We shall now apply the verification strategy for asserted predicates to show the claim {true}mergesortsort(L,M){sort(L,M)}. We will end with subclaims that are treated with the help of the basic verification strategy described in Subsection 9.1.5. These are verified using induction and the completion of the predicates defined in (1)-(12).

---

**Exercise 9.1**   Construct the completion for each of the predicates defined in (1)-(12).

**Exercise 9.2**   Show that the following 'funcionality axioms' for predicate append follows from the completion of predicate append together with structural induction on list L:

$\forall L \ \forall M \ \forall N1 \ \forall N2 \ ((append(L,M,N1) \wedge append(L,M,N2)) \rightarrow N1=N2)$

$\forall L \ \forall M \ \exists N \ append(L,M,N)$

---

Knowing that predicate append exhibits a functional character, it is recommended that functional notation be introduced. We write N=L•M instead of append(L,M,N). This will considerably simplify the presentation of proofs. The definition of append then turns into the following axioms:

$$[]\bullet L=L \text{ and } [X|L]\bullet M=[X|L\bullet M]$$

We apply the verification strategy for asserted predicates and equip the single rule from (7) that defines predicate mergesort with intermediate assertions as follows (note that variable L must be replaced by term [X,Y|L]):

{true}
      divide([X,Y|L],L1,L2)
{split([X,Y|L],L1,L2)}
      mergesort(L1,M1)
{sorted(M1)∧permute(L1,M1)}
      mergesort(L2,M2)
{sorted(M2)∧permute(L2,M2)}
      merge(M1,M2,M)
{sorted(M)∧permute([X,Y|L],M)}.

Then the following formulas have to be verified. (Although most of these sub-tasks are trivial we nevertheless list them all in order to demonstrate the enormous overhead of such trivialities in verification. This motivates the desire to have available a semi-automated system able to generate automatically all these subtasks and solve most of them.)

(a)    true → (sorted([])∧permute([],[]))

Results from 'consistency with the first fact for mergesort'. It is readily derived using corresponding facts from sorted and permute.

(b)    true→(sorted([X])∧permute([X],[X]))

Results from 'consistency with the second fact for mergesort'. It is readily derived using corresponding facts for sorted and permute.

(c)    {true} divide([X,Y|L],L1,L2) {split([X,Y|L],L1,L2)}

Results as a 'subgoal' from the rule for divide. It is immediately derived from the completions of predicates divide and split.

(d)    (true∧split([X,Y|L],L1,L2))→true

Results as a 'match pre'. Trivial.

(e)    (true∧split([X,Y|L],L1,L2)∧sorted(M1)∧permute(L1,M1))
→ (sorted(M1)∧permute(L1,M1))

Results as a 'match post'. Trivial.

(f)    (true∧split([X,Y|L],L1,L2)∧sorted(M1)∧permute(L1,M1))→true

Results as a 'match pre'. Trivial.

(g)    (true∧split([X,Y|L],L1,L2)∧sorted(M1)∧permute(L1,M1)∧
sorted(M2)∧permute(L2,M2)) → (sorted(M2)∧permute(L2,M2))

Results as a 'match post'. Trivial.

(h)         {true∧split([X,Y|L],L1,L2)∧sorted(M1)∧
            permute(L1,M1)∧sorted(M2)∧permute(L2,M2)}

                merge(M1,M2,M)

            {sorted(M)∧permute([X,Y|L],M)}

Results as a 'subgoal'. This is the only place where a non-trivial sub-task must be treated.

To show (h) we first derive the following asserted predicate:

            {sorted(M1)∧sorted(M2)}

                merge(M1,M2,M)

            {sorted(M)∧split(M,M1,M2)}

Then we show the following implication:

            (split(L,L1,L2)∧ permute(L1,M1)∧permute(L2,M2)∧

            split(M,M1,M2)) → permute(L,M)

Both together obviously solve (h). Using the completion of predicate split we may reformulate them as follows:

            {sorted(M1)∧sorted(M2)}

                merge(M1,M2,M)

            {sorted(M)∧permute(M,M1•M2)}

            (permute(L,L1L2)∧permute(L1,M1)∧permute(L2,M2)∧

            permute(M,M1•M2)) → permute(L,M)

The last formula expresses a simple relation between concatenation and permutation of lists. Though not difficult in principle, it requires a lot of lemmas and inductive proofs (in particular well-founded induction on the length of lists). We leave it to the reader to carry out a derivation, or to invoke an induction proof system, or to refer to a library of basic mathematical lemmas about lists. To show the former one, we add intermediate assertions into the body of the rules that define predicate merge:

*First rule*  Substitute ([X|L],[Y|M],[X|N]) for (M1,M2,M).

            {sorted([X|L])∧sorted([Y|M])}
                    less(X,Y)
            {less(X,Y)}
                    merge(L,[Y|M],N)
            {sorted([X|N])∧permute([X|N],[X|L]•[Y|M])}

*Second rule* Substitute ([X|L],[Y|M],[Y|N]) for (M1,M2,M).

{sorted([X|L])∧sorted([Y|M])}
    less(Y,X)
{less(Y,X)}
      merge([X|L],M,N)
{sorted([Y|N])∧permute([Y|N],[X|L]•[Y|M])}

*Third rule* Substitute ([X|L],[X|M],[X,X|N]) for (M1,M2,M).

{sorted([X|L])∧sorted([X|M])}
    merge(L,M,N)
{sorted([X,X|N])∧permute([X,X|N],[X|L]•[X|M])}

Concerning the first rule, the following formulas have to be derived (here we omit the overhead of trivial formulas resulting from requirements 'match pre' and 'match post' and list only the interesting ones):

(i)    (sorted(L)∧sorted([])) → (sorted(L)∧permute(L,L•[]))

Results from 'consistency with the first fact for merge'. Using structural induction on L one derives L•[]=L and permute(L,L), thus solving (i).

(j)    (sorted([])∧sorted(M)) → (sorted(M)∧permute(M,[]•M))

Results from 'consistency with the second fact for merge'. Use []•M=M and permute(M,M) from (i).

(k)    sorted([X|L])→sorted(L)

Results from 'match pre' for the first rule. Use the completion of predicate sorted and induction (where is induction required?).

(l)    (sorted([X|L])∧sorted([Y|M])∧less(X,Y)∧sorted(N)∧
permute(N,L•[Y|M]))
    → (sorted([X|N])∧permute([X|N],[X|L]•[Y|M]))

Results from 'match post' for the first rule. We decompose (l) into the following two sublemmas (m) and (n).

(m)    permute(N,L•[Y|M]) → permute([X|N],[X|L]•[Y|M])

First we derive [X|L]•[Y|M]=[X|L•[Y|M]]. Then, using the facts delete1([X|L],X,L) and delete1([X|L•[Y|M]],X,L•[Y|M]), our claim (m) follows immediately from the completion of predicate permute.

(n)    (sorted([X|L])∧sorted([Y|M])∧less(X,Y)∧sorted(N)∧
permute(N,L•[Y|M])) → sorted([X|N])

Here, we apply structural induction on N. If N is empty the result follows from sorted([X]). If N is non-empty, say N is [Z|K], we have to derive formula sorted([X|[Z|K]]), hence sorted([X,Z|K]), hence (less(X,Z)∨X=Z). This is done in the following steps:

| | |
|---|---|
| permute(N,L•[Y|M]) | assumption |
| permute([Z|K],L•[Y|M]) | since N=[Z|K] |
| element(Z,[Z|K]) | fact defining element |
| element(Z,L•[Y|M]) | rule for permute |
| element(Z,L)∨element(Z,[Y|M]) | simple lemma |
| element(Z,L)∨Z=Y∨element(Z,M) | rule for predicate element |
| less(X,Z)∨X=Z∨Z=Y∨less(Y,Z)∨Y=Z | use sorted([X|L]), sorted([Y|M]) and lemma (o) below |
| less(X,Z)∨X=Z | use less(X,Y) and (0) |

(o)    (sorted([X|L])∧element(Z,L)) → (less(X,Z)∨X=Z)

Simple structural induction on L.

The conditions 'match pre' and 'match post' that result from a treatment of the second and third rules above lead to just the same lemmas.

# Chapter 10
# Procedural Interpretation of Horn Logic and Steps Towards Prolog

In Chapter 7 we described the declarative semantics of Horn formulas. Now we turn Horn logic into a programming language by prescribing a **fixed order** for the generation of SLD-derivations. To do so it is necessary to order the program formulas in a logic program P. Hence, from now on, a logic program P will be a finite list of program formulas. Having such a logic program and a goal ?-$A_1,...,A_k$, search for an SLD-refutation of $P \cup \{$?-$A_1,...,A_k\}$ is sequentialized as follows:

- In the actual goal always process the left-most literal.

- Try to resolve the head of the first program formula in the list P with the leftmost literal in the actual goal.

We thus obtain an interpretation of Horn logic as a programming language, called **pure Prolog**. Then we introduce some 'non-logical features' aimed at the manipulation of the control structure (**cut**) or dynamic modifications of the executed logic program itself. These features are standard ingredients of the programming language Prolog. We do not present Prolog in detail both for space reasons and because this book is about *logic* and not programming languages.

## 10.1    On the indeterminism of SLD-resolution

So far we have regarded a logic program P as a means for defining a certain distinguished model $M_P$, the minimal Herbrand model of P. Considering goal ?-$A_1,...,A_k$ with variables $X_1,...,X_n$ means asking whether $P \models \exists X_1...\exists X_n (A_1 \wedge...\wedge A_k)$, and, in the case that this indeed is true, to determine a substitution $\sigma$ such that $P \models (A_1 \wedge...\wedge A_k)\sigma$. This constitutes the *declarative view* of logic programming. The algorithmic counterpart to this declarative view is SLD-resolution. By systematically generating all possible SLD-derivations from $P \cup \{$?-$A_1,...,A_k\}$ we always find an SLD-refutation of $P \cup \{$?-$A_1,...,A_k\}$, whenever $P \models \exists X_1...\exists X_n(A_1 \wedge...\wedge A_k)$. Moreover, the computed answer substitution constructed along such refutations yields every correct answer substitution $\sigma$ as an instance, restricted to the variables $X_1,...,X_n$.

SLD-resolution alone cannot be regarded as the ultimate procedural counterpart to the declarative view of logical programs, at least if we refer to the present state of computer architecture, namely sequential (von Neumann) machines. These machines allow only execution of one instruction at a time, so they are quite incompatible with the strong indeterminism involved in the generation of SLD-derivations. We are faced with the following two sorts of indeterminism:

- *Choosing the literal* $A_m$ in the actual goal ?-$A_1,...,A_k$ to be resolved with the head of a program formula in the considered logic program P.

- *Choosing the program formula* of logic program P whose head is to be resolved with $A_m$.

Let us first deal with the former indeterminism. The next theorem shows that with respect to *refutations* (but not arbitrary derivations) the choice of the literal $A_m$ of a goal to be processed next is of no importance, provided that indeed all possible choices of program formulas to be resolved with $A_m$ are systematically tried.

### 10.1.1    Exchange construction

Let P be a logic program. Consider the 2-step SLD-derivation from P in Figure 10.1 (the selected literals are in bold). Here,

(1)     $\mu$ is an mgu of $A_i$ and $C_{head}$.

(2)     $\rho$ is an mgu of $A_j\mu$ and $D_{head}$.

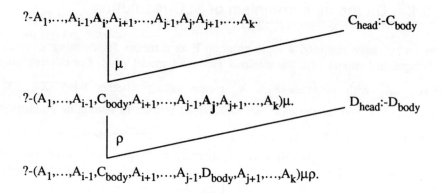

**Figure 10.1**

$C_{head}\text{:-}C_{body}$ and $D_{head}\text{:-}D_{body}$ are variants of formulas in P. By the variable disjointness condition for SLD-derivations, $D_{head}\text{:-}D_{body}$ does not contain variables that occur in $C_{head}$, $C_{body}$, $\mu$, $\rho$, $A_1,\ldots,A_{i-1},A_i,A_{i+1},\ldots,A_{j-1},A_j,A_{j+1}$, $\ldots,A_k$. Hence

(3)     $\mu$ does not act on $D_{head}$.

Using (2) and (3) we may conclude that $A_j\mu\rho=D_{head}\rho=D_{head}\mu\rho$, i.e. $\mu\rho$ unifies $A_j$ and $D_{head}$. Choose $\sigma$ and $\tau$ such that:

(4)     $\sigma$ is an mgu of $A_j$ and $D_{head}$.

(5)     $\mu\rho=\sigma\tau$.

By the way an mgu $\sigma$ of $A_j$ and $D_{head}$ is obtained, we know that the variables occurring in the pairs of $\sigma$ are taken from $A_j$ or $D_{head}$. Thus, by the variable disjointness condition again, we know that

(6)     $\sigma$ does not act on $C_{head}\text{:-}C_{body}$.

Using in succession (6), (5), (1) and (5) it follows that $C_{head}\tau=C_{head}\sigma\tau=C_{head}\mu\rho=A_i\mu\rho=A_i\sigma\tau$. Hence, $\tau$ unifies $C_{head}$ and $A_i\sigma$. Choose substitutions $\lambda$ and $\eta$ such that:

(7)     $\lambda$ is an mgu of $C_{head}$ and $A_i\sigma$.

(8)     $\tau=\lambda\eta$.

Now consider the 2-step SLD-derivation from P in Figure 10.2 (the reader should check that the variable disjointness condition is again fulfilled) resulting from the one above by first processing $A_j$ and then $A_i$.

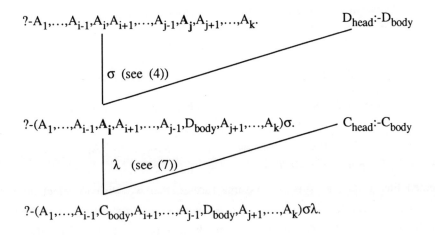

**Figure 10.2**

Let us compare the computed substitutions $\mu\rho$ and $\sigma\lambda$ in the two SLD-derivations above. First note that by (8) and (5), $\sigma\lambda\eta=\sigma\tau=\mu\rho$. Second note that $\sigma\lambda$ unifies $A_i$ and $C_{head}$, since by (7) and (6), $A_i\sigma\lambda=C_{head}\lambda=C_{head}\sigma\lambda$. Using (1) we may choose a substitution $\gamma$ such that

(9)     $\sigma\lambda=\mu\gamma$.

Applying (9), (4), (9) and (3) we may conclude that $A_j\mu\gamma=A_j\sigma\lambda=D_{head}\sigma\lambda=D_{head}\mu\gamma=D_{head}\gamma$. Hence $\gamma$ unifies $A_j\mu$ and $D_{head}$. Since, by (2), $\rho$ is an mgu of $A_j\mu$ and $D_{head}$, there exists a substitution $\delta$ such that

(10)     $\gamma=\rho\delta$.

So we finally obtain from (10) and (9) that $\mu\rho\delta=\mu\gamma=\sigma\lambda$.

Thus, the two computed answer substitutions $\mu\rho$ and $\sigma\lambda$ are correlated via $\mu\rho\delta=\sigma\lambda$ and $\sigma\lambda\eta=\mu\rho$, for suitable substitutions $\delta$ and $\rho$. We may of course assume that the pairs X/t in $\delta$ are such that no pair X/r occurs in $\mu\rho$, and X occurs in one of the terms substituted in $\mu\rho$ (if not, restrict $\delta$ to such pairs X/t). An analogous condition may be assumed for $\eta$ and $\sigma\lambda$. Since $\delta$ transforms $\mu\rho$ into $\sigma\lambda$, which is transformed back into $\mu\rho$ by $\eta$, and since $\delta$ and $\eta$ do not contain 'redundant' pairs w.r.t. the job just described, we may conclude that $\delta$ and $\eta$ are renaming substitutions and $\delta$ is the inverse substitution of $\eta$.

As a conclusion, the second SLD-derivation above computes, up to a renaming substitution, the same answer substitution as the first one.

**Theorem 10.1** (*reordering of SLD-refutations*)

Let an SLD-refutation be given. Then, any possible alternative sequence of choices for the literals of the actual goals to be resolved next can be achieved by an iterated application of the exchange construction, in such a way that the rearranged SLD-refutation computes the same answer substitution (up to renaming) as the original one. An analogous statement for SLD-derivations that do not end up with the empty goal is not true.

*Proof*   Consider the intermediate goal ?-$A_1,...,A_{i-1},A_i,A_{i+1},...,A_{j-1},A_j,$ $A_{j+1},...,A_k$ of an SLD-refutation at some stage s. Assume that the next selected literal at stage s is $A_i$. Assume further, that we wish to select instead of $A_i$ the literal $A_j$. Now observe that $A_j$ (with a certain substitution applied to it) will be the selected literal at some later stage t of the SLD-refutation (otherwise we would not arrive at the empty goal). Working backward from that later stage t up to the considered stage s, we may successively exchange the resolving of $A_j$ until it happens at stage s. Iterated application of this procedure allows any possible selection of literals to be achieved without changing the computed answer substitution (up to renaming).

It is clear that the argument above is not valid for infinite SLD-derivations and finite SLD-derivations which do not end up with the empty goal. The examples in the next sections will indeed show that the statement of the reordering theorem above is false for non-refutations.    ∎

# 10.2    Normal SLD-derivations and SLD-trees

Having seen that one 'selection rule' for the literal to be processed next is as good as any other (at least for the moment), we may fix a certain ordering of literals.

> **Definition 10.1:**   Let G be a goal, C be a program formula, and μ be a substitution. A goal G' is called **normal SLD-resolvent** of G and C via μ iff G' is an SLD-resolvent of G and C via μ such that the first literal in G was resolved with the head of a program formula. A **normal SLD-derivation (normal SLD-refutation)** is an SLD-derivation (SLD-refutation) all of whose steps are normal SLD-resolvent steps.

Graphically, we denote a normal SLD-derivation as in Figure 10.3 (often also vertically instead of horizontally). Here, $G_0,G_1,...,G_n,...$ is the sequence of generated goals, $C_0,C_1,...,C_n,...$ is the sequence of used program formulas, and $\mu_0,\mu_1,...,\mu_n,...$ is the sequence of applied substitutions.

$$G_0 \frac{C_0}{\mu_0} \quad G_1 \frac{C_1}{\mu_1} \quad G_2 \frac{C_2}{\mu_2} \quad \cdots \quad \frac{C_{n-1}}{\mu_{n-1}} \quad G_n \quad \cdots$$

**Figure 10.3**

***Corollary (Completeness of normal SLD-resolution)***

Let P be a logic program and G be a goal. If there exists an SLD-refutation of $P \cup \{G\}$, then there exists a normal SLD-refutation of $P \cup \{G\}$, too.

The next definition introduces a data structure that represents all possible normal SLD-derivations from $P \cup \{G\}$, for a logical program P and a goal G.

**Definition 10.2:** Let P be the logical program $C_1 C_2 ... C_k$, with program formulas $C_1,C_2,...,C_k$, and G a goal. The **SLD-tree** of P and G is defined (uniquely up to variants) as a finite or infinite tree SLD(P,G) with nodes which are labelled with goals, and edges which are labelled with rules of P and substitutions, in the following way:

- The root of SLD(P,G) is labelled with the goal G.

- Let $1 \leq i_1 < i_2 < ... < i_m \leq k$ be an enumeration of all indices $i \leq k$ such that the first literal in G can be resolved with a variant $C_i'$ of $C_i$ via mgu $\mu_i$. Furthermore, let $G_i$ be the normal SLD-resolvent obtained from G and $C_i'$ via $\mu_i$, for $i \in \{i_1,i_2,...,i_m\}$. If $\{i_1,i_2,...,i_m\}$ is empty, then SLD(P,G) consists of its root only. Otherwise, the root of SLD(P,G) possesses m successor nodes with labelled edges and subtrees as in Figure 10.4. There, the choice of the variants $C_i'$ of program formulas in P at some edge e of the SLD-tree is done in such a way that the variables of $C_i'$ do not occur elsewhere in the whole path from the root of the SLD-tree to the edge e.

Sometimes, we omit to label the edges of SLD-trees, when the program formulas and substitutions used are clear from the rest of the tree.

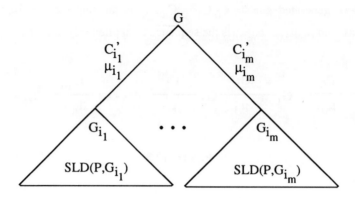

**Figure 10.4**

### Lemma 10.1

Every normal SLD-derivation from $P \cup \{G\}$ occurs (up to the renaming of variables) as a path of SLD(P,G). Conversely, every path of the SLD-tree of P and G represents a normal SLD-derivation from $P \cup \{G\}$.

*Proof*    Clear, since the choice of the variants in the construction of an SLD-tree guarantees that the variable disjointness condition for SLD-derivations is fulfilled along every path through the SLD-tree.    ∎

There are three sorts of paths in an SLD-tree:

- **Successful paths,** that is, paths of finite length ending with the empty goal.
- **Failure paths,** that is, paths of finite length ending with a non-empty goal.
- **Infinite paths,** that is, paths of infinite length.

Let us come back to our original concern, the study of the effects of the indeterminacy of the choice of literals and program formulas on the construction of SLD-derivations. Having fixed the choice of literals by the idea of a normal SLD-derivation, the former indeterminacy has been falsely resolved. It is the ordering of literals in the program formulas that allows several alternative selection rules to be simulated.

Hence, let us consider the effect on the SLD-tree of a *reordering of the literals* in the program formulas, and also the effect of a *reordering of the program formulas*. As we know from the reordering theorem above, a reordering of literals in program formulas of a logic program does not affect the set of all successful paths in an SLD-tree, although the position of a successful path in the

SLD-tree may change. What happens with infinite and failure paths is not clear, a priori. Let us illustrate this with an example (taken from Lloyd (1987)). Let P be the following logic program:

p(X,Z):-q(X,Y),p(Y,Z).

p(X,X).

q(a,b).

Consider the goal G=?-p(X,b). Then SLD(P,G) as presented in Figure 10.5 consists of two successful paths and one failure path.

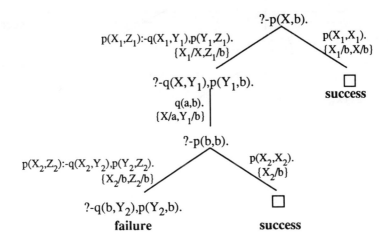

**Figure 10.5**

Next, let us see what happens if we reverse the ordering of literals in the program formulas of P (thus simulating a selection rule that always chooses the last literal in a goal). Then, the considered program Q is the following:

p(X,Z):-p(Y,Z),q(X,Y).

p(X,X).

q(a,b).

With ?-p(X,b) again as goal G, SLD(Q,G) is the tree presented in Figure 10.6. We should observe that the set of successful paths is unchanged (as was predicted), but that there are further failure paths branching off from a single infinite path. The presence of an infinite path is particularly problematic if we think of a depth-first search through this SLD-tree (thereby anticipating the usual Prolog refutation procedure). In case of logic program Q, we would be stuck with the infinite path, never reaching a successful one.

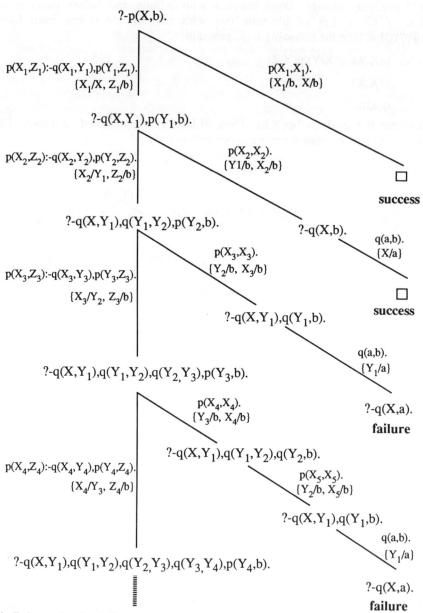

**Figure 10.6**

Now it is time to say something about the effect of the order of program formulas on SLD-trees. This is readily expressed by saying that a reordering of program formulas leads to a permutation of the paths of an SLD-tree, but does not affect the set of occurring paths. Thus, a suitable reordering of program formulas sometimes partially compensates for the drawbacks (namely that an infinite path comes in depth-first ordering of paths before every successful path) of a certain ordering of literals in the program formulas. It does not affect the number of successful, failure or infinite paths.

Such a compensation, which succeeded for our previous example, is not always possible, as the following example (also taken from Lloyd (1987)) shows. The considered logic program P is

(1)     p(a,b).

(2)     p(c,b).

(3)     p(X,Z):-p(X,Y),p(Y,Z).

(4)     p(X,Y):-p(Y,X).

Figure 10.7 shows a normal SLD-refutation of $P \cup \{?\text{-}p(a,c)\}$. We want to show that, regardless of the ordering of literals in the third rule of P and regardless of the ordering of facts and rules of P, the leftmost path of SLD(P,?-p(a,c)) will always be an infinite path. To show this we first prove that $\neg p(a,c)$ together with any three of the four elements of P possess a Herbrand model. For example, the set $\{\neg p(a,c),\ p(a,b),\ p(c,b),\ p(X,Z)\text{:-}p(X,Y),p(Y,Z)\}$ has Herbrand model $\{p(a,b),p(c,b)\}$.

---

**Exercise 10.1**   Construct Herbrand models for the remaining cases.

---

Now consider an arbitrary permutation Q of program P. Either rule (3) comes before rule (4) in Q, or conversely. Since the heads of rules (3) and (4) are unifiable with every literal, the leftmost path in SLD(Q,?-p(a,c)) either does not use rule (4) or rule (3), respectively. Since the used set of program formulas together with $\neg p(a,c)$ has a model, we may conclude that the considered leftmost path cannot be a successful path. Because of the permanent applicability of rule (3) or (4) we conclude that it is an infinite path.

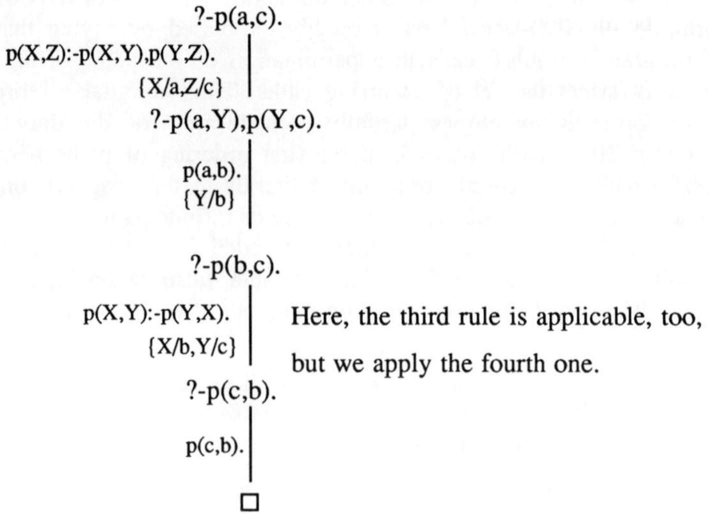

```
                              ?-p(a,c).
              p(X,Z):-p(X,Y),p(Y,Z).
                          {X/a,Z/c}
                              ?-p(a,Y),p(Y,c).

                          p(a,b).
                          {Y/b}

                     ?-p(b,c).

         p(X,Y):-p(Y,X).          Here, the third rule is applicable, too,
              {X/b,Y/c}
                                  but we apply the fourth one.
              ?-p(c,b).

         p(c,b).

              □
```

**Figure 10.7**

## 10.2.1   Summary

What should we keep in mind concerning the indeterminism in the selection of
a literal and a program formula in SLD-resolution?

(1)   Restriction of SLD-derivations to *normal derivations* is motivated by
      the exchange theorem. Strictly speaking, this restriction to normal
      SLD-derivations is more a matter of notational simplification than a se-
      vere restriction, since the reordering of literals allows us to simulate
      other selection rules.

Thus, our question should be posed more specifically as follows. In the context
of normal SLD-derivations, what is the effect of the *ordering of literals* in the
program formulas and the effect of the *ordering of program formulas* in a logic
program?

(2)   As long as we *systematically explore* every possible normal SLD-deri-
      vation, the ordering of literals and program formulas does not affect the
      reachability of successful paths. It may well affect the efficiency of arriv-
      ing at a successful path. Systematically exploring an SLD-tree could be
      implemented by a *breadth-first search* through such a tree.

(3)   If we search through SLD-trees in a *depth-first* manner (as we will do
      with the Prolog refutation procedure in the next section) then both the
      ordering of literals and the ordering of program formulas is important
      with respect to the reachability of successful paths.

(4)     This severe drawback of a depth-first search through SLD-trees may be mitigated by the following fact: if we can guarantee that, for a certain goal G and logic program P, there are only *finitely long normal SLD-derivations* from $P \cup \{G\}$, that is, that the *SLD-tree of P and G is finite*, then a depth-first search will indeed find a successful path if one exists.

## 10.3     The pure Prolog refutation procedure

Roughly speaking, the pure Prolog refutation procedure works as follows:

> Given a logic program P and a goal G, search through the SLD-tree of P and G in a depth-first manner until, eventually, a successful path is found. Then, the computed answer substitution along the found path will be passed back.

Besides the logic program P and goal G, we require further parameters for a proper formulation of the algorithm. First, the paths in an SLD-tree must fulfil the variable disjointness condition. This forces us to introduce the set of already consumed variables $V_{cons}$ as a parameter into our procedure. Second, when arriving at a success path we want to output the computed answer substitution along that path. For this reason we introduce as a parameter $\mu$ the **substitution build-up** until the actual node. Third, since we finally want to restrict the substitution built up along a successful path to the variables of the original goal G, we need one more parameter $V_{orig}$, which is instantiated with the set variables in the original goal G (a parameter which never changes). $V_{cons}$ is always a superset of $V_{orig}$.

> **Definition 10.3:**    The following algorithm is called the **pure Prolog refutation procedure** or **pure Prolog interpreter**. It is called with five parameters:
>
> (1)     a logic program P
> (2)     a goal G
> (3)     a finite set of variables $V_{orig}$
> (4)     a finite set of already consumed variables $V_{cons}$
> (5)     a substitution $\mu$
>
> It outputs a computed answer substitution for P and G, provided that such a substitution exists.
>
> The procedure is as follows:

- If G is the empty goal, then stop with success and pass back the substitution $\mu|_{V_{orig}}$ .

- If G is not the empty goal, then generate all program formulas $C_1,...,C_m$ of P such that there exist normal SLD-resolvents $G_1,...,G_m$ of G with suitable variants $C_1',...,C_m'$ via mgus $\mu_1,...,\mu_m$. (See the figure below. Note that m may be 0.) Variants are chosen in such a way that they do not contain variables from $V_{cons}$.

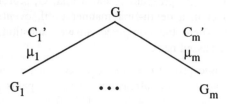

Then successively call the algorithm, for i=1,...,m, with actual parameters P, $G_i$, $V_{orig}$, $V_{cons} \cup Var(C_i')$, $\mu\mu_i$ until (eventually) for the first time one of these calls stops with success and passes back a computed answer substitution. (Note that parameters P and $V_{orig}$ are unchanged.) Then the whole procedure stops.

If one of these calls does not terminate before a successful path is reached, the whole procedure does not terminate.

If all of these subcalls terminate, but not with success, the whole procedure stops with failure. (This case occurs, for example, for m=0.)

It should be clear that the described Prolog refutation procedure, called with parameters P, G, $V_{orig}$=set of variables of G, $V_{cons}$=set of variables of G, $\mu$=empty substitution does nothing other than explore the SLD-tree of P and G in a depth-first manner until, eventually, a first successful path is found. Then, the answer substitution computed along this path is passed back. If we arrive at an infinite path before reaching a successful one, the procedure does not terminate. If all paths of SLD(P,G) are failure paths, the procedure halts and passes back the answer 'no'.

## 10.4    Jumping into Prolog

On basis of the pure Prolog interpreter described we can easily introduce some of the 'impure features' built into the programming language Prolog. The features we discuss here are

- the **cut** as a means of restricting search through SLD-trees;
- **assert** and **retract** as means of manipulating the rule base;
- **not** as a means of introducing negative information.

Since our theme is logic and not teaching full Prolog, we restrict our discussion to a short introduction of these concepts.

### 10.4.1    The cut

The purpose of using the cut predicate, usually written !, is to restrict a search through the SLD-tree of a goal. There are several reasons for abbreviating a complete search. We discuss a few of them.

#### Confirming the choice of a rule

Assume we have several rules defining a predicate p. There might be one rule, say p:-body, with the property that successful processing of body should rule out any possibility of alternatively satisfying p via another rule. For example, imagine a 2-person game whose rules are described by rules of the form:

$$move(State,NextState):-possibleMove_1(State,NextState)$$

.
.
.

$$move(State,NextState):-possibleMove_k(State,NextState)$$

It is then decided that for a certain move it should not be permissible to take it back and choose an alternative one. By writing

$$move(State,NextState):-possibleMove(State,NextState),!$$

such alternative choices are ruled out. In other words, having successfully processed the chosen rule, it confirms this choice to the interpreter and freezes it. As a second example, imagine a logic program that makes heavy use of list predicate element(X,L), but only with X and L instantiated with ground terms x and l. Now it might be possible that x occurs several times in list l. This leads to the possibility of having several success paths in the SLD-tree of ?-element(x,l), each of them leading to the same answer 'success'. In such a case it would be better to use another predicate, called elementcheck, that is defined as follows:

```
elementcheck(X,[X|L]):-!.
elementcheck(X,[Y|L]):-elementcheck(X,L).
```

Here, the use of ! prevents the interpreter from resatisfying goal elementcheck(x,l). Note the price that is paid with this new predicate: it can no longer be used as a predicate able to generate via backtracking successively all elements of a list.

### Confirming a satisfying instantiation of parameters

Consider a rule p(X,Y):-q(X),r(X,Y). It might be the case that it suffices to find a single successful instantiation of q(X) to recognize satisfiability of r(X,Y). In particular, this is the case if r(X,Y) does not depend on X. Then it would be desirable to prevent the interpreter from resatisfying p(X), once it has been successfully executed. Cut allows us to realize such a restriction of search by writing

```
p(X,Y):-q(X),!,r(X,Y).
```

Here, ! confirms to the interpreter the first satisfying instantiation for X in q(X). By the definition of cut, it will also confirm the choice of rule p(X,Y):-q(X),!,r(X,Y). This latter effect may be undesirable. It may be avoided by introduction of an auxiliary predicate s as follows:

```
p(X,Y):-s(X),r(X,Y).
s(X):-q(X),!.
```

### Saving multiple execution of a test

Consider the following definition of the maximum of two elements from a linearly ordered domain:

```
maximum(X,Y,Y):-less(X,Y).
maximum(X,Y,X):-less(Y,X).
maximum(X,X,X).
```

The disadvantage of this program is that it may require two executions of test less to find the appropriate case. This can be avoided as follows:

```
maximum(X,Y,Y):-less(X,Y),!.
maximum(X,Y,X).
```

Note that the use of cut allows the second rule to be chosen only if the test of the first one is not successful, that is, if X is greater than or equal to Y. Note the price that was paid with this version of maximum: considered from a declarative point of view (interpreting ! as a valid formula true), it is not a correct description of the maximum function.

**Definition 10.4 (Semantics of cut):**   From a declarative point of view the **cut** (written !) may be considered to be the 0-ary valid predicate true. Procedurally, its effect is to freeze certain decisions made in the exploration of SLD-trees. Let us consider a rule containing a cut:

$$A:-B_1,\ldots,B_m,!,C_1,\ldots,C_n.$$

Figure 10.8 explains the procedural effect of the involved cut: after the finite failure to satisfy $(C_1,\ldots,C_n,\text{rest})\mu\mu'$, backtracking leaves out the alternatives ∘ to satisfy $(B_1,\ldots,B_m)\mu$, as well as the alternatives * to satisfy A, and jumps back at once to the backtracking point •. In other words, the effect of ! is as follows:

> If ! is executed, then all the possible alternatives between the point where the rule $A:-B_1,\ldots,B_m,!,C_1,\ldots,C_n$ was chosen to satisfy A, and the point where ! was executed, are frozen (not searched via backtracking). Hence, if $B_1,\ldots,B_m$ become satisfied for the first time, all the choices made in order to satisfy $B_1,\ldots,B_m$, as well as the choice of rule $A:-B_1,\ldots B_m,!,C_1,\ldots,C_n$ are confirmed to the interpreter, telling it to not search for alternatives.

---

**Exercise 10.2**   Consider the following program:

```
A:-B,!,C.
A:-D.
```

Describe the effect of calling goal ?-A.

**Exercise 10.3**   Let fail be a predicate that can never be satisfied.

(a)     Can you envisage what such a predicate could be used for? Give an example.

(b)     Suppose we wanted to define a predicate that checks whether some person belongs to some group and is not a student. Consider the following program:

```
group(X) :- student(X), fail.

group(X):- ...
```

The first clause expresses the fact that a student does not belong to the group, the other clause(s) treat the other cases. Does this program fragment match our intuition? If so, prove your statement! If not, explain why not and amend the program.

(c)     Using cut and fail, program an n-ary predicate *negation* that behaves the same as *not* (when applied to n literals).

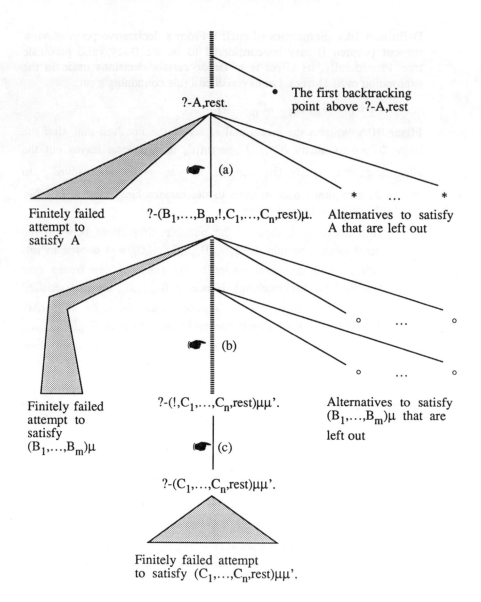

**Figure 10.8**

(a) introduction of !

(b) first path satisfying subgoal $(B_1,...,B_m)\mu$

(c) execution of ! ('point of no return')

---

**Exercise 10.4** The following program expresses the opinion of the ancient Greeks about spoken languages:

speak(greeks,greek) :- !.
speak(X,barbarian).

Greeks speak Greek and the rest speak barbarian (a somewhat subjective point of view!). Does the above program express this intuitive idea? If not, explain why and amend the program.

---

## 10.4.2    Manipulating the knowledge base

Prolog systems usually offer the user the possibility of modifying the executed program dynamically (that is, during execution) by providing 1-ary predicates asserta and retract (we concentrate on these two).

Roughly speaking, asserta(C) *adds* program formula C at the beginning of the actual Prolog program P, and thus proceeds execution with Prolog program CP, whereas retract(C) *deletes* the leftmost program formula in P that matches C, that is, is of the form $C\sigma$, for a substitution $\sigma$. In neither case is this side effect undone under backtracking. Also observe that there may be several substitutions $\sigma$ such that $C\sigma$ is a program formula of P. Then, retract(C) may be satisfied in several ways leading to a branching of the SLD-tree.

This rough description of the procedural effects of asserta and retract, usually presented in literature in such a form, is by no means sufficient for a full understanding of the effects of these predicates. Let us look at the following examples of a Prolog program together with a goal. Here, s=t succeeds if s and t are unifiable (in other words, we may look at predicate = as if it were defined by X=X).

(1)    p(a).

       ?-p(X),asserta(p(b)),X=b.

(2)    p(a).

       ?-retract(p(X)),asserta(p(b)),X=b.

(3)    q(a).

       q(b).

       p(a).

       ?-q(Y),retract(p(X)),asserta(p(b)),X=b.

(4)    p(a).

p(b).

?-retract(p(X)),retract(p(b)),X=b.

(5)    p(a).

p(b).

?-retract(p(X)),asserta(p(c)),asserta(p(b)),X=b.

The reader may predict what the answers should be and, eventually, test his/her supposition with an available Prolog system. The problems with the examples above are obvious. If, in the exploration of a node G of an SLD-tree, a finite number of successor nodes $G_1,...,G_m$ of G are introduced and the first one is further explored, what happens if, during the exploration of this first node, the underlying program is changed in such a way that $G_1,...,G_m$ no longer correctly reflect the set of choices possible at node G?

Another question is what happens if retract(C) can be satisfied in more than one way and, when trying to retract the second program formula matching C via backtracking, the program formula to be retracted has already been retracted, or is no longer the second program formula matching C? What we need is a more detailed description of the way a Prolog interpreter handles asserta and retract predicates.

**Definition 10.5 (Extended interpreter for logic programs with asserta and retract):**    Let us assume a logic program P containing asserta and retract, goal G, sets of initial and consumed variables $V_{init}$ and $V_{cons}$ and a substitution $\mu$. (Note that now the underlying program P may change during execution.)

(a)    *Interpretation of a goal G whose leftmost literal is a Horn logic literal*    Compute the list $C_1,...,C_m$ of program formulas in P such that there exist normal SLD-resolvents $G_1,...,G_m$ of G with suitable variants $C_1',...,C_m'$ via mgus $\mu_1,...,\mu_m$. Variants are chosen in such a way that they do not contain variables from $V_{cons}$. Then, the interpreter is successively called, for i=1,...,m, with program $Q_i$ that is dynamically computed by the preceding call ($Q_1$ is P), $G_i$, $V_{init}$, $V_{cons} \cup Var(C_i')$) and $\mu\mu_i$ until a success node is found. (Here it is laid down that the set of resolvents is statically fixed even if the executed subcalls may lead to further possible resolvents or disqualify some.)

(b)    *Interpretation of a goal with asserta(C) as leftmost literal*

?-asserta(C),G is executed by recursively calling the Prolog interpreter with new parameters CP (program formula C attached to the front of P), G, $V_{init}$, $V_{cons}$, $\mu$.

(c)     *Interpretation of a goal with retract(C) as leftmost literal*

?-retract(C),G is executed by computing the list $C_1,...,C_m$ of program formulas in P that match program formula C, that is, they are of the form $C\sigma_1,...,C\sigma_m$, for substitutions $\sigma_1,...,\sigma_m$. Introduce pointers $Ptr_1,...,Ptr_m$ pointing to the program formulas $C_1,...,C_m$ of P (see Figure 10.9). Then, the Prolog refutation procedure is successively called, for i=1,...,m, with the program $Q_i$ that is dynamically computed from the preceding call, new goal ?-delete($Ptr_i$),G, variable sets $V_{init}$, $V_{cons} \cup Var(\sigma_i)$ and substitution $\mu\sigma_i$, until a success node is found.

In (c) new 'literals' delete(Ptr) were used. So we must lay down how these are to be interpreted:

(d)     *Interpretation of a goal with delete(Ptr) as leftmost literal*

?-delete(Ptr),G is executed as follows. If Ptr points to a program formula C of the actual Prolog program then the program formula pointed to is deleted, and execution proceeds with the new program and goal G, and otherwise unchanged parameters. If Ptr does not point to a program formula of the actual program (this may occur if the program formula originally pointed to by Ptr is retracted at some earlier stage), then the interpretation of delete(Ptr) fails.

---

**Exercise 10.5** Execute the goals from examples (1)-(5) above. Explain the answers.

Actual program P (dynamically changing)

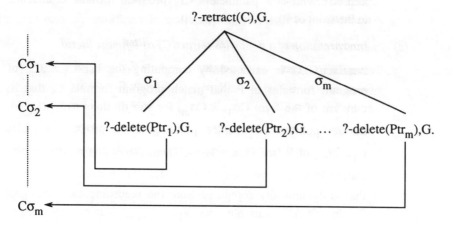

**Figure 10.9**

We shall now briefly discuss some applications of predicates asserta and retract.

### Fibonacci numbers and dynamic programming

The following 2-ary function fib on the set of natural numbers is called the Fibonacci function:

$$fib(0)=1$$
$$fib(1)=1$$
$$fib(x+2)=fib(x)+fib(x+1).$$

It may be readily implemented by the following logic program using 2-ary predicate fibo and a definition of addition:

fibo(0,succ(0)).
fibo(succ(0),succ(0)).
fibo(succ$^2$(X),Y):-fibo(X,U),fibo(succ(X),V),add(U,V,Y).

The disadvantage of this definition is that a call of goal  ?-fibo(succ$^m$(0),Y) requires $O(2^m)$ paths of the SLD-tree to be visited.

---

**Exercise 10.6**   Construct the SLD-tree of goal ?-fibo(succ$^6$(0),Y).

---

The same subgoal ?-fibo($succ^k(0)$,U), with k<m, may become resatis-fied over and over again. This is redundant since there are only m such sub-goals. This situation, namely that of a few subproblems being exponentially often processed, is a common situation in recursive programming. Usually, it is solved by replacing the top-down recursive solution by a bottom-up procedure that generates and solves subproblems, starting with small problems and pro-ceeding to larger ones. This strategy, called **dynamic programming**, may be realized with the asserta predicate in a quite simple way: whenever a subgoal is satisfied, its solution is stored in the rule base. For our example, this looks as follows:

    fibon(0,succ(0)).
    fibon(succ(0),succ(0)).
    fibon($succ^2$(X),Y):-
        fibon(succ(X),U),**asserta(fibon(X,U))**,fibon(X,V),add(U,V,Y).

Since asserta stores solutions to subproblems at the front of the rule base, these stored solutions are used instead of further calling the rule for fibon. The effect is that multiple satisfaction of the same subgoal is suppressed.

---

**Exercise 10.7**    Construct the SLD-tree of goal ?-fibon($succ^6$(0),Y).

---

### Path searching in a directed graph

Given a directed graph by a set of ground facts EDGE consisting of facts edge(a,b), we discussed in Section 7.2 a logic program defining a predicate path(X,Y) which expressed the fact that a path exists from node X to node Y in the considered graph:

    path(X,X).
    path(X,Y):-edge(X,Z),path(Z,Y).

Though being declaratively correct, this program is not well suited for execution since it admits infinite SLD-derivations if the graph contains cycles. So we replaced it in Section 8.9 by the following definition:

    path(X,Y):-restrictedPath(X,Y,[]).
    restrictedPath(X,X,L).
    restrictedPath(X,Y,L):-
        edge(X,Z),new(Z,L),restrictedPath(Z,Y,[X|L]).

The effect of restrictedPath(X,Y,[]) was to generate only cycle-free paths. Hence, for a finite graph, SLD-trees are finite. Nevertheless, from a complexity theory point of view, this alternative definition is still unsatisfying.

The reason is that there might exist fac(n) different cycle-free paths for a graph with n nodes, and thus the same number of paths within SLD-trees. Compared with the O(n) running time of good path searching algorithms, this is an unacceptable situation. The use of asserta offers a solution. Using a new predicate visited(X), we store all nodes X that have already been generated. Nodes that are visited are not explored again. Now look at Figure 10.10.

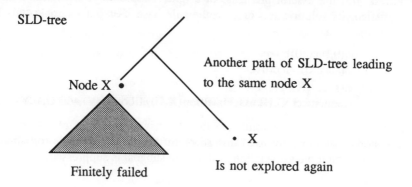

SLD-tree

Node X •

Another path of SLD-tree leading to the same node X

• X

Finitely failed        Is not explored again

**Figure 10.10**

Note that backtracking has forgotten the first occurrence of node X on arriving at the second one. It is because asserta is not undone under backtracking that we may store node X as being already visited. The solution looks as follows:

```
path(X,Y):-asserta(visited(X)),search(X,Y).

search(X,X).
search(X,Y):-
    edge(X,Z),unvisited(Z),asserta(visited(Z)),search(Z,Y).

unvisited(Z):-visited(Z),!,fail.
unvisited(Z).
```

In one sense, visited(Z) is comparable to a global variable in conventional programming. Note that unvisited(x) succeeds if visited(x) fails. It would be somewhat nicer to formulate this negation of predicate visited in a more direct way:

```
unvisited(X):-not visited(X).
```

The next section introduces negation.

## 10.4.3  Negation as finite failure

**Definition 10.6 (Procedural semantics of 'not'):**     Given a sequence $A_1,...,A_n$ of atomic formulas, Prolog systems generally allow us to call a goal ?-not $(A_1,...,A_n)$ (written '?-not $A_1$' in the case that n=1). Its procedural interpretation is as follows. Execute the auxiliary goal ?-$A_1,...,A_n$. Three outcomes are possible:

(1)    A successful path is found. This means that a substitution $\mu$ is found such that $P \models (A \wedge...\wedge A_n)\mu$, hence $P \models \exists(A_1 \wedge...\wedge A_n)$. Then subgoal ?-not $(A_1,...,A_n)$ fails.

(2)    After a finite number of failure paths Prolog runs into an infinite path. Then subgoal ?-not $(A_1,...,A_n)$ also loops.

(3)    The SLD-tree of goal ?-$A_1,...,A_n$ consists of failure paths only. Then subgoal ?-$A_1,...,A_n$ succeeds with empty substitution as the computed answer.

**Remark**

Let us summarize what the execution of 'not' means for a special case. Consider a logic program P (pure Prolog without the occurrence of not) that defines the predicates occurring in $A_1,...,A_n$ and a rule A:-not $(A_1,...,A_n)$. Assume that the predicate occurring in A does not occur in $A_1,...,A_n$ (so negation is not used within a recursion). Furthermore, assume that the SLD-tree of the auxiliary goal ?-$A_1,...,A_n$ is finite. Then

- ?-not A succeeds iff $P \not\models \exists(A_1 \wedge...\wedge A_n)$.

Assume, in addition, that $A_1,...,A_n$ are ground atoms. Then:

- ?-not A succeeds iff $P \not\models A_1 \wedge...\wedge A_n$.

In this restricted sense, 'not' models negation as finite failure. But never forget that 'not' mirrors negation only in the limited sense described above. As an example, try to explain what will be answered to a question ?-not not p(X). Or consider p:- not q, q:-p.

There is extensive literature on negation in logic programming, in particular concerning the question of how to make the use of 'not' more safe. The idea of preventing the use of not within a recursion has been seen to be of particular value. This idea leads to a special class of programs that use 'not' only in a restricted sense. These programs are called **stratified programs**. We shall give only a brief introduction here.

**Definition 10.7:**     **Logic programs with negation** are finite sets of strings of the form A or A:-$L_1$,...,$L_m$ with atomic formula A and formulas $L_1$,...,$L_m$ that are either atomic or of the form 'not B' with atomic formula B.

**Definition 10.8:**     Let P be a logic program with negation. P is **stratified** iff it can be decomposed into a disjoint sequence of subprograms $P_1$,...,$P_k$ with the following properties:

(1)     For every predicate p there is an index i=1,...n such that $P_i$ contains all program formulas that define p.

(2)     If predicate p is used in unnegated form within a program formula from $P_i$ then p is defined within one of $P_1$,...,$P_i$.

(3)     If predicate p is used in negated form within a program formula from $P_i$ then p is defined within one of $P_1$,...,$P_{i-1}$. In particular, $P_1$ is a logic program without an occurrence of 'not'.

The advantage of stratified programs is that we may assign to them a canonical semantics that reflects the procedural semantics described above in the same way as $M_P$ established the canonical semantics of a pure program P.

Having decomposed P into subprograms $P_1$,...,$P_k$ this canonical model is constructed in k steps as follows. We define $Q_1=P_1$ and start with the least Herbrand model of logic program $Q_1$. Next we consider the predicates that are used negatively in $P_2$ and interpret them (these are written in pure Prolog) according to the description given in the remark above. Then we create for every such predicate p a new one, called pComplement, that describes just the complement of predicate p, and insert all true ground facts p(a) as well as all ground facts pComplement(b) such that p(b) does not hold. (Note that there may be an infinite number of facts for pComplement; this does not disturb us, because the construction of canonical models is exactly as for finite programs.) Replacing occurrences of not p(t) in $P_2$ by pComplement(t) leads to a set (perhaps infinite) of Horn formulas $Q_2$ that can be assigned a canonical semantics via the $M$-operator. Proceeding this way, we arrive at a canonical model for the whole sequence $P_1$,...,$P_k$. It can be shown that this model is independent of the decomposition $P_1$,...,$P_k$ of P.

### 10.4.4    An extended example: blocks world

Assume we have some blocks piled up at some available places in some initial
order as shown in Figure 10.11. Imagine a robot able to grasp a block with no
block upon it (a **free block**) and to put it onto another free block or a free
place. How can a robot plan a sequence of such actions whose execution pro-
duces a certain goal situation?

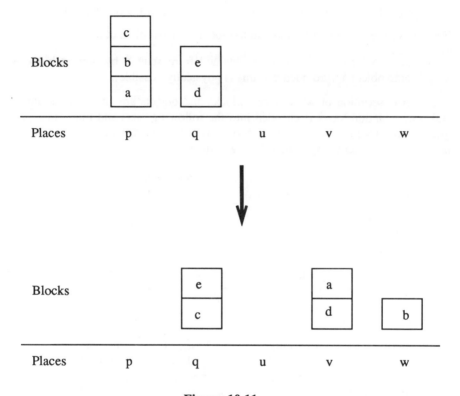

**Figure 10.11**

Let us now model this problem in logic. We proceed in a top-down
fashion, starting with a rather abstract problem description. The abstract version might be taken as the final program, but efficiency considerations will lead
us to more concrete realizations.

**First attempt: an abstract solution**

We use constants a,b,c,d,e,p,q,u,v,w,init,goal:

- a,b,c,d,e to denote blocks
- p,q,u,v,w to denote places
- init and goal to denote the initial and desired situations, respectively

We use variables R,S,T (**state variables** to describe the actual blocks world situation). Furthermore, we use predicates place, block, object, on and free with the following meanings:

- place(X) and block(X) to distinguish places from blocks
- object(X) collecting blocks and places
- on(S,X,Y) expressing that in state S object X is sitting on object Y
- free(S,X) expressing that in state S no block is sitting on object X

Finally, we use a function symbol put to model the movement of a block:

- put(S,X,Y) denotes the state obtained from state S by putting block X onto object Y (provided that this is physically possible).

A description of what places, blocks and objects are and how the initial and final situation looks is encoded into the following facts and rules (in a program with varying situations there should be a knowledge acquisition component asking the user to input these facts and rules):

|            |          |                    |
|------------|----------|--------------------|
| place(p).  | block(a).| object(X):-place(X).|
| place(q).  | block(b).| object(X):-block(X).|
| place(u).  | block(c).|                    |
| place(v).  | block(d).|                    |
| place(w).  | block(e).|                    |

Note that we do not use a cut after literal place(X) in the definition of predicate object. The reason is that we want object to act like a generator able to be successively satisfied with all considered places and blocks. Also, a cut would be rather superfluous, since an object cannot be both a block and a place.

|             |             |              |              |
|-------------|-------------|--------------|--------------|
| on(init,c,b). | free(init,c). | on(goal,e,c). | free(goal,p). |
| on(init,b,a). | free(init,e). | on(goal,c,q). | free(goal,e). |
| on(init,a,p). | free(init,u). | on(goal,a,d). | free(goal,u). |
| on(init,e,d). | free(init,v). | on(goal,d,v). | free(goal,a). |
| on(init,d,q). | free(init,w). | on(goal,b,w). | free(goal,b). |

Next we describe the effect of an application of put(S,X,Y) on a state S (see Figure 10.12). Here, we do not worry about the question of whether such an application is physically possible; this will be integrated into our program later.

| | |
|---|---|
| on(put(S,X,Y),X,Y). | ('local changes') |
| on(put(S,X,Y),U,V):-on(S,U,V),not U=X. | ('preservation') |
| free(put(S,X,Y),U):-on(S,X,U). | ('local changes') |
| free(put(S,X,Y),U):-free(S,U),not U=Y. | ('preservation') |

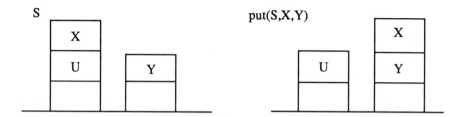

**Figure 10.12**

Let us now define a predicate possible(S) that expresses which states can be generated from the initial state by repeated applications of operation put, always taking care that these applications are indeed physically possible:

```
possible(init).
possible(put(S,X,Y)):-
        possible(S),block(X),free(S,X),object(Y),not X=Y,free(S,Y).
```

Now it is easy to formulate the problem of transforming state init into state goal as a path-searching problem in a directed graph. What are the nodes of this graph? Do not answer 'the possible states defined above'. The problem is that two such states may well encode the same situation. As an example, consider put(put(init,c,e),c,b) and init. We need thus a predicate same(S1,S2) that decides whether two states encode the same arrangement of blocks:

```
same(S1,S2):-not (block(X),object(Y),on(S1,X,Y),not on(S2,X,Y)).
```

This definition of predicate same requires some comments. First of all, the procedural semantics of not tells us that same(S1,S2) is to be interpreted as follows:

$$\neg\exists X\ \exists Y(block(X)\wedge object(Y)\wedge on(S1,X,Y)\wedge\neg on(S2,X,Y)), \text{ or}$$
$$\forall X\ \forall Y((block(X)\wedge object(Y))\rightarrow(on(S1,X,Y)\rightarrow on(S2,X,Y)))$$

Since every block X is sitting on exactly one object(Y), this implies that even:

$$\forall X\ \forall Y((block(X)\wedge object(Y))\rightarrow(on(S1,X,Y)\leftrightarrow on(S2,X,Y)))$$

This latter statement obviously expresses that S1 and S2 encode the same situation. Finally, we may apply our (efficient) path-searching programs discussed in Subsection 10.4.2 (taking care to avoid a repeated exploration of already visited nodes) to obtain the following program:

```
plan(P):-asserta(visited(init)),restricted_path(init,P).
restrictedPath(S,S):-same(S,goal).
```

```
restrictedPath(S,P):-
        possible(put(S,X,Y)),
        not (visited(Sbefore),same(put(S,X,Y),Sbefore)),
        asserta(visited(put(S,X,Y))),
        restrictedPath(put(S,X,Y),P).
```

Though path searching is done as well as possible, response times are absolutely unacceptable, even for our small example. How can this be explained?
The first reason is that the search space of our problem is quite large. For our
example of blocks world, in the worst case there are five choices for the block
x to be moved, and always four objects y to put it on. Estimating that at least
10 put operations are required to transform the initial state init into the final
state, we must cope with a search space of about $20^{10}$~$10^{13}$ different elements. Without good heuristics, even a good path-searching algorithm would
be overstrained with such a search space.

But there is another drawback to our solution, not as dramatic as that
concerning the dimension of the search space, but nevertheless not to be ig-
nored. To recognize this drawback let us take a look at the way predicate restrictedPath searches through the states of the blocks world. It generates (via
predicate possible) expressions $s=put(\dots put(init,x_1,y_1),\dots,x_n,y_n)$, with n that
may be considerably greater than the number k of objects and places. Each
such expression must be compared w.r.t. predicate same with goal. If we consider a more general blocks world with k places and blocks, the definition of
predicate same leads to the consideration of $k^2$ pairs (x,y) to be checked as to
whether on(s,x,y) and on(goal,x,y) are true. Each such check on(s,x,y) requires a recursive calculation of n similar further facts for states $s_i=put(put(\dots$
$put(init,x_1,y_1),\dots,x_i,y_i),x',y')$, with i running from n down to 0, until we arrive
at facts on(init,x'',y''), which may be true or false. Thus the comparison of s
with goal requires $k^2n$ operations.

This is unsatisfactory since the real physical situations are of size
bounded by parameter k, and thus independent of n, if we represent them in an
appropriate form as lists. Then the comparison of a state s with goal
(represented as lists) requires only the comparison of every element of list s
with every element of state goal, leading to $k^2$ operations.

## Second attempt: concrete representation of states by lists of stacks

It seems to be appropriate to represent a concrete situation of our blocks
world as a list l that contains as elements all the stacks at every place. As an
example, the given initial state init would be represented by the following list
of stacks:

[[c,b,a,p],[e,d,q],[u],[v],[w]].

Obviously, the order of objects makes the top element above some place directly available. Now we model our blocks world task alternatively as follows.

We use 1-ary predicates init(S) and goal(S) to describe initial and final states by facts:

init([[c,b,a,p],[e,d,q],[u],[v],[w]]).
goal([[p],[c,e,q],[u],[a,d,v],[w]]).

Then we use a predicate next(S,Snew) to express the possibilities of generating a successor state Snew of state S (see Figure 10.13). Hence, next is a substitute for the function symbol put(S,X,Y) and predicate possible from before:

```
next(S,Snew):-
          element([X,U|L],S),
          element([Y|M],S),
          not X=Y,
          delete(S,[X,U|L],S1),
          delete(S1,[Y|M],S2),
          Snew=[[U|L],[X,Y|M]|S2].
```

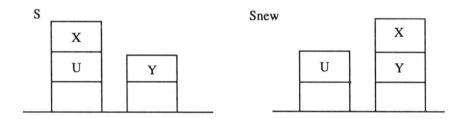

**Figure 10.13**

As before, we require a predicate same(S1,S2) to test whether two states represent the same situation. Here, this predicate simply compares lists w.r.t. to common elements.

```
same([],[]).
same([X|L],M):-
     element(X,M),delete(X,M,M1),delete(L,X,L1),same(L1,M1).
```

Using definitions for predicates element and delete as in previous examples, we arrive at the following program:

```
plan(P):-init(I),asserta(visited(I)),restricted_path(I,[I],P).

restrictedPath(S,L,L):-
          goal(G),same(S,G).
```

```
restrictedPath(S,L,P):-
    next(S,Snew),
    not (visited(Sbefore),same(Snew,Sbefore)),
    asserta(visited(Snew)),
    restrictedPath(Snew,[Snew|L],P).
```

Here, parameter L in predicate restrictedPath(S,L,P) maintains a sequence of intermediate states leading from the initial state to the actual state S. On success it is passed back to variable P. With a suitable further predicate action(P) we could then force a robot to transform the initial state into the final one.

Though our second solution is somewhat more efficient than the first one, the unchanged large search space again leads to execution times which are unacceptable.

What we finally try to do is to incorporate heuristics into our program. Fortunately, there are heuristics which cut down the search space considerably without omitting possible solutions. (Usually, heuristics tolerate that possible solutions are omitted for the sake of execution time; it may be rather difficult to guarantee that desirable solutions are not left out too often.)

### Third attempt: adding heuristics

The crucial point in our second attempt is the definition of predicate next controlling the choice of next state. To obtain a valuable heuristic guiding this selection, let us introduce the notion of a **partial** (or approximate) **solution**. This will be a stack L which occurs as a postfix of a stack in the final state $s_f$. The situation is shown in Figure 10.14.

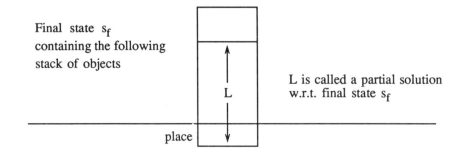

Final state $s_f$ containing the following stack of objects

L

place

L is called a partial solution w.r.t. final state $s_f$

**Figure 10.14**

Whenever such a partial solution L occurs as an element of some state

S, it is a good decision to *preserve* L. This is a good decision particularly because the presence of L by no means obstructs the reordering of the remaining blocks; putting blocks onto L is as good a possibility as putting them onto the place where L is positioned. Moreover, it seems wise to *extend* partial solutions to even longer, but still partial solutions. (Note that such a monotonicity is not always present in problems. Often we find that partial solutions must be again destroyed in order to obtain final solutions. As an example, consider the well known n×n-puzzles with one stone missing.) Whenever extensions are impossible, we choose an arbitrary next state but take care not to destroy a partial solution. Note that such a choice is always possible, unless the final state is reached. Let us first define the concept of a partial solution:

```
partialSolution(Stack):-
          final(F),element(Final_Stack,F),postfix(Stack,Final_Stack).

postfix(L,L).
postfix(L,[X|M]):-postfix(L,M).
```

The heuristic with highest priority, to extend further a partial solution to a partial solution, can be formulated as follows:

```
next(S,Snew):-
          element([Y|M],S),
          partialSolution([Y|M]),
          element([X,U|L],S),
          not X=Y,
          partialSolution([X,Y|M]),
          delete(S,[X,U|L],S1),
          delete(S1,[Y|M],S2),
          Snew=[[U|L],[X,Y|M]|S2].
```

Note that the second chosen list [X,U|L] is never a partial solution. If this rule is not applicable, then try the following:

```
next(S,Snew):-
          element([Y|M],S),
          element([X,U|L],S),
          not X=Y,
          not partialSolution([X,U|L]),
          delete(S,[X,U|L],S1),
          delete(S1,[Y|M],S2),
          Snew=[[U|L],[X,Y|M]|S2].
```

Since one of these rules is applicable, provided the final state is not already reached, we do not need a further rule for next. Of course, further heuristics might be incorporated into our program. Now a plan is quickly passed back to the user.

## Historical remarks and recommended reading for Part II

Logic programming began in the early 1970s and was based on previous work about logic and automated theorem proving (for example, Herbrand (1930), Davis and Putnam (1960)) and especially on resolution. The first authors who introduced logic programming were Colmerauer *et al.* (1973) and Kowalski (1979b). Their idea of using predicate logic (or some sublogic of it) was revolutionary, as until then logic was only used as a specification or deductive language in computer science.

The philosophy of logic programming is best described by Kowalski's famous equation Algorithm = Logic + Control in Kowalski (1979a): the ideal is that the programmer only describes what he or she wants the program to do, while execution is left to the machine. Unfortunately, current logic programming systems do not allow such a strict distinction. This is mirrored by the existence of non-logical features in Prolog, the most popular logic programming language today. But it should be noted that considerable effort is made to provide declarative semantics also for negation (see Lloyd (1987), Shepherdson (1985), (1987)) and dynamic program manipulation in Burt *et al.* (1990).

The semantic foundations of logic programming can be found in Kowalsky and van Emden (1976), Apt and van Emden (1982) and especially in Lloyd (1987), whereas implementation issues can be found in Campbell (1984) and Warren (1983).

Introductions to Prolog programming can be found in Clocksin and Mellish (1981) that established the language standard, Sterling and Shapiro (1986) or Bratko (1990). Aspects concerning the correctness and synthesis of logic programs are discussed in Hogger (1984).

One major line of current research deals with the combination of logic programming and databases. This has led to deductive databases which are generalizations of relational databases (see Gallaire and Minker (1978), Gallaire *et al.* (1984), Minker (1988)). One well-known system based on such ideas is Datalog (see Ceri *et al.* (1989)).

# Part III

# LOGIC OF EQUATIONS
# AND
# ABSTRACT DATA TYPES

Having recognized in Part II that sublogics of predicate logic like Horn logic deserve interest as languages being broad enough to be well suited for certain applications, and sufficiently narrow to exhibit special useful properties, we will study here another sublogic of predicate logic with equality, namely **logic of equations**.

First we analyze equations from several points of view. Proof-theoretic investigations concerning the portion of Hilbert's calculus necessary for deriving equations from other ones lead to **Birkhoff's calculus**. A model-theoretic analysis of a set E of equations leads to a specific model capturing the semantics of E, namely the **initial model** of E. A rewriting view of equations (equations naturally describe ways of replacing some objects by other ones) leads finally to **term rewriting mechanisms**. The inner relations between these approaches are developed in Chapter 11.

Then we study the way equations may be used for describing properties of data structures and interfaces of software systems. In Chapter 12 we thus use logic of equations as a specification language for abstract data types.

Interpretation of a set E of equations as a **directed term rewriting system** opens up the way under suitable conditions to consider E as an executable program. This idea, as well as questions concerning the expressive power of the logic of equations, is treated in Chapter 13.

# Chapter 11
# Logic of Equations

In this chapter we refer to predicate logic with equality over a signature $\Sigma$ consisting of function symbols only.

## 11.1    Birkhoff's calculus

We are interested in determining which part of Hilbert's calculus is necessary to prove that an equation follows from a set E of equations. In this section we shall define the so-called **Birkhoff's calculus**; it consists of the axioms and rules shown in Figure 11.1, with terms $r, s, t, s_1, \ldots, s_n, t_1, \ldots, t_n$, function symbols f of arity n, and substitutions $\sigma$.

Reflexivity axiom

$$\frac{}{t=t}$$

Symmetry rule

$$\frac{s=t}{t=s}$$

Transitivity rule

$$\frac{r=s, \; s=t}{r=t}$$

Congruence rule

$$\frac{s_1=t_1, \ldots, s_n=t_n}{f(s_1, \ldots, s_n)=f(t_1, \ldots, t_n)}$$

Substitution rule

$$\frac{s=t}{s\sigma=t\sigma}$$

**Figure 11.1**

Let E be a set of equations and l=r be an equation. Then we define:

- $\vdash_E$ l=r iff l=r can be derived from E in Birkhoff's calculus.

- l=$_E$r iff equation l=r logically follows from E, i.e. E $\models$ l=r.

(Note that $\vdash_E$ defines a congruence relation on terms). Obviously, Birkhoff's calculus is a restriction of Hilbert's calculus with equality:

$$\vdash_E \text{ l=r implies } E\vdash_H \text{ l=r, hence l=}_E\text{r}$$

Despite this restriction, we will be able to show completeness of the calculus for the case of equations. For the moment, let us look at a simple example. Let E be the set of equations

        pop(push(S,X))=S
        pop(nil)=nil
        top(nil)=0
        top(push(S,X))=X

(This is a method of describing the data type 'stack'; more about specifications of data types in the next chapter). We want to show $\vdash_E$ push(S,1)=push(pop(push(S,0)),1). The following is a derivation of the desired equation in Birkhoff's calculus:

| | |
|---|---|
| pop(push(S,X))=S | element of set E |
| pop(push(S,0))=S | substitution rule |
| 1=1 | reflexivity axiom |
| push(pop(push(S,0)),1)=push(S,1) | congruence rule |
| push(S,1)=push(pop(push(S,0)),1) | symmetry rule |

## 11.2    Sets of equations as term rewriting systems

If we take a closer look at Birkhoff's calculus, we notice that it is based on the idea of deriving equations between increasingly complicated terms from given equations. This is some kind of bottom-up approach.

Another variation that is very close to the intuitive content of equality would consist in applying equations on both sides of the equality to be proven, until we reach a common term. This approach is broadly used in elementary algebra, where expressions are often simplified according to certain rules. So, X-X may be simplified to zero.

**Definition 11.1:**    Let E be a set of equations and u,v terms. Then we define the following relations:

- $u \xrightarrow{E} v \Leftrightarrow$ there are strings a and b, a substitution $\sigma$ and an equation l=r in E such that u=a(l$\sigma$)b and v=a(r$\sigma$)b
- $u \xleftrightarrow{E} v \Leftrightarrow u \xrightarrow{E} v$ or $v \xrightarrow{E} u$

For example, push(**pop(push(S,0))**,1) $\xrightarrow{E}$ push(S,1). This example clearly shows that term rewriting is more comfortable than application of the rules of Birkhoff's calculus, since the former may be applied locally within a surrounding context, whereas the latter must build up such a context after replacement of a left-hand side l by a right-hand side r of an equation. Also note the difference between $\xrightarrow{E}$ and $\xleftrightarrow{E}$ : the former allows only replacement of left-hand sides of equations by corresponding right-hand sides, the latter allows also the converse replacement.

**Definition 11.2:**  We write $\xrightarrow{E}{}^n$, $\xrightarrow{E}{}^+$ and $\xrightarrow{E}{}^*$ for the n-th power, reflexive closure and reflexive and transitive closure of $\xrightarrow{E}$, respectively.  Corresponding relations for $\xleftrightarrow{E}$ are denoted $\xleftrightarrow{E}{}^n$, $\xleftrightarrow{E}{}^+$ and $\xleftrightarrow{E}{}^*$. (See appendix.)

The following theorem shows that term rewriting is actually equivalent to deriving in Birkhoff's calculus.

**Theorem 11.1 (Equivalence of Birkhoff's calculus and term rewriting)**

Let E be a set of equations and u=v be an equation. Then, $u \xleftrightarrow{E}{}^* v$ iff $\vdash_E u=v$.

*Proof*  Assume that $\vdash_E u=v$. By induction on the length l of a derivation of u=v in Birkhoff's calculus we show that $u \xleftrightarrow{E}{}^* v$. Let l be given and assume that our assertion has already been shown for derivations of length less than l. Take a derivation of u=v in Birkhoff's calculus of length l. We distinguish five cases according to the last step in the given derivation.

*Case 1*  The last step is an application of the reflexivity axiom. Then u and v are identical terms, hence $u \xleftrightarrow{E}{}^* v$.

*Case 2*  The last step is an application of the symmetry rule. This means that at some prior stage of our derivation the equation v=u must occur. By induction we may conclude that $v \xleftrightarrow{E}{}^* u$, hence $u \xleftrightarrow{E}{}^* v$.

*Case 3*  The last step is an application of the transitivity rule. This means that there is some term w such that at some prior stage of our derivation the equations u=w and w=v occur. By induction we conclude $u \xleftrightarrow{E}^* w$ and $w \xleftrightarrow{E}^* v$. Therefore also $u \xleftrightarrow{E}^* v$ holds.

*Case 4*　The last step is an application of the congruence rule. This means that u is of the form $f(u_1,\ldots,u_n)$ and v of the form $f(v_1,\ldots,v_n)$ with equations $u_1=v_1,\ldots,u_n=v_n$ which occur at some prior stage of our derivation. By induction hypothesis we know that $u_1 \xleftrightarrow{E}^* v_1,\ldots,\ u_n \xleftrightarrow{E}^* v_n$. Then we can conclude

$$f(u_1,u_2,u_3,\ldots,u_{n-1},u_n) \xleftrightarrow{E}^*$$
$$f(v_1,u_2,u_3,\ldots,u_{n-1},u_n) \xleftrightarrow{E}^*$$
$$f(v_1,v_2,u_3,\ldots,u_{n-1},u_n) \xleftrightarrow{E}^*$$

$$\cdot$$
$$\cdot$$
$$\cdot$$

$$f(v_1,v_2,v_3,\ldots,v_{n-1},u_n) \xleftrightarrow{E}^*$$
$$f(v_1,v_2,v_3,\ldots,v_{n-1},v_n).$$

*Case 5*　The last step is an application of the substitution rule. This means that there are terms p and q and a substitution $\sigma$ such that $u=p\sigma$, $v=q\sigma$ and the equation p=q occurs at some prior stage of our derivation. By induction hypothesis we know that $p \xleftrightarrow{E}^* q$. Then the desired conclusion, $p\sigma \xleftrightarrow{E}^* q\sigma$, follows immediately from the following simple fact which itself follows directly from the definitions:

For all terms p and q and substitutions $\sigma$, $p \xrightarrow{E} q$ implies $p\sigma \xrightarrow{E} q\sigma$.

Conversely, assume that $u \xleftrightarrow{E}^* v$. Since $\vdash_E$ is an equivalence relation it suffices to show the following statement:

For all terms u and v, $u \xrightarrow{E} v$ implies $\vdash_E u=v$.

To show this last claim, assume that $u \xrightarrow{E} v$. Let u be $a(l\rho)b$ and v be $a(r\rho)b$, with an equation l=r from E, a substitution $\rho$ and strings a,b. Using the substitution rule we obtain $\vdash_E l\rho=r\rho$. By induction on the length of string ab we show that $\vdash_E a(l\rho)b=a(r\rho)b$. This is trivially true if ab is empty, and uses the congruence rule otherwise. (Note that in the latter case, u is a composed term $f(t_1,\ldots,t_m)$ with $l\rho$ being a subterm of one of $t_1,\ldots,t_m$.)　∎

## 11.3    Free term algebra and initial model

In Sections 11.1 and 11.2 we interpreted a set E of equations in a purely syntactical way and, in particular, defined relations $\vdash_E$ and $\underset{E}{\longleftrightarrow}$ *. Since these relations capture the main features of equality we expect that Birkhoff's calculus is complete, and it will indeed be so. To prove this fact we have to look at a set E of equations from a model-theoretic point of view.

Let l=r be an arbitrary equation with variables from a set V. To show that $E \models l=r$ we have to consider every model of E and show that each such model is a model of l=r, too. One of the most important advantages of the logic of equations, though, is that it suffices to consider only one special model of E , the so-called **initial model of E over variable set V**. And if we are solely interested in ground terms l and r, it suffices to consider the initial model of E over the empty set of variables, called the **initial model** of E.

> **Definition 11.3:**    Let $\Sigma$ be a signature and V be a set of variables. Then the **free term algebra over** V, denoted $T(V)$, is defined as the $\Sigma$-interpretation with domain T(V)=the set of all $\Sigma$-terms with variables from V and the following interpretation of function symbols as free term constructors:
>
> $$f_{T(V)}(t_1,\ldots,t_n)=f(t_1,\ldots,t_n)$$
>
> Assume, in addition, that $\Sigma$ contains at least one constant. The subalgebra of $T(V)$ whose domain consists of all ground terms in the signature $\Sigma$ (which is a non-empty set by assumption) is denoted by $T$ and called the **free ground-term algebra**.

> **Definition 11.4:**    Let $\Sigma$ be a signature and E be a set of $\Sigma$-equations. Then $\vdash_E$ is a congruence relation on the free term algebra $T(V)$. For a term t, $[t]_{V,E}$ denotes the equivalence class of t in the set T(V) of all $\Sigma$-terms w.r.t $\vdash_E$ . The **initial model of E over variable set V** is defined as the quotient interpretation of the free term algebra $T(V)$ modulo the congruence relation $\vdash_E$ . We denote it by $T_E(V)$. For V=$\varnothing$, we obtain $T_E(\varnothing)$, called the   **initial model of E**. It is more simply denoted by $T_E$.

By the definition of the quotient interpretation, a function symbol f in $\Sigma$ is interpreted in $T_E(V)$ as follows:

$$f_{T_E(V)}([t_1]_{V,E},\ldots,[t_n]_{V,E})=[f(t_1,\ldots,t_n)]_{V,E}$$

for all terms $t_1,...,t_n$ with variables in V. Correspondingly,

$$f_{T_E}([t_1]_E,...,[t_n]_E)=[f(t_1,...,t_n)]_E$$

for all ground terms $t_1,...,t_n$.

---

**Exercise 11.1**  Show that $\vdash_E$ is a congruence relation on algebras $T(V)$ and $T$.

---

The following theorem summarizes the interesting properties of $T_E(V)$ and $T_E$.

### *Theorem 11.2: (Basic properties of initial algebras)*

(1)    For every term t with variables $X_1,...,X_n$, and every state sta over $T_E(V)$ with $sta(X_i)=[t_i]_{V,E}$, for $i=1,...,n$:

$$val_{T_E(V),sta}(t)=[t\{X_1/t_1,...,X_n/t_n\}]_{V,E}$$

(2)    For every term t with variables $X_1,...,X_n$, and every state sta over $T_E$ with $sta(X_i)=[t_i]_E$, for $i=1,...,n$:

$$val_{T_E,sta}(t)=[t\{X_1/t_1,...,X_n/t_n\}]_E$$

(3)    For every ground term t:

$$val_{T_E}(t)=[t]_E$$

(4)    $T_E(V)$ and $T_E$ are models of E.

(5)    Let $i:V\rightarrow T_E(V)$ be the function $i(X)=[X]_{V,E}$. Then for every model $A$ of E and every function $sta:V\rightarrow dom(A)$ there is a unique homomorphism h from $T_E(V)$ into $A$ such that $h\circ i=sta$. (See Figure 11.1.)

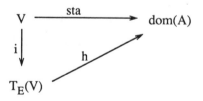

**Figure 11.1**

(6)    For every model $A$ of E there is a unique homomorphism h from $T_E$ into $A$. (This explains the name 'initial algebra over E'.)

(7)    The interpretation $T_E$ is, up to isomorphism, the only model of E with the property that for every model $A$ of E there is a unique homomorphism h from $T_E$ into $A$.

*Proof*

(1)    Let a state sta over $T_E(V)$ be given with $s(X_i)=[t_i]_{V,E}$, for i=1,...,n. The function h from $T(V)$ into $T_E(V)$ with $h(t)=[t]_{V,E}$ is a homomorphism. Define a state sta' over $T(V)$ by $sta'(X_i)=t_i$, for i=1,...,n, and $sta'(X)=$an arbitrary element of $sta(X)$ (that is, an arbitrary element of the equivalence class of $sta(X)$ in $T_E(V)$), for any other variable X. Then sta=h∘sta', so we obtain for every term t:

$$val_{T_E(V),sta}(t)$$
$$=val_{T_E(V),h\circ sta'}(t)$$
$$=h(val_{T(V),sta'}(t))$$
$$=(\text{simple induction on t}) \; h(t\{X_1/t_1,...,X_n/t_n\})$$
$$=[t\{X_1/t_1,...,X_n/t_n\}]_{V,E}$$

(2)    Omitting each occurrence of 'V' in the proof above yields a proof for (2).

(3)    Immediately follows from (2).

(4)    Let l=r be one of the equations in E containing variables $X_1,...,X_n$, and sta a state over $T_E(V)$ with $sta(X_i)=[t_i]_{V,E}$, for i=1,...,n. Then

$$T_E(V) \models_{sta} l=r \text{ iff}$$
$$val_{T_E(V),sta}(l)=val_{T_E(V),sta}(r) \text{ iff } \text{(use (1))}$$
$$[l\{X_1/t_1,...,X_n/t_n\}]_{V,E}=[r\{X_1/t_1,...,X_n/t_n\}]_{V,E} \text{ iff}$$
$$\models_E l\{X_1/t_1,...,X_n/t_n\}=r\{X_1/t_1,...,X_n/t_n\}$$

According to the substitution rule of Birkhoff's calculus the latter is true. The same argument shows that $T_E$ is a model of E.

(5)    Let i:V→ $T_E(V)$ be the function $i(X)=[X]_{V,E}$, let $A$ be a model of E and sta:V→A be an arbitrary function (note that for terms with variables among V, sta may be seen as a state). Define h from $T_E(V)$ into $A$ by $h([t]_{V,E})=val_{A,sta}(t)$. Since $A$ is a model of E the function h is uniquely

defined. Obviously it is a homomorphism from $T_E(V)$ into $A$. Furthermore, $h \circ i = sta$. We show that the function h just defined is the only homomorphism from $T_E(V)$ into $A$ with $h \circ i = sta$. Assume k is a further homomorphism. By structural induction on term t we show that $k([t]_{V,E}) = val_{A,sta}(t)$:

- $k([X]_{V,E}) = k(i(X)) = sta(X) = val_{A,sta}(X)$

- $k([f(t_1,...,t_n)]_{V,E})$

   $= k(f_{T_E(V)}([t_1]_{V,E},...,[t_n]_{V,E}))$

   =(since k is a homomorphism) $f_A(k([t_1]_{V,E}),...,k([t_n]_{V,E}))$

   =(inductive hypothesis) $f_A(val_{A,sta}(t_1),...,val_{A,sts}(t_n))$

   $= val_{A,sta}(f(t_1,...,t_n))$

(6)     Let $A$ be a model of E. The unique homomorphism from $T_E$ to $A$ is the function h defined by $h([t]_E) = val_A(t)$.

(7)     Let $A$ and $B$ be two models of E such that for every model $C$ of E there are unique homomorphisms $h:A \rightarrow C$ and $g:B \rightarrow C$. In particular, there are homomorphisms $h:A \rightarrow B$ and $g:B \rightarrow A$. Then $h \circ g$ is a homomorphism from $B$ into $B$, and $g \circ h$ is a homomorphism from $A$ into $A$. But the identity function $id_B$ and $id_A$ is a homomorphism from $B$ into $B$ or $A$ to $A$ respectively. Since we know from (6) that there is only one homomorphism from $B$ into $B$ or $A$ into $A$, we may conclude that $h \circ g = id_B$ and $g \circ h = id_A$. This implies that h is bijective with inverse function g, hence h is an isomorphism from $A$ to $B$.     ■

## 11.4     Completeness of Birkhoff's calculus and of term rewriting

In the previous sections we introduced three different views of a set E of equations. We already know that the two syntactical ones coincide. Here we shall bridge the gap to the semantical concepts of Section 11.3.

### Theorem 11.3

(1)     For every equation l=r with variables from V:

$$T_E(V) \models l{=}r \text{ iff } \vdash_E l{=}r$$

(2)     For every ground equation l=r:

$$T_E \models l{=}r \text{ iff } \vdash_E l{=}r$$

*Proof*

(1)    Assume that $\vdash_E l=r$. Then $l=_E r$. Since $T_E(V)$ is a model of E it is a model of $l=r$, too. Conversely, assume that $T_E(V)$ is a model of $l=r$. Take the state sta over $T_E(V)$ with $\text{sta}(X)=[X]_{V,E}$. Then, $\text{val}_{T_E(V),\text{sta}}(l)=\text{val}_{T_E(V),\text{sta}}(r)$. Using point (1) of Theorem 11.2, this implies that $[l]_{V,E}=[r]_{V,E}$, hence $\vdash_E l=r$.

(2)    Assume that $\vdash_E l=r$, for ground terms l and r. Then $l=_E r$. Since $T_E$ is a model of E it is a model of $l=r$, too. Conversely, assume that $T_E$ is a model of $l=r$. Then, by property (3) in Theorem 11.1:

$$[l]_E=\text{val}_{T_E}(l)=\text{val}_{T_E}(r)=[r]_E.$$

This implies that $\vdash_E l=r$.    ∎

### Theorem 11.4 (Completeness of Birkhoff's calculus)

For a set of equations E and an equation $l=r$ such that all the variables in E, l and r are contained in a set of variables V the following are equivalent:

(1)    $\vdash_E l=r$

(2)    $E \models l=r$

(3)    $T_E(V)$ is a model of $l=r$

For ground terms l and r, (1)-(3) are equivalent to

(4)    $T_E$ is a model of $l=r$

*Proof*    It is an easy consequence of Theorem 11.3 and the definitions of the involved notions.    ∎

The implication (4)⇒(1) of Theorem 11.4 does not hold for arbitrary terms l and r containing variables. This can be shown by the following example: in the signature with a constant 0, a 1-ary function symbol succ and a 2-ary function symbol + (with infix notation), consider the following set E of equations:

X+0=X
X+succ(Y))=succ(X+Y)

As can be easily shown, $T_E$ is isomorphic to the interpretation $(N,+)=(\mathbb{N},0,\text{succ},+)$ of natural numbers with zero, successor function and addition. $(N,+)$ is a model of X+Y=Y+X. We show that $E \not\models X+Y=Y+X$ by constructing a model $B$ of E which is not a model of X+Y=Y+X. As for $B$ we can take the interpretation $(\mathbb{N}\cup\{a,b\},0,s,+,=)$ with $s(i)=succ(i)$, for every $i\in\mathbb{N}$, $s(a)=a$ and $s(b)=b$, equality relation $=$, and the following definition of $+$:

| $+$ | j | a | b |
|---|---|---|---|
| i | i+j | b | a |
| a | a | b | a |
| b | b | b | a |

Obviously, **B** is a model of E. But **B** is not a model of X+Y=Y+X, since for example:

$a+b=a$

$b+a=b$

## Remarks

(1)   Concerning the logic of equations a quick side-glance at Horn logic is in order. Let E be a set of equations and l=r another equation such that $E \models l=r$. Let $X_1,...,X_n$ be the variables in l=r. Again, we consider logic with axiomatically defined equality as described in Chapter 4. This means that $E\cup Eq \models l=r$ in predicate logic without equality. The formulas in E and l=r are Horn formulas (facts). The set Eq consists of Horn formulas, too (fact and rules). So we could try to apply the Horn logic machinery to our situation. But then a difference becomes apparent. $E\cup Eq \models l=r$ stated in greater clearity means that $\forall(E\cup Eq) \models \forall(l=r)$, whereas the presentation of L=R as a goal to the logic program $E\cup Eq$ means proving that $\forall(E\cup Eq) \models \exists(l=r)$. Fortunately, because of the extraction of a *most general answer substitution* from the SLD-tree, SLD-resolution would nevertheless find the empty substitution as an answer substitution, and hence also prove that $\forall(E\cup Eq) \models \forall(l=r)$.

The question arises as to whether we should develop a separate theory of logic of equations instead of treating it as a sublogic of Horn logic. The answer is that because of the exclusive presence of functions, special requests and constructions come to light which, in the general context of Horn logic, either would not make sense or would not attract attention. This will become apparent in the next two chapters.

The other way around, expressing relations as (characteristic) functions and then simulating Horn formulas by equations (which is possible under certain circumstances) is not appropriate. Applications of a functional character are better described functionally using function symbols and equations, whereas those of a relational character are better described using relation symbols and Horn formulas. This guiding principle finds its expression in recent attempts to enrich Prolog by practically useful functional components which can be interpreted efficiently outside, but in harmony with the SLD-resolution mechanism. More information on this topic may be found in Goguen and Meseguer (1984).

(2)   Equations are the counterpart to facts in logic programming. So far, logic programming is more general in that it also admits rules. The natural counterparts of rules are **conditional equations**, i.e. implications of the form

$$(l_1=r_1 \wedge \ldots \wedge l_n=r_n) \rightarrow l=r$$

The logic we obtain is called the logic of conditional equations. Many parts of the theory developed so far for the logic of equations carries over word for word to the logic of conditional equations. In particular, the definition and main properties of initial models are easily generalized.

---

**Exercise 11.2**  Generalize the definition of initial algebras to the case of logic of conditional equations and prove the analogue to Theorem 11.2.

---

The generalization of other concepts like that of term rewriting require some more effort. For an introduction to the logic of conditional equations the reader should consult Kaplan (1984).

# Chapter 12
# Algebraic Specification of Abstract Data Types

In software engineering there is a growing demand for formal methods for the specification of software systems. The aim is to supply the user of a software module only with the absolute minimum amount of information needed for adequate use of such a module (**information hiding**) at a high level of abstraction (**data abstraction**). This not only improves the readability of software modules, but, since specifications are expressed in a formal language, also opens the way for formal verification of such modules.

What a user of a software module needs for its proper use consists, in general, of some knowledge of how the operations made available from the module are interrelated. A specification language which has proven to be well suited for describing such interrelations is equational logic, as presented in Chapter 11. In this chapter we will develop the basics of a theory for correct equational specifications of abstract data types. Emphasis is given to several examples (many-sorted logic will be freely used whenever appropriate) and general and practically useful methods for developing correct specifications in a stepwise manner (**incremental design**).

Since, in this chapter, we are mainly interested in particular algebras and not in all algebras over a certain signature, we shall use the same symbol to denote both a function symbol and its interpretation in the algebra under consideration.

## 12.1    Correctness of algebraic specifications

**Definition 12.1:**    Let E be a set of $\Sigma$-equations and $A$ be a $\Sigma$-algebra. E is called a **correct specification** of $A$, iff $A$ and the initial model $T_E$ of E are isomorphic.

## EXAMPLE 12.1

The empty set $\varnothing$ of equations is a correct specification of algebra $N=(\mathbb{N},0,\text{succ})$ that was introduced in Section 6.5. This is easily shown using the unique homomorphism $h:T_\varnothing \to \mathbb{N}$ defined by $h([t]_\varnothing)=\text{val}_N(t)$, for every ground term t. Obviously, h is surjective, since $h([\text{succ}^i(0)]_\varnothing)=i$, for all $i\in\mathbb{N}$. Furthermore, $\varnothing \nvDash \text{succ}^i(0)=\text{succ}^j(0)$, for $i\neq j$. This is clear since $N$ is an example of a model of $\varnothing$ which is not a model of $\text{succ}^i(0)=\text{succ}^j(0)$. Hence the equivalence class $[\text{succ}^i(0)]_\varnothing$ consists solely of the ground term $\text{succ}^i(0)$. As a consequence, h is also injective.

## EXAMPLE 12.2

For a finite alphabet $\Omega=\{a_1,...,a_n\}$ let **Word**$(\Omega)$ be the algebra $(\Omega^*,\varepsilon,\text{succ}_1,...,\text{succ}_n)$ with the set $\Omega^*$ of all finite strings over $\Omega$ as domain, the empty word $\varepsilon$ as a constant, and functions $\text{succ}_1,...,\text{succ}_n:\Omega^*\to\Omega^*$ with $\text{succ}_i(w)=a_i w$, for $i=1,...,n$ and a string w.

---

**Exercise 12.1**    As in Example 12.1 show that the empty set of equations is a correct specification of **Word**$(\Omega)$.

---

Both examples were quite simple and of the same type. The algebras under consideration were isomorphic to the free term algebra in the corresponding signature. Hence, different ground terms evaluate to different values. So, there is no need for any equations identifying ground terms. Free term algebras are always correctly specified by the empty set of equations. Let us now treat an algebra with the property that there are different ground terms evaluating to the same value.

## EXAMPLE 12.3

For a finite alphabet $\Omega = \{a_1, \ldots, a_n\}$ let $\textbf{\textit{Concat}}(\Omega)$ be the algebra $(\Omega^*, \varepsilon, a_1, \ldots, a_n, \bullet)$ with the set $\Omega^*$ of all finite words over $\Omega$ as domain, the empty word $\varepsilon$ and the alphabet symbols $a_1, \ldots, a_n$ as constants, and the concatenation function $\bullet$ (written in infix notation). Now, different ground terms may evaluate to the same element, for example $(b \bullet \varepsilon)$ and $b$, or $((b \bullet c) \bullet d)$ and $(b \bullet (c \bullet d))$. The task of a set of equations correctly specifying $\textbf{\textit{Concat}}(\Omega)$ is to identify exactly the ground terms with the same value. As such a set of equations E we take:

(1)    $(X \bullet \varepsilon) = X$

(2)    $(\varepsilon \bullet X) = X$

(3)    $(X \bullet (Y \bullet Z)) = ((X \bullet Y) \bullet Z)$

First observe that $\textbf{\textit{Concat}}(\Omega)$ is a model of E. Let $h: T_E \rightarrow \textbf{\textit{Concat}}(\Omega)$ be the unique homomorphism defined by $h([t]_E) = val_{\textit{Concat}(\Omega)}(t)$, for every ground term t. Since $val_{\textit{Concat}(\Omega)}((\ldots((c_1 \bullet c_2) \bullet c_3) \bullet \ldots \bullet c_n) = c_1 c_2 c_3 \ldots c_n$, for all $c_1, \ldots, c_n \in \Omega$, we may conclude that h is surjective. Let us next show that h is injective. For this purpose consider arbitrary ground terms s and t such that $h([s]_E) = h([t]_E)$, i.e. $val_{\textit{Concat}(\Omega)}(s) = val_{\textit{Concat}(\Omega)}(t)$. Using the equations (1) and (2) it is possible to eliminate successively occurrences of $\varepsilon$ in s and t, and obtain ground terms $s_1$ and $t_1$ with the following properties:

- $s_1$ either is $\varepsilon$ or it does not contain constant $\varepsilon$ as subterm
- $t_1$ either is $\varepsilon$ or it does not contain constant $\varepsilon$ as subterm
- $t =_E t_1$
- $s =_E s_1$

Using equation (3) we can successively shift brackets from the right to the left and obtain thus ground terms $s_2$ and $t_2$ of a certain 'normal form' with the following properties:

- $s_2$ is of the form $((\ldots((c_1 \bullet c_2) \bullet c_3) \bullet \ldots \bullet c_n)$
- $t_2$ is of the form $((\ldots((d_1 \bullet d_2) \bullet d_3) \bullet \ldots \bullet d_m))$,
- $t_1 =_E t_2$
- $s_1 =_E s_2$

with $c_1, \ldots, c_n, d_1, \ldots, d_m \in \Omega$ (with the understanding that $s_2$ is $\varepsilon$ for n=0, and $t_2$ is $\varepsilon$ for m=0). (The construction of terms $s_2$ and $t_2$ proceeds by structural in-

duction on terms $s_1$ and $t_1$ and is left as an exercise for the reader).    Since
*Concat*$(\Omega)$ is a model of E, we may conclude that

$$c_1 \ldots c_n$$
$$=\text{val}_{Concat(\Omega)}(s_2)$$
$$=\text{val}_{Concat(\Omega)}(s_1)$$
$$=\text{val}_{Concat(\Omega)}(s)$$
$$=\text{val}_{Concat(\Omega)}(t)$$
$$=\text{val}_{Concat(\Omega)}(t_1)$$
$$=\text{val}_{Concat(\Omega)}(t_2)$$
$$=d_1 \ldots d_m$$

hence $n=m$, $c_1=d_1,\ldots,c_n=d_n$. So $s_2$ and $t_2$ are identical ground terms. We conclude that $t=_E t_1=_E t_2=s_2=_E s_1=_E s$, that is, $[t]_E=[s]_E$. This shows injectivity of
h. Hence h is an isomorphism from $T_E$ onto *Concat*$(\Omega)$.

---

The proof presented in Example 12.3 is typical for a large collection of examples showing the correctness of specifications of certain basic algebras. We
extract the general method and formulate it as a theorem.

**Definition 12.2:**    A $\Sigma$-algebra $A$ is called **term generated**, iff for every
$a \in \text{dom}(A)$ there is a ground term t such that $a=\text{val}_A(t)$.

*Theorem 12.1*

Let $A$ be a $\Sigma$-algebra and E be a set of $\Sigma$-equations such that:
(1)     $A$ is a model of E.
(2)     $A$ is term generated.
(3)     There is a set C of ground terms (called terms in normal form)
        with the following properties:

>       (3.1)   **Reducibility to normal form**    For every ground term t
>               there is a ground term $s \in C$ such that $t=_E s$.
>
>       (3.2)   **Normal form property**    For all $s,t \in C$, $\text{val}_A(s)=\text{val}_A(t)$
>               iff s and t coincide as strings.

Then E is a correct specification of $A$.

*Proof*  Since $A$ is a model of E, we can make use of the unique homomorphism

$h:T_E \to A$ defined by $h([t]_E)=val_A(t)$, for every ground term t. Since $A$ is term generated, h is surjective. We show that h is also injective, hence an isomorphism. For this purpose let ground terms s and t be given such that $h([t]_E)=h([s]_E)$. Using property (3.1) we can choose ground terms $\underline{s}$ and $\underline{t}$ such that $\underline{s} \in C$, $\underline{t} \in C$, $s=_E\underline{s}$ and $t=_E\underline{t}$. Using again the fact that $A$ is a model of E, it follows that $val_A(\underline{s})=val_A(s)=h([s]_E)=h([t]_E)=val_A(t)=val_A(\underline{t})$. Using property (3.2) we conclude that $\underline{s}=\underline{t}$. Hence $[s]_E=[\underline{s}]_E=[\underline{t}]_E=[t]_E$.  ■

How can this theorem be used? The main work has to be done in determining the set C of ground terms in 'normal form'. C must be *rich enough* to contain a representative for each equivalence class $[t]_E$ (property (3.1)). On the other hand, it must be *sufficiently narrow* to guarantee that any two elements of C evaluate in $A$ to different values (property (3.2)). Recall the example of correctly specifying **Concat**$(\Omega)$. There, such a set C of ground terms in 'normal form' is readily constructed. We take as C the set of all ground terms of the form $((...((c_1 \bullet c_2) \bullet c_3) \bullet ... \bullet c_n)$ with $c_1,...,c_n \in \Omega$. Properties (3.1) and (3.2) obviously hold.

# EXAMPLE 12.4

We construct now the correct specification of algebra **Set**$(N)$ (see Section 6.5). Using a variable M of type set and variables X,Y of type nat we consider the following set E of equations:

(1)     insert(insert(M,X),X)=insert(M,X)

(2)     insert(insert(M,X),Y)=insert(insert(M,Y),X).

The first equation expresses the fact that multiple insertion of a number into a set has no effect, the second one states that the order of insertion of elements is irrelevant.

Using Theorem 12.1, we will show that E is a correct specification of **Set**$(N)$. Obviously, **Set**$(N)$ is a term generated model of E. So it remains only to construct suitable sets $C_{nat}$ and $C_{set}$ of ground terms in normal form. (Since we consider a 2-sorted algebra, we must of course construct a set of ground terms in 'normal form' for each of the involved sorts.)

As $C_{nat}$ we take the set of all ground terms of the form $succ^i(0)$, for $i \in \mathbb{N}$. As $C_{set}$ we take the set of all ground terms of the form

$$insert(insert(...(insert(\emptyset,succ^{i_1}(0)),succ^{i_2}(0)),...,succ^{i_k}(0)))$$

for all $k \in \mathbb{N}$ and $i_1<i_2<...<i_k$. Let us abbreviate this last term by

$$insert(\emptyset,i_1,i_2,...,i_k).$$

Now we know that:

$$\text{val}_{Set(N)}(\text{succ}^i(0))=i \text{ and}$$

$$\text{val}_{Set(N)}(\text{insert}(\varnothing,i_1,i_2,\ldots,i_k))=\{i_1,i_2,\ldots,i_k\}$$

This immediately gives us property (3.2) of Theorem 12.1. It now remains to show property (3.1). For this purpose let t be an arbitrary ground term of sort set (obviously, for sort nat there is nothing to show). By structural induction on t we construct a ground term $\underline{t} \in C_{set}$ such that $t =_E \underline{t}$. In the case that t is the constant $\varnothing$ we take $\varnothing$ as $\underline{t}$, too. Assume that t is of the form insert(r,succ$^i$(0)). By induction we assume that there are $k \in \mathbb{N}$ and $i_1<i_2<\ldots<i_k$ such that $r =_E \text{insert}(\varnothing,i_1,i_2,\ldots,i_k)$.

If $i_k<i$, then insert(r,succ$^i$(0)) is an element of $C_{set}$ and can be taken as normal form $\underline{t}$. Otherwise, there is a least m such that $1 \le m \le k$ and $i \le i_m$. Applying equation (2) k-m+1 times shows that

$$\text{insert}(r,\text{succ}^i(0))=_E\text{insert}(\varnothing,i_1,i_2,\ldots,i_{m-1},i,i_m,\ldots i_k)$$

If $i<i_m$ the latter term is a term in $C_{set}$, if $i=i_m$ one application of equation (2) shows that it can be further reduced to $\text{insert}(\varnothing,i_1,i_2,\ldots,i_{m-1},i_m,\ldots i_k)$.

---

For later use we will give an equivalent formulation for a set of equations to be a correct specification of an algebra $A$, which results from our constructions given so far in a simple way.

### Theorem 12.2

Let $A$ be an algebra and E be a set of equations. Then the following are equivalent:

(1)     E is a correct specification of $A$.

(2)     $A$ is a term-generated model of E and for all ground terms l, r:

$$A \models l=r \text{ iff } E \models l=r$$

*Proof*   Assume that E is a correct specification of $A$. Hence $A$ is a model of E, and the unique homomorphism $h:T_E \to A$ defined by $h([t]_E)=\text{val}_A(t)$, for all ground terms t, is even an isomorphism. Hence $A$ is term generated. Furthermore, for all ground terms l and r, $A \models l=r$ iff $T_E \models l=r$ iff $E \models l=r$.

Conversely assume that (2) holds. Since $A$ is a term generated model of E the unique homomorphism $h:T_E \to A$ defined by $h([t]_E)=\text{val}_A(t)$, for all ground terms t, is surjective. We can show that it is injective, too. For this purpose let l and r be ground terms such that $h([l]_E)=h([r]_E)$. It follows that $\text{val}_A(l)=\text{val}_A(r)$, so $A \models l=r$, hence by using (2) E $\models l=r$, that is, $[l]_E=[r]_E$.     ∎

## 12.2     Extensions of specifications

The algebras treated so far have in common that they all contain only functions constructing or generating the data objects in the domains under considerations (**constructor functions**). They all are term generated, but omission of one of the functions changes them into algebras which are no longer term generated. In commonly used data structures there are, in general, at least two further sorts of function, namely those that manipulate data objects (**modification functions**) and those that access the components which constitute a composed data object (**selection functions**).

There is a great need for a general method allowing enrichment of a given correct specification E of an algebra $A$ to a correct specification E' of an enriched algebra $(A,f_1,...,f_m)$ without having to deal with the fraction $A$ of $(A,f_1,...,f_m)$ again (**incremental design of correct specifications**). This will be achieved by the following theorem which is simple in use, but nevertheless widely applicable.

### Theorem 12.3 (Correctness of algebra extensions)

Let E be a correct specification of a $\Sigma$-algebra $A$. Let $(A,f_1,...,f_m)$ be the expansion of $A$ by new functions $f_1,...,f_m$. Denote by $\Sigma'$ the signature of this expansion $(A,f_1,...,f_m)$. Let E' be a set of $\Sigma'$-equations with the following properties:

(1)     **Validity of the new equations**  $(A,f_1,...,f_m)$ is a model of E'.

(2)     **Reducibility property of the new equations**  For every ground $\Sigma'$-term t' there is a ground $\Sigma$-term t such that $t'=_{E\cup E'}t$.

Then E∪E' is a correct specification of $(A,f_1,...,f_m)$.

*Proof*   E is a correct specification for $A$, so let  $h:T_E \rightarrow A$ be an isomorphism from $T_E$ onto $A$. From the uniqueness of a homomorphism from $T_E$ to $A$ we know that $h([t]_E)=\text{val}_A(t)$, for all ground $\Sigma$-terms t. The surjectivity of h implies that $A$ is a term-generated model of E. Thus, $(A,f_1,...,f_m)$ is term generated too, and, by (1), a model of E∪E'. Let $h':T_{E\cup E'} \rightarrow (A,f_1,...,f_m)$ be the unique homomorphism defined by $h'([t']_{E\cup E'})=\text{val}_{(A,f_1,...,f_m)}(t')$, for all ground $\Sigma'$-terms t'. Obviously, h is surjective. We can show that h is injective, too.

For this purpose consider ground $\Sigma'$-terms s' and t' such that $h'([s']_{E\cup E'})=h'([t']_{E\cup E'})$. Using (2) we choose ground $\Sigma$-terms s and t such that $s'=_{E\cup E'}s$ and $t'=_{E\cup E'}t$. Then we conclude that

$$val_{(A,f_1,...,f_m)}(s)$$
$$=val_{(A,f_1,...,f_m)}(s')$$
$$=h'([s']_{E\cup E'})$$
$$=h'([t']_{E\cup E'})$$
$$=val_{(A,f_1,...,f_m)}(t')$$
$$=val_{(A,f_1,...,f_m)}(t)$$

Since s and t do not contain any of the function symbols $f_1,...,f_m$, it follows that $val_A(s)=val_A(t)$. Hence $h([s]_E)=h([t]_E)$. Using the injectivity of h we obtain $[s]_E=[t]_E$, that is, $t=_E s$. This finally implies that $s'=_{E\cup E'}s=_E t=_{E\cup E'}t'$, hence $[s']_{E\cup E'}=[t']_{E\cup E'}$.  ∎

How is this theorem to be used? Having constructed a correct specification E for an algebra $A$, we obtain a correct specification for an expanded algebra $(A,f_1,...,f_m)$ by describing the new functions $f_1,...,f_m$ by further equations in an adequate way. Adequate means that these new equations must describe not only true properties of the new functions (this is requirement (1)) but also the new functions in a sufficiently complete way (this is requirement (2)) of Theorem 12.3.

## EXAMPLE 12.5

We consider the expansion $Nat=(N,+,*)$ of algebra $N$ (see Section 6.5). A correct specification of $Nat$ is given by the following equations:

| | |
|---|---|
| X+0=X | X*0=0 |
| X+succ(Y)=succ(X+Y) | X*succ(Y)=X+(X*Y) |

Requirements (1) and (2) of Theorem 12.3 are fulfilled. Apparently, $Nat$ is a model of these equations, and they allow reduction of any ground term t' containing + and * to a ground term t without + and *, as can easily be proved by structural induction on t'.

## EXAMPLE 12.6

We consider algebra $Stack(N)$ (see Section 6.5). We start with the reduct $A$

which results from **Stack(N)** by dropping the functions top and pop. Since different ground terms evaluate in **A** to different values, the empty set of equations is a correct specification for **A**. Now we extend **A** by functions top and pop. We claim that the following equations correctly specify **Stack(N)**:

| | |
|---|---|
| pop(emptystack)=emptystack | top(emptystack)=0 |
| pop(push(X,W))=W | top(push(X,W))=X |

Conditions (1) and (2) of Theorem 12.3 are easily seen to be true, hence our claim is shown.

## EXAMPLE 12.7

A correct specification of algebra **Boolean** (see Section 6.5) is provided by the following equations:

| | | |
|---|---|---|
| not(true)=false | and(true,true)=true | or(true,true)=true |
| not(false)=true | and(true,false)=false | or(true,false)=true |
| if(true,X,X)=X | and(false,true)=false | or(false,true)=true |
| if(false,X,Y)=Y | and(false,false)=false | or(false,false)=false |

---

**Exercise 12.2**  Show the correctness of the specification from Example 12.7.

**Exercise 12.3**  For a finite alphabet $\Omega = \{a_1,\dots,a_n\}$, recursively define lists over $\Omega$ as finite sequences as follows:

(i)     the empty sequence () is a list over $\Omega$

(ii)    $(a_i)$ is a list over $\Omega$, for all $i=1,\dots,n$

(iii)   $(L_1,\dots,L_m)$ is a list over $\Omega$, whenever $m\geq 1$ and $L_1,\dots,L_m$ are lists over $\Omega$

Let List($\Omega$) be the set of all lists over $\Omega$. Define (with m>0)

cons:List($\Omega$)×List($\Omega$)→List($\Omega$) by cons(($L_1,\dots,L_m$),L)=($L_1,\dots,L_m$,L)

car:List($\Omega$)→List($\Omega$) by car(())=() and car(($L_1,\dots,L_m$))=$L_m$

cdr:List($\Omega$)→List($\Omega$) by cdr(())=() and cdr(($L_1,\dots,L_m$))=($L_1,\dots,L_{m-1}$).

(a) Construct a correct specification for the algebra

**List(Ω)**=(List($\Omega$),nil,$a_1,\dots,a_n$,cons,car,cdr)

(Do not forget to show that **List(Ω)** is term generated.)

(b) Extend **List(Ω)** by functions

append:List($\Omega$)×List($\Omega$)→List($\Omega$) with

append(($L_1,\dots,L_p$),($M_1,\dots,M_q$))=($L_1,\dots,L_p,M_1,\dots,M_q$)

flat:List($\Omega$)→List($\Omega$) with

flat(L)=$(a_{i_1}, a_{i_2}, \ldots, a_{i_k})$ iff the string $a_{i_1} a_{i_2} \ldots a_{i_k}$ over $\Omega$ results from

list L, considered as a string over $\Omega \cup \{(,)\}$, by deleting all brackets

(in other words, flat(L) is the list of all occurrences of symbols of $\Omega$ in L written from left to right), and finally

reverse:List($\Omega$)→List(W) with reverse$((L_1, \ldots, L_m))=((L_m, \ldots, L_1))$

Construct correct specifications for all of these extensions.


**Exercise 12.4**

(a) Correctly specify the 2-sorted algebra *Bin_tree*

($\mathbb{N}$ ,{finite binary tree labelled with natural numbers},leaf,left,right,cons)

where leaf is a constructor function creating from a natural number a binary tree consisting solely of one node, left and right are functions constructing from a number and a tree another tree (their intuitive meaning is implied by their names), and cons a function that constructs from a number and two binary trees a new one.

(b) Extend *Bin_tree* by functions calculating the number of leaves, edges and nodes, as well as the height of a binary tree.

---

# 12.3  Hidden functions

So far, application of Theorem 12.3 always leads in a simple and straightforward way to correct specifications of the algebras under consideration. This might give the impression that any algebra could be correctly specified by a suitable *finite* set of equations. Or, more cautiously, it could lead to the question of which algebras could indeed be specified by a finite set of equations. The next theorem will show that sometimes the use of further functions besides the interesting ones, so called **hidden functions**, is unavoidable.


*Theorem 12.4*

Let $(N,\text{fac})$ be the expansion of the algebra $N$ by the factorial function fac defined by fac(0)=1, fac(1)=1, fac(X+2)=fac(X)+fac(X+1). Then the following holds:

(1)    $(N,\text{fac})$ can be correctly specified by an infinite set E of ground equations.

(2)    $(N,\text{fac})$ cannot be correctly specified by a finite set E of ground equations.

(3)    There is no finite set of equations which correctly specifies

(N,fac).

(4)     There is a finite set of equations which correctly specifies (N,+,fac).

*Proof*

(1)     Simply take as E the set of all ground equations holding in (N,fac). We use the proof method for correctness introduced in Section 12.1 to show that E correctly specifies (N,fac). For this purpose we choose as the set C of 'normal forms' the set of all ground terms $succ^i(0)$, for $i \in \mathbb{N}$. Requirements (1)-(3) of Theorem 12.3 then follow immediately. (The attentive reader will certainly have noticed that the argument just presented is applicable to any term generated algebra.)

(2)     Let E be a finite set of ground equations true in (N,fac). We show that E cannot be a correct specification of (N,fac) by providing a ground equation l=r which holds in (N,fac), but cannot be derived from E. It follows from the characterization of correct specifications in Section 12.1 that E is not a correct specification of (N,fac). To obtain such a ground equation l=r consider all subterms of terms in equations of E. Each such subterm evaluates in (N,fac) to a certain natural number. Let max be the maximum of these numbers. Let f be the factorial of max+1. We take as l=r the equation $fac(succ^{max+1}(0))=succ^f(0)$. By the definition of f, this equation holds in (N,fac).

But it does not follow from E: consider the algebra (N,fac') with fac' defined by fac'(n)=fac(n), for n≤max, and fac'(n)=0, for n>max. (N,fac') is a model of E by choice of max, since (N,fac) was a model of E and all the subterms and terms in E evaluate to the same value in (N,fac) as in (N,fac'). But (N,fac') is not a model of equation $fac(succ^{max+1}(0))=succ^f(0)$, since $fac(succ^{max+1}(0))$ evaluates in (N,fac') to zero.

(3)     We show that the only equations holding in (N,fac) are true *ground* equations or identities, that is, equations of the form t=t, for a term t. By (2) we may conclude that E is not a correct specification of (N,fac).

So let l=r be an equation which holds in (N,fac) and contains at least one variable. Terms are of the form $f_1(f_2(...f_m(0)...))$ or $f_1(f_2(...f_m(X)...))$, where X is a variable and $f_1,f_2,...,f_m \in \{succ,fac\}$. From this observation and the injectivity of succ and fac we may conclude that both l and r must contain a variable; moreover they must contain the same variable X. We show that l and r are identical terms by structural induction on l.

*Case 1* l and r are of the form f(l') and f(r'), for $f \in \{succ,fac\}$ and terms l' and r'. Using again injectivity of f we may conclude that l'=r' is also valid in (N,fac). By induction, l' and r' are identical terms, hence l and r are identical, too.

*Case 2* l is the variable X, and r is of the form f(r'), where

$f \in \{succ,fac\}$ and r' is a term (or symmetrically with l and r inter-changed). Interpreting X as 0 we immediately obtain that l=r cannot be valid in $(N,fac)$.

*Case 3* l is of the form succ(l') and r of the form fac(r') with terms l' and r' (or symmetrically with l and r interchanged). Validity of succ(l')=fac(r') in $(N,fac)$ enforces that succ(l') must contain an occurrence of fac, too: otherwise, succ(l') would only contain the function succ, so almost all numbers would be obtained when X runs through $\mathbb{N}$ (only some numbers at the beginning may not be included in the set). This means that the values of terms succ(l') in $(N,fac)$ with X (the variable occurring in l and r) running through all natural numbers form a set with finite complement, whereas the set of values of terms fac(r') in $(N,fac)$ with X running through all natural numbers form a set with an infinite complement). Hence, l is of the form $succ^{i+1}(fac(l''))$, for some $i \in \mathbb{N}$ and a term l'' containing the variable X. Since the gaps in the range of the factorial functions grow steadily we conclude that the equation $succ^{i+1}(fac(l''))=fac(r')$ cannot be valid in $(N,fac)$.

(4)    $(N,+,fac)$ can be correctly specified by the following equations:

fac(0)=succ(0)                    X+0=X

fac(succ(0))=succ(0)              X+succ(Y)=succ(X+Y)

fac(succ(succ(X))=fac(X)+fac(succ(X))                    ∎

Before we treat examples for specifications requiring the introduction of hidden functions we discuss how correct specifications may be obtained for the disjoint union of algebras.

### Theorem 12.5

Let $\Sigma$ and $\Sigma'$ be disjoint signatures, $A$ be a $\Sigma$-algebra, $A'$ be a $\Sigma'$-algebra, E be a correct specification of $A$, and E' be a correct specification of $A'$. By $(A,A')$ we denote the $\Sigma \cup \Sigma'$-algebra whose reduct to $\Sigma$ and $\Sigma'$ is $A$ and $A'$ respectively. Then $E \cup E'$ is a correct specification of $(A,A')$.

*Proof*    Apparently, $(A,A')$ is a term-generated model of $E \cup E'$. Applying the characterization theorem for correct specification presented in Section 12.1, it suffices to show that for all ground terms l and r, $(A,A') \models l=r$ iff $E \cup E' \models l=r$.

Assume that l and r are $\Sigma$-terms. Then $(A,A') \models l=r$ is equivalent to $A \models l=r$, and thus to $E \models l=r$. From this fact we can trivially conclude that $E \cup E' \models l=r$. Conversely, since $E \cup E' \models l=r$, we may conclude $E \models l=r$ as follows. Let an arbitrary model $B$ of E be given; then $(B,A')$ is a model of

E∪E'. Hence $(B,A')$ is a model of l=r. Since l and r are $\Sigma$-equations, it follows that $B$ is a model of l=r.

The case of $\Sigma'$-terms l and r is treated in the same way. ∎

## EXAMPLE 12.8

We now consider $(Set(N),\text{delete})$, the expansion of algebra $Set(N)$ by the function delete of type delete: set nat → set defined by delete(M,n)=M-{n}.

Let us try to construct a correct specification along the lines of the preceding examples. The idea should be to define delete(M,n) equationally by structural induction on M. Hence we start with delete(∅,n)=∅. But what about delete(insert(M,m),n)? If n=m we should output delete(M,n), otherwise insert(delete(M,n),m). This cannot be expressed equationally in terms of the given signature for two reasons. First, because there is no 'conditional' available in $(Set(N),\text{delete})$. Second, there is no function 'equal' that checks two numbers n and m for equality. These components must be added to our algebra. The first component, a conditional, could of course be added by expanding the considered algebra $(Set(N),\text{delete})$ by **Boolean**.  We do not follow this method as this would be a sort of overkill: neither do we require the new domain of truth values (numbers 0 and 1 as substitutes for false and true serve the same purpose), nor do we need functions not, and, and or. The expansion we propose is the following:

$(Set(N),\text{delete},\text{cond},\text{equal})$

with functions cond: nat nat nat → nat and equal: nat nat → nat defined as follows:

cond(t,x,y)=x for t>0, and cond(0,x,y)=y

equal(x,y)=1 for x=y, and equal(x,y)=0 otherwise

Now a correct specification of $(Set(N),\text{delete},\text{cond},\text{equal})$ is:

the equations for $Set(N)$ together with
cond(0,X,Y)=Y
cond(succ(T),X,Y)=X

equal(0,0)=succ(0)
equal(succ(X),0)=0
equal(0,succ(Y))=0
equal(succ(X),succ(Y))=equal(X,Y)

delete(∅,X)=∅
delete(insert(M,Y),X)=
$\qquad$ cond(equal(X,Y),delete(M,X),insert(delete(M,X),Y))

Correctness is again easily shown by application of Theorem 12.4.

**Exercise 12.5**   Consider the algebra (*Set(N)*,*BOOLEAN*,equal,element).

(a) Let E be the set of equations

    element(X,insert(M,X))=true
    element(X,insert(M,Y))=element(X,M)

as a reminder of the logic program:

    element(X,insert(M,X)).
    element(X,insert(M,Y)):-element(X,M).

Is E a correct specification of the function element? Show that P is a Prolog program correctly defining (in the sense of Herbrand semantics) the element predicate. Explain the different behaviour of the equations compared to Prolog rules.

(b) Correctly specify a function min of type min:set→nat defined by min($\varnothing$)=0 and min(M)=the least number in M, if M is non-empty.

## 12.4    Error handling

A common situation is having to deal with functions whose application on arbitrary objects should not be allowed. Division by zero is not allowed for rational numbers, square roots are not defined for negative real numbers, and so on. Situations like these can be treated in different ways. Firstly, we could introduce a notion of **partial algebra** by allowing partial, that is, not everywhere defined, functions. Such a notion has been investigated in the literature, but it brings with it conceptual and notational difficulties. Therefore, we try to avoid the use of partial functions as has already been done in the treatment of stacks.

## EXAMPLE 12.9

In the definition of pop and top in the algebra *Stack(N)* we made the conventions pop(emptystack)=emptystack and top(emptystack)=0. Defining functions like these on exceptional and unwanted inputs in an arbitrary way is not too unnatural. Thinking of the role undefinedness plays in computer science, we see that it is more related to the phenomenon of non-termination of a program for certain inputs, which is an undecidable property, than to the more innocent situation of avoiding having certain inputs assigned a value. In the examples mentioned, the latter can be easily decided: given a ground term t over *Stack(N)* of sort stack, we can easily decide whether it denotes the empty stack or not. Hence there is no real need for resorting to partiality of functions.

Nevertheless, pop(emptystack)=emptystack and top(emptystack)=0 are not very satisfactory definitions. It imposes all the burdens of the proper use of pop and top on the user, without providing any means of recognizing and analyzing any possible misuse of pop and top. A better treatment of such exceptional situations is to introduce special new objects, called **error objects**, as values of the involved functions for exceptional inputs. Instead of treating this problem generally, we will treat the algebra *Stack(N)*.

We shall illustrate the approach of error elements and its difficulties by extensively discussing the classical example of specifying stacks. First of all let us define an algebra *ErrorStack(N)* as follows. We choose new objects $error_{nat}$ and $error_{stack}$ and consider the extended domains:

$$\mathbb{N} \cup \{error_{nat}\} \text{ and } \mathbb{N}^* \cup \{error_{stack}\}$$

as domains of sort nat and stack. Then we redefine functions succ, push, pop and top on the extended domains as follows:

$succ(x)=x+1$
$succ(error_{nat})=error_{nat}$

$push(x_1...x_n,x)=x_1...x_nx$
$push(error_{stack},x)=error_{stack}$
$push(x_1...x_n,error_{nat})=error_{stack}$
$push(error_{stack},error_{nat})=error_{stack}$

$pop(emptystack)=error_{stack}$
$pop(x_1...x_n)=x_1...x_{n-1}$, for n>0
$pop(error_{stack})=error_{stack}$

$top(emptystack)=error_{nat}$
$top(x_1...x_n)=x_n$, for n>0
$top(error_{stack})=error_{nat}$

for $x_1,...,x_n,x \in \mathbb{N}$. Observe that three things have happened:

(1)    For the **exceptional input** emptystack, pop and top give out error elements of suitable sort.

(2)    On **normal inputs** which are not exceptional and not error objects, functions push, pop and top have their old values.

(3)    All other cases are treated via **error propagation**, that is, any of the functions is defined as an appropriate error object whenever one of its inputs is an error object.

We now come to the question of constructing a correct specification for the algebra *ErrorStack(N)*.

*First attempt*   Consider the following set of equations:

$$pop(emptystack)=error_{stack}$$
$$pop(push(Stack,X))=Stack$$

$$top(emptystack)=error_{nat}$$
$$top(push(Stack,X))=X$$

The reader will certainly recognize that this is not a correct specification for *ErrorStack(N)*. What is wrong with it is that we have forgotten to describe error propagation. For example, $push(emptystack,error_{nat})=error_{stack}$ is valid in *ErrorStack(N)*, but does not follow from the presented set of equations. The initial model of the proposed set of equations is far too large. Lots of terms occurring in different congruence classes should be contained in the same class. In particular we also forgot to treat error propagation for the successor function. So let us try to integrate error propagation into our specification.

*Second attempt*

$$succ(error_{nat})=error_{nat}$$

$$push(Stack,error_{nat})=error_{stack}$$
$$push(error_{stack},X)=error_{stack}$$

$$pop(emptystack)=error_{stack}$$
$$pop(push(Stack,X))=Stack$$
$$pop(error_{stack})=error_{stack}$$

$$top(emptystack)=error_{nat}$$
$$top(push(Stack,X))=X$$
$$top(error_{stack})=error_{nat}$$

This specification seems to be accurate, but is nevertheless not the desired one. The mistake now is, however, not as easily recognized as in our first attempt. But consider the following. Let t be an arbitrary term of type nat. Then the following reductions are possible on the basis of our set of equations:

$$t \underset{E}{\longleftrightarrow} top(push(error_{stack},t)) \underset{E}{\longleftrightarrow} top(error_{stack}) \underset{E}{\longleftrightarrow} error_{nat}$$

Correspondingly, for every term t of type stack:

$$t \underset{E}{\longleftrightarrow} pop(push(t,error_{nat})) \underset{E}{\longleftrightarrow} pop(error_{stack}) \underset{E}{\longleftrightarrow} error_{stack}$$

Hence the initial model of our set of equations collapses to the trivial model having exactly one element in each of its domains. Where is the mistake? Obviously, equations $pop(push(Stack,X))=Stack$ and $top(push(Stack,X))=X$

should be allowed only when Stack and X are not error objects. This observation leads us to the third attempt.

*Third attempt*   This requires a considerable extension of our algebra **ErrorStack(N)** by functions allowing the recognition of error elements. Furthermore, we need some boolean mechanisms. Unlike our first two attempts, we now strictly follow the strategy developed in the preceding sections. Let us start with algebra

$$A=(\mathbb{N}\cup\{error_{nat}\},\mathbb{N}*\cup\{error_{stack}\},0,error_{nat},succ,\varepsilon,error_{stack},push)$$

as the reduct of **ErrorStack(N)** to constants and constructor functions succ and push. Note that we already need equations to identify ground terms. As specification we try the following set of equations E:

(1)    $succ(error_{nat})=error_{nat}$

(2)    $push(Stack,error_{nat})=error_{stack}$

(3)    $push(error_{stack},X)=error_{stack}$

Note that variables X and Stack range over $\mathbb{N}\cup\{error_{nat}\}$ and $\mathbb{N}*\cup\{error_{stack}\}$ respectively. To prove correctness of this specification w.r.t. the considered algebra $A$, we apply the criterion described in Section 12.1. $A$ is a model of E, since succ and push obey error propagation. Furthermore, $A$ is term generated. It remains to construct sets $C_{stack}$ and $C_{nat}$ of terms in normal form such that conditions (3.1) and (3.2) of the criterion are fulfilled. We take:

$$C_{nat}=\{error_{nat}\}\cup\{succ^i(0) \mid i\in\mathbb{N}\}$$

$$C_{stack}=\{error_{stack}\}\cup$$

$$\{push(...(push(\varepsilon,succ^{i_1}(0),...,succ^{i_m}(0)) \mid m,i_1,...,i_m\in\mathbb{N}\}$$

Using equations (1)-(3) above we can reduce every ground term to a ground term in $C_{nat}$ or $C_{stack}$. This shows that condition (3.1) holds. Furthermore, different terms in $C_{nat}$ or $C_{stack}$ obviously evaluate in $A$ to different objects. This shows condition (3.2). Hence, $A$ is correctly specified by the three equations (1)-(3) above.

Next we expand the algebra $A$ by the boolean component **Boolean** to an algebra $(A,Boolean)$. A correct specification of $(A,Boolean)$ is obtained by putting together a correct specification of **Boolean** and the correct specification (1)-(3) of $A$ from above. (Remember Theorem 12.5.)

Next we extend $(A,Boolean)$ by functions

iserror$_{nat}$: nat → boole

iserror$_{stack}$: stack → boole

defined by

$$iserror_{nat}(x)= \begin{cases} true & for\ x=error_{nat} \\ false & for\ x\in \mathbb{N} \end{cases}$$

$$iserror_{stack}(st)= \begin{cases} true & for\ st=error_{stack} \\ false & for\ st\in \mathbb{N}* \end{cases}$$

A correct specification of $(A,\textbf{Boolean},iserror_{nat},iserror_{stack})$ is given by the following equations:

(0)    a correct specification of **Boolean**

(1)    $succ(error_{nat})=error_{nat}$

(2)    $push(Stack,error_{nat})=error_{stack}$

(3)    $push(error_{stack},X)=error_{stack}$

(4)    $iserror_{nat}(error_{nat})=true$

(5)    $iserror_{nat}(0)=false$

(6)    $iserror_{nat}(succ(X))=iserror_{nat}(X)$

(7)    $iserror_{stack}(error_{stack})=true$

(8)    $iserror_{stack}(emptystack)=false$

(9)    $iserror_{stack}(push(Stack,X))=or(iserror_{stack}(Stack),iserror_{nat}(X))$

This can easily be shown using the theorem on the correctness of expanded algebras. Now we expand our algebra by conditionals of type nat and stack. These are functions

$if_{nat}$: boole nat nat $\rightarrow$ nat

$if_{stack}$: boole stack stack $\rightarrow$ stack

and are defined as usual by $if_{anysort}(true,X,Y)=X$ and $if_{anysort}(false,X,Y)=Y$. The algebra $(A,\textbf{Boolean},iserror_{nat},iserror_{stack},if_{nat},if_{stack})$ is correctly specified by:

(0)-(9) +

(10)    $if_{nat}(true,X,Y)=X$

(11)    $if_{nat}(false,X,Y)=Y$

(12)    $if_{stack}(true,Stack,Stack')=Stack$

(13)    $if_{stack}(false,Stack,Stack')=Stack'$

So far we have a lot of preparations. We now come to the expansion of the algebra $(A,\textbf{Boolean},iserror_{nat},iserror_{stack},if_{nat},if_{stack})$ considered so far by the functions top and pop. The resulting algebra

$(A,\textbf{Boolean},iserror_{nat},iserror_{stack},\ if_{nat},if_{stack},top,pop)$

is correctly specified by the following set of equations:

(0)-(13) +

(14)    $pop(emptystack)=error_{stack}$

(15)    $pop(push(Stack,X))=$
$$if_{stack}(or(iserror_{stack}(Stack),iserror_{nat}(X)),error_{stack},Stack)$$

(16)    $pop(error_{stack})=error_{stack}$

(17)    $top(emptystack)=error_{nat}$

(18)    $top(push(Stack,X))=$
$$if_{stack}(or(iserror_{stack}(Stack),iserror_{nat}(X)),error_{nat},X)$$

(19)    $top(error_{stack})=error_{nat}$

---

**Exercise 12.6** Show that (0)-(19) above correctly specify the considered algebra. Apply the theorem on algebra expansions.

---

As the example above shows, the integration of error elements into an algebra leads to an enormous expense in the adaptation of equations. Treating errors correctly is a non-trivial task. Nevertheless, we can make the following pleasant observations. Consequent application of the theorems presented in Sections 12.1 and 12.2 to build more and more complicated specifications drastically reduces the possibility of obtaining incorrect specifications. Furthermore, treating error elements seems to be possible in a quite uniform way. Being able to distinguish error elements from normal elements by specifiable functions iserror with boolean values, we can readily obtain a correct specification as follows:

- Describe the effect of the involved functions for *normal* objects using tests 'iserror' for the detection of error elements.

- Describe the effect of the involved functions for *exceptional* cases.

- Describe *error propagation*.

This uniformity suggests the introduction of a special language for the description of function behaviour under the presence of error elements which should

- be easier to read than the example equations (0)-(19) above,

- concentrate on the main aspects of a specification (suppressing for example the always present error propagation),

- be easily translatable into specifications as treated so far.

Such languages have been investigated in the literature and are called **specification languages**. Typically, our example **ErrorStack(N)** would read as follows:

pop(emptystack)=error$_{stack}$

pop(push(Stack,X))=

**if** (Stack=error$_{stack}$ **or** X=error$_{nat}$) **then** error$_{stack}$ **else** Stack

top(emptystack)=error$_{nat}$

top(push(Stack,X))=

**if** (Stack=error$_{stack}$ **or** X=error$_{nat}$) **then** error$_{nat}$ **else** X

In the same way, the description of the function delete in the context of the algebra **Set(N)** might look as follows:

delete($\emptyset$,X)=$\emptyset$

delete(insert(M,Y),X)=

**if** X=Y **then** delete(M,X) **else** insert(delete(M,X),Y)

---

**Exercise 12.7**   We refer to algebra **Queue(N)** that was introduced in Section 6.5.

(a) Construct a correct specification for **Queue(N)**.

(b) Modify **Queue(N)** by dealing with the exceptional cases first(emptyqueue)=0 and rest(emptyqueue)=emptyqueue in a more adequate manner using error elements. Construct a correct specification for this modified version.

(c) Extend **Queue(N)** by functions for testing if a queue is empty, for the queue length and for removal of queue elements, and give a correct specification of this extension.

---

## 12.5   Related specification concepts

So far, we have used initial algebra semantics to specify a certain term generated algebra $A$ by a set of equations E uniquely up to isomorphism. Assume that we now want to prove certain assertions about such an algebra $A$ which is correctly specified by a set of equations E. (For example, in later chapters there will be programs running over $A$, and we will be interested in proving that certain programs exhibit a certain input-output behaviour.) How can the information present in E be used in such a proof?

First, we know that $A$ *is a model of* E, that is, all the equations in E, and thus all formulas following from E, are true within $A$. This information is easily applied since it is explicitly present. It is a sort of *positive information* in the sense that the set of all formulas following from E can be recursively enumerated.

But, in general, this cannot be all the information about $A$ that will be required to show the assertions under consideration. Think of the trivial 1-element algebra which is a model of every set of equations. Thus, as a second aspect, it will be sometimes necessary to use that $A$ is the *initial model* of E. This means that the *only* ground equations holding in $A$ are the ones which are logical consequences of E. Stated the other way around, ground equations which do not follow from E are not true in $A$, either. (Remember similar discussions in the section on logic programming under the title 'closed world assumption'.) So, in proving assertions about $A$, we may use the negation of all ground equations which are not entailed by E. This is a sort of *negative information*. The latter set may be undecidable, and thus not recursively enumerable (since its complement is always recursively enumerable). This negative information is often much harder to apply than positive information, since it is not as easily accessible. This motivates reformulation of specifications, in order to make explicit information about the algebra under consideration that is only present in this negative sense.

As a third aspect, initiality implicitly contains the information that the specified algebra is *term generated*. As we know from our discussion in Chapter 5 on non-standard models, it is impossible to capture completely the idea of term generatedness within predicate logic. The best that can be done is to incorporate *induction on the term structure* as an approximation to term generatedness. In later applications this will indeed be the way we deal with term-generatedness. Here, we will suppress this problem completely and restrict discussion to term-generated algebras.

Now let us come back to our plan to *turn negative information into positive*. If, in specifying a certain term-generated algebra $A$, we want to stay within the scope of equational logic or logic of conditional equations, the best way to achieve this is to obtain a set E of equations such that the only term-generated models of E are $A$ and the trivial, 1-element algebra, which we will henceforth denote by $\bot$. (In the case of a many-sorted signature, '1-element' of course means '1-element per sort'.)

> **Definition 12.3:**   We say that a set of conditional equations E **has few models** iff the only term-generated models of E, up to isomorphism, are the initial model $T_E$ and the trivial, 1-element algebra $\bot$ (which may coincide).

# EXAMPLE 12.10

Let us consider one of the examples treated earlier. The algebra $N$ of natural numbers with zero 0 and successor function succ was correctly specified by the empty set $\varnothing$ of equations. Of course, there are lots of non-isomorphic term generated models besides $N$ and $\bot$. Let us instead of $\varnothing$ consider the following set E consisting of the single conditional equation

$$\text{succ}(X)=\text{succ}(Y)\rightarrow X=Y$$

This conditional equation lays down that succ is to be interpreted as an injective function. Let us investigate the class of term-generated models of E. Such a model is of the form (A,a,f) with an element $a \in A$ and an injective function $f:A \rightarrow A$. Furthermore, since A is generated from a by iterated application of f, the possible models look like the following:

$$a \longrightarrow f(a) \longrightarrow f(f(a)) \longrightarrow f(f(f(a))) \longrightarrow \quad \ldots \quad \text{without cycles}$$

$$a \longrightarrow \quad \ldots \quad \longrightarrow f^n(a) \qquad \text{with a cycle of length n}$$

Hence, we still have an infinite number of non-isomorphic term generated models of E.

Let us expand algebra $N$ by the predecessor function pred into an algebra $(N,\text{pred})$ and consider the following set of equations E':

$$\text{pred}(0)=0$$

$$\text{pred}(\text{succ}(X))=X.$$

Now take a look at the term generated models $A$ of E'. First, injectivity of function $\text{succ}_A$ follows from E': $\text{succ}_A(X)=\text{succ}_A(Y)$ implies $\text{pred}_A(\text{succ}_A(X))=\text{pred}_A(\text{succ}_A(Y))$, so X=Y.

This means that $A$ is a model of the form discussed above. Let us see what sorts of cycles are possible. Assume that $\text{succ}_A^{n+m}(0)=\text{succ}_A^n(0)$, for some n,m. Applying n times the second equation for function pred, we obtain $\text{succ}_A^m(0)=0$. Applying the first equation for pred m-1 times, we obtain $\text{succ}_A(0)=0$. So, the only model containing cycles is the trivial one.

This shows that E' has few models. Observe that the reason that E' has few models is the introduction of some sort of *selector function* pred, that is, a function which is able to compute from a ground term t, using only the remaining functions, the subterms of t.

---

Sets of conditional equations with few models will not play a very important role in forthcoming chapters. It will be another notion, which goes one step further in turning negative information into positive and suppresses $\perp$ as a model, too. In our examples above, this can be easily achieved by extending

E' by the formula ¬succ(0)=0 to an axiom set E'':

The only term generated model (up to isomorphism) of E'' is the standard model (N,pred). But note that negation is required to achieve this. Likewise, we could extend E by the formula ¬succ(X)=0 to obtain an axiom system E''', whose only term generated model (up to isomorphism) is N.

> **Definition 12.4:**    For an algebra $A$, we define $A_{gen}$ to be the uniquely determined **term generated subalgebra** of $A$. We say that a set of first-order formulas Ax is a **rigid specification** iff for any two models $A$ and $B$ of Ax, $A_{gen}$ and $B_{gen}$ are isomorphic. If Ax is a rigid specification and $A$ is a term-generated model of Ax (the only one up to isomorphism), we say that Ax is a rigid specification of $A$.

Note that for an axiom system Ax consisting of universal formulas, Ax is a rigid specification iff any two term generated models of Ax are isomorphic. This is because validity of universal formulas is preserved when going from an algebra to a subalgebra.

## EXAMPLE 12.11

For a finite alphabet $\Omega=\{a_1,...,a_n\}$ consider the algebra $Word(\Omega)= (\Omega^*,\varepsilon,succ_1,...,succ_n)$, that was introduced in Section 12.1. A rigid specification of $Word(\Omega)$ is the following:

$$succ_i(X)=succ_i(Y)\rightarrow X=Y, \text{ for } i=1,...,n$$

$$\neg succ_i(X)=\varepsilon, \text{ for } i=1,...,n$$

$$\neg succ_i(X)=succ_j(Y), \text{ for } i,j=1,...,n \text{ with } i\neq j$$

If we had in mind an equational specification with few models, we could alternatively introduce the canonical selector function pred and axiomatize it as follows:

$$pred(\varepsilon)=\varepsilon$$
$$pred(succ_i(X))=X, \text{ for } i=1,...,n$$

We obtain a specification with few models. Proofs are left to the reader.

---

Now we turn our attention to the question of how we may prove that a given axiom system Ax is indeed a rigid specification of a term generated algebra $A$.

Since we already have some experience with correct initial specifications, it would be useful if this experience could be incorporated into sufficient criteria for the property studied here. This will indeed be possible.

### Theorem 12.6 (Criterion for the rigidness of specifications)

Let $A$ be a term generated algebra and Ax be a set of first-order formulas. Assume that we already have available a set E of equations such that $A$ is correctly specified by E (in the sense of initial algebra semantics). Then the following three conditions are sufficient to guarantee that Ax is a rigid specification of $A$:

(1)    $A$ is a model of Ax.

(2)    $\text{Ax} \models \text{E}$

(3)    $\text{val}_A(s) \neq \text{val}_A(t)$ implies $\text{Ax} \models \neg s = t$, for all ground terms s, t.

*Proof* Assume that (1)-(3) are fulfilled and $A$ is the initial model of E. Let $B$ be an arbitrary model of Ax. It suffices to show that $B_{\text{gen}}$ is isomorphic to $A$. By (2) we know that $B$ is a model of E, too. Thus, $B_{\text{gen}}$ is also a model of E. Let $h{:}A \rightarrow B_{\text{gen}}$ be the homomorphism defined by $h(\text{val}_A(t)) = \text{val}_{B_{\text{gen}}}(t)$, for all ground terms t. By definition of $B_{\text{gen}}$, h is surjective. We finally show that h is injective. To show this, let s and t be ground terms such that $\text{val}_A(s) \neq \text{val}_A(t)$. Then, it follows by (3) that $\text{Ax} \models \neg s = t$. Since $B$ is a model of Ax, we obtain that $B \models \neg s = t$, so $\text{val}_B(s) \neq \text{val}_B(t)$. This implies that $\text{val}_{B_{\text{gen}}}(s) \neq \text{val}_{B_{\text{gen}}}(t)$.

■

## EXAMPLE 12.12

Consider the algebra $T$ of binary trees with empty tree nil, constructor function cons and left and right subtree functions left and right. A correct specification of $T$ is provided by the following set of equations E:

left(nil)=nil

right(nil)=nil

left(cons(L,R))=L

right(cons(L,R))=R

Consider the following axiom system Ax:

(a)    E

(b)    cons(L,R)=cons(L',R')$\rightarrow$(L=L' $\wedge$ R=R')

(c)    $\neg$nil=cons(L,R)

Let us show that Ax is a rigid specification of $T$ by application of Theorem

12.6. Conditions (1) and (2) are trivially true. To show (3), let two ground terms s and t be given such that $val_T(s) \neq val_T(t)$. Since for every ground term s there exists a ground term s' consisting only of nil and cons such that $E \models s=s'$, we may assume that s and t are built up from nil and cons. By induction on s we show that $Ax \models \neg s=t$. If s=nil, then t must be of the form $cons(t_1,t_2)$, hence by axiom (c) we conclude that $Ax \models \neg s=t$. If $s=cons(st,s_2)$ and t=nil, then the result follows again from (c). Finally, assume that $s=cons(s_1,s_2)$ and $t=cons(t_1,t_2)$. Since $val_T(s) \neq val_T(t)$ we know that either $val_T(s_1) \neq val_T(t_1)$ or $val_T(s_2) \neq val_T(t_2)$. Applying the induction hypothesis we obtain that either $Ax \models \neg s_1=t_1$ or $Ax \models \neg s_2=t_2$. Applying (b) yields $Ax \models \neg cons(s_1,s_2)=cons(t_1,t_2)$.  ∎

---

**Exercise 12.8**  Prove that

$succ_i(X)=succ_i(Y) \rightarrow X=Y$, for i=1,...,n

$\neg succ_i(X)=\varepsilon$, for i=1,...,n

$\neg succ_i(X)=succ_j(Y)$, for i<j

is a rigid specification of **Word($\Omega$)**, where $\Omega$ is the finite alphabet $\{a_1,...,a_n\}$.

**Exercise 12.9**  Give rigid specifications for **Integer** and **Cardinal(max)**, the algebra of cardinal numbers less than a given bound max.

---

**Theorem 12.7 (How to show that a specification is rigid under the presence of boolean values)**

Let E be a correct (equational) specification in a signature with a sort boole and two constants true and false of an algebra **A**. Assume that for every sort s and ground terms $t_1,t_2$ of sort s with different values in **A** there exists a term b(X) of sort boole with a variable X of sort s such that the following is true:

$E \models b(t_1)=true$ and $E \models b(t_2)=false$, or conversely

$E \models b(t_1)=false$ and $E \models b(t_2)=true$.

Then $E \cup \{\neg true=false\}$ is a rigid specification of **A**.

*Proof*  Let **B** be an arbitrary term generated model of $E \cup \{\neg true=false\}$. Constants true and false are interpreted in **B** by different values. Consider the surjective homomorphism $h:A \to B$ defined by $h(\text{val}_A(t))=\text{val}_B(t)$, for all ground terms t. We show that h is injective. For this purpose, let $t_1$ and $t_2$ be ground terms such that $\text{val}_A(t_1) \neq \text{val}_A(t_2)$. Choose a term $b(X)$ of sort boole with a variable X of sort s such that $E \models b(t_1)=\text{true}$ and $E \models b(t_2)=\text{false}$. Then $\text{val}_B(t_1) \neq \text{val}_B(t_2)$, since otherwise $\text{val}_B(\text{true})=\text{val}_B(\text{false})$. ■

# EXAMPLE 12.13

Let us treat the algebra $(Set(N),\boldsymbol{Boolean},\text{equal},\text{element})$ with functions

equal: nat nat $\to$ boole
element: nat set $\to$ boole

defined by

equal(x,y)=true for x=y, and equal(x,y)=false otherwise
element(x,M)=true for $x \in M$, and element(x,M)=false otherwise

Consider the following specification E:

(0)     correct specifications of $Set(N)$ and $\boldsymbol{Boolean}$

(1)     equal(0,0)=true
(2)     equal(0,succ(Y))=false
(3)     equal(succ(X),0)=false
(4)     equal(succ(X),succ(Y))=equal(X,Y)

(5)     element(Y,$\emptyset$)=false
(6)     element(Y,insert(Set,X))=if(equal(X,Y),true,element(Y,Set))

Application of the criteria from preceding sections gives us at once that E correctly specifies algebra $(Set(N),\boldsymbol{Boolean},\text{equal},\text{element})$.

Let us show by application of Theorem 12.7 that $E \cup \{\neg\text{true}=\text{false}\}$ is a rigid specification of algebra $(Set(N),\boldsymbol{Boolean},\text{equal},\text{element})$. To show this, consider ground terms $t_1,t_2$ of sort nat with different values in the considered algebra. Then $t_1$ is $\text{succ}^n(0)$ and $t_2$ is $\text{succ}^m(0)$, with $n \neq m$. Consider the boolean term $b(X)=\text{equal}(X,t_1)$. Then n applications of (4) and one of (1) show that $Ax \models b(t_1)=\text{true}$, whereas $|n-m|$ applications of (4) and either one of (2) or (3) show that $Ax \models b(t_2)=\text{false}$.

Next consider ground terms $t_1,t_2$ of sort boole such that $E \not\models t_1=t_2$. Since $t_1$ and $t_2$ can be reduced via E to either true or false, there is nothing left to be proved in this case.

So consider finally ground terms $t_1,t_2$ of sort set such that $E \not\models t_1=t_2$. Let $t_1$ be the term insert(insert(…(insert($\varnothing$,succ$^{n_1}$(0)),…),succ$^{n_p}$(0))…) and $t_2$ be the term insert(insert(…(insert($\varnothing$,succ$^{m_1}$(0)),…),succ$^{m_q}$(0))…).

From $E \not\models t_1=t_2$ we know that $t_1$ and $t_2$ are interpreted differently in the initial model (*Set(N)*,*Boolean*,equal,element) of E. Thus, we may choose $x \in \{n_1,…,n_p\}-\{m_1,…,m_q\}$ (without loss of generality; the other case would be to choose $x \in \{m_1,…,m_q\}-\{n_1,…,n_p\}$). Let b(Z) be element(succ$^x$(0),Z). Applying equations (1)-(6), we obtain that $Ax \models b(t_1)$=true, whereas $Ax \models b(t_2)$=false.

## EXAMPLE 12.14

The specification presented in Example 12.7 for algebra *Boolean*, together with axiom ¬true=false is a rigid specification of the considered algebra. (Left as an exercise for the reader.)

## EXAMPLE 12.15

We present a rigid specification of algebra *Cardinal* (see Section 6.5). Let us start with an equational specification of cardinal (as initial model):

(0)     Correct specification of *Boolean, Nat* + specification of conditional if$_{nat}$:

if$_{nat}$(true,X,Y)=X
if$_{nat}$(false,X,Y)=Y

(1)     pred(0)=0
(2)     pred(succ(X))=X

(3)     X-0=X
(4)     X-succ(Y)=pred(X-Y)

(5)     equal(0,0)=true
(6)     equal(0,succ(Y))=false
(7)     equal(succ(X),0)=false
(8)     equal(succ(X),succ(Y))=equal(X,Y)

(9)     unequal(X,Y)=not(equal(X,Y))

(10)   less(X,0)=false
(11)   less(0,succ(Y))=true
(12)   less(0,0)=false
(13)   less(succ(X),succ(Y))=less(X,Y)

(14)   less_or_equal(X,Y)=or(less(X,Y),equal(X,Y))

(15)   greater(X,Y)=less(Y,X)

(16)   greater_or_equal(X,Y)=or(greater(X,Y),equal(X,Y))

(17)   X DIV 0=0
(18)   X DIV succ(Y)=
           $\text{if}_{nat}$(less(X,succ(Y)),0,1+((X-succ(Y)) DIV succ(Y)))

(19)   X MOD 0=0
(20)   X MOD succ(Y)=X-((X DIV succ(Y))*succ(Y))

Along the lines of earlier examples it is not difficult to show that E is indeed a correct specification of **Cardinal**. We show that (0)-(20), together with ¬true=false, are a rigid specification of algebra **Cardinal**. For this let ground terms $t_1$ and $t_2$ of type nat with different values in algebra **Cardinal** be given (as before terms of type boole are similarly treated). Since ground terms may be reduced via (0)-(20) to terms of the form $\text{succ}^n(0)$, we may assume that $t_1=\text{succ}^n(0)$ and $t_2=\text{succ}^m(0)$, for natural numbers n and m with n≠m. Now, for the boolean term $b(Z)=\text{equal}(Z,\text{succ}^m(0))$, $E \models b(t_1)=\text{false}$, whereas $E \models b(t_1)=\text{true}$. Hence, our criterion from above is applicable.

## 12.6    Summary of specification methods for an algebra

Let an algebra $A$ of cardinality $\kappa$ be given. So far we have studied three ways of describing $A$ *uniquely up to isomorphism* by logical axioms and further, non-logical features.

- $A$ may be the unique initial model (up to isomorphism) of a set of equations E. Initiality is the non-logical feature.

- $A$ may be the unique term generated model (up to isomorphism) of a set Ax of universal formulas. Here term generatedness, a weaker concept that initiality, is the non-logical feature.

- $A$ may be the unique model of cardinality $\kappa$ (up to isomorphism) of an axiom system Ax. It was the notion of $\kappa$-categoricity that was introduced to describe this situation. The non-logical feature used here is the cardinality of admitted models.

- *A* may be the unique model (up to isomorphism) of an axiom system Ax. This strongest way to fix a term generated algebra is applicable only in the case of finite algebras.


## 12.7    Parametrized specifications ✂

Until now we have investigated methods for specifying a single algebra. The purpose of this chapter is to introduce 'generic specifications', that is, specifications containing a parameter part which may be instantiated by a concrete algebra from a certain class of algebras. As an example we will discuss a specification *Set(Data)* with *Data* free for arbitrary instantiations. Allowing such instantiations is, of course, more economical than writing separate specifications for *Set(Nat)*, *Set(Int)*, and so on.


**Definition 12.5:**

(1)    A **parametrized specification** consists of a parameter specification $(S_{par}, \Sigma_{par}, Ax_{par})$ and an extension $(S, \Sigma, Ax)$ of $(S_{par}, \Sigma_{par}, Ax_{par})$. We write $(S_{par}, \Sigma_{par}, Ax_{par}) + (S, \Sigma, Ax)$ to indicate a parametrized specification.

(2)    A parametrized specification as above is called an **equational parametrized specification** iff Ax consists of equations. It is called a **completely equational parametrized specification** iff Ax and $Ax_{par}$ consist of equations. (This latter concept is not of great importance, since we will usually need more than equations to axiomatize the class of admitted parameters adequately.)

(3)    Let $(S_{par}, \Sigma_{par}, Ax_{par}) + (S, \Sigma, Ax)$ be a parametrized specification. The class of all **admissible parameters** for $(S_{par}, \Sigma_{par}, Ax_{par}) + (S, \Sigma, Ax)$ consists of all algebras P in the signature $(S_{par}, \Sigma_{par})$ that are models of Ax.


**Definition 12.6:**    Let $(S_{par}, \Sigma_{par}, Ax_{par}) + (S, \Sigma, E)$ be an equational parametrized specification and *P* be an admissible parameter for $(S_{par}, \Sigma_{par}, Ax_{par}) + (S, \Sigma, Ax)$. We define the **initial algebra** $T_E(P)$ **of** E **over** *P* as follows:

- Extend the signature $(S_{par},\Sigma_{par})+(S,\Sigma)$ by a new constant p, for each element p of $P$.
- Let $(P,p)_{p\in P}$ be the expansion of $P$ which interprets each of the new constants by itself.
- Extend E by all ground equations s=t over the extended signature which are true in $(P,p)_{p\in P}$.
- For this extended set of equations, form the (usual) initial model.
- Then define $T_E(P)$ to be the reduct of this initial model to the original signature $(S_{par},\Sigma_{par})+(S,\Sigma)$ (thus, forget the new constants).

## EXAMPLE 12.16

Consider the parameter specification

$(\{data\},\varnothing,\varnothing)$ (no function symbols and no axioms)

and the extension $(\{set\},\{\varnothing,insert\},E)$ with the following set E of equations:

insert(insert(S,D),D)=insert(S,D)

insert(insert(S,D),E)=insert(insert(S,E),D)

The class of admissible parameters consists of all algebras in the signature $(\{data\},\varnothing)$, that is, simply of all non-empty sets P. For such an admissible parameter P let us determine $T_E(P)$. Our goal will be to show that $T_E(P)$ is isomorphic to the algebra $Set(P)$. This will be achieved by applying the theorems of this chapter in just the same way as we applied them to examples without parameter part. Let us trace the construction presented in the definition of initial algebras for parametrized specifications.

First, P is equipped with new constants $p\in P$ to obtain the algebra $(P,p)_{p\in P}$. Then, E is extended by all valid ground equations s=t in the extended algebra. Now we consider the algebra $(Set(P),p)_{p\in P}$ and show that it is isomorphic to the initial model of our extended set of equations by showing the criteria presented in Section 12.1. Obviously, $(Set(P),p)_{p\in P}$ is a term generated model of the extended set of equations. We have to provide a set C of canonical forms of ground terms such that conditions (3.1) and (3.2) of Theorem 12.1 hold. As $C_{par}$ we take the set of all new constants. For the definition of $C_{set}$ we impose an arbitrary ordering < on P and define $C_{set}$ to consist of all terms of the form insert(...insert(insert($\varnothing,p_1),p_2),...,p_n$), for every finite subset $\{p_1,p_2,...,p_n\}$ of P with $p_1<p_2<...<p_n$. Now it is an easy exercise to check (3.1) and (3.2). Hence, $T_E(P)$ is isomorphic to $Set(P)$.

This shows that the same methods and theorems are applicable to parametrized specifications as to non-parametrized ones.

Now we enrich the above parametrized specification by a conditional $if_{set}$ selecting sets and a delete function. As in the non-parametrized case we need the presence of **Boolean** and an equality function $equal_{data}$. Thus, our parameter specification is $(\{data, boole\}, \{not, and, or, if, equal_{data}\}, Ax_{par})$ with the following set of axioms:

a correct specification of **Boolean** +

$equal_{data}(D, E) = true \leftrightarrow D = E$

(Note here that we use more than equations to specify adequately the parameter part.) Then the desired specification is:

$if_{set}(true, S, T) = S$

$if_{set}(false, S, T) = T$

$delete(emptyset, D) = emptyset$
$delete(insert(S, D), E) =$
    $if_{set}(equal_{data}(D, E), delete(S, E), insert(delete(S, E), D))$

Application of the theorem on the correctness of algebra extensions gives the desired result.

---

All previous definitions, results and proof for non-parametrized specifications carry over to parametrized specifications in a natural way. We discuss in the following the most important ones.

**Definition 12.7:**   Let a parameter specification $(S_{par}, \Sigma_{par}, Ax_{par})$ and a mapping which assigns to each model $P$ of $Ax_{par}$ over the signature $(S_{par}, \Sigma_{par})$ an expanded algebra $(P, A(P))$ in the signature of $(S_{par}, \Sigma_{par}) + (S, \Sigma)$ be given. We say that $(S_{par}, \Sigma_{par}, Ax_{par}) + (S, \Sigma, E)$, with a set of equations E, is a **correct parametrized specification** of this mapping iff $(P, A(P))$ is isomorphic to $T_E(P)$, for every admissible parameter $P$ of $(S_{par}, \Sigma_{par}, Ax_{par}) + (S, \Sigma, E)$.

So far, we have considered $(S_{par}, \Sigma_{par})$-algebras which are models of $Ax_{par}$ as admissible parameters of a parametrized specification $(S_{par}, \Sigma_{par}, Ax_{par}) +$

$(S,\Sigma,E)$. This might seem somewhat too restrictive. The following fact shows that we may admit models of $Ax_{par}$ in *arbitrary extension signatures* of $(S_{par},\Sigma_{par})$.

**Fact**    Let $(S_{par},\Sigma_{par},Ax_{par})+(S,\Sigma,E)$ be a correct parametrized specification of a mapping $P \mapsto (P,A(P))$ as described in Definition 12.7, with $P$ ranging over all models of $Ax_{par}$. Let $(S_{par}\cup S'_{par},\Sigma_{par}\cup\Sigma'_{par})$ be an extension of $(S_{par},\Sigma_{par})$ disjoint to $(S,\Sigma)$. Then $(S_{par}\cup S'_{par},\Sigma_{par}\cup\Sigma'_{par},Ax_{par})+(S,\Sigma,E)$ is a correct specification of the extension of the considered mapping to models $P$ of $Ax_{par}$ in the signature $(S_{par}\cup S'_{par},\Sigma_{par}\cup \Sigma'_{par})$.

*Proof*    Consider an algebra $P_1=(P,A',B',...,f',g',...)$ in the signature $(S_{par}\cup S'_{par},\Sigma_{par}\cup\Sigma'_{par})$, with an algebra $P$ in the signature $(S_{par},\Sigma_{par})$ and further domains $A',B',...$ and functions $f',g',...$ in $\Sigma'_{par}$. Assume that $P_1$ is a model of $Ax_{par}$. Define $P_2=(P_1,p)_{p\in P}$ and consider the definition of $T_E(P_2)$.

First of all, we enrich the given signature by a set C of new constants p, for each element p of $P$ and $A',B',....$ Let us at the moment forget about the new functions $f',g'$, and so on. Consider the set of equations E' consisting of all the equations in E together with all ground equations r=t which are true in $P_3=(P_1,p)_{p\in C}$. Obviously (if not obvious apply Theorem 12.1), the initial model of this set of equations is $(P,A',B',...,A(P),p)_{p\in C}$, provided that $T_E(P)=(P,A(P))$.

Now we extend the signature by the functions $F',G',...$ and consider the extension E'' of E' consisting of the equations of E' and all ground equations r=t which are true in $P_3$. An immediate application of the theorem on the correctness of algebra extensions shows that the initial model of E'' is $(P_3,A(P))$. We have thus shown the desired result.    ∎

**Definition 12.8:**    Let $(S_{par},\Sigma_{par},Ax_{par})+(S,\Sigma,Ax)$ be a parametrized specification and $P$ be an admissible algebra. An algebra $(P,A)$ in the signature $(S_{par},\Sigma_{par})+(S,\Sigma)$ is called **parameter generated** iff every element a of $A$ can be written as $val_{(P,A)}(t(p_1,...,p_n))$, for some term t with parameter variables $P_1,...,P_n$ and parameter values $p_1,...,p_n$. In other words, $(P,A)$ is parameter generated iff the extended algebra $(P,A,p)_{p\in P}$ is term generated. In an obvious way, the **parameter generated subalgebra** $(P,A_{par})$ of $(P,A)$ is defined by generalizing the notion of a term generated subalgebra.

**Definition 12.9:** A parametrized specification $(S_{par}, \Sigma_{par}, Ax_{par}) +$ $(S, \Sigma, Ax)$ is called a **rigid parametrized specification** iff for any two models $(P,A)$ and $(P,B)$ of Ax, with the same admissible parameter $P$, $(P,A_{par})$ and $(P,B_{par})$ are isomorphic. If $P \mapsto (P,A(P))$ is a mapping which assigns to each admissible parameter $P$ a parameter generated model $(P,A(P))$ of Ax, we say that $(S_{par}, \Sigma_{par}, Ax_{par}) + (S, \Sigma, Ax)$ is a rigid parametrized specification of this mapping.

## EXAMPLE 12.17

Proof that a parametrized specification is a correct or rigid specification of a mapping $P \mapsto (P,A(P))$ is done via the same theorems as applied in corresponding non-parametrized situations. As an example we show that the following parametrized specification is a rigid specification of the mapping **Set** that assigns to each admissible parameter **Data** the expansion **Set(Data)**.

The parameter specification is $(S_{par}, \Sigma_{par}, Ax_{par})$ with

$S_{par} = \{data, boole\}$

$\Sigma_{par}$ consists of function symbols

> $equal_{data}$: data data $\rightarrow$ boole
>
> true: $\rightarrow$ boole
>
> false: $\rightarrow$ boole
>
> if : boole boole boole $\rightarrow$ boole

$Ax_{par}$ consists of axioms

> $\neg true = false$
>
> if(true,B,C)=B
>
> if(false,B,C)=C,
>
> $equal_{data}(X,Y) = true \leftrightarrow X = Y$

Thus the parameter axioms lay down that only interpretations **Data** are admissible with a boolean-valued equality function. The parametrized specification we have in mind is obtained by the following extension:

$S = \{set\}$

$\Sigma$ consists of function symbols

> $\emptyset$: $\rightarrow$ set

insert: set data $\rightarrow$ set

element: data set $\rightarrow$ boole

Ax contains the usual axioms for insert and the following equations:

element(D,$\emptyset$)=false

element(D,insert(Set,E))=

if(equal$_{data}$(D,E),true,element(D,Set))

---

Finally, we carry over the concept of categoricity for parametrized specifications. This concept will play an important role in the discussion of data type constructors like *Array* [1..N] *of Data* and *Record* $sel_1:s_1,...,sel_n:s_n$ *of Data*.

**Definition 12.10:** A parametrized specification $(S_{par},\Sigma_{par},Ax_{par})+$ $(S,\Sigma,Ax)$ is called a **categorical parametrized specification** iff for any two models $(P,A)$ and $(P,B)$ of Ax, with the same admissible parameter $P$, $(P,A)$ and $(P,B)$ are isomorphic. If $P \mapsto (P,A(P))$ is a mapping which assigns to each admissible parameter $P$ a model $(P,A(P))$ of Ax, then we say that $(S_{par},\Sigma_{par},Ax_{par})+(S,\Sigma,Ax)$ is a categorical parametrized specification of this mapping.

The reader may wonder whether this definition makes sense in light of the results of Chapter 5, where we showed that infinite term generated algebras cannot be categorically axiomatized. But note that in the definition above, categoricity is introduced only *relative to a parameter algebra P*.

# EXAMPLE 12.18

Consider the following parameter specification for a subset $\{s_1,...,s_n\}$ of $S_{par}$:

$(S_{par},\{\perp_1:s_1,...,\perp_n:s_n\},\emptyset)$

Thus, every admissible parameter *Data* contains distinguished element $\perp_1,...,\perp_n$. Referring to these elements we may construct *Record* $sel_1:s_1,...,sel_n:s_n$ *of Data* as defined in Section 6.5. We construct a categorical parametrized specification of this mapping by the following extension:

$S=\{rec\}$

$\Sigma$ consists of the following function symbols (with $i=1,\ldots n$):

initrecord: $\rightarrow$ rec

$sel_i$: rec $\rightarrow s_i$

$assign_i$: rec $s_i \rightarrow$ rec

Ax is the following collection of equations:

(1)   $assign_i(assign_j(R,D),E)=assign_j(assign_i(R,E),D)$, for all $i \neq j$

(2)   $assign_i(assign_i(R,D),E)=assign_i(R,E)$

(3)   $sel_i(initrecord)=\perp_i$

(4)   $sel_i(assign_i(R,D))=D$

(5)   $assign_1(assign_2(\ldots(assign_n(R',sel_n(R)),\ldots,sel_2(R)),sel_1(R))=R$

Of course, **Record** $sel_1:s_1,\ldots,sel_n:s_n$ **of Data** is a model of (1)-(5). To show categoricity let **A** be an arbitrary model of (1)-(5). For simplicity, we denote the interpretations in **A** of all function symbols f by f, too. Consider the function

$$sel:A_{rec} \rightarrow A_{s_1} \times A_{s_2} \times \ldots \times A_{s_n} \text{ defined by } sel(R)=(sel_1(R),\ldots,sel_n(R))$$

We can show that sel is an injective function. Assume that $sel(R)=sel(S)$. Then by axiom (5):

$$R=assign_1(assign_2(\ldots assign_n(R,sel_n(R)),\ldots,sel_2(R)),sel_1(R))=$$

$$assign_1(assign_2(\ldots assign_n(S,sel_n(S)),\ldots,sel_2(S)),sel_1(S))=S$$

Next can we show that sel is surjective. Let $(a_1,\ldots,a_n) \in A_{s_1} \times A_{s_2} \times \ldots \times A_{s_n}$ be given. Consider $R=assign_1(\ldots(assign_n(R',a_n),\ldots),a_1)$, with an arbitrary $R' \in A_{rec}$. Then we obtain:

$$sel_i(R)$$

$$=sel_i(assign_1(\ldots(assign_n(R',a_n),\ldots),a_1))$$

$$= a_i$$

(The last equality is obtained by repeatedly applying (1), and then axiom (4).)

Finally, it is an easy exercise to show that sel is a homomorphism.

## EXAMPLE 12.19

Consider as parameter specification the following extension of a categorical specification of **Boolean**: $(\{data\},\{ \perp: \rightarrow data\},\emptyset)$. Thus, every admissible parameter **Data** contains a distinguished element $\perp_{\mathbf{Data}}$. Referring to this element we may construct **Array** [0..L] **of Data** as defined in Section 6.5. We construct a categorical parametrized specification of this mapping by the following extension:

S={index, array}
$\Sigma$ consists of the following function symbols:

> initarray: $\rightarrow$ array
> read: array index $\rightarrow$ data
> update: array index data $\rightarrow$ array
>
> $if_{data}$: boole data data $\rightarrow$ data
> $if_{array}$: boole array array $\rightarrow$ array
>
> 0: $\rightarrow$ index
> up: index $\rightarrow$ index
> down: index $\rightarrow$ index
> less: index index $\rightarrow$ boole
> $equal_{index}$: index index $\rightarrow$ boole

Ax = a categorical specification of the index algebra [0...N] with its functions together with a categorical specification of **Boolean** (note that such specifications exist since we deal with finite algebras) + usual specification of $if_{data}$ and $if_{array}$ + the following equations:

(1)    update(update(A,I,D),J,E)=
       $if_{array}(equal_{index}(I,J),update(A,I,E),update(update(A,J,E),I,D))$

(2)    read(initarray,I)=$\perp$

(3)    read(update(A,I,D),J)=$if_{data}(equal_{index}(I,J),D,read(A,J))$

(4)    A=update(...(update(initarray,0,read(A,0)),...),N,read(A,N))

It is left to the reader to show in an analogous manner as above that these equations completely axiomatize the array declaration above.

# Chapter 13
# Term Rewriting Systems

Consider once again a set E of equations. Imagine that E describes the properties of the export interface of a software module. Such an interface description is at first sight nothing more than a static description of the interrelations between the involved functions telling the *implementor* of a module, in a formal way, what he has to achieve when realizing these functions by procedures, and describing to the *user* the properties of the functions.

This use of equational specifications will be discussed in more detail later on, when verification of software modules is treated. There, it will become apparent that the process of partially automatizing such verification tasks requires special methods for showing that some equations l=r follow from a set E of equations.

The key to obtaining efficient methods to show the congruence of terms modulo E will be to treat E as a **directed term rewriting system**, meaning that replacements of left-hand sides of equations by right-hand sides are allowed but not vice versa (note that if replacing were also allowed in the opposite direction, then cyclic applications would be unavoidable). The main theme of this chapter will be to find out under what conditions the use of E as a directed term rewriting system is equivalent to its use as a symmetric one as in Chapters 11 and 12.

Having thus introduced the possibility of considering certain sets of equations not only as formal descriptions of interface properties in a *static description language*, but also as executable programs in a *programming language*, there opens up the possibility of executing interface specifications and observing their dynamic behaviour before they are implemented or used in other software modules. Owing to the fatal consequences of errors in early stages of software system design, this use of specifications as executable, although possibly not very efficient **rapid prototypes** ('rapid' does not refer to the efficiency of execution but to rapidity of obtaining a runnable version of a specification) is of enormous importance.

# 13.1    Simplification and the Church-Rosser property

In this section we shall introduce the concepts which will play an important role when we formulate the conditions under which a set of equations may be interpreted in a more efficient way and thus may be regarded as a program.

> **Definition 13.1:**    Let $\Sigma$ be a signature and E be a set of $\Sigma$-equations. A $\Sigma$-term s is called an **E-normal form** iff there is no $\Sigma$-term t such that $s \xrightarrow{E} t$. An **E-selection function** is a function
>
> $$\text{Sel}: \{t \in T(\Sigma) \mid t \text{ not an E-normal form}\} \to T(\Sigma)$$
>
> such that $t \xrightarrow{E} \text{Sel}(t)$, for all $\Sigma$-terms t which are not E-normal forms.
>
> For an E-selection function Sel we define a partial function $\text{Sel}^*:T(\Sigma) \to T(\Sigma)$ as follows:
>
> $$\text{Sel}^*(t) = \begin{cases} t & \text{if t is an E-normal form} \\ \text{Sel}^*(\text{Sel}(t)) & \text{otherwise} \end{cases}$$
>
> Hence $\text{Sel}^*(t)$ applies the function Sel as long as possible until eventually an E-normal form is reached. Then it stops with this E-normal form as result. Since it may be possible to apply Sel infinitely often without ever reaching a normal form, $\text{Sel}^*(t)$ may be undefined for certain terms t.
>
> A total function $\text{Simp}:T(\Sigma) \to T(\Sigma)$ is called an **E-simplifier** iff for all $\Sigma$-terms s and t:
>
> (1)     $\text{Simp}(t) =_E t$
>
> (2)     $s =_E t \Rightarrow \text{Simp}(s) = \text{Simp}(t)$

Let us briefly discuss these concepts. E-normal forms are terms at which the process of reduction via $\xrightarrow{E}$ comes to an end. For a given $\Sigma$-term t which is not an E-normal form, an E-selection function selects a term s among all the possibilities to reduce t into a term s via $\xrightarrow{E}$. Then $\text{Sel}^*(t)$ is defined iff there is a number m such that

- $Sel^i(t)$ is not an E-normal form, for all $i < m$

- $Sel^m(t)$ is an E-normal form

With a number m as above we know that

$$t \xrightarrow[E]{} Sel(t) \xrightarrow[E]{} Sel^2(t) \xrightarrow[E]{} \ldots Sel^{m-1}(t) \xrightarrow[E]{} Sel^m(t)$$

Hence, $t \xrightarrow[E]{}*Sel*(t)$, whenever $Sel*(t)$ is defined. In particular, $t \underset{E}{\longleftrightarrow}*$ $Sel*(t)$, whenever $Sel*(t)$ is defined. So, property (1) is fulfilled for $Sel*$ and terms t such that $Sel*(t)$ is defined.

Conditions (1) and (2) in the definition of E-simplifiers express the fact that an E-simplifier Simp picks up from each congruence class modulo $=_E$ of a $\Sigma$-term t exactly one representative Simp(t). Hence, the task of determining whether $s =_E t$, is reduced via Simp to the comparison of Simp(s) and Simp(t) as strings of symbols: $s =_E t \Leftrightarrow Simp(s) = Simp(t)$. Having available an effectively (efficiently) computable simplifier Simp leads thus to an effective (efficient) decision procedure for the relation $=_E$ on $\Sigma$-terms.

Our main interest will be in functions $Sel*$, for an E-selection function Sel, and the question of under what conditions on E such a function $Sel*$ is an E-simplifier. Requirement (1) will lead to the the notion of a **noetherian** set of equations, while (2) leads to the **Church-Rosser property**.

**Definition 13.2:**    A set E of equations is called **noetherian** iff there does not exist any infinite sequence $(t_i)_{i \in \mathbb{N}}$ of terms with $t_i \xrightarrow[E]{} t_{i+1}$, for all $i \in \mathbb{N}$.

Let E be a noetherian set of equations. Then for every E-selection function Sel, the corresponding function $Sel*$ is a total function and fulfils requirement (1) of Definition 13.1.

**Definition 13.3:**    A set of equations E has the **Church-Rosser property**, iff for any two terms s and t with $s =_E t$ there is a term r such that $s \xrightarrow[E]{}*r$ and $t \xrightarrow[E]{}* r$.

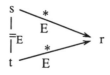

*Theorem 13.1*

Let E be a noetherian set of equations.

(1)    If there exists an E-selection function Sel such that Sel* is an E-simplifier, then E has the Church-Rosser property.

(2)    If E has the Church-Rosser property, then Sel* is an E-simplifier, for every E-selection function Sel.

*Proof*

(1)    Let Sel be an E-selection function such that Sel* is an E-simplifier. To show that E possesses the Church-Rosser property, let terms s and t be given such that $s =_E t$. From requirement (2) of Definition 13.1 we conclude that Sel*(s)=Sel*(t). Furthermore, we know by definition that $s \xrightarrow[E]{} {}^*Sel^*(s)$ and $t \xrightarrow[E]{} {}^*Sel^*(t)$. So, Sel*(t) is the desired term r with $s \xrightarrow[E]{} {}^*r$ and $t \xrightarrow[E]{} {}^*r$.

(2)    Assume that E has the Church-Rosser property and let Sel be an arbitrary E-selection function. To show that Sel* is an E-simplifier, we must verify condition (2) of Definition 13.1. For this purpose let s and t be terms such that $s =_E t$, that is, $s \underset{E}{\longleftrightarrow} {}^*t$. Since $s \xrightarrow[E]{} {}^*Sel^*(s)$ and $t \xrightarrow[E]{} {}^*Sel^*(t)$, we conclude that $Sel^*(s) \underset{E}{\longleftrightarrow} {}^*Sel^*(t)$, that is, $Sel^*(s) =_E Sel^*(t)$. Applying the Church-Rosser property to Sel*(s) and Sel*(t) we obtain a term r such that $Sel^*(s) \xrightarrow[E]{} {}^*r$ and $Sel^*(t) \xrightarrow[E]{} {}^*r$. Since Sel*(s) and Sel*(t) are E-normal forms, this is possible only when Sel*(s)=r and Sel*(t)=r. Hence, Sel*(s)=Sel*(t).    ∎

## 13.2    Proving the Church-Rosser property

We are now interested in establishing effective criteria for proving the Church-Rosser property. In three steps we will reduce the Church-Rosser property of a noetherian set of equations to increasingly simpler properties, namely:

- confluence
- local confluence
- critical pairs property

**Definition 13.4:**    A set of equations E is called **confluent** iff for any three terms u,s,t with $u \xrightarrow[E]{} {}^*s$ and $u \xrightarrow[E]{} {}^*t$ there is a term r such that $s \xrightarrow[E]{} {}^*r$ and $t \xrightarrow[E]{} {}^*r$.

This means that any divergence in reducing terms can be made convergent.

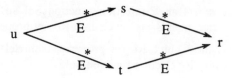

Proving a set E of equations to be confluent is simpler than showing the Church-Rosser property in the sense that we must show the convergence of terms s and t only for those resulting from a single term u by application of $\xrightarrow{E}*$ instead of considering all terms s and t with $s =_E t$. The following theorem shows that both properties are equivalent.

### Theorem 13.2 (Equivalence of confluence and Church-Rosser property)

Let E be an arbitrary (not necessarily noetherian) set of equations. Then E has the Church-Rosser property iff E is confluent.

*Proof*    Assume that E has the Church-Rosser property. Let terms u,s and t be given such that $u \xrightarrow{E} *s$ and $u \xrightarrow{E} *t$. Hence $u \xleftrightarrow{E} *s$ and $u \xleftrightarrow{E} *t$, that is, $s =_E t$. Applying the Church-Rosser property we obtain a term r with $s \xrightarrow{E} *r$ and $t \xrightarrow{E} *r$. So, E is shown to be confluent. Conversely, assume E to be confluent. To show the Church-Rosser property let terms s and t be given such that $s =_E t$, that is, $s \xleftrightarrow{E} *t$. Hence there is a number n such that $s \xleftrightarrow{E} {}^n t$. By induction on n we show that there is a term r with $s \xrightarrow{E} * r$ and $t \xrightarrow{E} *r$. If n=0 then s and t are identical terms and we can take s itself as the desired term r. Assume n>0. Then there is a term u such that $s \xleftrightarrow{E} u$ and $u \xleftrightarrow{E} {}^m t$, with m= n-1. By induction hypothesis applied to u and r there is a term v with $u \xrightarrow{E} *v$ and $t \xrightarrow{E} *v$. We distinguish two cases according to whether $s \xrightarrow{E} u$ or $u \xrightarrow{E} s$.

*Case 1*    $s \xrightarrow{E} u$. Then $s \xrightarrow{E} u \xrightarrow{E} *v$ and $t \xrightarrow{E} *v$. Hence, v is the desired term r.

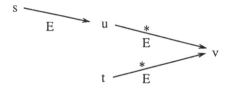

*Case 2* $u \underset{E}{\rightarrow} s$. Since also $u \underset{E}{\rightarrow} *v$, the confluence of E, applied to terms u,s and v, yields a term r such that $s \underset{E}{\rightarrow} *r$ and $v \underset{E}{\rightarrow} *r$. It follows that $s \underset{E}{\rightarrow} *r$ and $t \underset{E}{\rightarrow} *v \underset{E}{\rightarrow} *r$. Hence, r is the term we have been looking for.

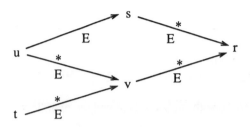

■

**Definition 13.5:**    A set of equations E is called **locally confluent** iff for any three terms u,s,t with $u \underset{E}{\rightarrow} s$ and $u \underset{E}{\rightarrow} t$ there is a term r such that $s \underset{E}{\rightarrow} *r$ and $t \underset{E}{\rightarrow} *r$. This means that any 1-step divergence in reducing terms via $\underset{E}{\rightarrow}$ can be made convergent.

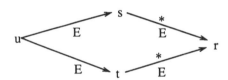

Obviously, it requires less effort to show local confluence of a set E than to show its confluence. Hence, confluence trivially implies local confluence. In the following we shall prove that for noetherian sets of equations, both notions coincide. To show this fact we have to introduce a special kind of proof principle in the context of noetherian sets of equations, or more generally of noetherian relations, called **noetherian induction**.

**Definition 13.6:**    A binary relation $\rightarrow$ on a set A is called **noetherian relation** on A iff there does not exist any infinite sequence $(a_i)_{i \in \mathbb{N}}$ such that $a_i \rightarrow a_{i+1}$, for all $i \in \mathbb{N}$. (In other words, the inverse relation of $\rightarrow$ is well-founded.)

A simple example of a noetherian relation the reader should have in mind for an understanding of the following two theorems is the relation $\rightarrow$ over $\mathbb{N}$ defined by n→m iff n=m+1. In this example, n→$^+$ m is equivalent to m<n.

### Theorem 13.3 (Noetherian induction)

Let → be a noetherian relation on a set A and X a subset of A. Assume that the following holds:

(Inductive step)  For every a∈ A, if {b∈ A | a→$^+$ b}⊆X then a∈ X.

Then we may conclude A=X.

*Proof*   Assume that X is a proper subset of A. Choose $a_0$∈ A such that $a_0$∉ X. Applying (Inductive step) with $a_0$ we can conclude that {b∈ A | $a_0$→$^+$ b} is not a subset of X. So we may choose some $a_1$ such that

$a_0$→$^+$ $a_1$  and  $a_1$∉ X

Applying (Inductive step) with $a_1$ we may conclude that {b∈ A | $a_1$→$^+$ b} is not a subset of X. So we may choose some $a_2$ such that

$a_1$→$^+$ $a_2$  and  $a_2$∉ X

Proceeding in this way we obtain an infinite sequence $(a_i)_{i\in \mathbb{N}}$ such that $a_i$→$^+$$a_{i+1}$ and $a_i$∉ X, for all i∈ $\mathbb{N}$. Since →$^+$ involves at least one application of →, we obtain an infinite sequence $(b_i)_{i\in \mathbb{N}}$ such that $b_i$→$b_{i+1}$, for all i∈ $\mathbb{N}$. This contradicts the fact that → was assumed to be noetherian.  ∎

Look again what (Inductive step) means for the natural numbers:

For every number a, if {b∈ $\mathbb{N}$ | b<a} is a subset of X, then a∈ X.

It is a variant of induction over the natural numbers. We are now prepared to show the equivalence between confluence and local confluence.

### Theorem 13.4 (Equivalence of confluence and local confluence for noetherian sets of equations)

Let E be a noetherian set of equations. Then E is confluent iff it is locally confluent.

*Proof*  Assume that E is locally confluent. We shall show that E is confluent. For this purpose let u be an arbitrary term. We have to show:

For all terms s and t, if u $\xrightarrow{E}$*s and u $\xrightarrow{E}$*t then
there exists a term r such that s $\xrightarrow{E}$*r and t $\xrightarrow{E}$*r.   (13.1)

This claim is shown by noetherian induction over u. Stated in terms of Theorem 13.3, we consider the set X of all terms u having property (13.1) and show that X consists of all terms u. According to the property (*Inductive step*) we assume that (13.1) has been already proved for all terms u' with $u \xrightarrow{E} u'$.

Now let terms s and t be given such that $u \xrightarrow{E} {}^*s$ and $u \xrightarrow{E} {}^*t$.

In the case that u is identical to one of s or t, nothing remains to be shown. So assume that u is different from s and t. Then there are terms $s_1$ and $t_1$ such that $u \xrightarrow{E} s_1 \xrightarrow{E} {}^*s$ and $u \xrightarrow{E} t_1 \xrightarrow{E} {}^*t$.

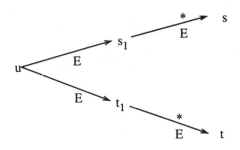

Since E is locally confluent, there is a term v with $s_1 \xrightarrow{E} {}^*v$ and $t_1 \xrightarrow{E} {}^*v$.

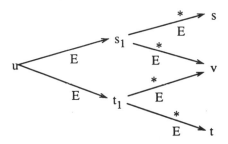

We may apply (13.1) to $s_1$. We obtain a term w with $s \xrightarrow{E} {}^*w$ and $v \xrightarrow{E} {}^*w$.

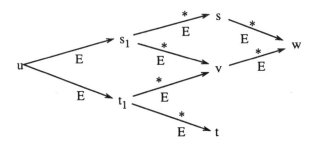

In the same way we may apply (13.1) to $t_1$. We obtain a term r with $w \xrightarrow{E} {}^*r$

and $t \xrightarrow[E]{} {}^{*}r$. So, $s \xrightarrow[E]{} {}^{*}r$ and $t \xrightarrow[E]{} {}^{*}r$.

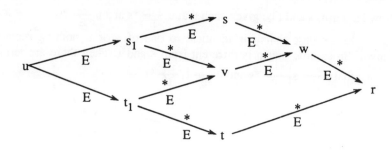

So far we have reduced the Church-Rosser property of a noetherian set of equations to the conceptually simpler property of local confluence. Nevertheless, in proving a set of equations to be locally confluent we are generally still left with the task of checking an infinite number of triples of terms u,s and t with $u \xrightarrow[E]{} s$ and $u \xrightarrow[E]{} t$. As a final step we will reduce this latter task to a finite one. Whereas the former reductions are possible for noetherian relations over an arbitrary set A, the following final step refers to the special case of a term rewriting relation $\xrightarrow[E]{}$.

## 13.3   Critical pairs

In the following let us investigate how a divergence $u \xrightarrow[E]{} s$, $u \xrightarrow[E]{} t$ in the application of $\xrightarrow[E]{}$ can result. This analysis will motivate the concept of critical pairs. Let some term u be given which we want to reduce via $\xrightarrow[E]{}$. Assume that there exist instances $l_1\sigma_1 = r_1\sigma_1$ and $l_2\sigma_2 = r_2\sigma_2$ of equations $l_1 = r_1$ and $l_2 = r_2$ in E such that $l_1\sigma_1$ and $l_2\sigma_2$ appear as different subterms of u. Without loss of generality we may assume that $l_1 = r_1$ and $l_2 = r_2$ do not contain common variables. (Otherwise rename the variables in $l_2 = r_2$ and adapt $\sigma_2$ to the renaming.) Assume that term s results from a replacement of subterm $l_1\sigma_1$ by $r_1\sigma_1$, and t results from a replacement of subterm $l_2\sigma_2$ by $r_2\sigma_2$. Can s and t be made convergent?

There is a simple case, namely that the considered occurrences of $l_1\sigma_1$ and $l_2\sigma_2$ in u do not overlap (Figure 13.1). In this case, obviously s and t converge.

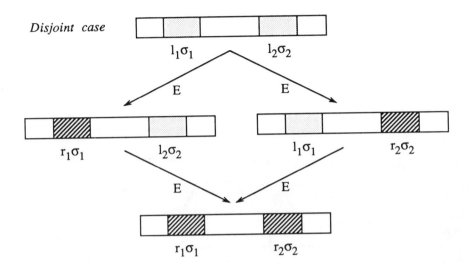

*Disjoint case*

$l_1\sigma_1$    $l_2\sigma_2$

E    E

$r_1\sigma_1$    $l_2\sigma_2$    $l_1\sigma_1$    $r_2\sigma_2$

E    E

$r_1\sigma_1$    $r_2\sigma_2$

**Figure 13.1**

What can be said in the case that $l_1\sigma_1$ and $l_2\sigma_2$ *overlap*? As we know from the uniqueness of term syntax, this is only possible if one of $l_1\sigma_1$ and $l_2\sigma_2$ is a subterm of the other. Let us assume that $l_2\sigma_2$ is a subterm of $l_1\sigma_1$. So we may decompose $l_1\sigma_1$ as a string into left($l_2\sigma_2$)right, with strings left and right. In order to show that s and t can be rewritten into a common term r it suffices to show that $r_1\sigma_1$ and left($r_2\sigma_2$)right can be rewritten into a common term r. The situation is shown in Figure 13.2.

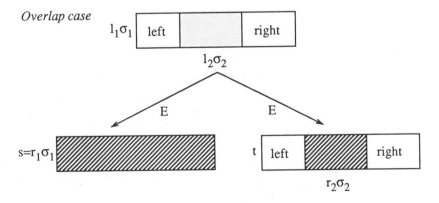

*Overlap case*

$l_1\sigma_1$ | left |       | right

$l_2\sigma_2$

E    E

s=$r_1\sigma_1$    t | left |       | right

$r_2\sigma_2$

**Figure 13.2**

The overlap case requires an investigation of the *minimal* (w.r.t. length) *subterm $l$ of $l_1$ whose image under substitution $\sigma_1$* covers the considered subterm $l_2\sigma_2$ of $l_1\sigma_1$. (See Figure 13.3. It may well be the case that $l_2\sigma_2$ is a proper subterm of $l\sigma_1$.)

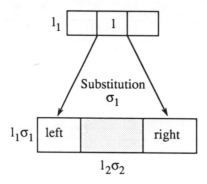

**Figure 13.3**

Two possibilities may occur corresponding to the cases that subterm $l$ may be a variable (the **variable overlap** case) or a composed term (the **critical overlap** case).

Assume that $l$ is a variable $X$. In this case we may decompose $l_1$ into $aXd$ and $l_1\sigma_1$ into $AB(l_2\sigma_2)CD$ with strings a,d,A,B,C and D such that $a\sigma_1=A$, $X\sigma_1=B(l_2\sigma_2)C$ and $d\sigma_1=D$. So, the situation is as in Figure 13.4.

Now define substitution $\sigma_3$ by $\sigma_3(X)=B(r_2\sigma_2)C$ and $\sigma_3(Y)=\sigma_1(Y)$ for all $Y\neq X$. Obviously, $\sigma_1(X)=B(l_2\sigma_2)C \xrightarrow[E]{} B(r_2\sigma_2)C=\sigma_3(X)$. It follows (see Exercise 13.1) that $s=r_1\sigma_1 \xrightarrow[E]{}{}^* r_1\sigma_3$.

We are done if we can show that t can also be rewritten into $r_1\sigma_3$. This is easily shown as follows: term t was obtained from $l_1\sigma_1$ by considering a *single occurrence* of variable $X$ in $l_1$ and replacing its image $B(l_2\sigma_2)C$ under substitution $\sigma_1$ by $B(r_2\sigma_2)C$. There may be further occurrences of $X$ in $l_1$. If we consider *all occurrences* of variable $X$ in $l_1$ and replace their images $B(l_2\sigma_2)C$ under substitution $\sigma_1$ by $B(r_2\sigma_2)C$, we obtain $l_1\sigma_3$. (Remember that $\sigma_1(X)=B(l_2\sigma_2)C$ and $\sigma_3(X)=B(r_2\sigma_2)C$.) This shows $l_1\sigma_1 \xrightarrow[E]{}{}^* l_1\sigma_3$. One further reduction step then leads from $l_1\sigma_3$ to $r_1\sigma_3$.

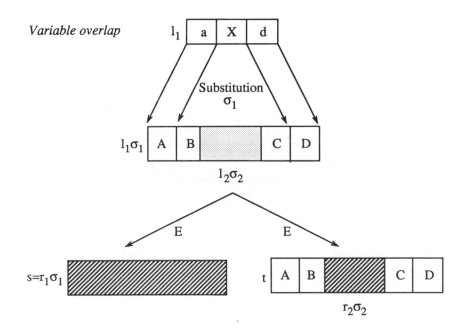

**Figure 13.4**

---

**Exercise 13.1** By induction on term r, show that $r\sigma \xrightarrow[E]{}^* r\rho$ provided that $\sigma(Z) \xrightarrow[E]{}^* \rho(Z)$, for all variables Z.

---

Now we come to the **critical overlap** case, that is, l is a composed term. Minimality of the length of subterm l ensures that $l\sigma_1$ coincides with $l_2\sigma_2$. So the situation is as in Figure 13.5 (for an arbitrary string w (not necessarily a term) consisting of variables and function symbols let us apply a substitution $\sigma$ to w and write $w\sigma$ for the string that results from w by simultaneous substitution of Z by $\sigma(Z)$, for all variables Z):

$l_1 = ald$

$l_1\sigma_1 = (a\sigma_1)(l_2\sigma_2)(d\sigma_1)$ with $l\sigma_1 = l_2\sigma_2$

$s = r_1\sigma_1$

$t = (a\sigma_1)(r_2\sigma_2)(d\sigma_1)$

*Critical overlap*

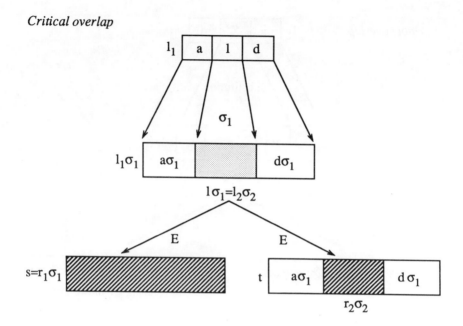

**Figure 13.5**

Note that until now, confluence of E has been reduced to the problem of showing that any two terms s and t as constructed above converge. The problem is that we still have to deal with infinitely many such pairs s and t, due to the presence of arbitrary substitutions $\sigma_1$ and $\sigma_2$. This latter difficulty is finally resolved by allowing only a most general unifier of $l$ and $l_2$ instead of arbitrary substitutions $\sigma_1$ and $\sigma_2$ with $l\sigma_1 = l_2\sigma_2$.

Since $\sigma_1$ and $\sigma_2$ do not act on common variables, we may define $\sigma = \sigma_1 \cup \sigma_2$ and conclude $l\sigma = l\sigma_1 = l_2\sigma_2 = l_2\sigma$. Hence $\sigma$ is a unifier of $l$ and $l_2$. Let $\mu$ be an mgu of $l$ and $l_2$ and $\rho$ a substitution with $\sigma = \mu\rho$. Now we consider terms $p = r_1\mu$ and $q = (a\mu)(r_2\mu)(d\mu)$. Pair $(p,q)$ is called a **critical pair** of E. Assume that $p$ and $q$ could be shown to converge to a common term $r$. Then, $p\rho$ and $q\rho$ converge to $r\rho$ (see Exercise 13.2). Since $p\rho = r_1\mu\rho = r_1\sigma = s$ and $q\rho = (a\mu)(r_2\mu)(d\mu)\rho = (a\sigma)(r_2\sigma)(d\sigma) = t$, it follows then that the original terms s and t also converge.

**Exercise 13.2** Show that $p\sigma \xrightarrow[E]{}{}^* q\sigma$ provided that $p \xrightarrow[E]{}{}^* q$. Use induction on the length of a derivation of q from p.

In the discussion above, the consideration of all critical pairs established to be sufficient for a test of local confluence of a set of equations E. Since, up to renaming, there is only a finite number of critical pairs, a finitary test has been achieved.

> **Definition 13.7:** Let E be a set of equations. A pair of terms (p,q) is called a **critical pair** of E (look at Figure 13.6) iff there exists an equation $l_1 = r_1$ in E, a non-variable subterm $l$ of $l_1$ that is unifiable via mgu $\mu$ with the left-hand side of a renamed version $l_2 = r_2$ of an equation in E that is variable disjoint from $l_1 = r_1$, such that p is the term $r_1\mu$ and q results from $l_1\mu$ by replacing an occurrence of $l_2\mu$ by $r_2\mu$.

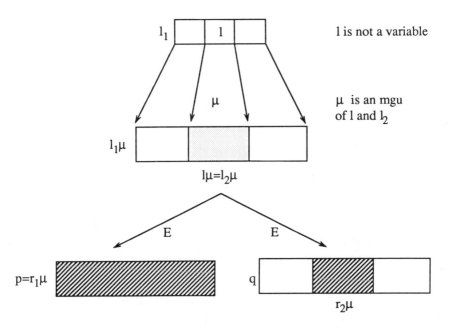

**Figure 13.6**

As a consequence of the development of the concept of critical pairs we obtain the following theorem.

### Theorem 13.5 (Equivalence of local confluence and critical pairs)

A set E of equations is locally confluent iff for every critical pair (p,q) of E there is a term r such that $p \xrightarrow[E]{} {}^*r$ and $q \xrightarrow[E]{} {}^*r$.

## EXAMPLE 13.1 _____

Let $\Sigma$ be a signature with a constant e and a 2-ary, infixedly written function symbol $\circ$. Consider the following set E of equations:

(1)    $((X \circ Y) \circ Z) = (X \circ (Y \circ Z))$    ($\circ$ is associative)

(2)    $(X \circ e) = X$    (e is a right unit)

(3)    $(e \circ Y) = Y$    (e is a left unit)

E is noetherian. This can be seen as follows: An application of (2) or (3) reduces the number of occurrences of e. Hence, (2) and (3) can be applied only finitely often to a term. An application of (1) shifts a closing bracket some positions to the right. Hence, equation (1) can be consecutively applied only finitely often.

E possesses the Church-Rosser property. This can be shown by examining all critical pairs. For this purpose, let us consider each one of the equations (1)-(3) and look at a non-variable subterm of the corresponding left-hand side unifiable with the left-hand side of one of the equations (1)-(3) (including the considered equation).

(a)    Consider equation $((X \circ Y) \circ Z) = (X \circ (Y \circ Z))$. There are two non-variable subterms of $((X \circ Y) \circ Z)$, namely $(X \circ Y)$ and $((X \circ Y) \circ Z)$ itself.

(a.1)    Let us first deal with the subterm $(X \circ Y)$. It can be unified with the left-hand side of each of the equations (1)-(3).

(a.1.1)    We unify $(X \circ Y)$ with the left-hand side of $((X' \circ Y') \circ Z') = (X' \circ (Y' \circ Z'))$ via mgu $\mu = \{X/(X' \circ Y'), Y/Z'\}$. This leads to the critical pair (a,b) with $a = ((X' \circ Y') \circ (Z' \circ Z))$ and $b = ((X' \circ (Y' \circ Z')) \circ Z)$. Terms a and b converge:

$$((X' \circ Y') \circ (Z' \circ Z)) \xrightarrow[E]{} (X' \circ (Y' \circ (Z' \circ Z)))$$

$$((X' \circ (Y' \circ Z')) \circ Z) \xrightarrow[E]{} (X' \circ ((Y' \circ Z') \circ Z)) \xrightarrow[E]{} (X' \circ (Y' \circ (Z' \circ Z)))$$

(a.1.2)    We unify $(X \circ Y)$ with the left-hand side of $(X' \circ e) = X'$ via mgu $\mu = \{X/X', Y/e\}$. This leads to the critical pair (a,b) with $a = ((X' \circ e) \circ Z)$ and $b = (X' \circ Z)$. Terms a and b converge, since

$((X'\circ e)\circ Z) \xrightarrow{E} (X'\circ Z).$

(a.1.3)   We unify $(X\circ Y)$ with the left-hand side of $(e\circ Y')=Y'$ via mgu $\mu=\{X/e,Y/Y'\}$. This leads to the critical pair (a,b) with $a=(e\circ(Y'\circ Z))$ and $b=(Y'\circ Z)$. Terms a and b obviously converge.

(a.2)   Let us deal with the subterm $((X\circ Y)\circ Z)$. It can be unified with the left-hand side of (1) and (2), but not of (3).

(a.2.1)   We unify $((X\circ Y)\circ Z)$ with the left-hand side of $((X'\circ Y')\circ Z')=(X'\circ(Y'\circ Z'))$ via mgu $\mu=\{X/X', Y/Y', Z/Z'\}$.This leads to the critical pair (a,b) with $a=(X'\circ(Y'\circ Z'))$ and $b=(X'\circ(Y'\circ Z'))$. Terms a and b converge. (The reader will have certainly noticed that cases like (a.2.1) are uninteresting in that they always lead to a critical pair (a,b) with identical terms a and b.)

(a.2.2)   We unify $((X\circ Y)\circ Z)$ with the left-hand side of $(X'\circ e)=X'$ via mgu $\mu=\{X'/(X\circ Y), Z/e\}$. This leads to the critical pair (a,b) with $a=(X\circ(Y\circ e))$ and $b=(X\circ Y)$. Obviously, terms a and b converge.

(b)   Consider the equation $(X\circ e)=X$. There are two non-variable subterms of $(X\circ e)$, namely e and $(X\circ e)$.

(b.1)   e is not unifiable with the left-hand side of one of (1)-(3).

(b.2)   $(X\circ e)$ is unifiable with the left-hand side of (1), (2) and (3).

(b.2.1)   We unify $(X\circ e)$ with the left-hand side of $((X'\circ Y')\circ Z')=(X'\circ(Y'\circ Z'))$. This leads to a situation as in case (a.2.2).

(b.2.2)   We unify $(X\circ e)$ with the left-hand side of $(X'\circ e)=X'$. This is a trivial case like (a.2.1).

(b.2.3)   We unify $(X\circ e)$ with the left-hand side of $(e\circ Y)=Y$ via mgu $\mu=\{X/e, Y/e\}$. This leads to the critical pair (a,b) with a=e and b=e.

(c)   Consider the equation $(e\circ Y)=Y$. There are two non-variable subterms of $(e\circ Y)$, namely e and $(e\circ Y)$.

(c.1)   e is not unifiable with the left-hand side of any of (1)-(3).

(c.2)   $(e\circ Y)$ is unifiable with the left-hand side of (2) and (3), but not with the left-hand side of (1).

(c.2.1)   We unify $(e\circ Y)$ with $(X\circ e)$. This leads to a situation like case (b.2.3).

(c.2.2)   We unify $(e\circ Y)$ with $(e\circ Y')$. This is a trivial case like (a.2.1).

Let us make some remarks based on this example. To compute all critical pairs in the simple example above was a trivial, but quite extensive, task involving several uninteresting cases. The reader will certainly feel the need for an automatic system able to generate all the relevant critical pairs and check their convergence. Such systems are indeed already available.

Let us now take a look at what a simplifier for the set E would look like. First we have to choose a selection function Sel. As such we define Sel(t), for a term t which is not an E-normal form, to be that term s resulting from a replacement of the leftmost subterm in t via $\xrightarrow{E}$ . Hence, for a ground term t we always have Sel*(t)=e. In the case that t contains occurrences of variables $X_1,X_2,...,X_n$ (denoted from left to right), Sel*(t) is the term $(X_1 \circ (X_2 \circ (... \circ (X_{n-1} \circ X_n))...))$. This can be easily seen, since terms of the latter form are E-normal forms and every term t can be E-reduced into such a term by successively applying the selection function Sel.

## EXAMPLE 13.2

Let $\Sigma$ be a signature with a constant e, a 2-ary infixedly written function symbol $\circ$ and a 1-ary function symbol i. Consider the following set E of equations (axioms of the theory of groups):

$$((X \circ Y) \circ Z)=(X \circ (Y \circ Z)) \qquad (X \circ e)=X \qquad (X \circ i(X))=e$$

As in the first example, it is easily shown that E is noetherian. But E does not possess the Church-Rosser property. Take the subterm $(X \circ Y)$ of $((X \circ Y) \circ Z)$ and unify it with the left-hand side of $(X' \circ i(X'))=e$ via mgu $\mu=\{X/X',Y/i(X')\}$. This leads to the critical pair (a,b) with $a=(X' \circ (i(X') \circ Z))$ and $b=(e \circ Z)$. Both are E-normal forms, and hence cannot be further reduced.

What can be done in a situation like this? We could try to extend E by new equations allowing reduction of critical pairs which do not converge so far to the same term r. Of course, the new equations should lead to a set of equations E' being logically equivalent to the given set E. This means that all new equations should follow from E. Furthermore, the property of being noetherian should be preserved. Such an extension is not always possible. Fortunately, in our example we are able to extend E by the equation

$$(X \circ (i(X) \circ Y))=(e \circ Y)$$

which follows from E. Now we can reduce both a and b from above into $(e \circ Z)$. But this is not the end of the story. With respect to the extended set E' of equations, there exists again a critical pair (a',b') such that a' and b' do not converge (the reader might find such a critical pair). As a final extension of E to a logically equivalent and noetherian set E' of equations with the Church-Rosser property, we would propose (following Buchberger and Loos (1980)):

| | |
|---|---|
| $(e \circ X)=X$ | $(i(X) \circ X)=e$ |
| $i(e)=e$ | $(i(X) \circ (X \circ Y))=Y$ |
| $(X \circ (i(X) \circ Y))=Y$ | $i(i(X))=X$ |
| $(i(X) \circ X)=e$ | $i((X \circ Y))=(i(Y) \circ i(X))$ |

The reader is encouraged to determine by hand all critical pairs. Let us finally formulate the famous **Knuth-Bendix completion procedure** which lays down how to extend successively a notherian set of equations to a set E' logically equivalent to E, which is hopefully still noetherian and possesses the Church-Rosser property.

## 13.4    The Knuth-Bendix completion procedure

Let a noetherian set E of equations be given. The following procedure, called the Knuth-Bendix completion procedure, tries successively to extend E to a logically equivalent set E' of equations which still is noetherian and, in addition, possesses the Church-Rosser property.

We must always take care that the considered extensions E' of E are noetherian. This implies that we can always effectively check whether two terms a and b converge: Simply build the tree of all possibilities to reduce a and b. Owing to the noetherian property of E', these trees consist only of paths of finite length. Thus, by König's lemma (see appendix), they are finite trees, allowing us to check in an effective way whether they contain leaves with the same E'-normal form.

The Knuth-Bendix completion procedure need not always be successful. It is not an effective algorithm, since it contains decisions which, in general, cannot be done algorithmically. Furthermore, it contains a non-deterministic element (the choice of the critical pair to be dealt with). Thus, it is more like a frame for a semi-automatic interactive system for the completion of noetherian sets of equations, or a frame to be refined and further automated for special cases which allow us to deal with the afore-mentioned decisions in an effective way.

```
E':=E ;
WHILE  the actual extension E' does not have the Church-Rosser
       property (to be checked using critical pairs)
   DO
       Choose a critical pair (a,b) of E' such that a and b do not
       converge ;
       IF E'∪{a=b} is noetherian THEN E':= E'∪{a=b}
       ELSE
           IF E'∪{b=a} is noetherian THEN E':= E'∪{b=a}
           ELSE stop with failure
           END
       END
END.
```

Whenever this procedure halts without failure, the obtained set E' is an extension of E which is logically equivalent to E, noetherian and possesses the Church-Rosser property.

Let us observe the running of the procedure in the case of the following set E of equations:

$$((X \circ Y) \circ Z) = (X \circ (Y \circ Z))$$

$$(X \circ e) = X$$

As a subset of the set of equations of our example from Section 13.4 we know that E is noetherian. We now compute the critical pairs. Consider the subterm $(X \circ Y)$ of the left-hand side of the first equation and unify it with the left-hand side of the equation $((X' \circ Y') \circ Z') = (X' \circ (Y' \circ Z'))$ via mgu $\mu = \{X/(X' \circ Y'), Y/Z'\}$. This leads to the critical pair (a,b) with $a = ((X' \circ Y') \circ (Z' \circ Z))$ and $b = ((X' \circ (Y' \circ Z')) \circ Z)$. Fortunately we obtain:

$$((X' \circ Y') \circ (Z' \circ Z)) \xrightarrow[E]{} (X' \circ (Y' \circ (Z' \circ Z)))$$

$$((X' \circ (Y' \circ Z')) \circ Z) \xrightarrow[E]{} (X' \circ ((Y' \circ Z') \circ Z)) \xrightarrow[E]{} (X' \circ (Y' \circ (Z' \circ Z)))$$

Hence, a and b converge and we do not need to extend E by the equation a=b. For the next critical pair, consider the subterm $(X \circ Y)$ of the left-hand side of the first equation and unify it with the left-hand side of the equation $(X' \circ e) = X'$ via mgu $\mu = \{X'/X, Y/e\}$. This leads to the critical pair (a,b) with $a = (X \circ (e \circ Z))$ and $b = (X \circ Z)$. Neither a nor b can be further reduced. Hence, one of a=b or b=a should be added to our set of equations. Obviously, E together with a=b is again noetherian. So our first extension of E is:

$$((X \circ Y) \circ Z) = (X \circ (Y \circ Z))$$

$$(X \circ e) = X$$

$$(X \circ (e \circ Z)) = (X \circ Z)$$

Again we compute critical pairs of our extension. Now we find that the terms in every critical pair converge (this is left as an exercise for the reader). Thus, the Knuth-Bendix completion procedure halts with success.

With this description of the Knuth-Bendix completion procedure we have only scratched the surface of an important and steadily growing field. Knuth-Bendix completion is used to: give decision algorithms for various axiomatic theories; synthesize programs; replace inductive proofs; integrate functional computing into logic programming and act as an interpreter for such languages (narrowing), and much more. For a good introduction into this actual field of research we refer the reader to Dershowitz (1989) and Bachmair *et al.* (1989).

## 13.5    The expressive power of noetherian and confluent specifications✂

In this section we investigate the question of which algebras can be correctly specified by noetherian and confluent sets of equations. As we have already seen, such specifications are of special importance, since they serve as executable rapid prototypes in software development.

### Theorem 13.6

Let E be a noetherian and confluent set of equations and $A$ be an algebra. If E correctly specifies $A$, then $A$ is term generated, and it is decidable whether two ground terms evaluate to the same value in $A$. Let us call algebras with these properties **computable algebras**.

*Proof*    Assume that E correctly specifies $A$. Then $A$ is trivially term-generated. Consider a pair (s,t) of ground terms. Then $val_A(s)=val_A(t)$ iff $s=_E t$ iff s and t converge w.r.t. E. By assumption on E, the latter property can be decided. ∎

What do computable algebras look like? Let us concentrate on one-sorted algebras. The following theorems give an answer to our question. We start with an investigation of algebras with finite domain. Even for this case, it is not as simple as it might seem at first sight.

### Theorem 13.7

Every term generated algebra with finite domain can be correctly specified by a finite set E which is noetherian and has the Church-Rosser property.

*Proof*    Let $A$ be a term generated algebra with finite domain. We define an equivalence relation ~ on ground terms as follows:

$$s\sim t \text{ iff } val_A(s)=val_A(t).$$

Obviously, ~ is a congruence relation on $A$. We carefully construct a finite set of representatives Rep of ground terms w.r.t. the congruence relation ~ with the following properties:

(1)    For every ground term t there is a ground term s∈ Rep with s~t.

(2)    For any two elements s and t of Rep, s~t iff s=t.

(3)    If $f(t_1,...,t_n)\in$ Rep, for an n-ary function symbol f and terms $t_1,...,t_n$, then $t_1,...,t_n\in$ Rep.

(These properties will later enable us to give a simple proof for the noetherian property of the set of equations to be constructed.) Rep will be the union of the following ascending sequence of sets of ground terms

$$\text{Rep}_0 \subseteq \text{Rep}_1 \subseteq \ldots \subseteq \text{Rep}_i \subseteq \text{Rep}_{i+1} \subseteq \ldots$$

defined as follows:

- *Basic case*    Consider $T_0$=the set of all constants. Let $\text{Rep}_0$ be a maximal subset of $T_0$ consisting of non-equivalent terms.

- *Inductive step*  Let $\text{Rep}_i$ be already defined. Consider

$$T_{i+1}=\{f(t_1,\ldots,t_n)| \text{ f n-ary, } t_1,\ldots,t_n \in \text{Rep}_i\}.$$

Let $M_{i+1}$ be a maximal subset of $T_{i+1}$ such that $\text{Rep}_i \cup M_{i+1}$ consists of non-equivalent terms. Then define $\text{Rep}_{i+1}=\text{Rep}_i \cup M_{i+1}$.

Property (1) is easily shown by structural induction on t. If t is a constant, existence of term $s \in \text{Rep}$ with $s \sim t$ follows from the maximality of $\text{Rep}_0$. If t is a composed term $f(s_1,\ldots,s_n)$, we first choose by induction $t_1,\ldots,t_n \in \text{Rep}$ with $s_1 \sim t_1,\ldots,s_n \sim t_n$. We fix i such that $t_1,\ldots,t_n \in \text{Rep}_i$. Then $t \sim f(t_1,\ldots,t_n)$ and $f(t_1,\ldots,t_n) \in T_{i+1}$. The existence of term $s \in \text{Rep}_{i+1}$ with $s \sim t$ follows from the maximality of $M_{i+1}$. Property (2) is fulfilled, since each set $\text{Rep}_i$ consists of non-equivalent terms. Property (3) immediately follows from the definition of $M_{i+1}$ as a subset of $T_{i+1}$.

Because of property (2) and the finiteness of the domain of $A$, Rep is a finite set of ground terms. Now consider the set E consisting of all equations $f(t_1,\ldots,t_n)=t$, for an n-ary function symbol f and ground terms $t_1,\ldots,t_n,t \in \text{Rep}$ such that $f(t_1,\ldots,t_n) \notin \text{Rep}$ and $f(t_1,\ldots,t_n) \sim t$.

Let us first show that E is a correct specification for $A$. In order to use Theorem 12.1, three conditions must be checked.

(a)    $A$ is a model of E. For this purpose let an equation $f(t_1,\ldots,t_n)=t$ in E be given. By definition of E, this implies that $f(t_1,\ldots,t_n) \sim t$, that is, $\text{val}_A(f(t_1,\ldots,t_n))=\text{val}_A(t)$. This means that $A$ is a model of $f(t_1,\ldots,t_n)=t$.

(b)    $A$ is term generated. This was part of the assumptions of our theorem.

(c)    As the required set C of 'normal forms' for condition (3) of Theorem 12.1 to be used, we take the set Rep of chosen representatives. Two conditions must be verified.

   (c1)    For every ground term s there exists a term $t \in \text{Rep}$ such that $t =_E s$. This is shown by structural induction on s. Assume s is written in the form $f(s_1,\ldots,s_n)$ with ground terms $s_1,\ldots,s_n$. By induction hypothesis we can choose terms $t_1,\ldots,t_n \in \text{Rep}$ such that

$s_1 =_E t_1,\ldots,s_n =_E t_n$. Then $s =_E f(t_1,\ldots,t_n)$. If $f(t_1,\ldots,t_n)$ is already an element of Rep, nothing remains to be done. Otherwise (using property (1) from above) there exists an equation $f(t_1,\ldots,t_n)=t$ in E with a ground term $t \in$ Rep. Then $t =_E s$.

(c2)   For all $s,t \in$ Rep, $s=t$ iff $val_A(s)=val_A(t)$. This follows from condition (2) above.

Next we show that E is noetherian. It suffices to consider reductions via $\xrightarrow[E]{}$ consisting solely of ground terms, since E consists of ground terms only. We show that any application of $\xrightarrow[E]{}$ to a ground term s decreases the number $\sigma(s)$ of subterms of s which are not elements of Rep. This can be seen as follows. Assume that term r results from term s by application of an equation $f(t_1,\ldots,t_n)=t$. This implies that $f(t_1,\ldots,t_n) \notin$ Rep and $t \in$ Rep. Hence, the subterm $f(t_1,\ldots,t_n)$ of s together with all its further subterms contribute at least 1 to the number $\sigma(s)$. The subterm $t \in$ Rep of r together with all its further subterms which all are elements of Rep by property (3) above contributes 0 to $\sigma(r)$.

Next, we have to consider all the K subterms of s containing the replaced occurrence of term $f(t_1,\ldots,t_n)$ as a subterm or all the K (with the same number K) subterms of r containing the considered occurrence of term t as a subterm. The former are all terms not in Rep by property (3) above, hence they all contribute to $\sigma(s)$. Some of the latter ones may be elements of Rep, some others may not. Anyhow, they do not contribute more to $\sigma(r)$ than the former ones contributed to $\sigma(s)$. Altogether we have that $\sigma(r)<\sigma(s)$.

Finally, we show that E has the Church-Rosser property. To do so, we compute all critical pairs of E. Take some equation $f(t_1,\ldots,t_n)=t$ from E. Then $f(t_1,\ldots,t_n) \notin$ Rep and $t,t_1,\ldots,t_n \in$ Rep. Consider another equation $g(s_1,\ldots,s_m)=s$ in E. Then $g(s_1,\ldots,s_m) \notin$ Rep and $s,s_1,\ldots,s_m \in$ Rep. This implies that none of the proper subterms of $f(t_1,\ldots,t_n)$ coincides with $g(s_1,\ldots,s_m)$ (note that unification is not possible, since all equations in E are ground equations). Furthermore, if $f(t_1,\ldots,t_n)$ coincides with $g(s_1,\ldots,s_m)$ then $t \sim s$, hence $t=s$. Thus, the only critical pairs of E are identities.     ∎

### Theorem 13.8 (Characterization of computable algebras with infinite domain)

Every computable algebra A with infinite domain is isomorphic to an algebra $(\mathbb{N},f_1,\ldots,f_m)$ with total recursive (computable and everywhere defined) functions $f_1,\ldots,f_m$.

*Proof*    Let $A$ be the algebra $(A, g_1, \ldots, g_m)$. Take an effective listing of all ground terms:

$$t_0, t_1, t_2, \ldots$$

Omit from this listing all terms $t_i$ such that there exists a term $t_j$ with $j < i$ and $\mathrm{val}_A(t_i) = \mathrm{val}_A(t_j)$. Since the latter property is decidable for our given computable algebra, we arrive at a second effective listing of ground terms

$$s_0, s_1, s_2, \ldots$$

such that two different terms have different values in $A$, and every element of $A$ is the value of one of these terms. Since A was assumed to be infinite, the listing consists of infinitely many terms. This implies that the function $h: \mathbb{N} \to A$ defined by $h(i) = \mathrm{val}_A(s_i)$ is bijective. Now define the functions $f_1, \ldots, f_m$ in such a way that h becomes an isomorphism from $(\mathbb{N}, f_1, \ldots, f_m)$ onto $A$. This means defining $f_k: \mathbb{N}^m \to \mathbb{N}$, in the case that $g_k: A^m \to A$ is an m-ary function, as follows:

$$f_k(i_1, \ldots, i_m) = h^{-1}(g_k(h(i_1), \ldots, h(i_m)))$$

It remains to show that $f_k$ is a computable function. This can be seen as follows:

$$f_k(i_1, \ldots, i_m)$$
$$= h^{-1}(g_k(h(i_1), \ldots, h(i_m)))$$
$$= h^{-1}(g_k(\mathrm{val}_A(s_{i_1}), \ldots, \mathrm{val}_A(s_{i_m})))$$
$$= h^{-1}(\mathrm{val}_A(g_k(s_{i_1}, \ldots, s_{i_m})))$$
$$= \text{the uniquely determined index i with } \mathrm{val}_A(s_i) = \mathrm{val}_A(g_k(s_{i_1}, \ldots, s_{i_m}))$$

The latter representation shows how to compute $f_k(i_1, \ldots, i_m)$.     ∎

Finally, we have to deal with algebras of the form $(\mathbb{N}, f_1, \ldots, f_m)$ with total recursive functions $f_1, \ldots, f_m$. As we already know from Section 12.3, it is not always possible to specify such an algebra correctly by a finite set of equations, but extension of $(\mathbb{N}, f_1, \ldots, f_m)$ by suitable functions sometimes helps us to find a finite equational specification. As we will soon see, it will always be possible to specify a suitable (computable) expansion of the given algebra correctly by a noetherian and confluent set of equations. The proof uses concepts and results from recursive function theory which were introduced in Section 8.3.

**Theorem 13.9 (Specifying total computable functions by noetherian and confluent sets of equations)**

For every total recursive function f there is an expansion of algebra $(\mathbb{N},0,\text{succ},f)$ that can be correctly specified by a confluent and noetherian set of equations.

*Proof* We show how a confluent and noetherian specification E of an algebra $(\mathbb{N},0,\text{succ},k_1,\dots,k_r)$ with total and computable functions $k_1,\dots,k_r$ can be extended into a confluent and noetherian specification E' of an expansion of $(\mathbb{N},0,\text{succ},k_1,\dots,k_r,f)$, whenever f is either a projection function or defined by composition, primitive recursion or minimization from $0,\text{succ},k_1,\dots,k_r$. Obviously, this proves our theorem. Equations used in specifications of algebras $(\mathbb{N},0,\text{succ},k_1,\dots,k_r)$ will have left-hand sides of the form $k_i(t)$, for $i=1,\dots,r$ and a term t.

So let a function f as described be given. The specifications constructed below can be verified to be correct w.r.t. the considered algebras by a simple application of Theorem 12.3. They are easily seen to be noetherian, with one exception which is treated in more detail. All of them are easily shown to be locally confluent by recognizing that the only critical pairs are identities (here we use the announced form of the left-hand sides of equations).

(1)     Assume that f is the projection function $p_{n,i}$. Obviously, a confluent and noetherian specification of the extension of $(\mathbb{N},0,\text{succ},k_1,\dots,k_r,f)$ is obtained by adding to E the following equation:

$$f(X_1,\dots,X_n)=X_i \qquad (13.2)$$

(2)     Assume that f defined by $f(x_1,\dots,x_n)=g(h_1(x_1,\dots,x_n),\dots,h_m(x_1,\dots,x_n))$ with functions $g,h_1,\dots,h_m$ occurring among the functions $0,\text{succ},k_1,\dots,k_r$. Then, a correct specification of $(\mathbb{N},0,\text{succ},k_1,\dots,k_r,f)$ is obtained by extending E by the equation

$$f(X_1,\dots,X_n)=g(h_1(X_1,\dots,X_n),\dots,h_m(X_1,\dots,X_n)) \qquad (13.3)$$

The obtained extension of E is noetherian and confluent.

(3)     Assume that f is defined recursively by $f(x_1,\dots,x_n,0)=g(x_1,\dots,x_n)$ and $f(x_1,\dots,x_n,y+1)=h(x_1,\dots,x_n,f(x_1,\dots,x_n,y),y)$ with functions $g,h$ occurring among the functions $0,\text{succ},k_1,\dots,k_r$. Then, a correct specification of the expanded algebra $(\mathbb{N},0,\text{succ},k_1,\dots,k_r,f)$ is obtained by extending E by the equations

$$f(X_1,\dots,X_n,0)=g(X_1,\dots,X_n) \qquad (13.4)$$

$$f(X_1,\ldots,X_n,\text{succ}(Y))=h(X_1,\ldots,X_n,f(X_1,\ldots,X_n,Y),Y) \qquad (13.5)$$

Obviously, the obtained extension of E is noetherian and confluent.

(4)     Assume that $f:\mathbb{N}^n\to\mathbb{N}$ defined by $f(x_1,\ldots,x_n)=$the least number y such that $g(x_1,\ldots,x_n,y)=0$ with a total function g occurring among the functions $0,\text{succ},k_1,\ldots,k_r$. Now we will require several additional functions in order to specify f. (Note that auxiliary functions have not been required so far.) A first idea of how to specify f correctly is obtained by regarding the way $f(x_1,\ldots,x_n)$ is computed step by step: we successively compute $g(x_1,\ldots,x_n,y)$, for $y=0,1,2,\ldots$, until a first y is found such that $g(x_1,\ldots,x_n,y)=0$ (such a y will be found, since f is assumed to be total). This suggests the introduction of a further parameter b measuring the position up to which the above search has proceeded. Then we search for a $y\geq b$ with $g(x_1,\ldots,x_n,y)=0$. If b already exceeds the first y with $g(x_1,\ldots,x_n,y)=0$, we output 0. This idea is formally captured by the function $f^*:\mathbb{N}^{n+1}\to\mathbb{N}$ defined by

$$f^*(x_1,\ldots,x_n,b)=\begin{cases} 0 \text{ if there exists } y<b \text{ with } g(x_1,\ldots,x_n,y)=0 \\ \text{the least } y\geq b \text{ with } g(x_1,\ldots,x_n,y)=0, \text{ otherwise} \end{cases}$$

Note that f can be expressed using $f^*$ as $f(x_1,\ldots,x_n)=f^*(x_1,\ldots,x_n,0)$. Let us now concentrate on specifying $f^*$. For this purpose, we first deal with the test defining the two cases in the definition of $f^*$. Define functions $\text{zero}:\mathbb{N}^{n+1}\to\mathbb{N}$ and $\text{if}:\mathbb{N}^3\to\mathbb{N}$ by

$$\text{zero}(x_1,\ldots,x_n,b)=\begin{cases} 0 \text{ if there exists } y<b \text{ with } g(x_1,\ldots,x_n,y)=0 \\ 1 \text{ otherwise} \end{cases}$$

$$\text{if}(t,x,y)=\begin{cases} x \text{ if } t=0 \\ y \text{ if } t>0 \end{cases}$$

We construct a correct specification for the expanded algebra $(\mathbb{N},0,s,k_1,\ldots,k_r,\text{if},\text{zero})$ by extending set E by the following equations:

$$\text{if}(0,X,Y)=X \qquad (13.6)$$

$$\text{if}(\text{succ}(T),X,Y)=Y \qquad (13.7)$$

$$\text{zero}(X_1,\ldots,X_n,0)=\text{succ}(0) \qquad (13.8)$$

$$\text{zero}(X_1,...,X_n,\text{succ}(B)) \tag{13.9}$$
$$=\text{if}(g(X_1,...,X_n,B),0,\text{zero}(X_1,...,X_n,B))$$

Obviously, $(\mathbb{N},0,\text{succ},k_1,...,k_r,\text{if},\text{zero})$ is a term generated model of these equations. Moreover, the presented extension of E fulfils the reduction property (3) of Theorem 12.3. So, it correctly specifies $(\mathbb{N},0,\text{succ},k_1,...,k_r,\text{if},\text{zero})$. Furthermore, it is locally confluent, since the only critical pairs are identities, and noetherian, because the last parameter B of zero decreases when going from left to right in the second equation for zero.

Finally, we have to deal with f*. First of all note that f* fulfils the following recurrence relation :

$f^*(x_1,...,x_n,b)=0$, if $\text{zero}(x_1,...,x_n,b)=0$

$f^*(x_1,...,x_n,b)=b$ if $\text{zero}(x_1,...,x_n,b)>0$ and $g(x_1,...,x_n,b)=0$

$f^*(x_1,...,x_n,b)=f^*(x_1,...,x_n,b+1)$, otherwise

Let us introduce a 5-ary conditional (the first two arguments s and t serve as test values, the other three arguments x, y and z as possible outputs whose choice depends on s and t) cond: $\mathbb{N}^5 \rightarrow \mathbb{N}$ by

cond(s,t,x,y,z)=x for s=0
cond(s,t,x,y,z)=y for s>0 and t=0
cond(s,t,x,y,z)=z otherwise

Function cond is easily specified by the equations

cond(0,T,X,Y,Z)=X
cond(succ(S),0,X,Y,Z)=Y
cond(succ(S),succ(T),X,Y,Z)=Z

Then we can recursively describe f* by adding to our set of equations

$f^*(x_1,...,x_n,b)$
$=\text{cond}(\text{zero}(x_1,...,x_n,b),g(x_1,...,x_n,b),0,b,f^*(x_1,...,x_n,b+1))$

This set of equations correctly specifies cond and f* (reduction property (3) of Theorem 12.3 follows from the assumption that f is total, hence there is always some y such that $g(x_1,...,x_n,y)=0$; by always first evaluating $\text{zero}(x_1,...,x_n,b)$, then, if necessary, $g(x_1,...,x_n,b)$, and finally $f^*(x_1,...,x_n,b+1)$, a reduction is always possible.). It is locally confluent, too (no critical pairs unless identities). But unfortunately, it is not noetherian. The reason is that the above-mentioned order of evaluation from left to right is not enforced by the mechanism of term rewriting. We could use the following order instead:

$f^*(x_1,\ldots,x_n,b) \xrightarrow{E}$

$cond(zero(x_1,\ldots,x_n,b),g(x_1,\ldots,x_n,b),0,b,f^*(x_1,\ldots,x_n,b+1)) \xrightarrow{E}$

$cond(zero(x_1,\ldots,x_n,b),g(x_1,\ldots,x_n,b),0,b,$

$\qquad cond(zero(x_1,\ldots,x_n,b+1),g(x_1,\ldots,x_n,b+1),0,b+1,$

$\qquad\qquad\qquad\qquad f^*(x_1,\ldots,x_n,b+2)))$

$\xrightarrow{E} cond(zero(x_1,\ldots,x_n,b),g(x_1,\ldots,x_n,b),0,b,$

$\qquad cond(zero(x_1,\ldots,x_n,b+1),g(x_1,\ldots,x_n,b+1),0,b+1,$

$\qquad\qquad cond(zero(x_1,\ldots,x_n,b+2),g(x_1,\ldots,x_n,b+2),0,b+2,$

$\qquad\qquad\qquad\qquad f^*(x_1,\ldots,x_n,b+3))))$

$\xrightarrow{E} \ldots$

which results in a non-terminating reduction. (The reader should compare this with the corresponding Prolog program from Section 9.2.2. There the evaluation strategy of Prolog, from left to right, enforces a terminating evaluation.) Can the correct evaluation order be enforced in the formalism of equations? The answer is yes. The way to do this uses n+2-ary auxiliary functions secondcondition and firstcondition defined by

$$secondcondition(t,x_1,\ldots,x_n,b)= \begin{cases} b \text{ if } t=0 \\ f^*(x_1,\ldots,x_n,b+1) \text{ if } t>0 \end{cases}$$

$firstcondition(t,x_1,\ldots,x_n,b)=$

$$\begin{cases} 0 \text{ if } t=0 \\ secondcondition(g(x_1,\ldots,x_n,b),x_1,\ldots,x_n,b) \text{ if } t>0 \end{cases}$$

Consider now the following algebra:

$$(\mathbb{N},0,succ,k_1,\ldots,k_r,cond,zero,f^*,firstcondition, secondcondition)$$

It is correctly specified by extending the set E+(13.6)-(13.9) obtained so far by the following equations:

$f^*(X_1,\ldots,X_n,B)$  (13.10)

$\qquad =firstcondition(zero(X_1,\ldots,X_n,B),X_1,\ldots,X_n,B)$

$firstcondition(0,X_1,\ldots,X_n,B)=0$  (13.11)

$firstcondition(succ(T),X_1,\ldots,X_n,B)$  (13.12)

$\qquad =secondcondition(g(X_1,\ldots,X_n,B),X_1,\ldots,X_n,B)$

$secondcondition(0,X_1,\ldots,X_n,B)=B$  (13.13)

$secondcondition(succ(T),X_1,\ldots,X_n,B)$  (13.14)

$$=f*(X_1,...,X_n,succ(B))$$

This algebra is a term generated model of E+(13.6)-(13.14), and it is easily seen that it is correctly specified. Also, the set of equations E+(13.6)-(13.14) is locally confluent. Finally it is noetherian. This can be seen as follows. In order to reduce further a term of the form firstcondition(zero($X_1$,...,$X_n$,B),$X_1$,...,$X_n$,B), we first must evaluate term zero($X_1$,...,$X_n$,B), since only then will one of the equations (13.11) or (13.12) becomes applicable. Correspondingly, to reduce further term secondcondition(g($X_1$,...,$X_n$,B),$X_1$,...,$X_n$,B), we must first evaluate g($X_1$,...,$X_n$,B), since only then will one of the equations (13.13) or (13.14) become applicable. Thus, we have perfectly realized the above-mentioned order of evaluation guaranteeing termination.     ∎

### Corollary

Given an algebra ($\mathbb{N}$,$f_1$,...,$f_m$) with total recursive functions $f_1$,...,$f_m$, we can always expand it by total recursive functions $g_1$,...,$g_l$ to an algebra ($\mathbb{N}$,$f_1$,...,$f_m$,$g_1$,...,$g_l$) which can be correctly specified by a finite, noetherian and confluent set of equations.

### Theorem 13.10

For an algebra $A$ the following are equivalent:

(1)     $A$ is a computable algebra.

(2)     A computable expansion of $A$ by a finite number of auxiliary functions can be correctly specified by a finite, noetherian and confluent set of equations.

# Historical remarks and recommended reading for Part III

Investigations of calculi for equational reasoning go back to Birkhoff (1935). The idea of abstract data types as a means of abstraction in software systems came up in the 1970s with Parnas (1972) and Guttag and Horning (1978). Initial semantics of algebraic specifications was first introduced and studied by the ADJ-group in Goguen *et al.* (1976) and Goguen *et al.* (1977). The standard book on algebraic specifications is Ehrig and Mahr (1985). The main step towards development of specification languages was done by Burstall and Goguen (1977). The semantics of their language CLEAR was presented in Burstall and Goguen (1980). Other interesting specification languages are ACT-TWO described in Fey (1986), OBJ2 in Futatsugi *et al.* (1985), PLUS in Gaudel (1984), Bidoit *et al.* (1987), and ASF in Bergstra *et al.* (1987). A theory of modules in algebraic specification can be found in Ehrig and Mahr (1990).

The Knuth-Bendix algorithm was introduced in Knuth and Bendix (1967). Using different variants of this procedure, it was possible to complete a great number of axiom systems; an overview may be found in Jouannaud and Kirchner (1986). Overviews for term rewriting and a starting point for further investigations are Buchberger and Loos (1982), Buchberger (1987), Dershowitz (1989) and Bachmair *et al.* (1989).

# Part IV

# PROGRAM VERIFICATION
# AND
# HOARE LOGIC

Program verification is the process of proving the correctness of programs. One may ask why this is important and necessary. In fact, in the first years of data processing nobody cared about correctness issues. The necessity for formal analysis of programs became apparent only when more and more programs did not behave in the way that had been expected. This crisis of the software industry was known as the **software crisis** and was the starting point for program verification.

But why isn't *testing* sufficient? This alternative to verification validates a program by letting it run on many 'critical' input data. As long as the program shows the expected behaviour confidence increases that the program is correct. Unfortunately this is not a correctness proof: testing can only show that a program is incorrect, never that it is correct! Perhaps the most convincing argument against testing instead of verification (but not testing in general!) is that every commercial program has been extensively tested, but almost none is free of errors. Thus, testing alone is not enough, especially in critical application areas.

So it is necessary to prove program correctness in a formal (logical) way. To do so we need a programming language with a formal description of the meaning of programs (program semantics) and a formal specification of what programs should do. In the following chapters we shall investigate programming languages containing some important program constructs. The specification language will be predicate logic, in the case of modules the logic of universal formulas, with equational logic as a sublogic that is often sufficient.

Chapter 18 is very important because it addresses a crucial problem of verification methods, namely applicability to realistic problems. It is the task of specifying, implementing and verifying large software systems. We feel that the key to solving these serious problems lies in **modularity**, where module interfaces not only contain syntactic description (as in current programming languages) but also semantical information. We show a way of splitting verification work into small pieces in a modular way.

# Chapter 14
# Program Correctness

## 14.1    The programming language of WHILE programs

In this section we shall define the programming language we want to discuss.
First we define program syntax, then we define program semantics by introducing the input-output and also the operational behaviour of programs over a
data structure.

The programming language we have in mind is an imperative programming language like Pascal or Modula-2. We start our investigation with the
simple language of WHILE programs which includes assignments, tests,
loops and composite instructions.

The reasons why we start with WHILE programs are:

- They exhibit a clear and simple semantics.
- They are contained in every higher procedural programming language.
- They are powerful enough to compute every computable function (see
  Section 8.1).
- They are well suited to implement interesting and realistic problems.

Then we extend the programming language by data type declarations and procedure calls including recursive procedures (Chapter 17), as well as a module
concept (Chapter 18).

> **Definition 14.1:**    For the definition of WHILE programs we use an in-
> finite set of variables Var and eight strings as special symbols, namely
>
> :=, IF, THEN, ELSE, END, WHILE, DO and ;

**Definition 14.2:**     If $\Sigma$ is a signature whose function and predicate symbols are different from the symbols in the vocabulary above, we define $\Sigma$-**programs** as follows: a $\Sigma$-program $\alpha$ is a composite instruction, that is, a string of the form $A_1;A_2;...;A_n$ where $n \in \mathbb{N}$ and $A_1,A_2,...,A_n$ are so-called $\Sigma$-instructions. For n=0 we have the **empty $\Sigma$-program**, written $\varepsilon$.

A $\Sigma$-**instruction** is either a $\Sigma$-assignment, a $\Sigma$-test or a $\Sigma$-loop. A $\Sigma$-**assignment** is a string of the form X:=t where X is a variable and t a $\Sigma$-term. A $\Sigma$-**test** is a string of the form IF B THEN $\beta$ ELSE $\gamma$ END where B is a $\Sigma$-formula without quantifiers and $\beta,\gamma$ are $\Sigma$-programs. A $\Sigma$-**loop** is a string of the form WHILE B DO $\beta$ END where B is a $\Sigma$-formula without quantifiers and $\beta$ is a $\Sigma$-program.

In program verification, formulas are also called **assertions**, and formulas without quantifiers **boolean expressions**.

In most of our examples we shall use the signature $\Sigma_{nat}$ of algebra *Cardinal* (see Section 6.5). For example, the following string is a program over the signature $\Sigma_{nat}$:

> X:=1;WHILE Z>0 DO IF (Z MOD 2)=0 THEN Y:=(Y*Y);
> Z:=(Z DIV 2) ELSE X:=(Y*X);Z:=(Z-1) END END

As usual, we write programs in a more readable form as follows:

```
X:=1;
WHILE Z>0 DO
        IF (Z MOD 2)=0 THEN Y:=(Y*Y);Z:=(Z DIV 2)
        ELSE  X:=(Y*X);Z:=(Z-1)
        END
END
```

*Theorem 14.1 (Uniqueness of the syntax of WHILE programs)*

A program is a string of the form $A_1;A_2;...;A_n$ with **uniquely determined** number n and instructions $A_1,A_2,...,A_n$. An instruction is of the form X:=t with uniquely determined variable X and term t, or of the form IF B THEN $\beta$ ELSE $\gamma$ END with uniquely determined boolean expression B and   programs   $\beta$ and $\gamma$, or of the form WHILE B DO $\beta$ END with uniquely determined boolean expression B and program $\beta$. The latter three cases exclude each other.

**Definition 14.3 (Input-output semantics of WHILE programs):**
Let $A$ be a $\Sigma$-algebra and sta, sta' be states over $A$. For a program $\alpha$ we define that $\alpha$ computes state sta' from state sta over algebra $A$ (or transforms state sta into state sta' over $A$, or terminates with final state sta' when starting in state sta over $A$) and denote it by sta$[\alpha]_A$sta'. $[\alpha]_A$ is thus a 2-ary relation over the set of states in $A$ and is defined in the following way:

- sta$[\varepsilon]_A$sta' $\Leftrightarrow$ sta'=sta

- sta$[X:=t]_A$sta' $\Leftrightarrow$ sta'=sta$(X/\text{val}_{A,\text{sta}}(t))$

- sta$[$IF B THEN $\beta$ ELSE $\gamma$ END$]_A$sta' $\Leftrightarrow$

  $(A \models_{\text{sta}} B$ and sta$[\beta]_A$sta'$)$ or $(A \not\models_{\text{sta}} B$ and sta$[\gamma]_A$sta'$)$

- sta$[$WHILE B DO $\beta$ END$]_A$sta' $\Leftrightarrow$ There are t$\in \mathbb{N}$ and states sta$_0$,sta$_1$,...,sta$_t$ such that

  sta=sta$_0$

  $A \models_{\text{sta}_i} B$ and sta$_i[\beta]_A$sta$_{i+1}$, for all i=0,...,t-1

  $A \not\models_{\text{sta}_t} B$

  sta$_t$ = sta'

- sta$[A_1;A_2;...;A_n]_A$sta', for n$\geq$2 $\Leftrightarrow$ There are states sta$_0$,sta$_1$,...,sta$_n$ such that

  sta=sta$_0$

  sta$_i [A_{i+1}]_A$sta$_{i+1}$, for all i=0,...,n-1

  sta$_n$=sta'

Let us look at some examples. They refer to algebra $A$=***Cardinal***.

(1)  Suppose sta$[X:=1;Y:=(X+1)]_A$sta'. Then there are states sta$_0$, sta$_1$ and sta$_2$ with sta$_0$=sta, sta$_1$=sta$_0$(X/1) and sta$_2$=sta$_1$(Y/val$_{A,\text{sta}_1}$(X+1))=sta$_1$(Y/sta$_1$(X)+1)=sta$_1$(Y/2)=sta(X/1,Y/2)=sta'.

(2)  Let us consider sta and sta' with sta$[$IF X>0 THEN X := X-1 ELSE $\varepsilon$ END$]_A$sta'. If sta(X)>0 then sta'=sta(X/sta(X)-1), else sta'=sta.

(3)  Suppose that sta$[$WHILE X$\neq$0 DO X:=(X-1) END$]_A$sta'. Then there exists t$\in \mathbb{N}$ and states sta$_0$,sta$_1$,...,sta$_t$ such that sta=sta$_0$, sta$_i[X:=(X-1)]_A$sta$_{i+1}$ (for i<t-1) and sta$_t$=sta'. It is an easy exercise

to show that   t=sta(X) and $sta_i$=sta(X/t-i), for all i≤t. This means that X takes all values from sta(X) down to zero.

---

**Exercise 14.1** Lay down the algebra **A=Cardinal**. For the following examples, express sta' in terms of sta.

(a)    sta [ IF X ≥Y THEN M:=X ELSE M:=Y END ]$_A$ sta'

(b)    sta [ X:=0; WHILE X≤Y DO X:=(X+1); Y:=(Y-1) END ]$_A$ sta'

(c)    sta [ WHILE X≤Y DO X:=(X+1); Y:=(Y-1) END ]$_A$ sta'.

---

### Lemma 14.1 (Determinism of WHILE programs)

Let α be a program and sta,sta',sta'' be states over *A*. If sta[α]$_A$sta' and sta[α]$_A$sta'' then sta'=sta''. So the relation [α]$_A$ is right unique. But in general it is not right complete, that is, there does not always exist a state sta' such that sta[α]$_A$sta', as the example WHILE 0=0 DO X:=X END over **Cardinal** shows.

*Proof*   Right uniqueness of [α]$_A$ is shown by structural induction on α.   ∎

For later use we introduce an alternative definition of the semantics of WHILE programs, the so-called **operational semantics**. Instead of defining the input-output effect of a program as was done above, the operational point of view defines how the instructions of a WHILE program are executed step by step from left to right by an interpreter function I. Thus, a program is considered as a stack of instructions with top instruction at the left.

### Definition 14.4 (Operational semantics of WHILE programs):
Let *A* be a Σ-algebra. We define an **interpreter function** I$_A$ which assigns to each pair (α,sta) consisting of a state sta over *A* and a Σ-program α a new pair (α',sta'). (Think of α as a stack of instructions that are executed from left to right and α' as the new stack after one step of execution.)

- I$_A$(ε,sta)=(ε,sta)

For a non-empty sequence of instructions $A_1;A_2;...;A_n$ (n=1 is possible) we distinguish three cases depending on the form of $A_1$:

- $I_A(X:=t;A_2;\ldots;A_n,\text{sta})=(A_2;\ldots;A_n,\text{sta}(X/\text{val}_{A,\text{sta}}(t)))$

- $I_A(\text{IF B THEN }\beta\text{ ELSE }\gamma\text{ END};A_2;\ldots;A_n,\text{sta})=$

$$\begin{cases} (\beta;A_2;\ldots;A_n,\text{sta}) \text{ if } A\models_{\text{sta}}B \\ (\gamma;A_2;\ldots;A_n,\text{sta}) \text{ if } A\not\models_{\text{sta}}B \end{cases}$$

- $I_A(\text{WHILE B DO }\beta\text{ END};A_2;\ldots;A_n,\text{sta})=$

$$\begin{cases} (\beta;\text{WHILE B DO }\beta\text{ END};A_2;\ldots;A_n,\text{sta}) \text{ if } A\models_{\text{sta}}B \\ (A_2;\ldots;A_n,\text{sta}) \qquad\qquad\qquad \text{ if } A\not\models_{\text{sta}}B \end{cases}$$

### Theorem 14.2 (Equivalence of input-output and operational semantics)

Let $A$ be a $\Sigma$-algebra and $\alpha$ be a $\Sigma$-program. Then, for all states sta over $A$ the following are true:

(1) If $\text{sta}[\alpha]_A\text{sta}'$, for a state sta', then a finite number of applications of $I_A$ transforms $(\alpha,\text{sta})$ into $(\varepsilon,\text{sta}')$.

(2) If there does not exist a state sta' with $\text{sta}[\alpha]_A\text{sta}'$, then iterated application of $I_A$ starting with $(\alpha,\text{sta})$ never leads to a pair whose first component is $\varepsilon$.

---

**Exercise 14.2**  Prove the equivalence of input-output and operational semantics.

---

Using the operational variant of the definition of the semantics of programs we will next provide an important construction expressing the effect of the execution of a program $\alpha$ within a fixed number T of steps.

### Definition 14.5 (Computation predicate with time bound):

Let $\alpha$ be a $\Sigma$-program with variables $X_1,\ldots,X_k$. Let $Y_1,\ldots,Y_k$ be fresh variables. For every natural number T we define a quantifier-free $\Sigma$-formula comp$_{\alpha,T}(X_1,\ldots,X_k,Y_1,\ldots,Y_k)$ with the following informal intended meaning over an arbitrary $\Sigma$-algebra $A$:

$(Y_1,...,Y_k)$ is computed from $(X_1,...,X_k)$ by $\alpha$ in at most T interpreter steps, including termination within at most T steps.

The definition of $comp_{\alpha,T}$ by induction on T is as follows:

*Induction base*

- $comp_{\varepsilon,0}(X_1,...,X_k,Y_1,...,Y_k)$ is $(Y_1=X_1\wedge...\wedge Y_k=X_k)$
- $comp_{\alpha,0}(X_1,...,X_k,Y_1,...,Y_k)$ is an arbitrary unsatisfiable formula if $\alpha$ is not empty

*Induction step*

- $comp_{\varepsilon,T+1}(X_1,...,X_k,Y_1,...,Y_k)$ is $(Y_1=X_1\wedge...\wedge Y_k=X_k)$

If $\alpha$ is a non-empty sequence of instructions $A_1;A_2;...;A_n$ we distinguish three cases depending on the form of $A_1$:

- $comp_{X_i:=t;A_2;...;A_n,T+1}(X_1,...,X_k,Y_1,...,Y_k)$ is
  $comp_{A_2;...;A_n,T}(X_1,...,X_{i-1},t,X_{i+1},...,X_k,Y_1,...,Y_k)$

- $comp_{\text{IF B THEN }\beta\text{ ELSE }\gamma\text{ END};A_2;...;A_n,T+1}(X_1,...,X_k,Y_1,...,Y_k)$ is
  $(B\wedge comp_{\beta;A_2;...;A_n,T}(X_1,...,X_k,Y_1,...,Y_k))\vee$
  $(\neg B\wedge comp_{\gamma;A_2;...;A_n,T}(X_1,...,X_k,Y_1,...,Y_k))$

- $comp_{\text{WHILE B DO }\beta\text{ END};A_2;...;A_n,T+1}(X_1,...,X_k,Y_1,...,Y_k)$ is
  $(B\wedge comp_{\beta;\text{WHILE B DO }\beta\text{ END};A_2;...;A_n,T}(X_1,...,X_k,Y_1,...,Y_k))$
  $\vee(\neg B\wedge comp_{A_2;...;A_n,T}(X_1,...,X_k,Y_1,...,Y_k))$

**Theorem 14.3 (Semantics of the computation predicate with time bound)**

Let *A* be a $\Sigma$-algebra and $\alpha$ be a $\Sigma$-program with variables $X_1,...,X_k$. Let sta and sta' be states over *A* with $sta(X_i)=a_i$ and $sta'(X_i)=b_i$, for i=1,...,k. Then the following are equivalent:

(1)     $sta[\alpha]_A sta'$

(2)     $A \models comp_{\alpha,T}(a_1,...,a_k,b_1,...,b_k)$, for some number T.

*Proof*   Assume that $sta[\alpha]_A sta'$. Choose a number T such that T applications of $I_A$ transform $(\alpha,sta)$ into $(\varepsilon,sta')$. Inductively on T it is easily shown that $A \models comp_{\alpha,T}(a_1,...,a_k,b_1,...,b_k)$. Likewise, the reversed implication is shown by induction on T.                                                    ■

### *Corollary*

Let $A$ be a $\Sigma$-algebra and $\alpha$ be a $\Sigma$-program with variables $X_1,...,X_k$. Let sta be a state over $A$ such that $sta(X_1)=a_1,...,sta(X_k)=a_k$. Then the following statements are equivalent:

(1)    There exists a state sta' with $sta[\alpha]_A$ sta'.

(2)    $A \models \exists Y_1...\exists Y_k \, comp_{\alpha,T}(a_1,...,a_k,Y_1,...,Y_k)$, for some $T \in \mathbb{N}$.

The formula $\exists Y_1...\exists Y_k \, comp_{\alpha,T}(X_1,...,X_k,Y_1,...,Y_k)$ from the corollary above thus serves as a sort of 'termination predicate with time bound T'. We will introduce an alternative termination predicate which is even quantifier free.

**Definition 14.6 (Termination predicate with time bound):**   Let $\alpha$ be a $\Sigma$-program with variables $X_1,...,X_k$. For every $T \in \mathbb{N}$ we define a quantifier-free $\Sigma$-formula $term_{\alpha,T}(X_1,...,X_k)$ by induction on T as follows:

- $term_{\varepsilon,0}(X_1,...,X_k)$ is an arbitrary valid formula
- $term_{\alpha,0}(X_1,...,X_k)$ is an arbitrary unsatisfiable formula, for every non-empty program $\alpha$
- $term_{\varepsilon,T+1}(X_1,...,X_k)$ is an arbitrary valid formula
- $term_{X_i:=t;A_2;...;A_n,T+1}(X_1,...,X_k)$ is the formula
    $$term_{A_2;...;A_n,T}(X_1,...,X_{i-1},t,X_{i+1},...,X_k)$$
- $term_{\text{IF B THEN } \beta \text{ ELSE } \gamma \text{ END};A_2;...;A_n,T+1}(X_1,...,X_k)$ is
    $$(B \wedge term_{\beta;A_2;...;A_n,T}(X_1,...,X_k)) \vee$$
    $$(\neg B \wedge term_{\gamma;A_2;...;A_n,T}(X_1,...,X_k))$$
- $term_{\text{WHILE B DO } \beta \text{ END};A_2;...;A_n,T+1}(X_1,...,X_k)$ is
    $$(B \wedge term_{\beta;\text{WHILE B DO } \beta \text{ END};A_2;...;A_n,T}(X_1,...,X_k)) \vee$$
    $$(\neg B \wedge term_{A_2;...;A_n,T}(X_1,...,X_k))$$

### *Theorem 14.4 (Semantics of the termination predicate)*

Let $A$ be a $\Sigma$-algebra and $\alpha$ be a $\Sigma$-program. Let sta be a state over $A$ and $T \in \mathbb{N}$. Then the following statements are equivalent:

(1)    $(\alpha,\text{sta})$ is transformed by at most T applications of $I_A$ into a pair with first component $\varepsilon$.

(2)    $A \models_{\text{sta}} \text{term}_{\alpha,T}(X_1,...,X_k)$

In particular: there exists a state sta' with $\text{sta}[\alpha]_A\text{sta}'$ (that is, $\alpha$ terminates over $A$ on state sta) iff there exists a number T such that $A \models_{\text{sta}} \text{term}_{\alpha,T}(X_1,...,X_k)$.

---

**Exercise 14.3**   Carry out the proof of Theorem 14.4.

---

An important consequence of the existence of a termination predicate as above is the following theorem.

### Theorem 14.5 (Programs with uniform termination time)

Let $\alpha$ be a $\Sigma$-program with variables $X_1,...,X_k$ and Ax be a set of predicate logic formulas. Assume that, for every model $A$ of Ax and every state sta over $A$, program $\alpha$ terminates over $A$ when started with input state sta (hence T applications of the interpreter function $I_A$ transform $(\alpha,\text{sta})$ into a pair $(\varepsilon,\text{sta}')$, with a number T which may, a priori, depend on $A$ and state sta). Then there is a constant number T such that, for every model $A$ of Ax and every state sta over $A$, T applications of $I_A$ transform $(\alpha,\text{sta})$ into a pair $(\varepsilon,\text{sta}')$.

*Proof*   Let $X_1,...,X_k$ be the variables occurring in $\alpha$. Assume that, for every model $A$ of Ax and every state sta over $A$, there is a number T such that T applications of the interpreter function $I_A$ transform $(\alpha,\text{sta})$ into a pair $(\varepsilon,\text{sta}')$. By contradiction, assume that the numbers T cannot be bound. Choose new constants $c_1,...,c_k$ and consider the following infinite set of predicate logic formulas:

$$M=\text{Ax} \cup \{\neg\text{term}_{\alpha,T}\{X_1/c_1,...,X_k/c_k\} \mid T\in \mathbb{N} \}$$

Since, by the second assumption above, for every number T there exists a model $A$ of Ax and a state sta over $A$ such that $\alpha$ requires more than T steps to terminate, when computing over $A$ with input state sta, we conclude (using the semantics of the termination predicate) that every finite subset of M has a model. By the compactness theorem, M has a model, too. Let $a_1,...,a_k$ be the values of the constants $c_1,...,c_k$ and $A$ be the reduct to the original signature (without the new constants) of such a model of M. Then, $A$ is a model of Ax and $\alpha$ does not terminate on a state sta with $sta(X_1)=a_1,...,sta(X_k)=a_k$ (again applying the semantics of our termination predicate). This contradicts the first assumption above. ∎

Informally stated, Theorem 14.5 states that a program which terminates over every model of an axiom system Ax for arbitrary input state must terminate within a constant number T of steps, independent of the considered model of Ax and the given input state. A program with this **uniform termination property** is surely rather uninteresting; it can readily be *unwound* in an obvious way into an equivalent loop-free one. Thus our theorem states that termination proofs for interesting programs will always require more than just the information delivered from a predicate logic axiom system.

## 14.2    Partial and total correctness assertions

Let us consider the program T:=X; X:=Y; Y:=T. Its meaning is apparently to interchange the values of X and Y using an auxiliary variable T. The name of the intermediate variable appears to be of no interest to the user of this piece of program. Therefore a reasonable specification of the program would look like this (to help the reader to distinguish between program text and comments inserted into program text we use a smaller type size for comments, that is, assertions; this is particularly helpful in the case of larger programs with lots of assertions):

$$\{(X=A \wedge Y=B)\}\ T:=X;\ X:=Y;\ Y:=T\ \{(X=B \wedge Y=A)\}$$

The first formula is called **precondition** and describes the input states that are of interest for the program. The second one is called **postcondition** and describes a property which is expected to hold after program execution. A program is correct if, after its execution on a state satisfying the precondition, the postcondition holds in the new state (if such a state exists). In many cases it is also interesting to find out whether the program always terminates. This is for our example trivial.

There are two points of view as far as correctness is concerned. The first one regards a program as correct if it is correct with respect to its specification. We call this correctness notion **partial correctness**. The second possibility is also to demand termination. This approach is called **total correctness**. In the above example it is easy to see that the program is partially and totally correct w.r.t. its specification.

At first the total correctness approach seems to be natural. What sense does a program make if we cannot expect it always to terminate and provide us with a correct result? There are, however, reasons why research is mainly done on partial correctness:

- In most cases termination can be easily shown separately. Partial correctness plus termination proof implies total correctness.
- As we shall see in Section 14.3 there are theoretical reasons indicating problems with total correctness. Partial correctness is more a logical property than termination.

**Definition 14.7:**   A **partial correctness assertion** (also called **Hoare formula**, abbreviated **pca**) is a string $\{\varphi\}\alpha\{\psi\}$ where $\alpha$ is a program and $\varphi, \psi$ are formulas. $\varphi$ is called the **precondition** and $\psi$ the **postcondition** of the partial correctness assertion $\{\varphi\}\alpha\{\psi\}$.

**Definition 14.8:**   We say that a pca $\{\varphi\}\alpha\{\psi\}$ is **valid in an algebra** $A$ **in state** sta (denoted by $A \models_{sta} \{\varphi\}\alpha\{\psi\}$) iff for all states sta' the following is true:

$$A \models_{sta}\varphi \text{ and } sta[\alpha]_A sta' \text{ implies } A \models_{sta'} \psi$$

We say that a pca $\{\varphi\}\alpha\{\psi\}$ is **valid in an algebra** $A$ (briefly denoted $A \models \{\varphi\}\alpha\{\psi\}$) iff it is valid in every state sta of $A$. Thus, $\{\varphi\}\alpha\{\psi\}$ is valid in $A$ iff for all states sta and sta':

$$A \models_{sta}\varphi \text{ and } sta[\alpha]_A sta' \text{ implies } A \models_{sta'} \psi$$

**Definition 14.9:**   A **total correctness assertion** is a string $[\varphi]\alpha[\psi]$ where $\alpha$ is a program and $\varphi, \psi$ are formulas. $\varphi$ is called the precondition and $\psi$ the postcondition of the total correctness assertion $[\varphi]\alpha[\psi]$. We say that $[\varphi]\alpha[\psi]$ is **valid in an algebra** $A$ **in state** sta (denoted by $A \models_{sta}[\varphi]\alpha[\psi]$) iff the following is true:

If $A \models_{sta}\varphi$ then there is a state sta' with

$sta[\alpha]_A sta'$ and $A \models_{sta'} \psi$.

We say that a total correctness assertion $[\varphi]\alpha[\psi]$ is **valid in algebra** $A$ (denoted by $A \models [\varphi]\alpha[\psi]$) iff it is valid in every state sta of $A$.

## EXAMPLE 14.1

{A≥0}
X:=0; Y:=1; Z:=1;
WHILE Z ≤A DO
        X:=(X+1); Y:=(Y+2); Z:=(Z+Y)
END
{( (X*X)≤A ∧ ((X+1)*(X+1))>A )}

is a partial correctness assertion which states the following when interpreted over the algebra *Cardinal*: if the program starts with input A≥0 and if it terminates (it is easily shown that it terminates) then $X^2 \leq A$ and $(X+1)^2 > A$ holds for the computed value X; this means that X is the integral part of the square root of A.

## EXAMPLE 14.2

{(A>0 ∧ B>0)}
X:=A; Y:=B ;
WHILE X≠Y DO
        IF X<Y THEN Y:=(Y-X) ELSE X:=(X-Y)
        END
END
{X = greatest common divisor of A and B}

states, when interpreted over *Cardinal*, that the program computes the greatest common divisor (gcd) of A and B (the program terminates for A>0 and B>0). But it should be pointed out that the postcondition above is not a formula but only an informal statement. Fortunately, it is easy to express it as a formula:

(X>0 ∧ (A MOD X)=0 ∧ (B MOD X)=0 ∧
∀Z ((Z>0 ∧ (A MOD Z)=0 ∧ (B MOD Z)=0) → Z≤X))

We shall prove the validity of the pca in algebra *Cardinal* in a subsequent chapter. Note that if the above program starts with values a>0 for A and 0 for B, or 0 for A and b>0 for B, then it does not terminate. But since we are not concerned with termination when investigating partial correctness, the following pca is also valid:

{(A>0 ∧ B=0) ∨ (A=0 ∧ B>0)}
X:=A ; Y:=B ;
WHILE X≠Y DO
        IF X<Y THEN Y:=(Y-X) ELSE X:=(X-Y)

        END
    END
    {X = gcd of A and B}

Note that the above expression written as a total correctness assertion would not be valid in **Cardinal**, since the program does not terminate on all inputs that satisfy the precondition. Finally, let us consider the following partial correctness assertion:

    {(A≥0 ∧ B≥0)}
    X:=A ; Y:=B ;
    WHILE X≠Y DO
        IF X<Y THEN Y:=(Y-X) ELSE X:=(X-Y)
        END
    END
    {X = gcd of A and B}

If the program starts with initial values A=0 and B=0 then it provides us with computed values A=0, B=0, X=0, Y=0. The above pca is thus valid in **Cardinal** iff we define gcd(0,0) as 0.


# EXAMPLE 14.3

    {(A≥0 ∧ B≥0)}
    X:=A ; Y:=B ; Z:=1 ;
    WHILE Y>0 DO
        IF (Y MOD 2)=0 THEN X:=(X*X) ; Y:=(Y DIV 2)
        ELSE Z:=(Z*X) ; Y:=(Y-1)
        END
    END
    {Z=A^B}

It is not easy to see that this pca is valid in **Cardinal**, but later on we shall prove this fact without any problem. Here we would like to point out that '$Z=A^B$' is not a formula over the considered signature. The question is whether the informal statement '$Z=A^B$' can be expressed by a formula as 'X = gcd of A and B' in the previous example. The reader should try to find an appropriate formula, and if unsuccessful, should wait until we discuss the expressive power of the assertion language in Chapter 16.

# EXAMPLE 14.4————————————————————————————————

Let us now turn our interest to arrays. For the following example we lay down the 3-sorted algebra $A=Array\ [0..L+1]\ of\ Cardinal$ (see Section 6.5).

The following pca states that its program decides whether the element X occurs among the elements of array A in the range between 0 and L. The occurring variables are variables of the following sorts: I of sort index, Found of sort boole, X of sort nat, A of sort array; 'true' is an arbitrary valid formula.

> {true}
> I:=0; Found:=false;
> WHILE I≤L ∧ Found=false DO
>> IF A[I]=X THEN Found:=true ELSE ε
>> END;
>> I:=I+1
> END
> {Found=true ↔ ∃J (J ≤L∧A[J]=X)}

In Chapter 15 we shall prove that this pca is valid in the considered algebra.

————————————————————————————————————————

**Exercise 14.4** What does it mean for a program $\alpha$ to be partially or totally correct w.r.t. the following specifications (pre- and postconditions)?

(a)    true, true

(b)    true, false

(c)    false, true

(d)    false, false

**Exercise 14.5** What does it mean that a program terminates w.r.t. true or false?

**Exercise 14.6** Assume that $(Cardinal,\text{fac}) \models [X=Y]\ \alpha\ [X=\text{fac}(Y)]$, where fac is the factorial function. Under which conditions does $\alpha$ indeed compute the factorial function?

**Exercise 14.7** Write a WHILE program calculating the factorial function and prove its partial correctness w.r.t. the specification of Exercise 14.6, using the definition of program semantics.

**Exercise 14.8** Is the pca {true}X:=t{X = t} (t is a term) valid? If not, construct a counter-example and give sufficient conditions for this property.

**Exercise 14.9** Give a specification for sorting programs with precondition true (that is, duplicate array elements are allowed).

————————————————————————————————————————

## 14.3     Computational and logical aspects of program correctness and termination

The purpose of this section is to investigate the computational difficulty of the concepts introduced so far. We ask whether termination of programs or validity of partial correctness are decidable or at least recursively enumerable (semi-decidable). The answer will be that these problems are neither decidable nor recursively enumerable, that is, there is not even a semi-decision procedure for them. First we present a well-known result from recursion theory.

### Theorem 14.6

In the signature $\Sigma$ underlying algebra $N$ define Tot=$\{\alpha \mid \alpha$ is a $\Sigma$-program which terminates over $N$ on every initial state$\}$ and Loop=$\{\alpha \mid \alpha$ is a $\Sigma$-program which does not terminate over $N$ on any input$\}$. Tot and Loop are not recursively enumerable.

### Theorem 14.7

The following sets PC, TE and TC are not recursively enumerable.

- PC = $\{((\varphi,\alpha,\psi) \mid N \models \{\varphi\}\alpha\{\psi\}\}$

- TE = $\{((\varphi,\alpha) \mid \alpha$ terminates in every state sta with $N \models_{sta}\varphi\}$

- TC = $\{((\varphi,\alpha,\psi) \mid N \models [\varphi]\alpha[\psi]\}$

*Proof* Immediately follows from Theorem 14.6 and the following reductions:

$\alpha \in$ Loop iff (true,$\alpha$,false)$\in$ PC
$\alpha \in$ Tot iff (true,$\alpha$)$\in$ TE
$\alpha \in$ Tot iff (true,$\alpha$,true)$\in$ TC     ∎

Note that Theorem 14.7 argues only about the special algebra $N$. It is in fact possible to extend these results considerably to algebras which contain the algebra of natural numbers with 0 and successor function as a subalgebra (**algebras with counters**). Most interesting algebras fall into this category.

---

**Exercise 14.10**   Let *A* be an algebra with finite domain. Show that PC, TC and TE for *A* are decidable.

---

Next, we shall investigate some logical questions concerning termination and total correctness. In particular we shall give theoretical reasons as to why partial correctness and termination are usually proved separately. The first result states that termination is not a first-order property. On the contrary, we shall prove that partial correctness is a first-order property.

**Definition 14.10:** A program $\alpha$ **terminates over an algebra $A$ w.r.t. a formula** $\varphi$ iff it terminates in every state sta over $A$ such that $\varphi$ is valid in $A$ in state sta.

As an example, the program WHILE $X \neq Y$ DO $X:=X+1$ END terminates over *Cardinal* w.r.t. the formula $X \leq Y$, but not w.r.t. the formula $0=0$.

### Theorem 14.8 (Termination is not first-order)

Termination of WHILE programs is not a first-order property, that is, there is a program $\alpha$ and elementary equivalent algebras $A$ and $B$ such that $\alpha$ terminates in $A$ w.r.t. the formula true but it does not terminate in $B$ w.r.t. true.

*Proof* Let $\alpha$ be a program and $A$ be an algebra such that $\alpha$ terminates over $A$ on every initial state, but with execution times that are unbounded. Then Theorem 14.5 implies that there exists an algebra $B$ that is elementary equivalent to $A$ such that $\alpha$ does not terminate over $B$ on every initial state. ∎

This result states that termination of a program is not only dependent on the first-order theory of an algebra, but also on its domain and algebraic structure. The following theorem states that partial correctness behaves differently, that is, it is a first-order property.

### Theorem 14.9 (Partial correctness is first-order)

$A \models \{\varphi\}\alpha\{\psi\}$ implies $\text{Th}(A) \models \{\varphi\}\alpha\{\psi\}$. In other words, $A \models \{\varphi\}\alpha\{\psi\}$ implies that $B \models \{\varphi\}\alpha\{\psi\}$, for every algebra $B$ that is elementary equivalent to $A$.

*Proof* Let $A \models \{\varphi\}\alpha\{\psi\}$ and $B$ be elementary equivalent to $A$. Assume, by contradiction, that $B \not\models \{\varphi\}\alpha\{\psi\}$. This means that there exist states sta and sta' over $B$ such that $B \models_{sta}\varphi$, sta$[\alpha]_B$sta' and $B \not\models_{sta'}\psi$. Let $X_1,...,X_k$ be the variables occurring in $\{\varphi\}\alpha\{\psi\}$, sta$(X_i)=a_i$ and sta'$(X_i)=b_i$, for $i=1,...,k$. Using the semantics of the computation predicate with time bound from Section 14.1 we conclude that $B \models \text{comp}_{\alpha,T}(a_1,...,a_k,b_1,...,b_k)$, for some number T. So the following closed formula is valid in $B$:

$$\exists X_1 ... \exists X_k \exists Y_1 ... \exists Y_k$$
$$(\varphi \wedge comp_{\alpha,T}(X_1,...,X_k,Y_1,...,Y_k) \wedge \neg \psi\{X_1/Y_1,...,X_k/Y_k\})$$

Since $B$ is elementary equivalent to $A$, this formula is also valid in $A$. Hence, there are elements $c_1,...,c_k$ and $d_1,...,d_k$ of the domain of $A$ such that $A \models$ $\varphi(c_1,...,c_k)$, $A \not\models \psi(d_1,...,d_k)$ and $A \models comp_{\alpha,T}(c_1,...,c_k,d_1,...,d_k)$. Choose states $sta_1$ and $sta_2$ over $A$ such that $sta_1(X_i)=c_i$ and $sta_2(X_i)=d_i$, for $i=1,...,k$. We obtain $A \models_{sta_1} \varphi$, $sta_1[\alpha]_A sta_2$ and $A \not\models_{sta_2} \psi$. This contradicts the assumption that $A \models \{\varphi\}\alpha\{\psi\}$. ∎

The results of Section 14.3 are summarized in Table 14.1.

## Table 14.1

|  | Termination of $\alpha$ over $A$ | Validity of pca in $A$ |
|---|---|---|
| First-order property | no | yes |
| Provable from an axiomatization of $A$, for example $Th(A)$ | only for loop-free programs | yes |
| Recursion theoretical status | not recursively enumerable | not recursively enumerable |

# Chapter 15
# Hoare Logic

Hoare (1969) presented a calculus for the derivation of partial correctness assertions. His influential work dominated the verification field for many years; his approach is still the most important in this area. Some advantages of Hoare's calculus are:

- It contains for each programming construct exactly one rule. This considerably reduces search space when trying to show program correctness.

- It can be used to define program semantics (**axiomatic semantics**).

- It is complete over the important algebra *Nat* of standard arithmetic.

In this chapter we shall introduce the calculus for the correctness of WHILE programs, prove its soundness, discuss questions concerning its practical applicability and present several examples. Theoretical results concerning completeness are discussed in Chapter 16.


## 15.1   The Hoare calculus

Let $\Sigma$ be a signature. Hoare's calculus over $\Sigma$ consists of the following axioms and rules, with assertions $\varphi$, $\psi$, $\chi$, $\varphi'$, $\psi'$, $\varphi_0$, $\varphi_1,\ldots,\varphi_n$, boolean expression B, programs $\alpha$, $\beta$, $\gamma$ and instructions $A_1,\ldots,A_n$:

(1)    **Assignment axiom**

$$\overline{\{\varphi\{X/t\}\}X:=t\{\varphi\}}$$    provided that $\varphi\{X/t\}$ is admissible

Note the role of the admissibility condition. Without it we could obtain the following false (over algebra $N$) partial correctness assertion as an instance of the assignment axiom: $\{\forall X\ X=X\}\ X:=Y\ \{\forall X\ X=Y\}$.

(2)    **Test rule**

$$\{(\varphi\wedge B)\}\beta\{\psi\}$$
$$\{(\varphi\wedge\neg B)\}\gamma\{\psi\}$$

$$\overline{\{\varphi\}\ \text{IF B THEN }\beta\text{ ELSE }\gamma\text{ END }\{\psi\}}$$

(3)    **Loop rule**

$$\{(\chi\wedge B)\}\beta\{\chi\}$$

$$\overline{\{\chi\}\text{WHILE B DO }\beta\text{ END}\{(\chi\wedge\neg B)\}}$$

$\chi$ is called **loop invariant**

(4)    **Composite instruction rule** (for n>1)

$$\{\varphi_0\}A_1\{\varphi_1\}$$
$$\{\varphi_1\}A_2\{\varphi_2\}$$
$$\ldots$$
$$\{\varphi_{n-1}\}A_n\{\varphi_n\}$$

$$\overline{\{\varphi_0\}\ A_1\ ;\ A_2\ ;\ldots;\ A_n\ \{\varphi_n\}}$$

(5)    **ε-axiom**

$$\overline{\{\varphi\}\varepsilon\{\varphi\}}$$

(6)    **Refinement rule**

$$(\varphi\rightarrow\varphi')$$
$$\{\varphi'\}\alpha\{\psi'\}$$
$$(\psi'\rightarrow\psi)$$

$$\overline{\{\varphi\}\alpha\{\psi\}}$$

**Definition 15.1:**    Let Ax be a **theory**, that is, a set of assertions that is closed under logical conclusion (only such axiom systems will be considered; note that Ax contains all valid predicate logic formulas over $\Sigma$). If a partial correctness assertion $\{\varphi\}\alpha\{\psi\}$ is derivable from Ax then we write $\text{Ax} \vdash_{\overline{\text{HC}}} \{\varphi\}\alpha\{\psi\}$. By **Hoare's calculus over an algebra $A$** we mean the calculus consisting of the rules above and all formulas of Th($A$) as axioms. The set of all partial correctness assertions that are derivable in Hoare's calculus from Th($A$) is denoted by HC($A$).

There are two ways to use Hoare's calculus to derive a partial correctness assertion from a set of formulas Ax:

- As a **deductive calculus** (*forward chaining, bottom up*)    We start with a set of formulas Ax and the pcas from rules (1) and (5), and derive via rules (2), (3), (4) and (6) more complicated correctness assertions, until the assertion to be shown is finally reached. This way cannot be recommended, since the search has no orientation point and a great number of pcas which are irrelevant to the derivation of the goal will be generated.

- As a **reduction calculus** (*backward chaining, top down, goal reduction*)    We reduce the problem of deducing a pca to the problem of deriving less complicated assertions (**subgoals**). This approach is, especially for Hoare's calculus, far more efficient than bottom up reasoning, because in rules (2) and (3) the subgoals are uniquely determined by the goal, and because it is easy to determine whether a partial correctness assertion has the form of rules (1) or (5). There are problems only with rules (4) and (6): a partial correctness assertion of the form $\{\varphi_0\}A_1;A_2;\ldots;A_n\{\varphi_n\}$ has to be reduced to pcas $\{\varphi_0\}A_1\{\varphi_1\}$, $\{\varphi_1\}A_2\{\varphi_2\},\ldots,\{\varphi_{n-1}\}A_n\{\varphi_n\}$ with suitable formulas $\varphi_1,\ldots,\varphi_{n-1}$, which must yet be found. The same problem exists when using rule (6). We shall see in many examples that these problems are far from being trivial.

Let us give a first example for using Hoare's calculus. The example is very easy, but the reader will have the opportunity to see more complex examples in subsequent sections and chapters.

# EXAMPLE 15.1

Here we want to show that the program T:=X; X:=Y; Y:=T indeed interchanges values of variables X and Y. The axiom system Ax consists of all valid predicate logic formulas.

$$\{X=A \wedge Y=B\}\ T:=X;\ X:=Y;\ Y:=T\ \{X=B \wedge Y=A\}.$$

As the program involved is a composite instruction we must provide intermediate assertions. We shall try to derive the following partial correctness assertions.

$$\{X=A \wedge Y=B\}\ T := X\ \{X=A \wedge Y=B \wedge T=A\} \tag{15.1}$$

$$\{X=A \wedge Y=B \wedge T=A\}\ X := Y\ \{X=B \wedge Y=B \wedge T=A\} \tag{15.2}$$

$$\{X=B \wedge Y=B \wedge T=A\}\ Y := T\ \{X=B \wedge Y=A \wedge T=A\} \tag{15.3}$$

Having derived 15.1-15.3, an application of rule (4) leads to

$$\{X=A \wedge Y=B\}T:=X;\ X:=Y;\ Y:=T\{X=B \wedge Y=A \wedge T=A\} \tag{15.4}$$

Since $((X=A \wedge Y=B) \to (X=A \wedge Y=B))$ and $((X=B \wedge Y=A \wedge T=A) \to (X=B \wedge Y=A))$
are valid formulas, we may apply rule (6) to derive the given partial correct-
ness assertion:

$$\{X=A \wedge Y=B\}T:=X;\ X:=Y;\ Y:=T\{X=B \wedge Y=A\} \tag{15.5}$$

So let us see how 15.1-15.3 may be derived. The assignment rule (1) allows
us to derive

$$\{X=A \wedge Y=B \wedge X=A\}T := X\ \{X=A \wedge Y=B \wedge T=A\} \tag{15.6}$$

Note that the assignment T:=X is mirrored in a precondition that is obtained
from the postcondition by application of substitution {T/X}. Now an obvious
application of rule (6) leads to pca 15.1. By the same way the assignment rule
allows us to derive

$$\{Y=B \wedge Y=B \wedge T=A\}\ X := Y\ \{X=B \wedge Y=B \wedge T=A\} \tag{15.7}$$

Once more applying rule (6) in an obvious way leads to pca 15.2. Finally, we
may derive

$$\{X=B \wedge T=A \wedge T=A\}Y := T\{X=B \wedge Y=A \wedge T=A\} \tag{15.8}$$

Together with (6) we obtain pca 15.3. A lot of work for a rather trivial example.

---

The first question concerning a calculus is always to establish its soundness,
that is, to show that everything derivable from an axiom system Ax also follows from Ax.

### Theorem 15.1 (Soundness of Hoare's calculus)

Let $\{\varphi\}\alpha\{\psi\}$ be a partial correctness assertion and Ax be a theory.
Then the following implication holds:

$$\text{Ax}\vdash_{\overline{HC}} \{\varphi\}\alpha\{\psi\} \Rightarrow \text{Ax} \models \{\varphi\}\alpha\{\psi\}$$

This means that every partial correctness assertion derivable from Ax
in Hoare's calculus follows from Ax. In particular, for an algebra $A$ the

following implication holds:

$$\text{Th}(A)\vdash_{\overline{\text{HC}}} \{\varphi\}\alpha\{\psi\} \Rightarrow A \vDash \{\varphi\}\alpha\{\psi\}$$

Hence every partial correctness assertion derivable from Th($A$) is valid in $A$.

*Proof*    By induction on the length of a derivation of $\{\varphi\}\alpha\{\psi\}$ from Ax we show that for every model $A$ of Ax and all states sta,sta' over $A$:

$$A \vDash_{\text{sta}}\varphi \text{ and sta}[\alpha]_A\text{sta' implies } A \vDash_{\text{sta'}} \psi$$

So let a model $A$ of Ax and states sta and sta' over $A$ be given such that $A \vDash_{\text{sta}}\varphi$ and sta$[\alpha]_A$sta', as well as a derivation of $\{\varphi\}\alpha\{\psi\}$ from Th($A$) of length m. By induction hypothesis we know that the statement above is true for all partial correctness assertions with derivations of length less than m. We have to show that $A \vDash_{\text{sta'}}\psi$. We consider the last rule which was applied in the derivation of $\{\varphi\}\alpha\{\psi\}$.

*Case 1*    The rule applied last was the assignment rule. This means that $\{\varphi\}\alpha\{\psi\}$ is of the form $\{\psi\{X/t\}\}X:=t\{\psi\}$ and $\psi\{X/t\}$ is admissible. Now we argue:

$A \vDash_{\text{sta}}\varphi$ and sta$[\alpha]_A$sta'                                  $\Rightarrow$

$A \vDash_{\text{sta}}\psi\{X/t\}$ and sta$[X:=t]_A$sta'              $\Rightarrow$ (substitution lemma)

$A \vDash_{\text{sta}(X/\text{val}_{A,\text{sta}}(t))}\psi$ and sta'=sta$(X/\text{val}_{A,\text{sta}}(t))$                $\Rightarrow$

$A \vDash_{\text{sta'}}\psi$

*Case 2*    The rule applied last was the test rule. This means that $\{\varphi\}\alpha\{\psi\}$ is a partial correctness assertion of the form $\{\varphi\}$ IF B THEN $\beta$ ELSE $\gamma$ END $\{\psi\}$, and there are derivations of $\{(\varphi\wedge B)\}\beta\{\psi\}$ and $\{(\varphi\wedge\neg B)\}\gamma\{\psi\}$ of length <m.

*Subcase 2a*  $A \vDash_{\text{sta}}B$. Then we argue as follows:

$A \vDash_{\text{sta}}\varphi$ and sta$[\alpha]_A$sta'                                  $\Rightarrow$

$A \vDash_{\text{sta}}(\varphi\wedge B)$ and sta$[\beta]_A$sta'              $\Rightarrow$ (induction hypothesis)

$A \vDash_{\text{sta'}}\psi$

*Subcase 2b*  $A \nvDash_{\text{sta}}B$. Then we argue as follows:

$A \vDash_{\text{sta}}\varphi$ and sta$[\alpha]_A$sta'                                  $\Rightarrow$

$A \models_{sta}(\varphi \wedge \neg B)$ and $sta[\gamma]_A sta'$ $\qquad \Rightarrow$ (induction hypothesis)

$A \models_{sta'}\psi$

*Case 3*    The rule applied last was the loop rule. This means that $\{\varphi\}\alpha\{\psi\}$ is a of the form $\{\varphi\}$ WHILE B DO $\beta$ END $\{(\varphi \wedge \neg B)\}$ and there is a derivation of $\{(B \wedge \varphi)\}\beta\{\varphi\}$ of length <m. We argue as follows:

$A \models_{sta}\varphi$ and $sta[\alpha]_A sta'$ $\qquad\qquad\qquad \Rightarrow$

$A \models_{sta}\varphi$ and there exist $t \in \mathbb{N}$ and states $sta_0, sta_1, \ldots, sta_t$ with:

$\qquad sta = sta_0,$

$\qquad A \models_{sta_i} B$ and $sta_i[\beta]_A sta_{i+1}$, for all $i = 0, \ldots, t-1,$

$\qquad A \not\models_{sta_t} B$

$\qquad sta_t = sta'$

Now we proceed as follows:

$A \models_{sta_0}(B \wedge \varphi)$ and $sta_0[\beta]_A sta_1$ $\qquad \Rightarrow$ (induction hypothesis)

$A \models_{sta_1}\varphi$ $\qquad\qquad\qquad\qquad\qquad \Rightarrow$

$A \models_{sta_1}(B \wedge \varphi)$ and $sta_1[\beta]_A sta_2$ $\qquad \Rightarrow$ (induction hypothesis)

$A \models_{sta_2}\varphi$ $\qquad\qquad\qquad\qquad\qquad \Rightarrow$

$A \models_{sta_2}(B \wedge \varphi)$ and $sta_2[\beta]_A sta_3$ $\qquad \Rightarrow$ (induction hypothesis)

$\qquad\qquad\qquad\qquad\qquad\qquad\qquad\qquad \Rightarrow$ (induction hypothesis)

$\qquad\qquad .$
$\qquad\qquad .$
$\qquad\qquad .$

$A \models_{sta_{t-1}}\varphi$ $\qquad\qquad\qquad\qquad\qquad \Rightarrow$

$A \models_{sta_{t-1}}(B \wedge \varphi)$ and $sta_{t-1}[\beta]_A sta_t$ $\qquad \Rightarrow$ (induction hypothesis)

$A \models_{sta_t}\varphi$ $\qquad\qquad\qquad\qquad\qquad \Rightarrow$

$A \models_{sta_t}(\varphi \wedge \neg B)$ $\qquad\qquad\qquad\qquad \Rightarrow$

$A \models_{sta'}\psi$

For the remaining cases see Exercise 15.1.

**Exercise 15.1** Complete the proof above by treating the rules for composite instructions, the empty program and the weakening rule.

Now that we know that Hoare's calculus cannot derive from Ax a partial correctness that does not follow from Ax, we may ask whether the opposite also holds. We restrict discussion to theories Ax=Th($A$), for an algebra $A$. In Chapter 16 we shall show that the answer heavily depends on the algebra which is laid down. But before we investigate such theoretical questions we turn our attention to aspects of the calculus' practical application.

## 15.2    Practical usage of Hoare's calculus

In this section we shall see how Hoare's calculus can and should be used in practice. First we present a lengthy example. An analysis of the points of the constructed derivation that disturb us leads to the notion of **correctly asserted programs**, a special denotation for derivations in Hoare's calculus which considerably simplifies the presentation of a derivation. Finally we present correctness proofs for many interesting programs.

The aim of the following example of a Hoare derivation is to frighten off the reader. But of course not from program verification! The example rather shows that Hoare's calculus in its raw form is not suitable for practical verification work. We shall take this example as a starting point to develop a more user friendly presentation of the rules of Hoare's calculus.

## EXAMPLE 15.1———————————————————————

We now lay down the algebra ***Cardinal***. Let us consider the following program which divides a natural number A by a number B>0 and gives as output the quotient T and the rest R; note that no use is made of the functions DIV and MOD of ***Cardinal***, since then the problem would be trivial.

    T:=0; R:=A;
    WHILE B≤R DO
        R:=(R-B); T:=(T+1)
    END

We claim that the following pca, which is obviously valid in algebra ***Cardinal***, is derivable in Hoare's calculus over algebra ***Cardinal***:

$\{(A{\geq}0 \wedge B{>}0)\}$
T:=0; R:=A ;
WHILE B≤R DO
     R:=(R-B); T:=(T+1)
END
$\{(A{=}((T{*}B){+}R) \wedge R{<}B)\}$

Our program is a sequence of three statements, so we have to give two intermediate formulas and derive three subgoals. Note that the intermediate formulas should express all properties that hold when program execution reaches the point where the formula stands. It is not very difficult to see that the following formulas and subgoals are suitable (this has of course yet to be shown):

*Subgoal 1*    $\{(A{\geq}0 \wedge B{>}0)\}$ T:=0 $\{((A{\geq}0 \wedge B{>}0) \wedge T{=}0)\}$

*Subgoal 2*    $\{((A{\geq}0 \wedge B{>}0) \wedge T{=}0)\}$ R:=A $\{(((A{\geq}0 \wedge B{>}0) \wedge T{=}0) \wedge R{=}A)\}$

*Subgoal 3*    $\{(((A{\geq}0 \wedge B{>}0) \wedge T{=}0) \wedge R{=}A)\}$
WHILE B≤R DO
     R:=(R-B); T:=(T+1)
END
$\{(A{=}((T{*}B){+}R) \wedge R{<}B)\}$

We further reduce these subgoals.

*For subgoal 1*   By the assignment rule (1) we know that the following pca is derivable:

$\{((A{\geq}0 \wedge B{>}0) \wedge 0{=}0)\}$ T:=0 $\{((A{\geq}0 \wedge B{>}0) \wedge T{=}0)\}$.

Applying the refinement rule (6) with the valid implication

$((A{\geq}0 \wedge B{>}0){\rightarrow}((A{\geq}0 \wedge B{>}0) \wedge 0{=}0))$

we obtain subgoal 1.

*For subgoal 2*   By the assignment rule (1) we know that the following pca is derivable:

$\{(((A{\geq}0 \wedge B{>}0) \wedge T{=}0) \wedge A{=}A)\}$
R:=A
$\{(((A{\geq}0 \wedge B{>}0) \wedge T{=}0) \wedge R{=}A)\}$

Applying the refinement rule (6) with the valid implication

$(((A{\geq}0 \wedge B{>}0) \wedge T{=}0){\rightarrow}(((A{\geq}0 \wedge B{>}0) \wedge T{=}0) \wedge A{=}A))$

we obtain subgoal 2.

*For subgoal 3*   Until now the game was boring, because it was quite trivial. Now it becomes more interesting.We have to find a suitable loop invariant $\chi$ and then apply rule (3). But how can such a loop invariant be found? As the name indicates, and as the loop rule requires, validity of the loop invariant must be preserved by each execution of the loop body. Let us see what is preserved by each execution of the loop body. Obviously, the loop successively subtracts B from the actual remainder R and increases the value of T by 1, until this is no longer possible, that is, until R<B. Therefore, it is reasonable to believe that the following formula is an appropriate loop invariant:

A=((T*B)+R)

In order to apply the loop rule with this invariant, we have to show the following subgoal:

*Subgoal 3.1*    {(A=((T*B)+R) ∧ B≤R)} R:=(R-B); T:=(T+1) {A=((T*B)+R)}

For the moment let us assume that subgoal 3.1 has already been derived. Loop rule (4) would then allow us to derive the following pca:

{A=((T*B)+R)}
**WHILE B≤R DO**
       R:=(R-B); T:=(T+1)
**END**
  {(A=((T*B)+R) ∧ ¬B≤R)}

Using this pca and the refinement rule (6), we may reduce subgoal 3 to the following two subgoals:

*Subgoal 3.2*    ((((A≥0 ∧ B>0) ∧ T=0) ∧ R=A)→ A=((T*B)+R))

*Subgoal 3.3*    ((A=((T*B)+R) ∧ ¬B≤R)→(A=((T*B)+R) ∧ R<B))

The validity in algebra **Cardinal** of formulas 3.2 and 3.3 is obtained as a quite trivial consequence of number theoretic results. We finally handle subgoal 3.1. It is a sequence of two instructions, so we have to give a suitable intermediate formula. We try the following:

A=(((T+1)*B)+R).

Subgoal 3.1 is then reduced to the derivation of the following two assertions:

*Subgoal 3.1.1* {(A=((T*B)+R) ∧ B≤R)} R:=(R-B) {A=(((T+1)*B)+R)}

*Subgoal 3.1.2* {A=(((T+1)*B)+R)} T:=(T+1) {A=((T*B)+R)}

Assertion 3.1.2 follows directly from the assignment rule. In order to derive 3.1.1 we first apply the assignment rule to derive

{A=(((T+1)*B)+(R-B))} R:=(R-B) {A=(((T+1)*B)+R)}

Finally the puzzle is solved using the refinement rule (6) with the following implication that is easily seen to be valid in *Cardinal*:

$$((A=((T*B)+R) \land B \leq R) \rightarrow A=(((T+1)*B)+(R-B)))$$

It is easy to see that this is the case. So the proof is completed.

---

Having worked out the example above, some observations and remarks are in order.

First, when reducing a goal to subgoals, creativity is only needed to find suitable loop invariants. When handling sequences of instructions, intermediate formulas are also necessary, but their choice is quite forced.

Second, when during the reduction of the original goal or some subgoals a predicate logic formula was obtained in our example, we were able to stop at that point because the validity of the formulas was quite trivial. This is not always so. The formulas yet to be proven could be definitely non-trivial; we shall see in Chapter 16 that the remaining formulas may even be as hard to prove as the original partial correctness assertion!

Nevertheless, we draw a line at this point. We are concerned with the proof as long as programs occur, and assume that the remaining formulas are either trivial or can be proven by a mathematician. An interactive verification system could give out the formulas which the goal was reduced to and ask the user whether validity of these formulas is accepted. An alternative would be to supply an automated theorem prover with these formulas and obtain an answer about their validity. We presented formal methods for proving formulas in Part I.

Note that if at some point we are unable to prove a formula, then it might be the case that some loop invariant or intermediate assertion was badly chosen; in this case we have to find other formulas and repeat the proof as far as these formulas are concerned. Another case is that the program is incorrect with respect to the given pre- and postcondition. Then we have to find out whether our specification is not the intended one or the program must be changed. Finally, it might be the case that the program is correct w.r.t. its specification, but this cannot be derived in Hoare's calculus. This case addresses the question of the calculus' completeness which will be thoroughly treated in the next chapter.

Thirdly, nobody would like to verify further programs in the style of Example 15.2, especially for more complicated programs. But which are the points of the above process that can be criticized? Here are some suggestions for improvement:

(1)    When handling sequences of assignments, the situation is usually so simple that we might try to do the verification in one step instead of providing lots of intermediate formulas. For this situation we shall define a more comfortable rule.

(2)     It is quite an urgent task to simplify formula notation in order to improve readability. A considerable problem is the amount of parentheses, particularly in arithmetic expressions.

(3)     It is very disturbing to write the program piece to be verified each time a rule is applied. Think of a program with some thousands of program lines, most parts of which must be written over and over again. For this reason we shall introduce the idea of correctly asserted programs.

To illustrate point (3) let us consider a program of the following form:

```
WHILE B1 DO
     Z1; WHILE B2 DO
          Z2; WHILE B3 DO
               Z3; WHILE B4 DO
                    .
                              .
                                        .
                              WHILE B(n-1) DO Zn
                              END
                                        .
                    .
               END
          END

          END
END

     END
END
```

Let us assume that B1,...,Bn are boolean expressions which fit on one monitor line; also let Z1,...,Zn be similarly short assignments. Then the program above consists of 2n+1 lines. To write down a derivation of a correct pca containing the program above we have to write at least the 2n+1 lines of program text plus an appropriate intermediate formula after Z1 plus a derivation of the second loop. For the second loop we need 2(n-1)+1 lines plus the derivation of the third loop, and so on. So the whole derivation would at least need

$$2n+1+2(n-1)+1+2(n-2)+1+...+2(3-1)+1+2(2-1)+1$$

$$=2(n+(n-1)+...+3+2+1)+n$$

$$=n(n+1)+n$$

$$=n^2+2n$$

lines of program text. If the program consists of 10 monitor pages with fifty lines each, then a derivation requires more than 1200 monitor pages! And this is primarily for unnecessarily writing program pieces over and over again.

**Definition 15.2 (More comfortable rules)**

**(7) Rule for assignment sequences**

For each non-empty sequence of assignments $X_1:=t_1;\dots;X_n:=t_n$ and for all assertions $\varphi$ and $\psi$ such that $\psi\{X_n/t_n\}\dots\{X_2/t_2\}\{X_1/t_1\}$ is admissible

$$\frac{(\varphi\rightarrow\psi\{X_n/t_n\}\dots\{X_2/t_2\}\{X_1/t_1\})}{\{\varphi\}\ X_1:=t_1;\dots;X_n:=t_n\ \{\psi\}}$$

(Note that the substitutions above have to be applied to $\psi$ in the given order, first $\{X_n/t_n\}$ and finally $\{X_1/t_1\}$. This is not the same as applying $\{X_n/t_n,\dots,X_2/t_2,X_1/t_1\}$ on $\psi$. Also note that the occurrence of a substitution in the postcondition fits much better with our intuitive understanding of the effect of an assignment than the occurrence of a substitution in the precondition in the original assignment rule.)

**(8) Rule for sequences of non-empty programs**

For all $n\geq2$, non-empty programs $\alpha_1,\dots,\alpha_n$ and assertions $\varphi_0,\varphi_1,\dots,\varphi_n$:

$$\{\varphi_0\}\alpha_1\{\varphi_1\}$$
$$\{\varphi_1\}\alpha_2\{\varphi_2\}$$
$$\cdot$$
$$\cdot$$
$$\cdot$$
$$\{\varphi_{n-1}\}\alpha_n\{\varphi_n\}$$

$$\frac{}{\{\varphi_0\}\alpha_1\ ;\ \alpha_2\ ;\dots;\ \alpha_n\ \{\varphi_n\}}$$

*Theorem 15.2 (Soundness of the new rules)*

Each application of rule (7) or (8) in a derivation from a theory Ax can be replaced by a finite number of applications of rules (1)-(6).

*Proof*    An application of rule (7)

$$(\varphi\rightarrow\psi\{X_n/t_n\}\dots\{X_2/t_2\}\{X_1/t_1\})$$
$$\dots\text{ (other proof steps)}$$
$$\{\varphi\}\ X_1:=t_1;\dots;X_n:=t_n\ \{\psi\}$$

in a deduction can be replaced by

$(\varphi\to\psi\{X_n/t_n\}...\{X_2/t_2\}\{X_1/t_1\})$

· ... (other proof steps)

$\{\psi\{X_n/t_n\}...\{X_2/t_2\}\{X_1/t_1\}\} X_1:=t_1 \{\psi\{X_n/t_n\}...\{X_2/t_2\}\}$    (rule (1))

$\{\psi\{X_n/t_n\}...\{X_2/t_2\}\} X_2:=t_2 \{\psi\{X_n/t_n\}...\{X_3/t_3\}\}$    (rule (1))

...

$\{\psi\{X_n/t_n\}\} X_n:=t_n \{\psi\}$    (rule (1))

$\{\psi\{X_n/t_n\}...\{X_2/t_2\}\{X_1/t_1\}\} X_1:=t_1;X_2:=t_2;...;X_n:=t_n \{\psi\}$    (rule (4))

$(\psi\to\psi)$    (valid)

$\{\varphi\} X_1:=t_1;...;X_n:=t_n \{\psi\}$    (rule (6))

Every application of rule (8) can obviously be simulated by rule (4). ■

**Notational simplifications for formulas** (in particular formulas in the signature of *Cardinal*)

(1)    Instead of $(\varphi_1\wedge\varphi_2\wedge\varphi_3\wedge...\varphi_n)$ we simply write: $\varphi_1,\varphi_2,\varphi_3,...,\varphi_n$.

(2)    Referring to the signature $\Sigma_{nat}$ we define an **arithmetic expression** to be a *sum of products* $P_1 op_1 P_2 op_2...P_n$, where $op_1,op_2,...\in\{+,-\}$ and $P_1,P_2...,P_n$ are products. A **product** is a string of form $F_1 op_1 F_2 op_2...F_m$, where $op_1,op_2,...\in\{*,DIV,MOD\}$ and $F_1,F_2,...,F_m$ are factors. A **factor** is a variable, a constant or a string of form (E), where E is again an arithmetic expression. By $P_1 op_1 P_2 op_2 P_3...P_n$ we understand the arithmetic expression $((...(P_1 op_1 P_2) op_2 P_3)...P_n)$; the same for $F_1 op_1 F_2 op_2...F_m$. This means that we use left bracketing and stronger binding of *,DIV,MOD than + and -. Finally we write $\forall X rel E \varphi$ instead of $\forall X(X rel E\to\varphi)$, where $rel\in\{>,<,\geq,\leq\}$.

**Definition 15.3:**   **Correctly asserted programs** (briefly **cap**) **over a theory** Ax are programs that are in addition equipped with assertions (comments) at certain places, fitting together in a specific way. They are formally defined as the strings derivable from Ax in the calculus whose rules are presented in Figure 15.1 (with assertions $\varphi$, $\psi$, $\chi$, $\varphi'$, $\psi'$, $\varphi_0$, $\varphi_1$, ..., $\varphi_n$, boolean expression B, programs $\alpha$, $\beta$, $\gamma$ and non-empty programs $\alpha_1$, $\alpha_2$, ..., $\alpha_n$). A correctly asserted program over an algebra *A* is a correctly asserted program over Th(*A*).

### Asserting assignment sequences

$$\frac{(\varphi \to \psi\{X_n/t_n\}\ldots\{X_2/t_2\}\{X_1/t_1\})}{\{\varphi\}X_1:=t_1;\ldots;X_n:=t_n\{\psi\}}$$

provided that $n \geq 1$ and $\psi\{X_n/t_n\}\ldots\{X_2/t_2\}\{X_1/t_1\}$ is admissible

### Asserting a test

$$\frac{\{((\varphi \wedge B)\}\beta\{\psi\}}{\{((\varphi \wedge \neg B))\}\gamma\{\psi\}}$$

$\{\varphi\}$ IF B THEN $\{((\varphi \wedge B)\}\beta\{\psi\}$ ELSE $\{((\varphi \wedge \neg B))\}\gamma\{\psi\}$ END $\{\psi\}$

### Asserting a loop

$$\frac{\{(\chi \wedge B)\}\beta\{\chi\}}{\{\chi\}\text{ WHILE B DO }\{(\chi \wedge B)\}\beta\{\chi\}\text{ END }\{(\chi \wedge \neg B)\}}$$

### Asserting a sequence of programs

$$\{\varphi_0\}\alpha_1\{\varphi_1\}$$
$$\{\varphi_1\}\alpha_2\{\varphi_2\}$$
$$.$$
$$.$$
$$.$$
$$\frac{\{\varphi_{n-1}\}\alpha_n\{\varphi_n\}}{\{\varphi_0\}\alpha_1;\{\varphi_1\}\alpha_2;\{\varphi_2\}\ldots\alpha_{n-1};\{\varphi_{n-1}\}\alpha_n\{\varphi_n\}}$$

### Asserting the empty program

$$\frac{(\varphi \to \psi)}{\{\varphi\}\varepsilon\{\psi\}}$$

### Refinement rule

$$\frac{\begin{array}{c}(\varphi \to \varphi')\\(\psi' \to \psi)\\\{\varphi'\}\alpha\{\psi'\}\end{array}}{\{\varphi\}\{\varphi'\}\ \alpha\ \{\psi'\}\{\psi\}}$$

**Figure 15.1**

For a correctly asserted program $\{\varphi\}\alpha\{\psi\}$ we denote by $\alpha^*$ the program which is obtained from $\alpha$ by deleting all occurrences of substrings $\{\varphi\}$ for assertions $\varphi$. Note that in the rule for program sequences $\alpha_1$, $\alpha_2$, ..., $\alpha_n$ are allowed to be arbitrary non-empty programs, that is sequences of instructions.

The following result states that correctly asserted programs over Ax are nothing more than a comfortable way of denoting derivations from Ax in Hoare's calculus.

### Theorem 15.3 (Correspondence between caps and Hoare's calculus)

If $\{\varphi\}\alpha\{\psi\}$ is a correctly asserted program over Ax, then the partial correctness assertion $\{\varphi\}\alpha^*\{\psi\}$ is derivable from Ax in Hoare's calculus. In particular, if $\{\varphi\}\alpha\{\psi\}$ is a correctly asserted program over $A$ then $\{\varphi\}\alpha^*\{\psi\}$ is derivable in HC($A$).

*Proof*  Inductively over the length of a derivation of cap $\{\varphi\}\alpha\{\psi\}$; this is left as a simple exercise for the reader.  ∎

The definition of correctly asserted programs might appear at first sight to be quite complicated, but it is actually something very natural, which is in fact used by every programmer. Caps are programs containing formulas as comments at distinguished points of the program text ('check points'). Every programmer uses informal comments to describe what is valid when execution of a program reaches specific program points. Caps mirror this usage of comments in a formal manner.

How can we equip a given partial correctness assertion $\{\varphi\}\alpha\{\psi\}$ with additional comments such that a correctly asserted programs results? Either bottom up, that is, working from inside to outside, or top down, working from outside to inside. We shall usually make use of the second approach. Then we start with a pca $\{\varphi\}\alpha\{\psi\}$ and go into the structure of $\alpha$, thereby inserting suitable comments. If, for example, we start with $\{\varphi\}$ IF B THEN $\beta$ ELSE $\gamma$ END $\{\psi\}$, then we have to insert comments before and after $\beta$ and $\gamma$ as follows:

$$\{\varphi\} \text{ IF B THEN } \{(\varphi\wedge B)\}\beta\{\psi\} \text{ ELSE } \{(\varphi\wedge\neg B)\}\gamma\{\psi\} \text{ END } \{\psi\}$$

Then we must proceed to assert $\{(\varphi\wedge B)\}\beta\{\psi\}$ and $\{(\varphi\wedge\neg B)\}\gamma\{\psi\}$ correctly. If on the other hand we start with $\{\varphi\}$ WHILE B DO $\beta$ END $\{\psi\}$, then we have to guess an appropriate loop invariant $\chi$ and insert three comments:

$$\{\varphi\}\{\chi\} \text{ WHILE B DO } \{(\chi\wedge B)\}\beta\{\chi\} \text{ END } \{(\chi\wedge\neg B)\}\{\psi\}$$

Then we have further to assert $\{(\chi\wedge B)\}\beta\{\chi\}$ correctly. When this process is completed we only have to check whether for all sequences of assignments the corresponding implications hold, and whether for subsequent comments $\{\varphi\}\{\psi\}$ the implication $(\varphi\rightarrow\psi)$ is an element of axiom set Ax.

## 15.3    Examples of correctly asserted programs

For Examples 15.3-15.6 we lay down the algebra *Cardinal*.

### EXAMPLE 15.3  Natural number division

Our well-known example of natural number division with rest can be turned
into a cap as follows:

{A≥0 , B>0}
$$\text{T:=0; R:=A;}$$
{A≥0, B>0, T=0, R=A)}  ☞(a)
{A=T∗B+R}  ☞(b)
WHILE B≤R DO
{B≤R, A=T∗B+R}
$$\text{R:=R-B; T:=T+1}$$
{A=T∗B+R }  ☞(c)
END
{¬B≤R, A=T∗B+R}
{A=T∗B+R, R<B)}  ☞(d)

Points  ☞(a)-(d) indicate the implications whose validity in *Cardinal* must
be shown.

(a)    A≥0, B>0→A≥0, B>0, 0=0, A=A                     (trivial)

(b)    A≥0, B>0, T=0, R=A→ A=T∗B+R                     (trivial)

(c)    B≤R, A=T∗B+R→A=(T+1)∗B+(R-B)                     (simple)

(d)    ¬B≤R, A=T∗B+R→A=(T∗B)+R, R<B                     (trivial)

### EXAMPLE 15.4  Greatest common divisor

Let gcd(a,b) be the greatest common divisor of numbers a>0 and b>0. We
claim that the following is a correctly asserted program:

{A>0, B>0}
$$\text{X:=A; Y:=B;}$$
{A>0, B>0, X=A,Y=B}  ☞(a)
{A>0, B>0, gcd(X,Y)=gcd(A,B)}  ☞(b)
WHILE X≠Y DO
{X≠Y, A>0, B>0,gcd(X,Y)=gcd(A,B)}
IF X<Y THEN
{X<Y,X≠Y, A>0, B>0, gcd(X,Y)=gcd(A,B)}

Y:=Y-X
{A>0, B>0, gcd(X,Y)=gcd(A,B)} ☞(c)
    ELSE
        {¬X<Y, X≠Y, A>0, B>0, gcd(X,Y)=gcd(A,B)}
        X:=X-Y
        {A>0, B>0, gcd(X,Y)=gcd(A,B)} ☞(d)
    END
    {A>0, B>0, gcd(X,Y)=gcd(A,B)}
END
{¬X≠Y, A>0, B>0, gcd(X,Y)=gcd(A,B)}
{gcd(A,B)=X} ☞(e)

Here we have to show that the following formulas are valid in *Cardinal*:

(a)    A>0, B>0→A>0, B>0, A=A,B=B        (trivial)

(b)    A>0, B>0, X=A,Y=B
        →A>0, B>0, gcd(X,Y)=gcd(A,B)        (trivial)

(c)    X<Y,X≠Y, A>0, B>0, gcd(X,Y)=gcd(A,B)
        →A>0, B>0, gcd(X,Y-X)=gcd(A,B)        (simple)

(d)    ¬X<Y, X≠Y, A>0, B>0, gcd(X,Y)=gcd(A,B)
        →A>0, B>0, gcd(X-Y,Y)=gcd(A,B)        (simple)

(e)    ¬X≠Y, A>0, B>0, gcd(X,Y)=gcd(A,B)
        →gcd(A,B)=X        (as gcd(X,X)=X)

## EXAMPLE 15.5  Square root

In this example we shall verify a program that calculates the (integer) square root using only addition. The idea of the computation of the square root of X is the following:

Successively calculate the squares of 0,1,2,3,..., until for the first time the square is greater than or equal to X. Given $I^2$, we calculate $(I+1)^2$ as $I^2+I+I+1$. Based on this formula, the program manages triplets $(U,V,W)=(I,I+I+1,I^2)$ with increasing I as follows: given a triple $(U,V,W)$, the next one can be easily calculated as $(U+1,V+2, W+V+2)$.

Using this idea we can show that the following is a correctly asserted program:

{0=0}
U:=0; V:=1; W:=1;
{U=0,V=1,W=1} ☞(a)
{U*U≤X, V=U+U+1, W=(U+1)*(U+1)} ☞(b)

WHILE W≤X DO

$\{W{\le}X,\ U{*}U{\le}X,\ V{=}U{+}U{+}1,\ W{=}(U{+}1){*}(U{+}1)\}$

U:=U+1; V:=V+2; W:=W+V

$\{U{*}U{\le}X,\ V{=}U{+}U{+}1,\ W{=}(U{+}1){*}(U{+}1)\}$  ☞(c)

END

$\{\neg W{\le}X,\ U{*}U{\le}X,\ V{=}U{+}U{+}1,\ W{=}(U{+}1){*}(U{+}1)\}$

$\{U{*}U{\le}X,\ (U{+}1){*}(U{+}1){>}X\}$  ☞(d)

We have to show that the following formulas are valid in *Cardinal*:

(a)     $0{=}0{\rightarrow}0{=}0,1{=}1,1{=}1$                                             (trivial)

(b)     $U{=}0,V{=}1,W{=}1$

      $\rightarrow\ U{*}U{\le}X,\ V{=}U{+}U{+}1,\ W{=}(U{+}1){*}(U{+}1)$       (simple)

(c)     $W{\le}X,\ U{*}U{\le}X,\ V{=}U{+}U{+}1,\ W{=}(U{+}1){*}(U{+}1)$

      $\rightarrow\ (U{+}1){*}(U{+}1){\le}X,\ V{+}2{=}(U{+}1){+}(U{+}1){+}1,$

      $W{+}V{+}2{=}((U{+}1){+}1){*}((U{+}1){+}1$

(we shall show this formula below)

(d)     $\neg W{\le}X,\ U{*}U{\le}X,\ V{=}U{+}U{+}1,\ W{=}(U{+}1){*}(U{+}1)$

      $\rightarrow\ U{*}U{\le}X,(U{+}1){*}(U{+}1){>}X$                         (easy)

It remains to prove that the following formulas are valid in *Cardinal*:

    $W{\le}X,\ U{*}U{\le}X,\ V{=}U{+}U{+}1,\ W{=}(U{+}1){*}(U{+}1)\ \rightarrow\ (U{+}1){*}(U{+}1){\le}X$
(clear, because $(U{+}1){*}(U{+}1){=}W{\le}X$)

    $W{\le}X,\ U{*}U{\le}X,\ V{=}U{+}U{+}1,\ W{=}(U{+}1){*}(U{+}1)\ \rightarrow\ V{+}2{=}(U{+}1){+}(U{+}1){+}1$
(clear, because $V{+}2{=}U{+}U{+}1{+}2$)

    $W{\le}X,\ U{*}U{\le}X,\ V{=}U{+}U{+}1,\ W{=}(U{+}1){*}(U{+}1)$

        $\rightarrow\ W{+}V{+}2{=}((U{+}1){+}1){*}((U{+}1){+}1)$
(since $((U{+}1){+}1){*}((U{+}1){+}1){=}(U{+}1){*}(U{+}1){+}(U{+}1){+}(U{+}1){+}1{=}W{+}V{+}2$)

## EXAMPLE 15.6  Exponential function

Let us assume that we can express the assertion $A^B{=}C$ as a formula (this will be shown in Chapter 16). Then we claim that the following is a correctly asserted program:

$\{0{=}0\}$

X:=A; Y:=B; Z:=1

$\{X{=}A,\ Y{=}B,\ Z{=}1\}$  ☞ (a)

$\{X^Y{*}Z{=}A^B\}$  ☞(b)

WHILE Y>0 DO

$$\{Y>0, X^Y*Z=A^B\}$$
IF Y MOD 2=0 THEN
$$\{Y \text{ MOD } 2=0, Y>0, X^Y*Z=A^B\}$$
$$Y:=Y \text{ DIV } 2; \ X:=X*X$$
$$\{X^Y*Z=A^B\} \quad \text{☞(c)}$$
ELSE
$$\{\neg Y \text{ MOD } 2=0, Y>0, X^Y*Z=A^B\}$$
$$Y:=Y-1; \ Z:= Z*X$$
$$\{X^Y*Z=A^B\} \quad \text{☞(d)}$$
END
$$\{X^Y*Z=A^B\}$$
END
$$\{\neg Y>0, X^Y*Z=A^B\}$$
$$\{Z=A^B\} \quad \text{☞(e)}$$

We have to show validity in *Cardinal* of the following formulas:

| | | |
|---|---|---|
| (a) | $0=0 \rightarrow A=A, B=B, 1=1$ | (trivial) |
| (b) | $X=A, Y=B, Z=1 \rightarrow X^Y*Z=A^B$ | (clear) |
| (c) | $Y \text{ MOD } 2=0, Y>0, X^Y*Z=A^B \rightarrow (X*X)^{Y \text{ DIV } 2}*Z=A^B$ | (easy) |
| (d) | $\neg Y \text{ MOD } 2=0, Y>0, X^Y*Z=A^B \rightarrow X^{Y-1}*(Z*X)=A^B$ | (easy) |
| (e) | $\neg Y>0, X^Y*Z=A^B \rightarrow Z=A^B$ | (trivial) |

---

**Exercise 15.2** Write and verify a program in *Nat* with + but without * (that is, no multiplication is used) that calculates the product of two numbers.

---

## EXAMPLE 15.7   Searching an element on an array

We lay down the algebra *Array*[0...L+1] *of Cardinal* (see Section 6.5). The purpose of the following program is to decide whether number X occurs among the array elements A[0],...,A[L], for an array A. Besides variables X of type nat and A of type array it uses a variable Found of sort boole and a variable I of sort index. Note that we search for X only in the range between 0 and L. Index L+1 serves as an endmarker. Using such an endmarker often simplifies programs working on arrays. We claim that the following is a correctly asserted program. For better reading, we write $X \in A[L..R)$ for the formula $\exists J\ (L \leq J\ \wedge\ J<R \wedge A[J]=X)$.

{true}

I:=0; Found:=false;

{I=0, Found=false} ☞(a)

{$X \in A[0..I) \leftrightarrow$Found=true} ☞(b)

WHILE I≤L ∧ Found=false DO

{I≤L, Found=false, $X \in A[0..I) \leftrightarrow$Found=true}

IF A[I]=X THEN

{A[I]=X, I≤L, Found=false, $X \in A[0..I) \leftrightarrow$Found=true}

Found:=true

{I≤L, $X \in A[0..I+1) \leftrightarrow$Found=true} ☞(c)

END; ☞(d)

{I≤L, $X \in A[0..I+1) \leftrightarrow$Found=true}

I:=I+1

{I≤L+1, $X \in A[0..I) \leftrightarrow$Found=true} ☞(e)

END

{¬I≤L, $X \in A[0..I) \leftrightarrow$Found=true}

{$X \in A[0..L+1) \leftrightarrow$Found=true} ☞(f)

We use instruction IF B THEN α END as an abbreviation for IF B THEN α ELSE ε END. An appropriate cap rule for such instructions is the following:

$$\frac{\{(\varphi \wedge B)\}\beta\{\psi\} \quad\quad (\varphi \wedge \neg B)\to\psi}{\{\varphi\}\ \text{IF B THEN}\{(\varphi \wedge B)\}\beta\{\psi\}\ \text{END}\ \{\psi\}}$$

Contrary to the subgoal $\{(\varphi \wedge B)\}\beta\{\psi\}$ the implication $(\varphi \wedge \neg B)\to\psi$ does not become visible in the notation of a cap. So do not forget to derive it when using the instruction IF B THEN α END. In the example above, pointer ☞(d) reminds us of this task. Let us now show the validity of the corresponding formulas at the check points (a)-(f).

(a)    Trivial.

(b)    Clear since both $X \in A[0..0)$ and Found=true are not valid.

(c)    Clear since I≤L and A[I]=X imply $X \in A[0..I+1)$.

(d)    Clear since I≤L, $\neg X \in A[0..I)$ and ¬A[I]=X imply $\neg X \in A[0..I+1)$.

(e)    Obvious.

(f)    Obvious.

# EXAMPLE 15.8   Maximum element of an array———————

We lay down the algebra *Array*[0...L] *of Cardinal*. The purpose of the following program is to compute the maximum array element between indices 0 and L. Note that we do not use an endmarker as in Example 15.7. We write Max=maximum(A,L,R) for the formula

$$\forall J((L \leq J \wedge J \leq R) \to A[J] \leq Max) \wedge \exists J((L \leq J \wedge J \leq R) \wedge A[J] = Max)$$

So, Max=maximum(A,L,R) says that Max is the maximum array element of array A between array indices L and R.

```
{true}
I:=0; Max:=A[0];
{I=0, Max=A[0]} ☞(a)
{Max=maximum(A,0,I)} ☞(b)
WHILE I<L DO
          {I<L, Max=maximum(A,0,I)}
          IF A[I+1]>Max THEN
                    {A[I+1]>Max, I<L, Max=maximum(A,0,I)}
                    Max:=A[I+1]
                    {I<L, Max=maximum(A,0,I+1)} ☞ (c)
          END; ☞(d)
          {I<L, Max=maximum(A,0,I+1)}
          I:=I+1
          {Max=maximum(A,0,I)} ☞(e)
END
{¬I<L, Max=maximum(A,0,I)}
{Max=maximum(A,0,I)} ☞(f)
```

---

**Exercise 15.3**   Show the validity of the formulas at check points (a) - (f).

---

After these examples the reader may have the strong feeling that every pca valid in *Cardinal* or *Array* [0..L] *of Cardinal* is indeed derivable by Hoare's calculus over *Cardinal* or *Array* [0..L] *of Cardinal* respectively. This refers us to the question of the calculus' completeness, which is treated in the next chapter.

---

**Exercise 15.4**   Write and verify a program calculating the arithmetical mean of an array.

**Exercise 15.5** In the sorting algorithm *insertion sort* the element A[I] is added into the sorted segment A[0],...,A[I-1] at the correct position. Write and verify a program over algebra *Array* [0..N+1] *of Cardinal* that implements this algorithm.

**Exercise 15.6** Extend the programming language by a REPEAT loop of the form REPEAT $\alpha$ UNTIL $\varphi$ with the meaning $\alpha$; WHILE $\neg\varphi$ DO $\alpha$ END. Give a proof rule for this construct and prove its soundness.

**Exercise 15.7** Extend the programming language by a FOR loop of the form

FOR I := $t_1$ TO $t_2$ DO $\alpha$ END

Define appropriate semantics for this loop according to its intuitive meaning and extend Hoare's calculus by a proof rule for this construct. Establish a sufficient condition for the fact that a FOR loop always terminates, provided that a is a total program. For programs containing only such loops, partial and total correctness coincides.

**Exercise 15.8** Prove the partial correctness assertion

```
{0=0}
Min := A[1]; Sum := A[1];
FOR K := 2 TO L DO
      Sum := Sum+A[I];
      IF Min>A[I] THEN Min := A[I] END
END
{∀I (I≤L→Min≤A[I])∧Sum≥L*Min}
```

either by using your rule from Exercise15.7 or by bringing the program into the form of a WHILE program and verifying the result by Hoare's calculus.

**Exercise 15.9** Verify the following program $\alpha$ over algebra *Array* [0..L] *of Cardinal* which uses the sieve of Eratosthenes to find all prime numbers between 2 and L:

```
FOR I := 2 TO L DO
    A[I] := 1;
    I := 2;
    WHILE I²≤L DO
        J := I²;
        IF A[I] = 1 THEN REPEAT A[J] := 0; J := J+I UNTIL J>L ELSE ε
        END;
        I := I+1
    END
END
```

Verify $\{0=0\}\alpha\{\forall K(2\leq K\leq L\to(A[K]=1\leftrightarrow prime(K)))\}$, where prime is a predicate which holds iff its argument is a prime number.

**Exercise 15.10** Show that it is always possible to verify a pca $\{\varphi\}\alpha\{\psi\}$ using only intermediate formulas and loop invariants with free variables among the free variables of $\varphi$ and $\psi$ and the variables of $\alpha$.

# Chapter 16
# Completeness of Hoare Logic

In this chapter we shall prove that Hoare's calculus is complete over **expressive algebras**. These algebras have the property that all intermediate assertions necessary for a derivation in Hoare's calculus can be expressed in first-order logic. We shall introduce a method for showing that an algebra is expressive and demonstrate its usage on many algebras, including among others the standard algebras of natural numbers *Nat* and *Cardinal*.

Then we shall turn our attention to negative results concerning completeness. It will even be possible to give very simple valid partial correctness assertions which cannot be verified by Hoare's calculus over the algebra under consideration. Nevertheless, the situation is not hopeless, since it is possible to expand every algebra to an expressive one.

## 16.1   Expressiveness

In this section we shall define the class of **expressive algebras**. First we introduce some technical issues that play a central role in the definition of expressiveness and in the completeness proof.

**Definition 16.1:**   Let $A$ be a $\Sigma$-algebra. For each program $\alpha$ over the signature of $A$ and assertions $\varphi$ and $\psi$ we define the following sets of states :

- $spc_A(\varphi,\alpha)$ is the set of all states sta' such that there is a state sta with $A \models_{sta}\varphi$ and $sta[\alpha]_A sta'$.

- $wlp_A(\alpha,\psi)$ is the set of all states sta such that, for all states sta', $sta[\alpha]_A sta'$ implies $A \models_{sta'} \psi$.

$spc_A(\varphi,\alpha)$ is called the **strongest postcondition** for $\varphi$ and $\alpha$ over $A$, $wlp_A(\alpha,\psi)$ is called the **weakest liberal precondition** for $\alpha$ and $\psi$ over $A$.

Thus, $spc_A(\varphi,\alpha)$ consists of all states that can be computed by $\alpha$ from a state satisfying $\varphi$, $wlp_A(\alpha,\psi)$ consists of all states sta that are transformed by $\alpha$ into a state satifying $\psi$, provided that $\alpha$ terminates on the given state sta.

Let us illustrate these concepts with some examples. We lay down the algebra $A=$***Cardinal***.

(1)    $wlp_A(X:=X+1,X>2)=\{sta \mid sta(X)\geq 2\}$. This is obviously the weakest possible precondition to assure validity of $X>2$ after execution of the instruction.

(2)    $spc_A(X:=X+1,X>2)=\{sta \mid sta(X)>3\}$, because for each such sta, sta'= $sta(X/sta(X)-1)$ is a state with $A \models_{sta}X>2$ and $sta'[X:=X+1]_A sta$. This information is obviously everything we know that holds after execution of the assignment.

(3)    $wlp_A(X:=0;$WHILE $X<Y$ DO $X:=X+1$ END, $X=10)=\{sta \mid sta(Y)=10\}$. After program execution $X=Y$ holds (why?). Therefore, $X$ and $Y$ have the value 10. Since $Y$'s value is not manipulated by the program, $Y$ must already have value 10 before program execution.

(4)    $wlp_A($WHILE $0=0$ DO $\varepsilon$ END, $\varphi)=\{sta \mid sta$ is an arbitrary state over algebra $A\}$, because for no sta does there exist a state sta' with $sta[$WHILE $0=0$ DO $\varepsilon$ END$]_A sta'$.

(5)    $spc_A($WHILE $0=0$ DO $\varepsilon$ END, $\varphi) = \varnothing$.

For the following definitions and theorems we remind the reader of the concepts of definability of state sets introduced in Chapter 5. There it was laid down that a set S of states over an algebra $A$ is definable in $A$, if there exists a formula $\sigma$ defining S in $A$, that is, for all states sta over $A$: $A \models_{sta}\sigma \Leftrightarrow sta\in S$.

For example, the set of all states over **Cardinal** is defined by formula 0=0, the empty set of states over **Cardinal** is defined by 0=1 and the set {sta | sta(X)>0} is defined by the formula X>0.

The following theorem describes the most important properties of the weakest liberal preconditions and the strongest postconditions.

### Theorem 16.1 (Basic properties of wlp and spc)

(1)    $\text{sta} \in \text{wlp}_A(\alpha,\psi)$ and $\text{sta}[\alpha]_A \text{sta'} \Rightarrow A \models_{\text{sta'}} \psi$

(2)    $A \models \{\varphi\}\alpha\{\psi\}$ and $A \models_{\text{sta}} \varphi \Rightarrow \text{sta} \in \text{wlp}_A(\alpha,\psi)$

(3)    $A \models_{\text{sta}} \varphi$ and $\text{sta}[\alpha]_A \text{sta'} \Rightarrow \text{sta'} \in \text{spc}_A(\varphi,\alpha)$

(4)    $A \models \{\varphi\}\alpha\{\psi\}$ and $\text{sta'} \in \text{spc}_A(\varphi,\alpha) \Rightarrow \text{then } A \models_{\text{sta'}} \psi$

*Proof*

(1)    Follows directly from the definition of $\text{wlp}_A$.

(2)    Let $\{\varphi\}\alpha\{\psi\}$ be valid in $A$ and sta be a state such that $A \models_{\text{sta}} \varphi$. We have to show that $\text{sta} \in \text{wlp}_A(\alpha,\psi)$. For this reason let sta' be a state with $\text{sta}[\alpha]_A \text{sta'}$ (if no such state exists, then nothing is to be shown). From the validity of $\{\varphi\}\alpha\{\psi\}$ we may conclude $A \models_{\text{sta'}} \psi$. So sta is an element of $\text{wlp}_A(\alpha,\psi)$ by definition.

(3)    Follows directly from the definition of $\text{spc}_A$.

(4)    Let $\{\varphi\}\alpha\{\psi\}$ be valid in $A$ and sta' be a state in $\text{spc}_A(\varphi,\alpha)$. We have to show $A \models_{\text{sta'}} \psi$. By definition of the strongest postcondition there is a state sta such that $A \models_{\text{sta}} \varphi$ and $\text{sta}[\alpha]_A \text{sta'}$. It follows from the validity of $\{\varphi\}\alpha\{\psi\}$ that $A \models_{\text{sta'}} \psi$.    ∎

### Corollary

For an algebra $A$, program $\alpha$ and assertions $\varphi$ and $\psi$, let $\text{wlp}_A(\alpha,\psi)$ and $\text{spc}_A(\varphi,\alpha)$ be defined in $A$ by formulas $\omega(\alpha,\psi)$ and $\pi(\varphi,\alpha)$, respectively. Then the following hold:

(1)    $\{\omega(\alpha,\psi)\}\alpha\{\psi\}$ is valid in $A$.

(2)    For each partial correctness assertion $\{\xi\}\alpha\{\psi\}$ that is valid in $A$, the formula $(\xi \rightarrow \omega(\alpha,\psi))$ is valid in $A$. Thus, $\omega(\alpha,\psi)$ is in-

deed the weakest formula $\xi$ such that $\{\xi\}\alpha\{\psi\}$ is valid in $A$.

(3)     $\{\varphi\}\alpha\{\pi(\varphi,\alpha)\}$ is valid in $A$.

(4)     For each partial correctness assertion $\{\varphi\}\alpha\{\zeta\}$ that is valid in $A$ the formula $(\pi(\varphi,\alpha)\rightarrow\zeta)$ is also valid in $A$. Thus, $\pi(\varphi,\alpha)$ is indeed the strongest formula $\zeta$ such that $\{\varphi\}\alpha\{\zeta\}$ is valid in $A$.

**Definition 16.2:** An algebra $A$ is called **expressive** iff for each program $\alpha$ and each formula $\psi$ the set of states $wlp_A(\alpha,\psi)$ is definable in $A$.

---

**Exercise 16.1** Let $\alpha$ be the program WHILE $\neg X=Y$ DO X:=succ(X) END over $N$. Compute $wlp_N(\alpha,X=Y)$, $wlp_N(\alpha,\neg X=Y)$, $spc_N(0=0,\alpha)$, $spc_N(X>Y,\alpha)$.

**Exercise 16.2** Determine $wlp_A(\alpha,\varphi)$ and $spc_A(\varphi,\alpha)$ (in terms of $\varphi$) for the programs X := X and WHILE true DO X := X END.

**Exercise 16.3** Show the following properties of the wlp and spc operator:

$$wlp(\alpha_1;\alpha_2,\varphi) = wlp(\alpha_1;wlp(\alpha_2,\varphi)) \text{ and } spc(\varphi,\alpha_1;\alpha_2) = spc(spc(\varphi,\alpha_1),\alpha_2)$$

Give similar equations for test instructions. What is the problem for the case of loop?

**Exercise 16.4** Let $A$ be an algebra, $\alpha$ be a program over $A$ and $\varphi$ and $\psi$ be formulas. Further let $X_1,\dots,X_n$ be all variables occurring in $\alpha$ or $\varphi$, $\psi$. Choose fresh variables $Y_1,\dots,Y_n$. Let $\omega$ be a formula that defines $wlp_A(\alpha,\neg(X_1=Y_1\wedge\dots\wedge X_n=Y_n))$ in $A$. Show that $spc_A(\varphi,\alpha)$ is defined in $A$ by

$$\exists Y_1\dots\exists Y_n(\varphi\{X_1/Y_1,\dots,X_n/Y_n\} \wedge \neg\omega\{X_1/Y_1,\dots,X_n/Y_n,Y_1/X_1,\dots,Y_n/X_n\})$$

Let $\pi$ be a formula that defines $spc_A((X_1=Y_1\wedge\dots\wedge X_n=Y_n),\alpha)$ in $A$. Show that $wlp_A(\alpha,\psi)$ is defined in $A$ by

$$\exists Y_1\dots\exists Y_n(\pi\{X_1/Y_1,\dots,X_n/Y_n,Y_1/X_1,\dots,Y_n/X_n\} \rightarrow \psi\{X_1/Y_1,\dots,X_n/Y_n\})$$

Conclude that an algebra $A$ is expressive iff for each program $\alpha$ and formula $\varphi$ the set $spc_A(\varphi,\alpha)$ is definable in $A$.

**Exercise 16.5** Show that algebras with finite domain are expressive.

## 16.2     Completeness of Hoare's calculus over expressive algebras

**Definition 16.3:**     Hoare's calculus is called **complete over an algebra** $A$ iff $Th(A)\vdash_H \{\phi\}\alpha\{\psi\}$ for each partial correctness assertion $\{\phi\}\alpha\{\psi\}$ that is valid in $A$.  ∎

*Theorem 16.2 (Completeness of Hoare's calculus over expressive algebras)*

Hoare's calculus is complete over every expressive algebra $A$. Moreover, for every pca $\{\phi\}\alpha\{\psi\}$ valid in $A$ there is a correctly asserted program $\{\phi\}\beta\{\psi\}$ over $A$ such that $\beta^*=\alpha$. This means that every valid pca $\{\phi\}\alpha\{\psi\}$ can be equipped with assertions in such a way that a correctly asserted program $\{\phi\}\beta\{\psi\}$ over $A$ results.

*Proof*     By structural induction on $\alpha$.

*Case 1*     $\alpha$ is an assignment X:=t. Then we first rename quantified variables in $\psi$ in such a way that we obtain a formula $\psi'$ such that $(\psi\leftrightarrow\psi')$ is valid in $A$ and $\psi'\{X/t\}$ is admissible. Then $\{\phi\}X:=t\{\psi'\}$ is also valid in $A$. This implies that the formula $(\phi\rightarrow\psi'\{X/t\})$ is valid in $A$. So, $\{\phi\}\{\psi'\{X/t\}\}$ X:=t $\{\psi'\}\{\psi\}$ is a correctly asserted program over $A$.

*Case 2*     $\alpha$ is a test IF B THEN $\gamma$ ELSE $\delta$ END. Then the partial correctness assertions $\{(B\wedge\phi)\}\gamma\{\psi\}$ and $\{(\neg B\wedge\phi)\}\delta\{\psi\}$ are valid in $A$. By induction hypothesis we may choose caps over $A$ of the form $\{(B\wedge\phi)\}\xi\{\psi\}$ and $\{(\neg B\wedge\phi)\}\zeta\{\psi\}$ with $\xi^*=\gamma$ and $\zeta^*=\delta$. Then the following is also a cap over $A$:

$\{\phi\}$ IF B THEN $\{B\wedge\phi)\}\xi\{\psi\}$ ELSE $\{(\neg B\wedge\phi)\}\zeta\{\psi\}$ END $\{\psi\}$

and (IF B THEN $\{B\wedge\phi)\}\xi\{\psi\}$ ELSE $\{(\neg B\wedge\phi)\}\zeta\{\psi\}$ END)* coincides with the given program IF B THEN $\gamma$ ELSE $\delta$ END.

*Case 3*     $\alpha$ is the empty program $\epsilon$. Then $(\phi\rightarrow\psi)$ is valid in $A$. So $\{\phi\}\epsilon\{\psi\}$ is the desired correctly asserted program over $A$.

*Case 4*     $\alpha$ is a sequence of instructions $A_1;A_2;...;A_m$ with m≥2. We choose a formula $\omega$ that defines $wlp_A(A_2;...;A_m,\psi)$ in $A$. By the properties of wlp discussed in Section 16.1 we know that the pca $\{\omega\}A_2;...;A_m\{\psi\}$ is valid in $A$. Next we show that $\{\phi\}A_1\{\omega\}$ is also valid in $A$. For this purpose let sta and sta' be states such that $A \models_{sta} \varphi$ and $sta[A_1]_A sta'$. If sta'' is a state with $sta'[A_2;...;A_m]_A sta''$, then it follows that $sta[A_1;A_2;...;A_m]_A sta''$ and $A \models_{sta''} \psi$. Hence, sta' is an element of $wlp_A(A_2;...;A_m, \psi)$, and $A \models_{sta'} \omega$. By induction hypothesis we may choose caps $\{\varphi\}\xi\{\omega\}$ and $\{\omega\}\zeta\{\psi\}$ over $A$ such that $\xi*=A_1$ and $\zeta*=A_2;...;A_m$. By definition of caps, $\{\varphi\}\xi\{\omega\};\zeta\{\psi\}$ is also a cap over $A$, and $(\xi\{\omega\};\zeta)*$ coincides with $A_1;A_2;...;A_m$.

*Case 5* α is the loop WHILE B DO γ END. Let χ be a formula defining $wlp_A(\alpha,\psi)$ in $A$. For a better understanding of what follows note that χ may informally be read as follows: 'After execution of α formula ψ holds'.

(a)    By the basic properties of wlp presented in Section 16.1, $(\varphi \to \chi)$ is valid in $A$. Note that this implication may be read informally as follows:

> If φ holds before execution of α, then ψ holds after execution of α.

It expresses nothing more than validity of $\{\varphi\}\alpha\{\psi\}$.

(b)    We show that the pca $\{(B\wedge\chi)\}\gamma\{\chi\}$ is valid in $A$. Informally, this pca expresses the following intuitive statement:

> If test B holds and if after execution of the whole loop α the formula ψ holds (this was the intuitive meaning of χ), then after execution of the loop body γ (which must actually be executed at least once since B is valid) it is true that after succeeding complete execution of the whole loop α the formula ψ holds.

The corresponding formal proof looks as follows. Let sta and sta' be states such that $A \models_{sta} B$, $sta \in wlp_A(\alpha,\psi)$ and $sta[\gamma]_A sta'$. We shall prove that $sta' \in wlp_A(\alpha,\psi)$. To do so, let sta'' be a state such that $sta'[\alpha]_A sta''$. Then $sta[\gamma;\alpha]_A sta''$. Since $A \models_{sta} B$ we know that $[\alpha]_A=[\gamma;\alpha]_A$, so $sta[\alpha]_A sta''$ follows. Because of $sta \in wlp_A(\alpha,\psi)$ we finally obtain $A \models_{sta''} \psi$.

(c)    We show that $((\neg B \wedge \chi) \to \psi)$ is a formula that is valid in $A$. Informally stated this formula reads as follows:

> If test B does not hold and if after execution of the loop α (which by definition terminates at once without a single execution of the loop body) ψ holds, then ψ holds.

An intuitively clear statement. In the following we give a corresponding formal proof of the validity of our formula. Let sta be a state such that $A \models_{sta}(\neg B \wedge \chi)$. We have to show $A \models_{sta}\psi$. Since $A \models_{sta}\chi$, we know that $A \models_{sta'}\psi$ holds for each state sta' with $sta[\alpha]_A sta'$. Since $A \models_{sta}\neg B$, we conclude that $[\alpha]_A=[\varepsilon]_A$, that is, $sta[\alpha]_A sta$, for every state sta. Our claim follows from this fact.

Combining result (b) and the induction hypothesis for $\gamma$ (note that $\gamma$ is a shorter program) we know that there exists a cap $\{(B \wedge \chi)\}\delta\{\chi\}$ over $A$ with $\delta*=\gamma$. Because of results (a) and (c), the following is also a cap over $A$:

$$\{\varphi\}\{\chi\} \text{ WHILE B DO } \{(B \wedge \chi)\}\delta\{\chi\} \text{ END}\{(\neg B \wedge \chi)\}\{\psi\}$$

Since $(\{\chi\} \text{ WHILE B DO } \{(B \wedge \chi)\}\delta\{\chi\} \text{ END } \{(\neg B \wedge \chi)\})*$ coincides with WHILE B DO $\gamma$ END, the proof is finished.    ∎

As was proved in Chapter 14, validity of a partial correctness assertion in an algebra $A$ is a first-order property, that is, it depends only on Th($A$). Obviously, completeness of Hoare's calculus over an algebra $A$ also depends only on Th($A$). This observation allows us to prove that expressiveness of an algebra, which was shown to be a sufficient criterion for completeness of Hoare'c calculus, is not necessary for completeness.

### Theorem 16.3

There is an inexpressive algebra $A$ such that Hoare's calculus is complete over $A$.

*Proof*  Consider any non-standard model $A$ of arithmetic, that is, an algebra $A$ that is elementary equivalent to algebra $Nat=(\mathbb{N},0,succ,+,*)$, but not term generated.  As we shall show in a subsequent section, Hoare's calculus is complete over $Nat$, and therefore also over the elementary equivalent algebra $A$.

On the other hand $A$ is not expressive. Consider, for example, the following program $\alpha$:

X:=0; WHILE ¬X=Y DO X:=X+1 END

Obviously, $wlp_A(\alpha,false)$ consists of all states sta with $sta(Y) \notin \{succ_A^n(0_A)|$ $n \in \mathbb{N}\}$. This subset of dom($A$) was shown in Chapter 5 to be not definable in $A$. It follows that $wlp_A(\alpha,false)$ is not definable, either.    ∎

Now we turn our attention to the question of how to *prove* expressiveness for concrete algebras. Obviously, a direct check of the definition does not help in practice. We would have to consider all programs $\alpha$ and sets of states $wlp_A(\alpha,\psi)$ and show that these sets are expressible in $A$. For this reason we introduce in the next section a sufficient condition for expressiveness, which can be used in many practical cases.

## 16.3    Arithmetical algebras

**Definition 16.4:**    A one-sorted algebra $A$ is called an **arithmetical algebra** iff the following two properties are fulfilled:

(1)    **Definability of counters**    An isomorphic copy of the algebra $(N,<)$ of natural numbers with zero, successor function and ordering can be defined within $A$.

For simplicity of presentation we assume that $(N,<)$ itself can be defined within $A$, in particular that the set of natural numbers is a subset of dom($A$). Let nat(X), $\varphi_0$(Y), $\varphi_{succ}$(X,Y), $\varphi_<$(X,Y) be formulas that define the subset of natural numbers, number zero, successor function and ordering of numbers within $A$.

(2)    **Encoding property for finite sequences**    There is a function dec:dom($A$)×$\mathbb{N}$ →dom($A$) definable within $A$ such that for each finite sequence $a_0,a_1,...,a_n$ (of arbitrary length) of elements of dom($A$) there exists some (code) element u of dom($A$) such that dec(u,i)=$a_i$, for i≤n.

Let $\varphi_{dec}$(U,I,A) be a formula defining dec within $A$.

The encoding property for finite sequences states that it is possible to encode a finite sequence $a_0,a_1,...,a_n$ of arbitrary length as an element u∈dom($A$) in such a way that each element $a_i$ can be accessed by a definable function dec(u,i), for i=0,...,n. Emphasis is put on definability of dec. Mere presence of an encoding mechanism for finite sequences together with a decoding function is trivially shown for an algebra with the set of natural numbers as subset.

**Definition 16.5:**    A **many-sorted algebra** $A$ over signature (S,Σ) with domains $A_s$, for s∈ S, is called an **arithmetical algebra** iff the following two properties are fulfilled:

(1)    **Definability of counters**    An isomorphic copy of the algebra $(N,<)$ of natural numbers with zero, successor function and ordering can be defined within $A$.

As in the one-sorted case we assume that $(N,<)$ itself can be defined within $A$, in particular that the set of natural numbers is a subset of $A_s$, for one of the sorts s∈ S.

(2)    **Encoding property for finite sequences**    For every sort s∈ S there is a sort r∈ S and a function $dec_s$:$A_r$×$\mathbb{N}$ →$A_s$ definable

within $A$ such that for each finite sequence $a_0, a_1, \ldots, a_n$ of elements of $A_s$ there exists some (code) element u of $A_r$ with $\text{dec}(u,i) = a_i$, for $i \leq n$. (Thus, elements of $A_r$ serve as code elements for finite sequences over $A_s$).

Let $\varphi_{\text{dec},s}(U,I,A)$ be a formula defining $\text{dec}_s$ within $A$.

## Notational simplification

For an arithmetical algebra $A$ with defining formulas $\text{nat}(X)$, $\varphi_0(Y)$, $\varphi_{\text{succ}}(X,Y)$, $\varphi_<(X,Y)$ and $\varphi_{\text{dec},s}(U,I,A)$ as in the definitions above, we will usually make use of **constants and function symbols** 0, succ and $\text{dec}_s$ as well as **relation symbol** < instead of using the corresponding defining formulas. This considerably improves the readability of succeeding formulas.

As an example, a use of term succ(t) in a formula $\psi(\text{succ}(t))$ is to be replaced by the formula $\exists Y (\varphi_{\text{succ}}(t,Y) \wedge \psi(Y))$ where Y is a new variable.

---

**Exercise 16.6** Show, in general, how the usage of 0, succ, dec and < in a formula can be simulated by an appropriate use of defining formulas $\varphi_0(Y)$, $\varphi_{\text{succ}}(X,Y)$, $\varphi_<(X,Y)$ and $\varphi_{\text{dec},s}(U,I,A)$.

---

As a first typical application of the expressive power provided by the function dec we show how addition + on $\mathbb{N}$ can be defined in an arithmetical algebra $A$. For all $a,b,c \in \text{dom}(A)$:

$a+b=c \Leftrightarrow$

$a,b,c \in \mathbb{N}$ and there is a sequence $a_0, a_1, \ldots, a_n$ such that $n \geq b$, $a_0 = a$ and

$a_{i+1} = \text{succ}(a_i)$ for all $i < b$, and $a_b = c \Leftrightarrow$

$a,b,c \in \mathbb{N}$ and there is a code element $u \in A$ such that $\text{dec}(u,0) = a$,

$\text{dec}(u,i+1) = \text{succ}(\text{dec}(u,i))$ for all $i < b$, and $\text{dec}(u,b) = c \Leftrightarrow$

$A \models (\text{nat}(a) \wedge \text{nat}(b) \wedge \text{nat}(c) \wedge \exists U\ (\text{dec}(U,0)=a \wedge \varphi \wedge \text{dec}(U,b,c)))$

with formula $\varphi = \forall I\ ((\text{nat}(I) \wedge I<b) \rightarrow \text{dec}(U,\text{succ}(I)) = \text{succ}(\text{dec}(U,I)))$

---

**Exercise 16.7** Define multiplication, the Fibonacci series and the exponential function in an arithmetical algebra.

---

Since addition and multiplication on $\mathbb{N}$ are definable in an arithmetical algebra, we shall freely use $+$ and $*$ as 2-ary function symbols in formulas in the same way as we have already used function symbols 0, succ and dec. This will prevent subsequent formulas from being more complicated.

### Theorem 16.4 (Definability of the input-output relation of programs over arithmetical algebras)

Let $A$ be an arithmetical algebra as in the definition above and $\alpha$ be a program, in which only variables $X_1,...,X_n$ occur. Let $Y_1,...,Y_n$ be fresh variables not occurring in $\alpha$. Then we may construct a formula $comp_\alpha(X_1,...,X_n,Y_1,...,Y_n)$ such that for all states sta over $A$ and all $a_1,...,a_n,b_1,...,b_n \in dom(A)$ the following statement are equivalent:

(1)    $sta(X_1/a_1,...,X_n/a_n)[\alpha]_A sta(X_1/b_1,...,X_n/b_n)$

(2)    $A \models comp_\alpha(a_1,...,a_n,b_1,...,b_n)$

(Compare this result with the **time bounded** computation predicate $comp_{\alpha,T}(X_1,...,X_n,Y_1,...,Y_n)$ discussed in Chapter 14.)

*Proof* Inductively over the structure of $\alpha$. Note that in the following n is a fixed number; the constructed formula $\varphi_\alpha$ will depend on n and $X_1,...,X_n,Y_1,...,Y_n$.

*Case 1*    $\alpha$ is $X_i:=t$, where $1 \leq i \leq n$ and t is a term. We take as $comp_\alpha$

$(Y_1=X_1 \wedge ... \wedge Y_{i-1}=X_{i-1} \wedge Y_i=t \wedge Y_{i+1}=X_{i+1} \wedge ... \wedge Y_n=X_n)$

*Case 2*    $\alpha$ is IF B THEN $\gamma$ ELSE $\delta$ END. We take as $comp_\alpha$ the formula

$((B \wedge \varphi_\gamma(X_1,...,X_n,Y_1,...,Y_n)) \vee (\neg B \wedge \varphi_\delta(X_1,...,X_n,Y_1,...,Y_n)))$

*Case 3*    $\alpha$ is the empty program $\varepsilon$. Then we take as $comp_\alpha$ the formula

$(Y_1=X_1 \wedge ... \wedge Y_i=X_i \wedge ... \wedge Y_n=X_n)$

*Case 4*    $\alpha$ is $A_1;A_2;...;A_m$ with $m \geq 2$. Then $comp_\alpha$ is the formula

$$\exists Z_0 \ldots \exists Z_{n-1} \ \exists Z_n \ldots \exists Z_{2n-1} \ \cdots \ \exists Z_{mn} \ldots \exists Z_{(m+1)n-1}$$

$$(X_1 = Z_0 \wedge \ldots \wedge X_n = Z_{n-1}$$

$$\wedge \ \varphi_{A_1}(Z_0, \ldots, Z_{n-1}, Z_n, \ldots, Z_{2n-1})$$

$$\wedge \ \varphi_{A_2}(Z_n, \ldots, Z_{2n-1}, Z_{2n}, \ldots, Z_{3n-1})$$

$$\wedge$$

$$\cdot$$
$$\cdot$$
$$\cdot$$

$$\wedge \ \varphi_{A_m}(Z_{(m-1)n}, \ldots, Z_{mn-1}, Z_{mn}, \ldots, Z_{(m+1)n-1}) \ \wedge$$

$$\wedge \ Z_{mn} = Y_1 \wedge \ldots \wedge Z_{(m+1)n-1} = Y_n)$$

*Case 5*   $\alpha$ is WHILE B DO $\gamma$ END. As comp$_\alpha$ we take the formula

$$\exists T \ \exists U_1 \ldots \exists U_n \ (\text{nat}(T) \wedge \varphi_1 \wedge \varphi_2 \wedge \varphi_3 \wedge \varphi_4 \wedge \varphi_5)$$

with the following subformulas:

$\varphi_1$ is the formula $(X_1 = \text{dec}(U_1, 0) \wedge \ldots \wedge X_n = \text{dec}(U_n, 0))$

$\varphi_2$ is $\forall I \ ((\text{nat}(I) \wedge I < T) \rightarrow B\{X_1/\text{dec}(U_1, I), \ldots, X_n/\text{dec}(U_n, I)\} \ )$

$\varphi_3$ is the formula $\forall I \ ((\text{nat}(I) \wedge I < T) \rightarrow$

$\qquad \varphi_\gamma(\text{dec}(U_1, I), \ldots, \text{dec}(U_n, I), \text{dec}(U_1, \text{succ}(I)), \ldots, \text{dec}(U_n, \text{succ}(I))))$

$\varphi_4$ is the formula $\neg B\{X_1/\text{dec}(U_1, I), \ldots, X_n/\text{dec}(U_n, I)\}$

$\varphi_5$ is $(Y_1 = \text{dec}(U_1, T) \wedge \ldots \wedge Y_n = \text{dec}(U_n, T))$

T denotes the number of loop executions until the program terminates, and $U_i$ encodes the sequence of $T+1$ intermediate values of variable $X_i$ (for each $i = 1, \ldots, n$). Hence, $(\text{dec}(U_1, I), \ldots, \text{dec}(U_n, I))$ are the values of $(X_1, \ldots, X_n)$ after the I-th run through the body of the loop.

It is easy to see that the formulas above have the desired properties. The proof is therefore left as an exercise for the reader.   ■

### Theorem 16.5 (Arithmetical algebras are expressive)
Every arithmetical algebra is expressive.

*Proof*   Let $A$ be an arithmetical algebra, $\alpha$ be a program and $\psi$ be a formula. Let $X_1, \ldots, X_n$ be all variables occurring in $\alpha$. Also let $Y_1, \ldots, Y_n$ be variables not occurring in $\alpha$ or $\psi$. We claim that $\text{wlp}_A(\alpha, \psi)$ is expressed in $A$ by the following formula:

$$\forall Y_1...\forall Y_n (comp_\alpha(X_1,...,X_n,Y_1,...,Y_n)\rightarrow\psi\{X_1/Y_1,...,X_n/Y_n\})$$

This requires us to show that the states in $wlp_A(\alpha,\psi)$ are exactly those that satisfy this formula. Let sta be a state in $wlp_A(\alpha,\psi)$. We want to show

$$A\models_{sta}\forall Y_1...\forall Y_n(comp_\alpha(X_1,...,X_n,Y_1,...,Y_n)\rightarrow\psi\{X_1/Y_1,...,X_n/Y_n\})$$

In the following let us write $a_1,...,a_n$ for $sta(X_1),...,sta(X_n)$. We take elements $b_1,...,b_n \in dom(A)$ such that

$$A\models_{sta(Y_1/b_1,...,Y_n/b_n)}comp_\alpha(X_1,...,X_n,Y_1,...,Y_n)$$

that is,

$$A\models comp_\alpha(a_1,...,a_n,b_1,...,b_n)$$

It follows from Theorem 16.4 that

$$sta(X_1/a_1,...,X_n/a_n)\ [\alpha]_A\ sta(X_1/b_1,...,X_n/b_n)$$

As $sta=sta(X_1/a_1,...,X_n/a_n)$ is an element of $wlp_A(\alpha,\psi)$, it follows that $A\models_{sta(X_1/b_1,...,X_n/b_n)}\psi$, and so we obtain

$$A\models_{sta(Y_1/b_1,...,Y_n/b_n)}\psi\{X_1/Y_1,...,X_n/Y_n\}$$

Conversely let sta be a state such that

$$A\models_{sta}\forall Y_1...\forall Y_n(comp_\alpha(X_1,...,X_n,Y_1,...,Y_n)\rightarrow\psi\{X_1/Y_1,...,X_n/Y_n\})$$

We have to show that sta is an element of $wlp_A(\alpha,\psi)$. For this purpose let sta' be a state such that $sta[\alpha]_A sta'$. In the following we denote $sta(X_1),...,sta(X_n)$ by $a_1,...,a_n$ and $sta'(X_1),...,sta'(X_n)$ by $b_1,...,b_n$. Then

$$sta(X_1/a_1,...,X_n/a_n)[\alpha]_A sta(X_1/b_1,...,X_n/b_n)$$

thus

$$A\models comp_\alpha(a_1,...,a_n,b_1,...,b_n)$$

Applying

$$A\models_{sta}\forall Y_1...\forall Y_n(comp_\alpha(X_1,...,X_n,Y_1,...,Y_n)\rightarrow\psi\{X_1/Y_1,...,X_n/Y_n\})$$

with $b_1,...,b_n$ for $Y_1,...,Y_n$ we obtain

$$A\models_{sta(Y_1/b_1,...,Y_n/b_n)}\psi\{X_1/Y_1,...,X_n/Y_n\}$$

hence $A\models_{sta(X_1/b_1,...,X_n/b_n)}\psi$. Therefore $A\models_{sta'}\psi$. So we have shown that sta is an element of $wlp_A(\alpha,\psi)$.    ∎

## 16.4    Examples of arithmetical algebras

In this section we want to show for several algebras that they are arithmetical. From the results of the previous sections it follows that Hoare's calculus is complete over these algebras.

The starting point of our investigations is the algebra *Nat* of natural numbers with zero, successor function, addition and multiplication. In order to prove that this algebra is arithmetical we need some technical results from number theory. We develop a mechanism for encoding finite sequences of numbers of variable length by one number in such a way that it is possible to access the coded elements.

### Lemma 16.1

Let $b_0, b_1, \ldots, b_t$ be positive numbers such that every two of them are relative prime. Let $m = b_0 * \ldots * b_t$ and

$$f: \{0,1,\ldots,m-1\} \to \{0,1,\ldots,b_0-1\} \times \{0,1,\ldots,b_1-1\} \times \ldots \times \{0,1,\ldots,b_t-1\}$$

be the function defined by $f(a) = (a \text{ MOD } b_0,\ a \text{ MOD } b_1, \ldots,\ a \text{ MOD } b_t)$, for all $a = 0, 1, \ldots, m-1$. It follows that f is a bijective function.

*Proof*    Since both the domain and range of f contain exactly m elements, only injectivity of f has to be shown. Let a and a' be elements of $\{0,1,\ldots,m-1\}$ such that $f(a)=f(a')$. Then, by definition of f, $a \text{ MOD } b_i = a' \text{ MOD } b_i$, for $i=0,\ldots,t$, that is, $b_i$ is a divisor a-a'. Since this is true for all $i=0,\ldots,t$ and since every two of $b_0, b_1, \ldots, b_t$ are relative prime, it follows that also $m=b_0 * \ldots * b_t$ is a divisor of a-a'. But we know that a-a' lies in the interval between -m+1 and m-1, so this is only possible if a=a'.    ∎

### Lemma 16.2

Let $k_0, k_1, \ldots, k_t$ be arbitrary natural numbers. Define

$m := \max\{t, k_0, k_1, \ldots, k_t\}$
$b = fac(m)$ (the factorial of m)
$b_i = 1+(i+1)*b$, for $i=0,\ldots,t$

Then every two elements of $b_0, b_1, \ldots, b_t$ are relatively prime.

*Proof*    Let $i,j \leq t$ with $i \neq j$ be given. We have to show that no prime number p divides both $b_i$ and $b_j$. Let us suppose that this were the case. Then p would be divisor of $1+(i+1)*b$ and $1+(j+1)*b$, so it would also divide $1+(i+1)*b-1-(j+1)*b$, that is, $(i-j)*b$. Hence p would have to be a divisor of i-j or of b (because p is a prime number). We know that $|i-j| \leq t \leq m$ and $b=fac(m)$. So if p were a divisor of i-j, then it would also be of b. In both cases p would divide b. Then p would also be a divisor of $(i+1)*b$, and therefore it would also divide $b_i-(i+1)*b$, which means it would divide 1. This is a contradiction.     ■

**Definition 16.6:** We define a ternary function $\beta$ as follows:

$\beta(a,b,i)=a \text{ MOD } (1+(i+1)*b)$

*Lemma 16.3*

Let $(k_0,k_1,...,k_t)$ be a sequence of t+1 numbers. Then there are numbers a and b such that

$(k_0,k_1,...,k_t)=(\beta(a,b,0),\beta(a,b,1),...,\beta(a,b,t))$

The function $\beta$ can be defined in *Nat* by a formula Beta(A,B,I,X).

*Proof*     Let $(k_0,k_1,...,k_t)$ be a sequence of t+1 numbers. We define m= $\max\{t,k_0,k_1,...,k_t\}$ and b:=fac(m). With $b_i:=1+(i+1)*b$, for i=0,...t, we consider the function f defined above. The function f is bijective. Also $(k_0,k_1,...,k_t)\in \{0,1,...,b_0-1\}\times\{0,1,...,b_1-1\}\times...\times\{0,1,...,b_t-1\}$.     Therefore,     there     exists     a number a such that $f(a)=(k_0,k_1,...,k_t)$. We claim that the numbers a and b defined in this way have the desired properties. This is true, because a MOD $b_i=k_i$ (for i=0,...,t), that is $\beta(a,b,i)=k_i$, for i=0,...,t. Finally, as Beta(A,B,I,X) we may take the formula $X=A \text{ MOD } (1+(I+1)*B)$.     ■

In order to show that *Nat* is an arithmetical algebra, we only have to bridge a small gap between the achieved results and the definition of arithmetical algebras. We have so far developed a method for encoding arbitrary finite sequences of numbers by *two* numbers, whereas in arithmetical algebras this has to be done by a *single* number.

The reader might ask why we defined arithmetical algebras in such a strict way. In fact we could have used two, or even an arbitrary (but fixed) number. of numbers to encode sequences. This would only have the drawback that many of the formulas used in several earlier proofs would have become more complicated. Therefore we have chosen the approach presented above and must now invest a little work in order to show that *Nat* is arithmetical.

We make use of the following bijective coding function which is based on Cauchy's diagonal numbering of number pairs:

$<x,y>:=(x+y)*(x+y+1) \text{ DIV } 2$

We also use decoding functions z/1 and z/2 defined by

$<x,y>/1=x$

$<x,y>/2=y$

$<z/1,z/2>=z$

Since it is clear that these functions are definable in *Nat*, we shall make use of functions $<\ >$, $/1$ and $/2$ in formulas as if they were function symbols of the underlying signature.

### Theorem 16.6

*Nat* is an arithmetical algebra. Hence, also *Cardinal* is arithmetical.

*Proof*    $(N,<)$ and function $dec(u,i)=\beta(u/1,u/2,i)$ can be defined within *Nat* by the following formulas $nat(X)$, $\varphi_0(Y)$, $\varphi_{succ}(X,Y)$, $\varphi_<(X,Y)$  and  $\varphi_{dec}(U,I,A)$ respectively:

$$X=X,\ Y=0,\ Y=succ(X),\ \exists Z\ Y=X+succ(Z),\ BETA(U/1,U/2,I,A)\qquad\blacksquare$$

In the rest of this section we shall show that some algebras important for computer science are arithmetical.

### Theorem 16.7

*Int* is an arithmetical algebra.

*Proof*    We must construct a formula $nat(X)$ which defines $\mathbb{N}$ in *Int*. This can be done, because every natural number can be written as sum of four quadratic numbers. Conversely, every such sum is of course a natural number. So we are able to express $nat(X)$ by the following formula:

$$\exists A\exists B\exists C\exists D\ X=A*A+B*B+C*C+D*D$$

Furthermore, constant 0, function succ and ordering relation $<$ on $\mathbb{N}$ may be defined in *Int* by formulas $Y=0$, $(nat(X)\wedge Y=succ(X))$ and $(nat(X)\wedge nat(Y)\wedge \exists Z\ (nat(Z)\wedge\ Y=X+succ(Z))$. Finally we have to encode finite sequences of integers and give a definable decoding function $dec_{int}$. Here we use the decoding function $dec$ used above for natural numbers and the pairing function $<>$. For an integer z, let $|z|$ be the absolute value of z. Also define $sgn(z)=0$ if $z<0$, and $sgn(z)=1$, if $z\geq 0$. Now let $z_0,z_1,...,z_m$ be a finite sequence of integers. We consider the sequence of pairs $<sgn(z_i),|z_i|>$, for $i=0,...,m$, and choose an element $u\in\mathbb{N}$ such that $dec(u,i)=<sgn(z_i),|z_i|>$ for $i=0,...,m$. Now we define

$$dec_{int}(u,i)=\begin{cases} dec(u,i)/2, & \text{if } dec(u,i)/1=1 \\ -dec(u,i)/2, & \text{if } dec(u,i)/1=0 \end{cases}$$

Obviously $dec_{int}(u,i)=z_i$ for $i=0,...,m$. Finally we show that $dec_{int}$ is definable in *Int*. We know that functions dec, /1 and /2 are definable in *Nat* by formulas DEC(U,I,X), and so on. We need such definitions in algebra *Int*. This requires a means of talking about algebra *Nat* within the extension *Int*. Such a means is relativization as introduced in Chapter 5. The following formula $DEC_{int}(U,I,X)$ defines function $dec_{int}$ in *Int*:

$$nat(U) \wedge nat(I) \wedge nat(X) \wedge [DEC]^{nat}(U,I,X)$$

where $[DEC]^{nat}(U,I,X)$ is the relativization of formula DEC(U,I,X) to formula nat(X). Now it is not difficult to construct a formula $DEC_{int}(U,I,X)$ that defines function $dec_Z$ in algebra *Int*. ∎

### Theorem 16.8

*Set(Nat)* is an arithmetical algebra.

*Proof* First note that $(N,<)$ can be defined within *Set(Nat)*. Next we describe how a finite sequence consisting of finite sets of numbers can be encoded by a single set and give a definable decoding function. Let $Q_0,Q_1,...,Q_m$ be a sequence of finite sets of numbers. We define a set

$$Q:=\{<i,q> \mid i{\le}m \text{ and } q{\in} Q_i\}$$

(Recall that $<i,q>$ is the encoding of pair (i,q).)

If we take $dec_{set}(Q,i)=\{q \mid <i,q>{\in} Q\}$ as a decoding function, then we have $dec_{set}(Q,i)=Q_i$. It remains to show that function $dec_{set}$ is definable in *Set(Nat)*. First we note that the element relation $q{\in} R$ can be defined in *Set(Nat)* by the formula Insert(R,q)=R. It follows that

$$dec_{set}(Q,i)=R \Leftrightarrow$$

$$Set(Nat) \models \forall q(q{\in} R \leftrightarrow <i,q>{\in} Q) \Leftrightarrow$$

$$Set(Nat) \models \forall q(insert(R,q)=R \leftrightarrow insert(<i,q>,Q)=Q) \qquad ∎$$

### Theorem 16.9

For $L{\ge}0$ and a one-sorted arithmetical algebra *Data,* the 3-sorted algebra *Array* [0..L] *of Data* (See section 6.5) is arithmetical, too.

*Proof* The only task to be done is the encoding and decoding of a sequence $B_0,B_1,...,B_m$ of arrays, as well as the encoding of a finite sequence $i_0,i_1,...,i_m$ of indices (though the domain of indices is finite we may not forget this latter task). So let a sequence $B_0,B_1,...,B_m$ of arrays with $B_i=(b_{i0},...,b_{iL})$, for i=0,…,m, be given. First, we encode the elements of array $B_i$ by an object $u_i$ of dom(***Data***) (for i=0,…,L) using the encoding machinery of ***Data***, with decoding function $dec_{data}$. In a second step we encode the sequence of code objects $u_0,u_1,\ldots,u_m$ again by an object u of dom(***Data***). The situation is illustrated by Figure 16.1.

$b_{00}\ b_{01}\ \cdots\ b_{0L} \longrightarrow u_0$

$b_{10}\ b_{11}\ \cdots\ b_{1L} \longrightarrow u_1$ with $dec_{data}(u_i,k)=a_{ik}$, for $i\le m$, $k\le L$

$\vdots$

$b_{m0}\ b_{m1}\ \cdots\ b_{mL} \longrightarrow u_m$

$\downarrow$

u with $dec_{data}(u,i)=u_i$, for $i\le m$

**Figure 16.1**

Now we define $dec_{array}(u,i)$ as

$$(dec_{data}(dec_{data}(u,i),0),dec_{data}(dec_{data}(u,i),1),\ldots,dec_{data}(dec_{data}(u,i),L))$$

and obtain $dec_{array}(u,i)=B_i$, for i=0,...,m. Function $dec_{array}$ is definable in the algebra ***Array*** [0…L] ***of Data*** on the basis of the following equivalence:

$$dec_{array}(u,i)=B \Leftrightarrow \text{For all indices } j,\ read(B,j)=dec_{data}(dec_{data}(u,i),j)$$

Encoding of a finite sequence $i_0,i_1,\ldots,i_m$ of indices is even simpler. We simply use the **type conversion function** conv that assigns to each object i of type index the corresponding natural number i as an object of type data and observe that it is trivially definable in ***Array*** [0…L] ***of Data*** by the following formula CONV(I,N) (here we must take care to distinguish between the constant $0_{index}$ of type index and the definable constant $0_{data}$ of the arithmetical algebra ***Data***, as well as between the function symbol $succ_{index}$ of type index→index and the definable function $succ_{data}$):

$$(I=0_{index}\wedge N=0_{data})\vee(I=succ_{index}(0_{index})\wedge N=succ_{data}(0_{data}))\ldots\vee$$
$$(I=succ_{index}{}^{L}(0_{index})\wedge N=succ_{data}{}^{L}(0_{data}))$$

Function conv allows us to transfer the encoding machinery for natural numbers to indices. ■

## 16.5   Incompleteness results

All algebras with complete Hoare logic that we have seen so far contain a copy of natural numbers as a definable subalgebra and the other parts of the algebras are firmly connected to this copy. In this section we want to present some algebras with incomplete Hoare logic. They are obviously neither expressive nor arithmetical. A first algebra with incomplete Hoare logic was presented by Wand (1978). The algebra he used was quite artificial though. Here we shall consider natural algebras. The proof scheme will be the same for all our examples, so we want to describe it first. It makes use of some results from computability theory.

> **Definition 16.7:**    An algebra $A$ is called **algebra with counters** iff there is a ground term $t_0$ and a term $t_{succ}(X)$ with a single variable $X$ such that the following algebra (Nat,a,f) is isomorphic to the algebra $N$:
>
> $$a=val_A(t_0)$$
> $$f(b)=val_A(t_{succ}(b))$$
> $$Nat=\{f^i(a)|\ i\in \mathbb{N}\ \}$$

Note that isomorphy with algebra $N$ means that iterated application of f to $val_A(t_0)$ generates an infinite sequence of objects. Also note the difference between an algebra $A$ with counters and an algebra $A$ such that an isomorphic copy of $N$ can be defined within $A$. The former concept is weaker than the latter in the sense that it does not require a definition of $\mathbb{N}$ in $A$, but stronger in the sense that it requires definitions of 0 and succ by terms instead of formulas.

> ### Lemma 16.4
>
> Let $A$ be an algebra with counters as in Definition 16.7. For each closed formula $\varphi$ of the form $\forall X_1 ... \forall X_n\ p=q$ over the algebra **Nat** (thus p and q are nothing more than polynomials) we construct a partial correctness assertion pca($\varphi$) of the form
>
> > $\{true\}$
> >> $test(X_1);...;test(X_n);$
> >> $comp_p(X_1,...,X_n,X_p);$
> >> $comp_q(X_1,...,X_n,X_q)$
> > $\{X_p=X_q\}$
>
> such that the following statements are equivalent:

(1)    $Nat \models \forall X_1...\forall X_n\ p=q$

(2)    $A \models pca(\varphi)$.

As program test(X) we may take :

$$Z:=t_0; WHILE\ \neg Z=X\ DO\ Z:=t_{succ}(Z)\ END$$

It terminates over $A$ in state sta iff sta(X) belongs to $\mathbb{N}$, and it does not alter the value of X.

As programs $comp_p(X_1,...,X_n,X_p)$ and $comp_q(X_1,...,X_n,X_q)$ we take programs that compute the values of polynomials p and q for inputs $X_1,...,X_n$ respectively by simulating additions and multiplications occurring in p and q by appropriate subprograms using $t_0$ and $t_{succ}$.

*Proof*    $comp_p(X_1,...,X_n,X_p)$ can be easily defined by induction over term p. (See Exercise 16.8.) Then we proceed, using a 'dummy state' sta over $A$, as follows:

$Nat \models \forall X_1...\forall X_n\ p=q \Leftrightarrow$

for all $a_1,...,a_n \in \mathbb{N}$, if $sta(X_1/a_1,...,X_n/a_n)[comp_p(X_1,...,X_n,X_p)]_A$ sta' and $sta(X_1/a_1,...,X_n/a_n)[comp_q(X_1,...,X_n,X_q)]_A$ sta'' then sta'$(X_p)$=sta''$(X_q)$ $\Leftrightarrow$

for all $a_1,...,a_n \in dom(A)$, if programs test$(X_1)$, ..., test$(X_n)$ terminate over $A$ on $sta(X_1/a_1,...,X_n/a_n)$, and if $sta(X_1/a_1,...,X_n/a_n)[comp_p(X_1,...,X_n,X_p)]_A$ sta' and $sta(X_1/a_1,...,X_n/a_n)[comp_q(X_1,...,X_n,X_q)]_A$ sta'' then sta'$(X_p)$=sta''$(X_q)$ $\Leftrightarrow$

$A \models pca(\varphi)$

Note how the relativization of quantifiers $\forall X_1...\forall X_n$ to elements $a_1,...,a_n \in \mathbb{N}$ is simulated by the semantics of test$(X_1)$;...; test$(X_n)$ in the pca above.    ∎

---

**Exercise 16.8** Give a definition of program $comp_p(X_1,...,X_n,X_p)$ used in the proof of Lemma 16.4.

---

### Theorem 16.10

Let $A$ be an algebra with counters. Then the set of all partial correctness assertions that are valid in $A$ is not recursively enumerable.

*Proof*    Lemma 16.4 gives a reduction of validity of polynomial equations in *Nat* to the set of valid pcas in $A$. The claim follows from a well-known fact of recursion theory that the former set is not recursively enumerable.    ■

### Lemma 16.5

The following algebras are algebras with counters and have a decidable first-order theory:

(1)    $(N,+)$

(2)    Every free term algebra $T_\Sigma$, for an arbitrary finite signature $\Sigma$ containing only function symbols, among them at least one constant symbol and at least one n-ary function symbol with n>0.

(3)    *Set(N)*

*Proof*

(1)    Th($N$,+) is known under the name of Presburger arithmetic. It is a well-known result that its theory is decidable.

(2)    This was shown in Chapter 5 as an application of $\omega_1$-categoricity.

(3)    Follows from the canonical reduction of Th(*Set(N)*) to a decidable theory known under the name of Büchi arithmetic (see Büchi (1960)).    ■

Using these results we may conclude the following incompleteness results.

### Theorem 16.11

For all algebras $A$ listed in Lemma 16.5, Hoare's calculus is incomplete over $A$.

*Proof*    In each case the set of pcas derivable by Hoare's calculus from Th($A$) is a recursively enumerable set. But all considered algebras are algebras with counters, so the set of pcas valid in them is not recursively enumerable. This means that Hoare's calculus over these algebras is not complete.    ■

---

**Exercise 16.9** Show that algebras *Stack(N)* and *Queue(N)* have incomplete Hoare logic. (*Hint*: First consider the algebras above without the selector functions pop, top, first and rest and apply the theorem on free term algebras. Then introduce the selector functions.)

---

After proving these incompleteness results, the following question aris-es. Is it perhaps true, that Hoare's calculus is able to derive all interesting pcas and is only incomplete because it cannot derive some 'pathological' as-sertions? The following result shows that this is definitely not true.

### Theorem 16.12

Let $A$ be the algebra $(\mathbb{N},0,\text{succ},\text{pred})$ of natural numbers with zero, successor function succ and predecessor function pred. The partial cor-rectness assertion

$\{X=0,\ Y=Z\}$
$\quad$ WHILE $\neg Y=0$ DO $X:=\text{succ}(X);\ Y:=\text{pred}(Y)$ END
$\{X=Z\}$

is valid in $A$, but not derivable in Hoare's calculus over $A$.

*Proof* Validity of the pca in $A$ is quite trivial. Assume by contradiction that the pca is derivable in Hoare's calculus. Then there is a formula $\varphi$ such that the following implications are valid in $A$. (The existence of $\varphi$ with (1)-(3) re-quires a moment's thought.)

(1)    $(X=0 \wedge Y=Z) \rightarrow \varphi$

(2)    $(\neg Y=0 \wedge \varphi) \rightarrow \varphi\{X/\text{succ}(X),Y/\text{pred}(Y)\}$

(3)    $(Y=0 \wedge \varphi) \rightarrow X=Z$

We may assume without loss of generality that $\varphi$ contains only variables X, Y and Z, and conclude that the following formulas (4)-(6) are also valid in $A$.

(4)    $\varphi(0,Y,Y)$

(5)    $\varphi(X,\text{succ}(Y),Z) \rightarrow \varphi(\text{succ}(X),Y,Z)$

(6)    $\varphi(X,0,Z) \rightarrow X=Z$

We then show that the formula $\varphi(X,Y,Z)$ defines in $(\mathbb{N},0,\text{succ},\text{pred})$ the addi-tion function. Let x,y,z be given such that x+y=z. By (4) we obtain $(\mathbb{N},0,\text{succ},\text{pred}) \models \varphi(0,x+y,z)$. Applying validity of (5) x times we obtain $(\mathbb{N},0,\text{succ},\text{pred}) \models \varphi(x,y,z)$. Conversely, let numbers x,y,z be given such that $(\mathbb{N},0,\text{succ},\text{pred}) \models \varphi(x,y,z)$. Applying formula (5) y times we obtain $(\mathbb{N},0,\text{succ},\text{pred}) \models \varphi(x+y,0,z)$. By (6) this implies that x+y=z. Now a contra-diction is achieved. Using this definition of addition it follows that the set of even numbers can also be defined in $(\mathbb{N},0,\text{succ},\text{pred})$. Since pred can be de-fined in $(\mathbb{N},0,\text{succ})$ we finally obtain that the set of even numbers can be de-fined in $(\mathbb{N},0,\text{succ})$. But as we know from Chapter 5, in $(\mathbb{N},0,\text{succ})$ only finite sets and their complements can be defined.    ∎

The reason for this incompleteness result was that we were not able to express a suitable loop invariant of the program loop in first-order logic.

---

**Exercise 16.10**  Let us consider the disjoint union **Nat+Nat** defined as the 2-sorted algebra $(\mathbb{N}_1, \mathbb{N}_2, 0_1, succ_1, +_1, *_1, 0_2, succ_2, +_2, *_2)$ consisting of two disjoint copies of the standard model of arithmetic. Show that Hoare's calculus is incomplete over **Nat+Nat**. (*Hint*: Note that there is no interrelation between the two copies. So try to find a program (running on the disjoint union) such that every loop invariant would need some correlation between the two copies.)

---

## 16.6    Resolving incompleteness

Assume that we want to show that $A \models \{\varphi\}\alpha\{\psi\}$, for an algebra $A$ and a pca $\{\varphi\}\alpha\{\psi\}$. Imagine that the expressive power of the underlying assertion language is too weak to provide the necessary loop invariants. A common attempt to resolve this weakness is to expand the considered algebra $A$ by suitably chosen functions to an algebra $B$. Then these new functions may be used to express stronger loop invariants. In the case of a concrete valid but unprovable pca from the last section, expansion of the algebra $(\mathbb{N}, 0, succ, pred)$ by the addition function would serve this purpose. Thus, if the originally considered partial correctness assertion can be derived over algebra $B$ we may conclude that it is true within $B$, hence also within $A$.

### Fact

Every algebra $A$ whose domain is denumerable can be expanded to an arithmetical algebra: add a copy of natural numbers and use any bijection between Nat and finite sequences of domain elements; finally add the natural decoding function w.r.t. this bijection to the algebra. Therefore we may always expand $A$ to an algebra $B$ in such a way that Hoare's calculus over $B$ allows us to derive $\{\varphi\}\alpha\{\psi\}$.

This is not so simple if we want to show that a pca $\{\varphi\}\alpha\{\psi\}$ follows from an axiom set Ax:

$$\text{Ax} \models \{\varphi\}\alpha\{\psi\}$$

A solution, namely to use a conservative extension of Ax to derive $\{\varphi\}\alpha\{\psi\}$, is investigated in Bergstra nd Klop (1984). (For the notion of a conservative extension see Chapter 5.)

The following theorem establishes a kind of soundness result for

Hoare's calculus in a conservative extension.

### Theorem 16.13

Let a set Ax' of $\Sigma$'-formulas be a conservative extension of a set Ax of $\Sigma$-formulas and $\{\varphi\}\alpha\{\psi\}$ be a partial correctness assertion over signature $\Sigma$. If Ax'$\vdash_{\overline{HC}}\{\varphi\}\alpha\{\psi\}$ then Ax $\models\{\varphi\}\alpha\{\psi\}$.

*Proof*   Assume that Ax'$\vdash_{\overline{HC}}\{\varphi\}\alpha\{\psi\}$. Then, Ax' $\models\{\varphi\}\alpha\{\psi\}$. To show Ax $\models\{\varphi\}\alpha\{\psi\}$, consider a model $A$ of Ax. Choose a $\Sigma$-algebra $B$ and an expansion $B_{exp}$ of $B$ to a $\Sigma$'-algebra such that $B$ is elementary equivalent to $A$ and $B_{exp}$ is a model of Ax'. We conclude that $B_{exp}\models\{\varphi\}\alpha\{\psi\}$. Since $\{\varphi\}\alpha\{\psi\}$ is a partial correctness assertion over signature $\Sigma$ we obtain that $B\models\{\varphi\}\alpha\{\psi\}$, that is, Th($B$)$\models\{\varphi\}\alpha\{\psi\}$. It follows Th($A$)$\models\{\varphi\}\alpha\{\psi\}$ and $A\models\{\varphi\}\alpha\{\psi\}$.   ∎

Thus, in order to show that Ax $\models\{\varphi\}\alpha\{\psi\}$, we must try to find a conservative extension Ax' of Ax such that Ax'$\vdash_{\overline{HC}}\{\varphi\}\alpha\{\psi\}$. In most cases it will be possible to show that Ax' is indeed a conservative extension of Ax by proving condition (4) of Theorem 5.6. The question arises as to whether every true statement Ax $\models\{\varphi\}\alpha\{\psi\}$ can be obtained by deriving $\{\varphi\}\alpha\{\psi\}$ from a conservative extension Ax' of Ax. This is indeed the case, even in the stronger sense that a conservative extension Ax' of Ax can be found independently of $\varphi$ and $\psi$.

### Theorem 16.14

For every axiom Ax in a signature $\Sigma$ there is a conservative extension Ax' in a signature $\Sigma$' such that for all partial correctness assertions $\{\varphi\}\alpha\{\psi\}$ over $\Sigma$' the following is true:

$$Ax' \models \{\varphi\}\alpha\{\psi\} \Leftrightarrow Ax' \vdash_{\overline{HC}} \{\varphi\}\alpha\{\psi\}$$

Ax' is called a **logically complete theory**.

*Proof*   See Bergstra and Klop (1984).   ∎

### Corollary

If Ax $\models\{\varphi\}\alpha\{\psi\}$ then there exists a conservative extension Ax' of Ax such that Ax'$\vdash_{\overline{HC}}\{\varphi\}\alpha\{\psi\}$.

**Exercise 16.11**    Let $\Sigma$ be a signature containing a constant 0 and 1-ary function symbols succ and pred. Consider the theory Ax that consists of all formulas that logically follow from the following axioms (a fragment of arithmetics):

$\forall X$ pred(succ(X))=X
pred(0)=0
$\forall X$ ($\neg X=0 \rightarrow X$=succ(pred(X)))
$\forall X \neg 0$=succ(X)

Consider the following partial correctness assertion $\rho$:

{X=0, Y=Z} WHILE $\neg$Y=0 DO X:=succ(X) ; Y:=pred(Y) END {X=Z}.

Extend $\Sigma$ by a 3-ary predicate symbol plus and define Ax' as the theory defined by the four axioms above together with the following ones:

$\forall X \forall Z$ (plus(X,0,Z)$\leftrightarrow$Z=X)
$\forall Y \forall Z$ (plus(0,Y,Z)$\leftrightarrow$Z=Y)
$\forall X \forall Y \forall Z$ (plus(X,succ(Y),succ(Z))$\leftrightarrow$plus(X,Y,Z))
$\forall X \forall Y \forall Z$ (plus(succ(X),Y,succ(Z))$\leftrightarrow$plus(X,Y,Z))
$\forall X \forall Y$ (plus(X,Y,0)$\rightarrow$(X=0$\wedge$Y=0))

(a)    What do models of Ax look like (in particular models that are not term generated)?

(b)    Show that Ax $\models \rho$.

(c)    Show that not Ax $\vdash_{HC} \rho$. Use the statement of Theorem 16.12 concerning algebra Th($\mathbb{N}$,0,succ,pred).

(d)    Prove that Ax' is a conservative extension of Ax.

(e)    Show that Ax' $\vdash_{HC} \rho$.

## 16.7    Program inclusion and program equivalence ✂

In this section we will introduce an application of Hoare's rules to the questions of program equivalence and program inclusion that were discussed in Bergstra and Klop (1984). Program equivalence plays an important role in some fields of computer science, such as program optimization.

**Definition 16.8:**    Let $\alpha$ and $\beta$ be $\Sigma$-programs and $A$ be a $\Sigma$-algebra. We say that $\alpha$ **is included in** $\beta$ w.r.t. $A$ iff for all states sta and sta' over $A$ the following holds:

If sta$[\alpha]_A$sta' then sta$[\beta]_A$sta'.

This property is denoted by $A \models \alpha{\subseteq}\beta$. Informally stated, it means that $\beta$ yields the same effect as $\alpha$ on all states sta such that $\alpha$ terminates on sta, but $\beta$ may terminate on further states than only those on which $\alpha$ terminates. We say that $\alpha{\subseteq}\beta$ **follows** from a set of first-order formulas Ax (denoted Ax $\models \alpha{\subseteq}\beta$) iff $A \models \alpha{\subseteq}\beta$ is true for every model $A$ of Ax. We say that $\alpha$ is **equivalent** to $\beta$ w.r.t. $A$ (denoted $A \models \alpha{\leftrightarrow}\beta$) iff $A \models \alpha{\subseteq}\beta$ and $A \models \beta{\subseteq}\alpha$. We say that Ax $\models \alpha{\leftrightarrow}\beta$, iff Ax $\models \alpha{\subseteq}\beta$ and Ax $\models \beta{\subseteq}\alpha$.

Let us look at some simple examples:

(1)    All programs include WHILE True DO $\alpha$ END, since this program terminates on no state.

(2)    Program IF B THEN $\alpha$ ELSE WHILE True DO $\beta$ END END is included in program IF B THEN $\alpha$ ELSE $\beta$ END.

(3)    Let $\alpha$ be the program X:=a; Y:=b and $\beta$ be the program X:=a. Then neither $\alpha$ is included in $\beta$, nor $\beta$ in $\alpha$.

(4)    Programs WHILE B DO $\alpha$ END and IF B THEN WHILE B DO $\alpha$ END END are equivalent.

(5)    Programs X:=f(Y);X:=g(Y) and X:=g(Y) are equivalent in every algebra.

(6)    The examples we have seen so far were quite independent of algebras. Now we show a domain-specific example. Programs X:=Y and X:=0;WHILE X$\neq$Y DO X:=succ(X) END and are equivalent in the algebra $N$. But this equivalence does not follow from Th($N$), since the former program does not terminate in a non-standard model when Y has a non-standard value. Note though that Th($N$) $\models\alpha{\subseteq}\beta$.

Now let us turn our attention to the question of how to prove Ax $\models \alpha{\subseteq}\beta$ in a general setting. The following simple observation gives us a hint.

### *Corollary*

Assume that Ax $\models \alpha{\subseteq}\beta$. Then, for all assertions $\varphi$ and $\psi$, Ax $\models \{\varphi\}\beta\{\psi\}$ implies Ax $\models \{\varphi\}\alpha\{\psi\}$.

(Note the order of programs.)

*Proof*  Follows immediately from the definitions.    ■

The question now arises of whether the converse is true, that is, whether the statement $Ax \models \alpha{\subseteq}\beta$ can be concluded from the following property:

For all assertions $\varphi$ and $\psi$, $Ax \models \{\varphi\}\beta\{\psi\}$ implies $Ax \models \{\varphi\}\alpha\{\psi\}$.

A short look at this question tells us that this cannot be expected to be true in general. The reason is that the property above might well be true by lack of interesting assertions $\varphi$ and $\psi$ with $Ax \models \{\varphi\}\beta\{\psi\}$, although $Ax \models \alpha{\subseteq}\beta$ does not hold. This problem might be overcome by considering extended signatures and stronger axiom systems which allow more detailed decriptions of the behaviour of $\alpha$ and $\beta$ by partial correctness assertions $\{\varphi\}\beta\{\psi\}$ and $\{\varphi\}\alpha\{\psi\}$. Observe that the following is true, too.

### Theorem 16.15

If $Ax \models \alpha{\subseteq}\beta$ then for every extension $Ax'$ of $Ax$ in an extended signature $\Sigma'$ and all assertions $\varphi$ and $\psi$ over $\Sigma'$:

$$Ax' \models \{\varphi\}\beta\{\psi\} \text{ implies } Ax' \models \{\varphi\}\alpha\{\psi\}$$

Now, Bergstra and Klop (1984) contains a reversed statement. It suffices to consider conservative extensions only.

### Theorem 16.16

Assume that $Ax' \models \{\varphi\}\beta\{\psi\}$ implies $Ax' \models \{\varphi\}\alpha\{\psi\}$, for all conservative extensions $Ax'$ of $Ax$ and assertions $\varphi$ and $\psi$ over an extended signature $\Sigma'$. Then $Ax \models \alpha{\subseteq}\beta$.

We are of course interested in *proving* program inclusion and equivalence, and it is natural to ask whether we can use Hoare's calculus for this purpose. So let us next discuss the question of whether $Ax' \models$ in Theorem 16.16 may be replaced by $Ax'\vdash_{\overline{HC}}$.

### Corollary

Assume that $Ax'\vdash_{\overline{HC}} \{\varphi\}\beta\{\psi\}$ implies $Ax'\vdash_{\overline{HC}} \{\varphi\}\alpha\{\psi\}$, for all conservative extensions $Ax'$ of $Ax$ and all assertions $\varphi$ and $\psi$ over an extended signature $\Sigma'$. Then $Ax \models \alpha{\subseteq}\beta$.

*Proof*   Follows directly from results in this section and Section 16.6.     ■

The problem is that we have now become too strong, that is, the originally considered easy direction, starting from the assumption that $Ax \models \alpha \subseteq \beta$, no longer holds. The reason is that an axiom system Ax' might be sufficiently strong to allow the derivation of $\{\varphi\}\beta\{\psi\}$ via Hoare's rules (thus $Ax' \models \{\varphi\}\beta\{\psi\}$, and hence $Ax' \models \{\varphi\}\alpha\{\psi\}$), but it might still be too weak to allow a derivation of $\{\varphi\}\alpha\{\psi\}$ via Hoare's rules. The corollary at the end of Section 16.6 suggests what might be the final version of a theorem characterizing program inclusion.

### Theorem 16.17

Let $\alpha$ and $\beta$ be $\Sigma$-programs and Ax a set of $\Sigma$-formulas. Then the following are equivalent:

(1)     $Ax \models \alpha \subseteq \beta$

(2)     For every conservative extension Ax' of Ax there is a conservative extension Ax'' of Ax' such that, for all assertions $\varphi$ and $\psi$, $Ax'' \vdash_{HC} \{\varphi\}\beta\{\psi\}$ implies $Ax'' \vdash_{HC} \{\varphi\}\alpha\{\psi\}$.

*Proof*   See Bergstra and Klop (1984).     ■

# Chapter 17
# Recursive Procedures and Data Type Declarations

In this chapter we make an essential step from the simple WHILE programs studied so far towards Pascal or Modula-2-like programs. This step will be the introduction of recursive procedures.

We shall refer here to the well-known result from recursive function theory stating that every partial recursive function can be computed by a WHILE program over the algebra $N$ of natural numbers with zero and successor function. This raises the question of what a discussion of recursive procedures, which requires considerable effort if a complete treatment is aspired to, might be good for. The answer is twofold.

First, the presence of procedures is a first step towards a modular design of more extended programs, and is thus indispensable from a software engineering point of view. In particular, recursive procedures are the natural programming language feature to implement not only the divide-and-conquer algorithm design paradigms but also the usage of recursively defined data structures like trees. Thus, recursive procedures are indispensable also from an algorithm design point of view.

These practical aspects might be already sufficient to justify the introduction of recursive procedures. But there is a second, more theoretical aspect which shows that recursive procedures are really inevitable in obtaining a reasonable theory of programming over arbitrary algebras. The above-mentioned universality of WHILE programs over the algebra of natural numbers does not carry over to arbitrary algebras, not even if we restrict the discussion to the computable ones treated in Chapter 13. There are simple algebras, for example the algebra of binary trees with empty tree nil and constructor function cons, with the property that every total WHILE program (that is, program that terminates for every input) is equivalent to a loop-free program, and thus to a program of a rather simple form. It is the introduction of recursive procedures (of course with formal parameters for input and output) that, together with other trivial matter, yields a programming language able to implement over an arbitrary algebra $A$ every function which is computable over $A$ (see Kfoury and Urzyczyn (1985). Note that we have defined computability only over algebras of natural numbers; for a generalization of this notion see also Kfoury and Urzyczyn (1985).

The trivial matter is the presence of a **counting facility**, that is, $A$ must be an algebra with counters. Most of the standard data structures used in computer science are equipped with such a counter structure. The simplest way of providing a counting structure explicitly is, of course, to build into the considered algebra $A$ the algebra $N$ as a part. This will lead us to the introduction of declarations of simple built-in data types like BOOLEAN, INTEGER and CARDINAL. Because of its usefulness in practical applications, we also introduce the possiblity of declaring RECORD and ARRAY data types. We will not provide more sophisticated types like POINTER types, nor will we introduce user-defined abstract data types. The latter are introduced in the following chapter on modules.

## 17.1  Syntax and semantics of recursive procedures

Let us first give an outline of the proposed procedure concept. Our intention will not be to discuss procedures completely (an enormous task), but rather to restrict ourselves to as simple a procedure concept as possible (thus simplifying verification), which is nonetheless sufficiently expressive for applications (both with respect to modules in Chapter 18 and to practical applications). The reader interested in a more comprehensive treatment of procedures should consult Olderog (1981).

What is part of our procedure concept and what is omitted?

- A **declaration environment** consists of a collection of procedures with mutual calls within the procedure bodies allowed. In particular, **recursive calls** are admitted.

- We do not use **nested procedure declarations**.

- Procedures communicate with their environment via **formal parameters**. We use a strict distinction of formal parameters between **in parameters** and **out parameters**. The former serve the purpose of reading input values, the latter pass back results. Each procedure has exactly one out parameter. In parameters are not altered in the body of a procedure. By calling a procedure with actual in parameters, we ensure that any other variable occurring in the procedure body receives a value before its first usage. Thus, a procedure will define a partial function that assigns a value under its out parameter to each tuple of actual in parameters.

- The restriction to only one out parameter serves only to simplify the presentation of proofs, but all of our results carry over to procedures with several formal out parameters.

- All variables in the body of a procedure are treated as **local variables**; therefore there are no **global variables**.

- We do not use **procedures as parameters** of other procedures. (This affects the usefulness of procedures as a means of obtaining modularity. But since we will introduce modularity in the next chapter by an explicit module concept containing a parameter part, this restriction is not severe.)

- When calling a procedure $P(\text{in } X_1,\ldots,X_n; \text{out } Y)$ with expressions (terms) $t_1,\ldots,t_n$ as actual in values and a variable U as the variable receiving the output, we assume that U does not occur within $t_1,\ldots,t_n$.

**Definition 17.1:**  We use an infinite set of procedure names. Usually, we will use words of the English language as mnemotechnical names for procedures. A **procedure head** over a many-sorted signature $(S,\Sigma)$ looks like this:

PROCEDURE $P(\text{in } X_1{:}s_1,\ldots,X_n{:}s_n; \text{out } Y{:}s)$

Here, P is a procedure name, $n \geq 0$, $s_1,\ldots,s_n,s$ are sorts from S, $X_1,\ldots,X_n,Y$ are different variables of sort $s_1,\ldots,s_n,s$ respectively. $X_1,\ldots,X_n$ are the **formal in parameters** of the considered procedure head, Y is its **formal out parameter**. The programming language of WHILE programs over $(S,\Sigma)$ is extended to the language of **recursive programs** over $(S,\Sigma)$ by introduction of a new instruction, called **procedure call** over $(S,\Sigma)$, which looks as follows:

CALL $P(t_1,\ldots,t_n,U)$

with a procedure name P, $(S,\Sigma)$-terms $t_1,\ldots,t_n$ and a variable U not occurring in $t_1,\ldots,t_n$. There is a further instruction that is used only in the definition of the semantics of a procedure call to deallocate auxiliary storage space that is no longer used. It is written

dispose(Z)

and its effect is to release variable Z from the used storage space.

Before we introduce the semantics of recursive programs we need a technical concept concerning the role of variables occurring in a recursive program as instantiated and computed variables respectively. We introduce thus the notions:

- $\alpha$ **computes on a set of variables** V
- **computed**$(\alpha,V)=V'$, for sets of variables V and V'

where $\alpha$ is a recursive program and V a set of some variables occurring in $\alpha$.

The first statement expresses that $\alpha$ may be executed provided that the variables in V are supplied with values, and the second tells us which variables possess values after execution of $\alpha$. The definition of these concepts orients itself at the syntax of the given program $\alpha$, and does not refer to its execution. Thus, these definitions are of a purely static nature, leading to decidable notions. They are more strict than the corresponding dynamic concepts would be, but are nevertheless sufficient for applications. To illustrate what is meant by a 'static definition' let us look at two examples:

> computed(IF test THEN Y:=X ELSE Z:=X END,{X}) contains neither Y nor Z, unless Y and Z coincide, since we cannot independently from the value of X and the underlying algebra conclude that Y is instantiate with a value after execution of the IF statement, nor can we conclude this for Z. Likewise, computed(WHILE test DO Y:=X END,{X}) does not contain Y, since it might be the case that the loop body is not executed.

**Definition 17.2:**    Let $\alpha$ be a recursive program and V be a subset of the set of all variables occurring in $\alpha$. We introduce the the concepts $\alpha$ **computes on** V and **computed**$(\alpha,V)=V'$ as follows (by induction on the structure of $\alpha$):

- X:=t computes on V iff V contains all variables occurring in t.
- computed(X:=t,V)=$V\cup\{X\}$

- CALL $P(t_1,\ldots,t_n,U)$ computes on V iff V contains all variables occurring in $t_1,\ldots,t_n$.
- computed(CALL $P(t_1,\ldots,t_n,U),V)=V\cup\{U\}$

- dispose(Z) computes on arbitrary set V.
- computed(dispose(Z),V)=$V-\{Z\}$

- IF B THEN $\beta$ ELSE $\gamma$ END computes on V iff V contains all variables occurring in B, and $\beta$ and $\gamma$ compute on V.
- computed(IF B THEN $\beta$ ELSE $\gamma$ END,V)
  $\qquad$ =computed$(\beta,V)\cap$computed$(\gamma,V)$

- WHILE B DO $\beta$ END computes on V iff V contains all variables occurring in B, and $\beta$ computes on V.
- computed(WHILE B DO $\beta$ END,V)=V

- $A_1;A_2;...;A_n$ computes on V (for n>1) iff $A_1$ computes on V and $A_2;...;A_n$ computes on computed($A_1$,V).
- computed($A_1;A_2;...;A_n$,V)
    =computed($A_2;...;A_n$,computed($A_1$,V))

**Definition 17.3:**     A **procedure declaration** over $(S,\Sigma)$ looks as follows:

PROCEDURE P(in $X_1:s_1,...,X_n:s_n$;out Y:s)
BEGIN
    $\alpha$
END

PROCEDURE P(in $X_1:s_1,...,X_n:s_n$;out Y:s) is called its **head**, $\alpha$ is called its **body**. It is assumed that

- $\alpha$ is a recursive program over $(S,\Sigma)$ (without dispose calls)
- $\alpha$ computes on $\{X_1,...,X_n\}$
- $Y \in$ computed($\alpha,\{X_1,...,X_n\}$)
- $\alpha$ does not contain instructions $X_i:=t$ or CALL $Q(t_1,...,t_m,X_i)$, that is, it does not alter in parameters.

A **procedure environment** $\Delta$ over $(S,\Sigma)$ is a finite set of procedure declarations of procedures with different procedure names and the following property:

- If instruction CALL $P(t_1,...,t_n,U)$ occurs in the body of one of the procedures in $\Delta$, then $\Delta$ contains a procedure declaration with head PROCEDURE P(in $X_1:s_1,...,X_n:s_n$;out Y:s) such that $t_1,...,t_n,U$ are of sort $s_1,...,s_n,s$, respectively.

The simplest way to realize that $\alpha$ computes on $X_1,...,X_n$ and Y is an element of set computed($\alpha,\{X_1,...,X_n\}$) is to ensure that $\alpha$ contains a prefix consisting of assignments $Z:=t(X_1,...,X_n)$, one for each variable Z that occurs in $\alpha$ and is different from $X_1,...,X_n$. Usually, our procedures indeed look like this.

Next we want to define the semantics of recursive programs over a given algebra $A$ with respect to a procedure environment $\Delta$. Here, the presence of recursive calls prevents a simple input-output semantics as defined for the WHILE programs. One way of defining the semantics of recursive programs, the most elegant one from a mathematical point of view, is to use complete partial orders, continuous functionals and least fixed points. We will not follow this possibility (the interested reader can read Loeckx and Sieber (1987).

An alternative is to define the semantics of recursive programs via an interpreter, which executes recursive programs over a given state step by step, thereby using a stack for the implementation of procedure calls. This alternative way, which might be better grasped for a reader aware of the implementation of higher programming languages, and which has been already introduced for WHILE programs in Chapter 14, is adopted here. It is primarily used to justify a proof rule (that is, to show its correctness and completeness) for recursive procedures. There will be a further theoretical advantage in the use of stack semantics, namely the possibility of defining and expressing weakest liberal preconditions and strongest postconditions over an arithmetical algebra along the lines of Section 16.3. For reasons which will soon become apparent, we cannot define states as in the case of WHILE programs as functions from the set of all variables into the domain of an algebra. What we need here, are *finite* functions. Of course, this is a modification of only minor importance, as both concepts are equivalent for WHILE programs.

**Definition 17.4:**   Let $\Delta$ be a procedure environment over signature $(S,\Sigma)$, $\alpha$ be a recursive program with procedure calls of procedures from $\Delta$, and $A$ be an $(S,\Sigma)$-algebra. A **state** sta of $\alpha$ over $A$ is a function such that

- dom(sta) (the domain of function sta) is a *finite* set of variables
- sta(X) is an element of $A_s$ if $X \in$ dom(sta) is a variable of sort s
- $\alpha$ computes on dom(sta)

For a state sta over $A$, a variable Z of sort s and an element b of $A_s$, define sta(Z/b) to be the state sta' with sta'(Z0=b and sta'(X)=sta(X), for all $X \in$ dom(sta)-{Z}.

Note that semantical concepts like $val_{A,sta}(t)$ and $A \models_{sta} \varphi$ can be introduced in a natural manner to the case of finite functions sta as states, provided that dom(t) contains all variables occurring in term t and all variables occurring free in formula $\varphi$.

**Definition 17.5:**    Let $\Delta$ be a procedure environment over signature $(S,\Sigma)$. We define an **interpreter function** $I_{A,\Delta}$ operating on pairs $(\alpha,sta)$, for a recursive program $\alpha$ with procedure calls from $\Delta$ and a state sta of $\alpha$ over $A$. Note that all the statements below are indeed defined due to the fact that sta is a state of $\alpha$. Also note that, for $I_{A,\Delta}(\alpha,sta)=(\alpha',sta')$, sta' is again a state of $\alpha'$.

- $I_{A,\Delta}(\varepsilon,sta)=(\varepsilon,sta)$

- $I_{A,\Delta}(X:=t;A_2;...;A_n,sta)=(A_2;...;A_n,sta(X/val_{A,sta}(t)))$

- $I_{A,\Delta}(\text{IF } B \text{ THEN } \beta \text{ ELSE } \gamma \text{ END};A_2;...;A_n,sta)=$

$$\begin{cases} (\beta;A_2;...;A_n,sta) \text{ if } A \models_{sta} B \\ (\gamma;A_2;...;A_n,sta) \text{ if } A \not\models_{sta} B \end{cases}$$

- $I_{A,\Delta}(\text{WHILE } B \text{ DO } \beta \text{ END};A_2;...;A_n,sta)=$

$$\begin{cases} (\beta;\text{WHILE } B \text{ DO } \beta \text{ END};A_2;...;A_n,sta) \text{ if } A \models_{sta} B \\ (A_2;...;A_n,sta) \hspace{3.2cm} \text{if } A \not\models_{sta} B \end{cases}$$

- $I_{A,\Delta}(\text{CALL } P(t_1,...,t_n,U);A_2;...;A_m,sta)=$

$(X_1':=t_1;...; X_n':=t_n; \beta'; U:=Y';$

$...\text{dispose}(Z');...;A_2;...;A_m,sta)$

provided that

(1)   $\Delta$ contains a procedure declaration PROCEDURE P(in $X_1:s_1,..., X_n:s_n;$out $Y:s_{out}$) BEGIN $\beta$ END.

(2)   Dashing indicates a renaming of variables in such a way that dashed variables does not occur in $t_1,...,t_n,U,A_2;...;A_m,$dom(sta).

(3)   For every used dashed variable Z an instruction dispose(Z') occurs in the right-hand side of the clause above.

- $I_{A,\Delta}(\text{dispose}(Z);A_2;...;A_m,sta)=(A_2;...;A_m,sta')$

where sta' is the restriction of sta to dom(sta)-$\{Z\}$.

Note that the concept of an activation stack that implements procedure calls is incorporated into the sequence of statements and their execution from left to right. Also note that according to the interpretation of all variables in a procedure as local variables, a call of $P(t_1,...,t_n,U)$ leads to an allocation of local storage space for the execution of the body of P which is deallocated at the end of execution of the procedure body (see Figure 17.1).

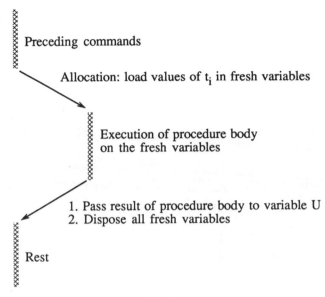

**Figure 17.1**

**Definition 17.6:** Let $\Delta$ be a procedure environment over signature $(S,\Sigma)$ and sta be a state of recursive program $\alpha$ over algebra $A$. We define $sta[\alpha]_{A,\Delta}sta'$ iff iterated application of $I_{A,\Delta}$ transforms pair $(\alpha,sta)$ into $(\varepsilon,sta')$.

Imagine the semantics of WHILE programs had been defined using finite states. Then, the input-output semantics $[\alpha]_A$ defined for WHILE programs is related to our interpreter $I_{A,\Delta}$ according to the following obvious result.

*Corollary*

Let $\Delta$ be the empty procedure environment, $\alpha$ be a WHILE program over signature $(S,\Sigma)$ and $A$ be an $(S,\Sigma)$-algebra. Then the following are true:

(1)    If $sta[\alpha]_A sta'$, then there is a number t such that t applications of interpreter function $I_{A,\Delta}$ lead from $(\alpha,sta)$ to $(\varepsilon,sta')$.

(2)    If there does not exist a state sta' such that $sta[\alpha]_A sta'$, then iterated application of $I_{A,\Delta}$ does not lead from $(\alpha,sta)$ to a pair of the form $(\varepsilon,sta')$.

We have not introduced a generalization of input-output semantics here. An attempt to do so might look as follows. Let $\Delta$ be a procedure environment over a signature $\Sigma$ and $A$ be an $(S,\Sigma)$-algebra. If $\Delta$ contains the procedure declaration

$$\text{PROCEDURE } P(\text{in } X_1{:}s_1,\ldots,X_n{:}s_n;\text{out } Y{:}s)$$
$$\text{BEGIN}$$
$$\beta$$
$$\text{END},$$

then we define

$$sta[\text{CALL } P(t_1,\ldots,t_n,U)]_A\, sta' \text{ iff states } sta_1 \text{ and } sta_2 \text{ exist with}$$

- $sta_1 = \{X_1/val_{A,sta}(t_1),\ldots,X_n/val_{A,sta}(t_n)\}$
- $sta_1[\beta]_A\, sta_2$
- $sta' = sta(U/sta_2(Y))$

For the remaining sorts of instructions, we carry over the definitions for WHILE programs from Chapter 14. But note that the definition for recursive programs presented here is incomplete. The reason is that it is not a simple inductive definition as in the case of WHILE programs, with the obvious guarantee to be well-founded. Our reduction of $sta[\text{CALL } P(t_1,\ldots,t_n,U)]_A\, sta'$ to $sta_1[\beta]_A\, sta_2$ may lead to an infinite regress. This is the point where we would need complete partial orders, continuous functionals and least fixed points to turn the above definition into a meaningful one. Nevertheless, we can give an analogy of input-output semantics in terms of our interpreter. This reads as follows:

If $sta_1 = \{X_1/val_{A,sta}(t_1),\ldots,X_n/val_{A,sta}(t_n)\}$ and iterated application of $I_{A,\Delta}$ leads from $(\beta,sta_1)$ to $(\varepsilon,sta_2)$, then iterated application of $I_{A,\Delta}$ leads from pair $(\text{CALL } P(t_1,\ldots,t_n,U),sta)$ to pair $(\varepsilon,sta(U/sta_2(Y)))$.

This is obvious from the definition of the way $I_{A,\Delta}$ interprets a procedure call: note that in order to obtain this result, the deallocation of no longer used local storage space is essential. Finally, we define a termination predicate as for WHILE programs and establish the corresponding result on uniform termination.

**Definition 17.7 (Termination predicate with time bound):** Let $\Delta$ be a procedure environment, $\alpha$ be an $(S,\Sigma)$-program with variables $X_1,\ldots,X_k$ without dispose calls, and $T \in \mathbb{N}$. Then we define an $(S,\Sigma)$-formula $term_{\Delta,\alpha,T}(X_1,\ldots,X_k)$ by supplementing the definition of the corresponding formula $term_{\alpha,T}(X_1,...,X_k)$ for WHILE programs by the following clause:

- $term_{\Delta,\ CALL\ P(t_1,...,t_n,X_i);A_2;...;A_m,T+1}(X_1,...,X_k)$ is the formula
  $term_{\Delta,\ \beta,T}(X_1,...,X_k, X_1,X_1,...,X_1)$

where $\beta$ is obtained in the following way. Assume that the procedure head of P is PROCEDURE P(in $U_1:s_1,...,U_n:s_n$;out V:s), its body is $\gamma$, and the variables occurring in P are $U_1,...,U_n,V,W_1,...,W_p$. Then $\beta$ is the program

$$U_1':=t_1;...;U_n':=t_n; \gamma'; X_i :=V'; A_2;...;A_m$$

with fresh variables $U_1',...,U_n',V', W_1',...,W_p'$ and a renamed version $\gamma'$ of $\gamma$.


In Definition 17.7 we disregard dispose instructions since these have no effect on termination. Furthermore, variables $U_1',...,U_n',V',W_1',...,W_p'$ are initialized by the value of $X_1$; in fact, they could be initialized in an arbitrary manner, since by the semantics definition for procedure calls these variables are assigned values in program $\beta$. So we could have defined $term_{\Delta,\alpha,T+1}(X_1,...,X_k)$ alternatively as

$$\exists U_1'...\exists U_n'\exists V' \exists W_1'...\exists W_p'$$
$$term_{\Delta,\ \beta,T}(X_1,...,X_k,U_1',...,U_n',V',W_1',...,W_p')$$


### Theorem 17.1 (Semantics of the termination predicate)

Let $A$ be an $(S,\Sigma)$-algebra and $\alpha$ be an $(S,\Sigma)$-program. Let sta be a state over $A$. Then there exists a state sta' with $sta[\alpha]_A sta'$ iff there exists $T \in \mathbb{N}$ such that $A \models_{sta} term_{\Delta,\alpha,T}$.

---

**Exercise 17.1**   Prove Theorem 17.1.

---

An important consequence of Theorem 17.1 is the following corollary that is shown as the corresponding one for WHILE programs.

*Corollary (Programs with uniform termination time)*

Let $\Delta$ be a procedure environment, $\alpha$ be a recursive $(S,\Sigma)$-program (without dispose calls) and Ax be a set of predicate logic formulas. Assume that, for every model $A$ of Ax and every state sta over $A$, there is a number T (which may, a priori, depend on $A$ and state sta) such that T applications of interpreter function $I_{A,\Delta}$ transform $(\alpha,\text{sta})$ into a pair $(\varepsilon,\text{sta'})$. Then there is a fixed number T such that, for every model $A$ of Ax and every state sta over $A$, at most T applications of interpreter function $I_{A,\Delta}$ transform $(\alpha,\text{sta})$ into a pair $(\varepsilon,\text{sta'})$.

Finally, we carry over the concepts of partial correctness assertions and their validity in algebras.

**Definition 17.8:**    A **partial correctness assertion** (pca) over procedure environment $\Delta$ is now a formula $\{\varphi\}\alpha\{\psi\}$, with a recursive program $\alpha$ (without dispose calls) with procedure calls taken from $\Delta$ and assertions $\varphi$ and $\psi$. An algebra $A$ is a **model** of a partial correctness assertion $\{\varphi\}\alpha\{\psi\}$ w.r.t. $\Delta$ (denoted $A \models_\Delta \{\varphi\}\alpha\{\psi\}$) iff, for all states sta and sta' over $A$ the following holds:

If sta is a state of $\alpha$ over $A$, dom(sta) contains all free variables of $\varphi$ and $\psi$, $A \models_{\text{sta}} \varphi$ and sta$[\alpha]_{A,\Delta}$sta' then $A \models_{\text{sta'}} \psi$.

A partial correctness assertion $\{\varphi\}\alpha\{\psi\}$ **follows** from an axiom set Ax w.r.t. a procedure environment $\Delta$ (denoted Ax $\models_\Delta \{\varphi\}\alpha\{\psi\}$) iff, for every model $A$ of Ax, $A \models_\Delta \{\varphi\}\alpha\{\psi\}$.

# 17.2 Correctly asserted recursive programs

**Definition 17.9:**  An **asserted procedure head** $\pi$ looks as follows:

PROCEDURE   $\{\varphi(X_1,...,X_n)\}$

$\qquad\qquad\qquad$ P(in $X_1$:$s_1$,...,$X_n$:$s_n$;out Y:s)

$\qquad\qquad$ $\{\psi(X_1,...,X_n,Y)\}$

with assertions $\varphi(X_1,...,X_n)$ and $\psi(X_1,...,X_n,Y)$ containing at most $X_1,...,X_n$ resp. $X_1,...,X_n,Y$ as free variables.

An **asserted procedure environment** $\Omega$ consists of asserted procedure heads together with procedure bodies, that is, entities of the form:

PROCEDURE   $\{\varphi(X_1,...,X_n)\}$

$\qquad\qquad$ P(in $X_1$:$s_1$,...,$X_n$:$s_n$;out Y:s)

$\qquad$ $\{\psi(X_1,...,X_n,Y)\}$

BEGIN
$\quad$ $\beta$
END

with the same restrictions as in the definition of procedure environments. The semantical concepts introduced for procedure environments in Section 17.1 carry over to asserted procedure environments.

**Definition 17.10:**   Let $\pi$ be the asserted procedure head

PROCEDURE   $\{\varphi(X_1,...,X_n)\}$

$\qquad\qquad$ P(in $X_1$:$s_1$,...,$X_n$:$s_n$;out Y:s)

$\qquad$ $\{\psi(X_1,...,X_n,Y)\}$

Then the following rule is called the **rule associated with** $\pi$:

$$\frac{\begin{array}{c}\varphi^* \to \varphi^\wedge\{X_1/t_1,...,X_n/t_n\}\\ (\exists U \varphi^* \wedge \psi^\wedge\{X_1/t_1,...,X_n/t_n,Y/U\}) \to \psi^*\end{array}}{\{\varphi^*\}\ \text{CALL } P(t_1,...,t_n,U)\ \{\psi^*\}} \quad (\pi)$$

where $\varphi^\wedge$ and $\psi^\wedge$ are arbitrary renamed versions of $\varphi$ and $\psi$ such that $\varphi^\wedge\{X_1/t_1,...,X_n/t_n\}$ and $\psi^\wedge\{X_1/t_1,...,X_n/t_n,Y/U\}$ are admissible substitutions. By $HC_\Omega$ we denote Hoare's calculus for WHILE programs enriched by the procedure call rules for all asserted procedure heads occurring in $\Omega$.

**Remarks**

(1)    What is the intuitive meaning of the proof rule associated with an asserted procedure head as above? Assume, for the moment, that pre- and postconditions $\varphi(X_1,...,X_n)$ and $\psi(X_1,...,X_n,Y)$ indeed correctly describe the behaviour of procedure P. Also assume that we want to prove the partial correctness assertion $\{\varphi^*\}$ CALL $P(t_1,...,t_n,U)$ $\{\psi^*\}$.

How can such a proof be obtained? What we know when entering the procedure call is that $\varphi^*$ is true. Furthermore, we know that $\psi(X_1,\ldots,X_n,Y)$ (and thus $\psi^\wedge(X_1,\ldots,X_n,Y)$) defines the value Y computed by our procedure call from input values $X_1,\ldots,X_n$, provided that $\varphi(X_1,\ldots,X_n)$ (equivalently $\varphi^\wedge(X_1,\ldots,X_n)$) holds. If U does not occur in $t_1,\ldots,t_n$ then $\psi^\wedge\{X_1/t_1,\ldots,X_n/t_n,Y/U\}$ defines the computed output U for actual parameters $t_1,\ldots,t_n$, provided that $\varphi^\wedge\{X_1/t_1,\ldots,X_n/t_n\}$ holds. In order that this latter condition is fulfilled we should know that $\varphi^*\to\varphi^\wedge\{X_1/t_1,\ldots,X_n/t_n\}$ holds. This is just the first premise of the rule above. All free variables in $\varphi^*$ and $\psi^*$ besides U are left unchanged by the procedure call, so they denote the same value in $\varphi^*$ and $\psi^*$. So, the best that we know about these variables after execution of the call is $\exists U\varphi^*$. (Read this as: Once there was some U, namely the one before execution of the procedure call, with $\varphi^*$.)

(2)    In most applications, U will not occur free in $\varphi^*$. Then $\exists U\varphi^*$ is equivalent to $\varphi^*$ and we just use the unchanged information about the input state to obtain information about the output state. Hence, in this case the rule above simplifies to the following form (with renamings of bound variables via $\wedge$ as before):

$$\frac{\varphi^*\to\varphi^\wedge\{X_1/t_1,\ldots,X_n/t_n\} \qquad (\varphi^*\wedge\psi^\wedge\{X_1/t_1,\ldots,X_n/t_n,Y/U\})\to\psi^*}{\{\varphi^*\}\ CALL\ P(t_1,\ldots,t_n,U)\ \{\psi^*\}}\ (\pi)$$

**Definition 17.11:**    Let $\Omega$ be an asserted procedure environment and Ax be a set of formulas that is closed under logical conclusion. We say that $\Omega$ is a **correctly asserted procedure environment over Ax** iff for each asserted procedure in $\Omega$ of the form

PROCEDURE $\{\varphi(X_1,\ldots,X_n)\}$

P(in $X_1$:$s_1$,\ldots,$X_n$:$s_n$;out Y:s)

$\{\psi(X_1,\ldots,X_n,Y)\}$

BEGIN
$\beta$
END

the pca $\{\varphi(X_1,...,X_n)\}\beta\{\psi(X_1,...,X_n,Y)\}$ can be derived from Ax in $HC_\Omega$. The notion of **correctly asserted recursive programs over environment $\Omega$ and theory Ax** is defined in a similar way as for WHILE programs.

## Remarks

(1)    There is an important remark about this definition. Assume that the body $\beta$ of our procedure above contained a recursive call of P. What does it mean that the partial correctness assertion $\{\varphi(X_1,...,X_n)\}\beta$ $\{\psi(X_1,...,X_n,Y)\}$ can be derived by Hoare's rules together with the deduction rule associated with the procedure head of P? In particular, how do we cope with the problem that recursive calls of P in the body of P must be verified? The availability of the deduction rule associated with the procedure head gives the answer. Owing to the form of their premises, we implicitly assume that the intended behaviour of P, expressed by $\varphi(X_1,...,X_n)$ and $\psi(X_1,...,X_n,Y)$, is already established for the recursive calls within the body of P. Based on this information we then deduce the intended behaviour of the enclosing call of P, too. Induction on computation time (number of interpreter steps up to termination) will show that the procedure P indeed exhibits the behaviour described by pre- and postconditions $\varphi(X_1,...,X_n)$ and $\psi(X_1,...,X_n,Y)$. So there is no infinite regression. As an example, consider an asserted procedure of the following form:

    PROCEDURE $\{\varphi(X)\}$ P(in X; out Y) $\{\psi(X,Y)\}$
    BEGIN
        IF test(X) THEN $\beta$ ELSE CALL P(E,U)
        END
    END

with a WHILE program $\beta$. In verifying the partial correctness assertion

    $\{\varphi(X)\}$ IF test(X) THEN $\beta$ ELSE CALL P(E,U) END $\{\psi(X,Y)\}$

we must prove $\{\varphi(X), test(X)\}$ $\beta$ $\{\psi(X,Y)\}$ via application of the usual Hoare calculus (the rule associated with the asserted procedure head of P is not applicable, as $\beta$ does not contain a call of P). Then we must verify $\{\varphi(X), \neg test(X)\}$ CALL P(E,U) $\{\psi(X,Y)\}$ via the associated deduction rule, that is, with the implicit hypothesis that CALL P(E,U) exhibits the intended behaviour.

(2)    Note what it means that $\Omega$ is a correctly asserted procedure environment w.r.t Ax, in the special case that the bodies of the procedures in $\Omega$ do not contain procedure calls. Then, derivation of

$\{\varphi(X_1,...,X_n)\}\beta\{\psi(X_1,...,X_n,Y)\}$, for an element of $\Omega$ of the form

PROCEDURE   $\{\varphi(X_1,...,X_n)\}$

$\qquad\qquad\qquad P(\text{in } X_1{:}s_1,...,X_n{:}s_n;\text{out } Y{:}s)$

$\qquad\qquad\qquad \{\psi(X_1,...,X_n,Y)\}$

$\qquad$ BEGIN

$\qquad\qquad \beta$

$\qquad$ END

cannot use the rules associated with the asserted procedure heads. Thus, we may look at this situation as if we had simply established a certain behaviour of a procedure body $\beta$ by deriving the partial correctness assertion $\{\varphi(X_1,...,X_n)\}\beta\{\psi(X_1,...,X_n,Y)\}$, and then passed this information to a potential user of the procedure by asserting

PROCEDURE   $\{\varphi(X_1,...,X_n)\}$

$\qquad\qquad\qquad P(\text{in } X_1{:}s_1,...,X_n{:}s_n;\text{out } Y{:}s)$

$\qquad\qquad\qquad \{\psi(X_1,...,X_n,Y)\}$

Before we proceed with the theoretical investigations concerning soundness and completeness of the proposed calculus, let us give an example.

## EXAMPLE 17.1 MacCarthy's 91-function ⎯⎯⎯⎯⎯⎯⎯⎯

We lay down the algebra *Cardinal*.

PROCEDURE
$\qquad$ {true}
$\qquad\qquad$ MacCarthy(in X:nat ; out Y:nat)
$\qquad$ {X>100→Y=X-10, X≤100→Y=91}
$\qquad\qquad$ BEGIN
$\qquad\qquad$ {true}
$\qquad\qquad\qquad$ IF X>100 THEN
$\qquad\qquad\qquad\qquad$ {X>100}
$\qquad\qquad\qquad\qquad$ Y:=X-10
$\qquad\qquad\qquad\qquad$ {X>100→Y=X-10, X≤100→Y=91} ☞ (a)
$\qquad\qquad\qquad$ ELSE
$\qquad\qquad\qquad\qquad$ {X≤100}
$\qquad\qquad\qquad\qquad$ MacCarthy(X+11,U);
$\qquad\qquad\qquad\qquad$ {X≤100, X+11>100→U=X+11-10, X+11≤100→U=91}
$\qquad\qquad\qquad\qquad$ ☞ (b1), (b2)

MacCarthy(U,V);

{X≤100, X+11>100→U=X+11-10, X+11≤100→U=91,

U>100→V=U-10, U≤100→V=91} ☞ (c1), (c2)

Y:=V

{X>100→Y=X-10, X≤100→Y=91} ☞ (d)

END

END

{X>100→Y=X-10, X≤100→Y=91}

First note that the use of fresh variables U and V allows us to apply the simplified proof rule from remark (2) above. What has to be shown? Validity in *Cardinal* of the following assertions:

(a)    X>100→((X>100→X-10=X-10)∧(X≤100→X-10=91))

(b1)   X≤100→true

(b2)   (X≤100∧(X+11>100→U=X+11-10)∧(X+11≤100→U=91))

    →    (X≤100∧(X+11>100→U=X+11-10)∧(X+11≤100→U=91))

(c1)   (X≤100∧(X+11>100→U=X+11-10)∧(X+11≤100→U=91))→true

(c2)   (X≤100∧(X+11>100→U=X+11-10)∧(X+11≤100→U=91)∧

    (U>100→V=U-10)∧(U≤100→V=91))

    →    (X≤100∧(X+11>100→U=X+11-10)∧(X+11≤100→U=91)∧

    (U>100→V=U-10)∧(U≤100→V=91))

All these assertions are trivially valid in *Cardinal*.

(d)    (X≤100∧(X+11>100→U=X+11-10)∧(X+11≤100→U=91)∧

    (U>100→V=U-10)∧(U≤100→V=91))

    →    ((X>100→V=X-10)∧(X≤100→V=91))

It is point (d) where correctness of the procedure above must be shown. So let us show (d). We start with the following information:

(1)   X≤100

(2)   X+11>100→U=X+1

(3)   X+11≤100→U=91

(4)   U>100→V=U-10

(5)   U≤100→V=91

Because of (1), X>100→V=X-10 is trivially valid. Next we show V=91.

    *Case 1* X>89. Then X+11>100, and so by (2) U=X+1>90. By (1) we know that U≤101. So, if U>100 we know that U=101 and by (4) also V=u-10=101-10=91. If U≤100 we know by (5) that also V=91.

    *Case 2* X≤89. Then X+11≤100, so by (3) U=91, and by (5) V=91.

Note also that our proof does not contain information about termination of MacCarthy's procedure, which is the real difficulty with this procedure.

---

## 17.3    Soundness and completeness of the proof rule for recursive procedures

### Soundness Theorem 17.2

Let $\Omega$ be a correctly asserted procedure environment over a theory Ax. Then for every recursive program $\alpha$ over $\Omega$ and arbitrary assertions $\varphi*$ and $\psi*$, if $Ax \vdash_{HC(\Omega)} \{\varphi*\}\alpha\{\psi*\}$ then $Ax \models_{\Omega} \{\varphi*\}\alpha\{\psi*\}$.

*Proof* Assume that sta is a state of $\alpha$ over $A$ such that $\alpha$ computes on dom(sta), dom(sta) contains all free variables of $\varphi*$ and $\psi*$, and

(1)     $Ax \vdash_{HC(\Omega)} \{\varphi*\}\alpha\{\psi*\}$

(2)     $A \models Ax$

(3)     $A \models_{sta} \varphi*$

(4)     $I^t_{A,\Omega}(\alpha,sta)=(\varepsilon,sta_1)$

Since $\alpha$ does not contain dispose calls we know that dom($sta_1$) contains all free variables of $\psi*$. We show that $A \models_{sta_1} \psi*$. The proof is by induction on t.

We distinguish six cases depending on the form of $\alpha$. If $\alpha$ is empty, a compound statement consisting of more than one instruction, an assignment, test or loop instruction, then the proof proceeds word for word as for WHILE programs. (It is left as an exercise for the reader to convince him/herself that this latter claim is true.) We are left with the case of a procedure call. So assume that $\alpha$ is CALL $P(t_1,...,t_n,U)$ with a procedure P from $\Omega$ with body $\beta$ and asserted head $\{\varphi(X_1,...,X_n)\}$ P(in $X_1,...,X_n$; out Y) $\{\psi(X_1,...,X_n,Y)\}$. We know that:

(5)     U does not occur in $t_1,...,t_n$

We define $a_i=val_{A,sta}(t_i)$, for i=1,...,n. By definition of $I_{A,\Omega}$ there exists a number s<t, a state $sta_2$ and b$\in$ dom($A$) such that:

(6)    $I^s_{A,\Omega}(\beta,sta(X_1/a_1,...,X_n/a_n))=(\ \epsilon sta_2(X_1/a_1,...,X_n/a_n,Y/b))$

(7)    $sta_1=sta(U/b)$

(Here we have used the fact that the body of a procedure does not alter its in parameters.) Since $\Omega$ was assumed to be correctly asserted we know that:

(8)    $Ax\vdash_{\overline{HC(\Omega)}}\ \{\varphi(X_1,...,X_n)\}\beta\{\psi(X_1,...,X_n,Y)\}$

By definition of the proof rule associated with P this means that

(9)    $Ax\vdash (\varphi^*\rightarrow\varphi^\wedge\{X_1/t_1,...,X_n/t_n\}$

(10)   $Ax\vdash ((\exists U\varphi^*\wedge\psi^\wedge\{X_1/t_1,...,X_n/t_n,Y/U\})\rightarrow\psi^*)$

with renamed versions $\varphi^\wedge$ and $\psi^\wedge$ of $\varphi$ and $\psi$ such that all substitutions occurring in (9) and (10) are admissible. (This will be used in the following when applying the substitution lemma.) From (3), (9), the substitution lemma and validity of renamings we conclude:

(11)   $A\models_{sta(X_1/a_1,...,X_n/a_n)}\varphi(X_1,...,X_n)$

Now we use (8), (2), (11), (6) and induction hypothesis (note that s<t) to conclude:

(12)   $A\models_{sta_2(X_1/a_1,...,X_n/a_n,Y/b)}\psi(X_1,...,X_n,Y)$

From (12) we conclude:

(13)   $A\models_{sta(U/b)}\psi^\wedge\{X_1/t_1,...,X_n/t_n,Y/U\}$

To show (13) observe that, for i=1,...,n, $val_{A,sta(U/b)}(t_i)=val_{A,sta}(t_i)$ as U does not occur in $t_i$. Furthermore, $val_{A,sta(U/b)}(U)=b$. By the substitution lemma and validity of renaming of bound variables, (13) is equivalent to $A\models_{sta(U/b)(X_1/a_1,...,X_n/a_n,Y/b)}\psi(X_1,...,X_n,Y)$. Since $X_1,...,X_n,Y$ are the only free variables of $\psi(X_1,...,X_n,Y)$, this latter claim follows from (12). From (3) we conclude:

(14)   $A\models_{sta(U/b)}\exists U\varphi^*$

Using (10), (14), (13) and (2) we obtain $A\models_{sta(U/b)}\psi^*$, that is, (by (7)) the desired conclusion $A\models_{sta_1}\psi^*$.                           ∎

### Corollary

Let $\Omega$ be a correctly asserted procedure environment over a theory Ax.

Assume that $\{\varphi(X_1,\ldots,X_n)\}$ P(in $X_1,\ldots,X_n$; out Y) $\{\psi(X_1,\ldots,X_n,Y)\}$ occurs as procedure head of a procedure in $\Omega$. Then, for all terms $t_1,\ldots,t_n$ and variable U not occurring in $t_1,\ldots,t_n$:

$$Ax \models \{\varphi(t_1,\ldots,t_n)\} \text{ CALL } P(t_1,\ldots,t_n,U) \{\psi(t_1,\ldots,t_n,U)\}$$

In particular:

$$Ax \models \{\varphi(X_1,\ldots,X_n)\} \text{ CALL } P(X_1,\ldots,X_n,Y) \{\psi(X_1,\ldots,X_n,Y)\}$$

*Proof*  Apply the proof rule associated with P's head to derive the partial correctness assertions under consideration.  ∎

### Remarks

(1)  The corollary above is wrong if U appears in one of $t_1,\ldots,t_n$. To see this, consider the correctly asserted procedure environment

PROCEDURE {true} P(in X; out Y) {Y=X+1};
BEGIN Y:=X+1 END

and the partial correctness assertion {true} CALL P(X,X) {X=X+1}. This pca is not valid in algebra *Nat* though it is derivable from Th(*Nat*). This shows that the requirement that U does not occur in $t_1,\ldots,t_n$, for an instruction CALL $P(t_1,\ldots,t_n,U)$, is essential for the correctness of our proof rule.

(2)  The corollary is also wrong if we allow more free variables than $X_1,\ldots,X_n$ resp. $X_1,\ldots,X_n,Y$ in the pre- and post-condition of an asserted procedure head

$$\{\varphi(X_1,\ldots,X_n)\} \text{ P(in } X_1,\ldots,X_n; \text{ out Y) } \{\psi(X_1,\ldots,X_n,Y)\}.$$

To see this, consider the correctly asserted procedure environment

PROCEDURE {X=U} P(in X; out Y) {Y=U+1};
BEGIN Y:=X+1 END

and the invalid partial correctness assertion {X=U} CALL P(X,U) {U=U+1} that can be derived by an application of the proof rule associated with the head of P.

We now turn our attention to the question of completeness. Consider an asserted procedure environment $\Omega$ consisting of asserted procedures

PROCEDURE $\{\varphi(X_1,...,X_n)\}$

$\qquad$ P(in $X_1$:$s_1$,...,$X_n$:$s_n$;out Y:s)

$\qquad \{\psi(X_1,...,X_n,Y)\}$

BEGIN
$\quad \beta$
END

Assume that $\text{Ax} \models_\Omega \{\varphi(X_1,...,X_n)\}$ CALL $P(X_1,...,X_n,Y)$ $\{\psi(X_1,...,X_n,Y)\}$. Can we prove that $\Omega$ is a correctly asserted procedure environment, that is, can we show that $\{\varphi(X_1,...,X_n)\}\beta\{\psi(X_1,...,X_n,Y)\}$ can be derived from Ax in $\text{HC}_\Omega$? Furthermore, can we derive from Ax in $\text{HC}_\Omega$ every partial correctness assertion which follows from Ax w.r.t. $\Omega$? The answer to both questions is: this is generally impossible. This answer, which is discouraging at first sight, is, however, not as bad as it seems. The explanation is that our question was posed in a totally foolish way. To recognize this, imagine we had implemented the mergesort algorithm discussed in Chapter 8 by a procedure Mergesort. Consider the following asserted procedure head:

$\quad$ {R-L is odd}
$\qquad$ Mergesort(in A:array, L,R:nat; out B:array)
$\quad$ {B results from A by sorting between L and R}.

With a suitable axiomatization Ax of arrays of natural numbers, this partial correctness assertion certainly follows from Ax. But we cannot expect that $\Omega$ is a correctly asserted procedure environment, since the body of our procedure will probably contain a recursive call with array bounds L' and R' such that R'-L' is even. Since the proposed asserted procedure head does not contain information about actual parameters A,L,R for mergesort with even R-L, the deduction rule associated with our asserted procedure head is not applicable. The same argument shows us that we cannot always derive from Ax in $\text{HC}_\Omega$ a partial correctness assertion which follows from Ax w.r.t. $\Omega$. This latter failure is even more dramatically demonstrated by the precondition false and postcondition false. But now it becomes clear what went wrong with our question.

$\quad$ Since it might be the case that the recursive calls of P within the body of P may be called with every possible list of actual parameters, we must equip the procedure head of P with a precondition as weak as possible, and a postcondition as strong as possible, in order to be prepared for every possible call of P. The weakest precondition is readily chosen, namely true. The strongest postcondition then is chosen as 'the strongest postcondition w.r.t. true and the body of P'.

$\quad$ In the following we will restrict the discussion of the completeness of our proof rule for procedures to the case of a fixed arithmetical algebra $A$. Thus, our axiom system Ax consists of the first-order theory of $A$.

**Definition 17.12:**   Let $\Delta$ be a procedure environment, $A$ be an algebra, $\varphi$ and $\psi$ be assertions, and $\alpha$ be a recursive program.

(1)    The **strongest postcondition** $spc_{A,\Delta}(\varphi,\alpha)$ of $\alpha$ and $\varphi$ w.r.t. $A$ and $\Delta$ is the set of all states sta' such that there is a state sta with $A \models_{sta}\varphi$ and $sta[\alpha]_{A,\Delta}sta'$.

(2)    The **weakest liberal precondition** $wlp_{A,\Delta}(\alpha,\psi)$ of $\alpha$ and $\psi$ w.r.t. $A$ and $\Delta$ is the set of all states sta such that for all states sta', $sta[\alpha]_{A,\Delta}sta'$ implies $A \models_{sta'} \varphi$.

### Theorem 17.3 (Arithmetical algebras are expressive)

Every arithmetical algebra $A$ is expressive for our programming language, that is, every set $spc_{A,\Delta}(\varphi,\alpha)$ and every set $wlp_{A,\Delta}(\alpha,\psi)$ can be defined in $A$ by first-order formulas $\xi$ and $\zeta$:

$$spc_{A,\Delta}(\varphi,\alpha)=\{sta| A \models_{sta}\xi\}$$
$$wlp_{A,\Delta}(\alpha,\psi)=\{sta| A \models_{sta}\zeta\}$$

*Proof*   The proof uses an arithmetization of the interpreter function $I_{A,\Delta}$. Such an arithmetization is obtained after having fixed a suitable encoding of all the ingredients of our language of recursive programs by natural numbers. What must be encoded are variables, terms, statements, statement sequences and states. This extensive, but essentially simple task is left as an exercise for the reader.    ∎

### Completeness Theorem 17.4

Let $A$ be an algebra that is expressive for recursive programs, and $\Delta$ be a procedure environment. We equip every procedure

PROCEDURE P(in $X_1,...,X_n$; out Y);

BEGIN $\beta$ END

as follows with pre-condition true as $\varphi(X_1,...,X_n)$ and a post-condition $\psi(X_1,...,X_n,Y)$ that defines $spc_{A,\Delta}(true,$ CALL $P(X_1,...,X_n,Y))$ in algebra $A$. (Since sta $\in spc_{A,\Delta}(true,$ CALL $P(X_1,...,X_n,Y))$ depends only on $sta(X_1),...,sta(X_n)$ and sta(Y), we may assume that $\psi$ contains at most free variables $X_1,...,X_n,Y$.) Then, for every partial correctness assertion $\{\varphi^*\}\alpha\{\psi^*\}$ with a recursive program $\alpha$ over $\Delta$, if $A \models_{\Omega}\{\varphi^*\}\alpha\{\psi^*\}$ then $\text{Th}(A)\vdash_{\overline{HC(\Omega)}} \{\varphi^*\}\alpha\{\psi^*\}$. In particular, the resulting asserted procedure environment $\Omega$ is correctly asserted.

*Proof* We argue by induction on $\alpha$. If $\alpha$ is empty, a compound statement consisting of more than one instruction, an assignment, test or loop instruction, then the proof proceeds word for word as for WHILE programs. (Exercise: Convince yourself that this latter claim is true.) We are left with the case of a procedure call. So assume that $\alpha$ is CALL $P(t_1,...,t_n,U)$, with a procedure P from $\Delta$ with body $\beta$ and head $P(\text{in } X_1,...,X_n; \text{out } Y)$ and a variable U not occurring in terms $t_1,...,t_n$. Further assume that:

(1)    $A \models_{\Omega}\{\varphi^*\}$ CALL $P(t_1,...,t_n,U)$ $\{\psi^*\}$

It suffices to show that the premisses of the deduction rule with conclusion $\{\varphi^*\}$CALL $P(t_1,...,t_n,U)\{\psi^*\}$ are valid in $A$. So we must show:

(2)    $A \models (\varphi^*\rightarrow\text{true})$

(3)    $A \models ((\exists U\varphi^*\wedge\psi^\wedge\{X_1/t_1,...,X_n/t_n,Y/U\})\rightarrow\psi^*)$

where the renamed version $\psi^\wedge$ of $\psi$ is chosen in such a way that the substitution $\psi\{X_1/t_1,...,X_n/t_n,Y/U\}$ is admissible. (2) is trivial. To show (3) consider a state sta such that dom(sta) contains all free variables of $\exists U\varphi^*$, $\psi^\wedge\{X_1/t_1,...,X_n/t_n,Y/U\}$ and $\psi^*$. Assume that:

(4)    $A \models_{\text{sta}}\exists U\varphi^*$

(5)    $A \models_{\text{sta}}\psi^\wedge\{X_1/t_1,...,X_n/t_n,Y/U\}$

We have to show $A \models_{\text{sta}}\psi^*$. Using (4) we may choose some $u\in \text{dom}(A)$ with

(6)    $A \models_{\text{sta}(U/u)}\varphi^*$

We may assume that dom(sta) contains all of the variables occurring in $t_1,...,t_n$ (if not, extend sta in an arbitrary manner). Define $a_i=\text{val}_{A,\text{sta}}(t_i)$, for $i=1,...,n$, and $a=\text{sta}(U)$. By (5) and the substitution lemma we know that

$$A \models_{\text{sta}(X_1/a_1,...,X_n/a_n,Y/a)}\psi^\wedge(X_1,...,X_n,Y)$$

so also

$$A \models_{\text{sta}(X_1/a_1,...,X_n/a_n,Y/a)}\psi(X_1,...,X_n,Y)$$

By choice of $\psi$ and definition of spc we conclude that there exists a state sta' with

(7)     $sta'(X_1/a_1,...,X_n/a_n)[CALL\ P(X_1,...,X_n,Y)]_{A,\Delta}sta(X_1/a_1,...,X_n/a_n,Y/a)$

Note that we used the fact that in parameters $X_1,...,X_n$ are not changed by execution of CALL $P(X_1,...,X_n,Y)$. Since element a is the value computed by the procedure call CALL $P(X_1,...,X_n,Y)$ on initial state $sta'(X_1/a_1,...,X_n/a_n)$ and $a_i=val_{A,sta(U/u)}(t_i)$, for i=1,...,n, we conclude that also

(8)     $sta(U/u)[CALL\ P(t_1,...,t_n,U)]_{A,\Delta}sta(U/a)$

Since sta(U/a)=sta we conclude from (1) and (6) that $A \models_{sta}\psi^*$.     ∎

---

**Exercise 17.2** Extend the procedure concept by admitting several out parameters. Define a proof rule and give a soundness and completeness proof along the lines of Section 17.3.

**Exercise 17.3** Write and verify a recursive program calculating the Fibonacci numbers.

**Exercise 17.4** Write and verify a recursive program implementing quicksort.

**Exercise 17.5** The 8-Queens Problem is the task of placing 8 queens on a chessboard, so that none attacks any of the others. Write and verify a recursive program solving this problem.

---

## 17.4     Recursive programs with data type declarations

The purpose of this section is to equip recursive programs with data type declarations of simple basic data types BOOLEAN, INTEGER and CARDINAL, and to introduce the possibility of building in RECORD and ARRAY types over a finite index set from the types available in the underlying signature. Discussion will be restricted to these simple data types and type constructors. We thus omit discussion of REAL and POINTER types. The large field of user-defined data types is treated in Chapter 18 on modules. So, the full power of data abstraction is achieved in the next chapter. The discussion here serves only to provide us with some type declarations which are often used later. This frees us from introducing these standard data types by hand as user-defined data types.

We will keep the discussion as simple as possible and restrict data type declarations over the many-sorted signature (S,Σ) to the following cases:

(1)     BOOLEAN, INTEGER and CARDINAL

(2)     RECORD

> selector$_1$:s$_1$;
> selector$_2$:s$_2$;
>
> .
>
> .
>
> .
>
> selector$_n$:s$_n$

END

(3)     ARRAY [0..N] OF s

with sorts s,s$_1$,s$_2$,...,s$_n$ in S, different function symbols selector$_i$, for i=1,...,n, and a natural number N. In the case of an ARRAY declaration as above we assume the presence of a distinguished constant $\perp$ of sort s. In the case of a RECORD declaration as above we assume the presence of distinguished constants $\perp_1$,...,$\perp_n$ of sorts s$_1$,...,s$_n$.

The discussion of *nested type declarations*, as well as the use of other types as index types for arrays is left to the reader as a simple generalization.

**Definition 17.13 (Semantics of declarations):**

Assume that $\delta$ is a finite set of data type declarations over a signature $(S,\Sigma)$. With every $(S,\Sigma)$-algebra $A$ we associate an expanded algebra $A(\delta)$ which results from $A$ by expanding it by the canonical interpretations of the declared data types in $\delta$. The **canonical interpretations** of data type declarations BOOLEAN, INTEGER and CARDINAL are algebras *Boolean*, *Integer* or *Cardinal*. RECORD and ARRAY declarations are interpreted by appropriate *Record* and *Array* algebras over $A$ as discussed in Section 6.5.

Next, assume that $\alpha$ is a program over a signature $(S,\Sigma)$ with procedure calls from a procedure environment $\Delta$ and data type declarations from a finite set $\delta$ as described above. Assume that $\alpha$ uses the sorts and function symbols made available by the data type declarations in $\delta$. Furthermore, let Ax be a set of $(S,\Sigma)$-formulas and $\{\varphi\}\alpha\{\psi\}$ be a partial correctness assertion. Then we define that $\{\varphi\}\alpha\{\psi\}$ **follows** from Ax with respect to procedure environment $\Delta$ and set of data type declarations $\delta$ iff $A(\delta) \models_\Delta \{\varphi\}\alpha\{\psi\}$, for every model $A$ of Ax. We denote this property by $Ax \models_{\Delta,\delta} \{\varphi\}\alpha\{\psi\}$.

So, the concept of logical conclusion of a partial correctness assertion from an axiom set Ax is generalized to programs using declared data types by building into the models of Ax the canonical interpretations of the declarations in a fixed way. It should be clear how to generalize other semantical concepts in a similar manner.

Having defined what $\text{Ax} \models_{\Delta,\delta} \{\varphi\}\alpha\{\psi\}$ means, the question arises of how such a conclusion might be derived in a calculus, without explicitly referring to the semantical construction of $A(\delta)$, for a model $A$ of Ax. In other words, this is the question of adequate axiomatizations of the built-in data types.

Assume we had an axiom system $\text{Ax}(\delta)$ that fixes the canonical interpretations of the data types declared in $\delta$ up to isomorphy. Then the property

$$\text{Ax} \models_{\Delta,\delta} \{\varphi\}\alpha\{\psi\}$$

could be equivalently replaced by

$$\text{Ax} \cup \text{Ax}(\delta) \models_{\Delta} \{\varphi\}\alpha\{\psi\}.$$

This latter claim could then be attacked via methods discussed previously (conservative extensions, Hoare's calculus). As we already know from earlier sections (specification of abstract data types, categoricity, rigid specifications, and so on) such an axiom system does not always exist. For the different sorts of data type declarations we obtain the following picture.

## Axiomatizing RECORD and ARRAY type declarations

Here, the situation is as good as it can be. We know that the mapping that assigns to an algebra $A$ the algebra *Array* [0..N] *of A* or *Record* $sel_1:s_1,...,sel_n:s_n$ *of A* can be completely described by a categorical parametrized specification (see Chapter 12). Thus, such a specification admits as models exactly the algebras isomorphic to $A(\delta)$ for an algebra $A$.

## Axiomatizing BOOLEAN

We axiomatize BOOLEAN by the rigid specification discussed in Chapter 12. Thus, in any model, the structure of the term-generated submodel with {true, false} as domain is uniquely fixed up to isomorphism. We could even go one step further and extend this rigid specification by the axiom $\forall B(B=\text{true} \lor B=\text{false})$ obtaining a categorical specification of the standard model of BOOLEAN.

## Axiomatizing CARDINAL and INTEGER

As in the case of BOOLEAN we may axiomatize these data structures by the rigid specifications discussed in Chapter 12. In this way, we fix the structure of the term generated subalgebra of an arbitrary model of these specifications. But as we know from model theory, the proposed specifications are by no means categorical, or even ω-categorical. Even worse, there is no ω-categorical axiomatization at all. So, what could be proposed as a practically feasible way to deal with declarations of CARDINAL or INTEGER? We propose two answers.

- First, if a program (or later an interface of a module) is such that it either uses *only the term generated part* of a model of CARDINAL or INTEGER or depends only on this term generated part, then the proposed *rigid specifications* suffice as axiomatizations. It is to be clarified separately what it means that a situation depends only on the term generated part of a model of CARDINAL or INTEGER. In the case of using CARDINAL in a program, this is fulfilled, for example, if every variable Z of type CARDINAL is instantiated with a ground term of data type CARDINAL before its first usage.

- Second, if the former situation is not present we might hope that the use of *induction together with a rigid specification* suffices to treat most practical problems. Although we know from model theory that the resulting axiomatization is not categorical, we propose this second variant as the way to deal with data types CARDINAL and INTEGER. Note that, when a proof of a partial correctness assertion succeeds on the basis of a rigid specification together with induction, it also succeeds w.r.t. the standard model of the considered specification as the only admitted model.

# Chapter 18
# Verification of Modules

As a final step we enrich our programming language by a module concept. Support of modularity is nowadays recognized as the key to the applicability of verification methods to realistic software systems. In this chapter we shall be concerned with proving correctness of modules. Work on four fields of computer science converges in our approach:

- Our concept is guided by the module concept of imperative programming languages like Modula-2 or Ada. It allows the development of software parts independently of each other. Modular design is essential for practical program development.

- Modula-2-like modules are equipped with logical specifications of their import and export interfaces consisting of universal formulas (often equations are sufficient).

- Correctness of such modules is defined and expressed (under some weak assumptions) as a partial correctness assertion. Thus, Hoare logic will serve as a candidate calculus for a proof theory to verify modules of this form.

- Considering specifications not only as declarative descriptions of the behaviour of module interfaces but also as (more-or-less efficiently) executable term rewriting systems (provided necessary properties like confluence and the noetherian property are fulfilled), we may also view interface descriptions as **rapid prototypes** available and executable, and thus observable, at an early stage of software development.

Since syntax and semantics of modules require rather voluminous definitions, we start this chapter with an extended collection of examples that are discussed in an informal way. We suggest that the reader quickly reads through these examples in order to get a first impression of what modules are and how they operate. Then he or she can read the formal definitions of syntax and semantics of modules and, in parallel or afterwards, come back to the presented examples.

## 18.1    Motivation of the module concept

A first glance at modules offers the following view which recognizes modules as entities consisting of three parts (see Figure 18.1).

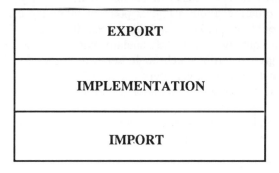

**Figure 18.1**

The **export interface** EXPORT consists of a collection of sorts and operations, together with an axiom system describing certain interrelations between these operations. EXPORT describes for a user of the module which operations are made available and how they are related to each other. Beyond the pure syntactical information about the functionality of the operation, which is the only information of module interfaces in current imperative programming languages like Modula-2 or Ada, our modules will contain semantical information, too. The restriction of visibility to the operations present in the export interface realizes thus the well-known software engineering principle of **information hiding**. Use of predicate logic as a very high-level description language for the semantics of the visible operations realizes what is known as **data abstraction**.

The **import interface** IMPORT looks just the same as the export interface. It tells the implementor of a module what data and operations are available when implementing the operations of the export interface, together with some information about the semantical behaviour of these operations. Stated differently, it expresses what the implementor may use as an implementation basis for the export interface; a subordinate implementor then has to take the

import interface just considered as the export interface of a subordinate module realizing it on the basis of an import interface of a still lower level, and so
on. Thus, by putting together modules we may realize a **hierarchical development** of large modular systems.

Finally, a module's IMPLEMENTATION part contains a procedural
implementation of the operations at the export interface on basis of the import
interface.

So far, we have not modelled a feature that is of enormous importance
with respect to the reusability of modules, namely **generic modules**. Imagine
a module whose purpose is to provide a sorting procedure for lists at its export
interface. It would be rather uneconomical to develop separate sorting modules, one for lists of integers, one for lists of characters, and so on. It is much
more convenient to develop a sorting module with an open parameter called
'linearly ordered domain' that can be instantiated by any concrete data type
with a linear ordering. From a logical point of view, the class of admitted instantiations could be perfectly well described by a 2-ary relation symbol <
(together with equality =) and the standard axioms for linear order. But then,
there are no ground terms which could serve as names for the available data.
Such **freely instantiable parameters** are modelled by a component called PA-
RAMETER which allows instantiation of arbitrary models of the axioms at
this part.

Thus, under a slightly more detailed view, a module presents itself as a
construct with four components (see Figure 18.2).

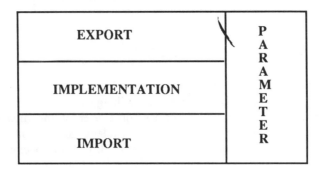

**Figure 18.2**

At the EXPORT and IMPORT interfaces and the IMPLEMENTATION
part (but not at the PARAMETER part) of a module we will freely use the
simple built-in data types INTEGER, CARDINAL, BOOLEAN, finite AR-
RAYS and RECORDS discussed in Chapter 17. In particular, the implementation of relations and equality will always use the data type BOOLEAN. These
built-in data types are always axiomatized by their canonical rigid or even categorical specifications. These specifications are then part of the module's interfaces where the corresponding data types are used. Since a module will use
only the parameter generated parts of the data types occurring at the IMPORT

interface and IMPLEMENTATION part, the intended data type is, in each case, indeed the only model (up to isomophism) of the used specification.

## EXAMPLE 18.1 A module implementing finite sets of data by arbitrary queues

Below we give a complete presentation of the interfaces and implementation part of a module implementing finite sets of data by queues of data.

PARAMETER consists of a universal specification of a data type *linearly ordered domain* with a sort *data*, distinguished constant ⊥ of type → *data* serving as a sort of error element, and a relation symbol < of type *data data*. Its axioms are the usual axioms of linear order:

$$\neg D<E, \; (D<E \wedge E<F) \rightarrow D<F, \; (D<E \vee D=E \vee E<D)$$

IMPORT makes available a parametrized data type *Queue(data)* with sort *queue*, constant [] of type → *queue* (the empty queue), functions [d|q] of type *data queue* → *queue* (appends a data element d at the left of a queue q), *first* of type *queue* → *data* (reads the rightmost element of a list), and *rest* of type *queue* → *queue* (removes the rightmost element of a list). We use the denotation for finite lists as introduced in Section 8.2. The specification below is a rigid specification of parametrized data type *Queue(data)*. (The reader should attempt to show this as an exercise using methods developed in Chapter 12. Why is the axiom ¬[D|L]=[] necessary to obtain a rigid specification?) The use of a rigid specification is particularly important since the semantics of a module will enforce that only the parameter-generated parts of an IMPORT algebra will be used. Hence, such a rigid specification uniquely fixes the desired parametrized data type *Queue(data)*.

¬[D|L]=[]

| | |
|---|---|
| first([])=⊥ | rest([])=[] |
| first([D| [] ]=D | rest([D| [] ]=[] |
| first([D‖[E|Q]])=first([E|Q]) | rest([D|[E|Q]])=[D| rest([E|Q])] |

EXPORT defines the data type (*Set(data)*,element,delete,max), with the usual interpretation of functions delete and max and element relation. It supplies the user with a sort *set*, a constant *emptyset* of type → *set*, functions *insert* and *delete* of type *set data* → *set*, relation *element* of type *data set*, and function *max* of type *set* → *data*. The specification is as follows:

insert(insert(M,D),E)=insert(insert(M,E),D)
insert(insert(M,D),D)=insert(M,D)

$\neg$insert(M,D)=emptyset

delete(emptyset,D)=emptyset
delete(insert(M,D),D)=delete(M,D)
$\neg$D=E$\rightarrow$delete(insert(M,D),E)=insert(delete(M,E),D)

$\neg$element(D,emptyset)
element(D,insert(M,E))$\leftrightarrow$(D=E$\vee$element(D,M))

max(emptyset)=$\perp$
max(insert(emptyset,D))=D
$\neg$D<E$\rightarrow$max(insert(insert(M,D),E)))=max(insert(M,D))
D<E$\rightarrow$max(insert(insert(M,D),E)))=max(insert(M,E))

The axioms at the EXPORT interface are **universal formulas**. The reader preferring equational specifications may replace them by equational axioms (this is a simple exercise). We use universal formulas here since they often read a little bit better than the corresponding equational specifications. Especially at the PARAMETER part, use of equational logic often leads to somewhat inconvenient formulations (try to transform the specification of linear ordering in the next example into equational form). It is easily shown that the only (up to isomorphism) parameter generated model over parameter algebra D of the axioms above is the standard model of finite subsets of D with the usual interpretation of function and relation symbols.

Now we come to the IMPLEMENTATION part of our example module realizing finite sets by queues. First, it contains a component REP laying down which import sorts (actually sorts of the IMPORT and PARAMETER part) are used to represent export sorts. In our example there is not much choice of how to represent sets: we define REP(set)=queue. Note that REP lays down the representation of export data by import data *only on the level of sorts*, that is, it only expresses how export sorts will be represented by import sorts in the procedures in the following program part. There is nothing said about the way single export objects are concretely represented by import objects. This concrete representation is implicitly contained in the procedures to be discussed next.

The IMPLEMENTATION part also contains a **PROC** part consisting of a procedure $P_f$, for every function symbol f of the export part that does not already appear in the PARAMETER part (these latter functions are simply carried over from the IMPORT to the EXPORT interface), with a procedure head being compatible with the functionality of the function symbol f and the representation function REP. So, if f is a function symbol of type $s_1 s_2 \ldots s_n \rightarrow s$, then the procedure head of $P_f$ is $P_f$(in $Y_1$:REP($s_1$);...; $Y_n$:REP($s_n$); out $Y_F$:REP(s)). We give a complete list of the procedures used.

```
PROCEDURE  P_emptyset(out Y_emptyset: queue);
BEGIN Y_emptyset:=[]
END

PROCEDURE  P_insert(in L: queue, D: data; out Y_insert: queue);
BEGIN Y_insert:=[D|L]
END

PROCEDURE  P_delete(in L: queue, D: data; out Y_delete: queue);
BEGIN
    Y_delete:=[] ; L':=L ;
    WHILE ¬L'=[] DO
        IF ¬first(L')=D THEN
        Y_delete:=[first(L')|Y_delete]
        END ;
        L':=rest(L')
    END
END

PROCEDURE  P_element(in D: data, L: queue ; out Y_element:boole);
BEGIN
    Y_element:=false ; L':=L;
    WHILE ¬L'=[] DO
        IF first(L')=D THEN Y_element:=true
        ELSE L':=rest(L')
        END
    END
END

PROCEDURE  P_max(in  L: queue ; out Y_max:data);
BEGIN
    IF L=[] THEN Y_max:=⊥
    ELSE  Y_max:=first(L) ;
          L':=L ;
          WHILE ¬L'=[] DO
              IF Y_max<first(L') THEN Y_max:=first(L')
              END;
              L':=rest(L')
          END
    END
END
```

Let us see now which IMPORT data occur as the computed values of these procedures, when called with parameter elements of an arbitrary model of the IMPORT specification. We observe that by calling $P_{emptyset}$ and then repeatedly $P_{insert}$ over a model $A$ of the IMPORT interface of our module, it is possible to *generate an arbitrary queue* of the form $[d_1,d_2,...,d_n]$, with data elements $d_1,d_2,...,d_n$ from the PARAMETER part of $A$. All queues $[d_1,d_2,...,d_n]$ thus appear as representations for finite sets.

The implementation part of our module is not finished yet. Already at this stage of the discussion it should be intuitively clear that the presented procedures of the implementation part, executed on a model of the import interface, do not lead to a data type realizing the required properties of the export interface. The explanation is simple: different queues may fairly well represent the same set, for example $[d,e,e,f]$ and $[d,f,d,f,e,e,e]$. More generally, any two queues with the same elements occurring in them, regardless of the order or multiple occurrences, represent the same set. This phenomenon, called **multiple representation** of export data by import data, should not be regarded as a defect of a module. Rather, it is a mechanism well suited for simplifying implementations. Nevertheless, we have to take multiple representation into account by identifying some import data. This happens in a third component of a module's implementation part, called the IDENTIFICATION part. It contains the following procedure expressing how two queues are to be identified when looking at them as sets:

```
PROCEDURE set-Equal(in L₁,L₂:queue; out B:boole);
BEGIN
    IF L₁=[] THEN
        IF L₂=[] THEN B:=true
        ELSE B:=false
        END
    ELSE
        CALL P_element(first(L₁),L₂,B₁);
        IF B₁=false THEN B:=false
        ELSE  CALL P_delete(first(L₁),L₁,M₁);
              CALL P_delete(first(L₁),L₂,M₂);
              CALL queue-Equal(M₁,M₂,B)
        END
    END
END
```

PROC and IMPLEMENTATION part together form a procedure environment over the signature of the IMPORT interface. In particular, arbitrary mutual calls of the occurring procedures are allowed.

## EXAMPLE 18.2 A module implementing finite sets of data by sorted queues————————————————————————————

The interfaces of our next module are the same as for the module in Example 18.1. Also, REP(set)=queue. But now we use only **sorted queues**, that is, queues $[d_1,d_2,...,d_n]$ with $d_1<d_2<...<d_n$ as representatives for finite sets. Where is this restriction to sorted queues laid down? Of course in the PROC part. The procedures are now as follows:

PROCEDURE $P_{emptyset}$(out $Y_{emptyset}$: queue);
BEGIN $Y_{emptyset}$:=[]
END


PROCEDURE $P_{insert}$(in L: queue, D: data; out $Y_{insert}$: queue);
BEGIN {*ordered insertion of D into queue L*}
    L':=L ; $Y_{insert}$:=[] ;
    WHILE ¬L'=[] ∧ D<first(L') DO
       $Y_{insert}$:=[first(L')|$Y_{insert}$] ; L':=rest(L')
    END ;
    IF L'=[] THEN $Y_{insert}$:=[D|$Y_{insert}$]
    ELSE
       IF first(L')<D THEN $Y_{insert}$:=[D|$Y_{insert}$]
       END ;
       WHILE ¬L'=[] DO
          $Y_{insert}$:=[first(L')|$Y_{insert}$] ; L':=rest(L')
       END
    END
END


PROCEDURE $P_{delete}$(in L: queue, D: data; out $Y_{delete}$: queue);
BEGIN
    L':=L; $Y_{delete}$:=[] ;
    WHILE ¬L'=[] ∧ D<first(L') DO
       $Y_{delete}$:=[first(L')|$Y_{delete}$]; L':=rest(L')
    END;
    IF ¬L'=[] ∧ D=first(L') THEN L':=rest(L')
    END ;

```
        WHILE ¬L'=[] DO Y_insert:=[first(L')|Y_insert]; L':=rest(L')
        END
    END

    PROCEDURE P_element(in D: data, L: queue ; out Y_element:boole);
    BEGIN
        WHILE ¬L'=[] ∧ D<first(L') DO L':=rest(L')
        END;
        IF L'=[] ∨ first(L')<D THEN Y_element:=false
        ELSE Y_element:=true
        END
    END

    PROCEDURE P_max(in L: queue ; out Y_max:data);
    BEGIN
        IF L=[] THEN Y_max:=⊥
        ELSE Y_max:=first(L)
        END
    END
```

Looking at these procedures we recognize that $P_{emptyset}$ outputs a sorted queue, and the remaining procedures with queue as output sort again pass back a sorted queue when called with a sorted queue as input. Thus, only sorted queues are **reachable** for a user of the module. As a side-effect, multiple representation of export objects by import objects does not occur. Thus, the IDENTIFICATION part does not contain any procedures. Also note that the functions realized by procedures $P_{emptyset}$, $P_{insert}$, and so on, fulfil the EXPORT equations only when restricted to sorted inputs. (Give an example of an equation that is not true for unsorted input.)

# EXAMPLE 18.3 A module implementing stacks by infinite arrays and a counter——————————————————————

Let us, as a final example, discuss an array implementation of a data type of stacks. Here, a stack is implemented by an infinite array with natural numbers as indices, together with a natural number pointing to the first position after the head of a stack (see Figure 18.3). Since we do not use a finite index set we cannot use the built-in data types ARRAY[0..L] OF data that were discussed in Chapter 17.

stack:                    $d_0 d_1 \ldots d_{n-1}$ with top element $d_{n-1}$

array with pointer ptr:

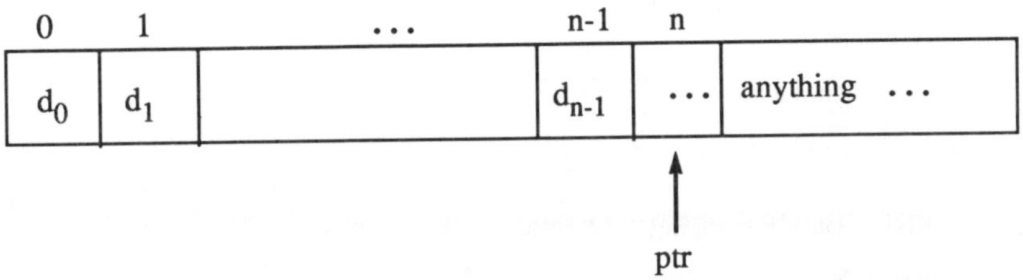

Figure 18.3

We do not specify what 'anything' in Figure 18.3 is, nor do we 'delete' any array element when simulating stack operation pop. This leads, as in Example 18.1, to multiple representation of EXPORT data by IMPORT data, which we must take into account within the module's IDENTIFICATION part. Two pairs (A,ptr) and (A',ptr') represent the same stack iff ptr=ptr' and A[i]=A'[i], for all i<ptr.

At the PARAMETER part we specify that we need a data type *data* with a distinguished constant $\bot$.

EXPORT contains the standard specification of data type *Stack(data)* with sort *stack* as introduced in Example 12.6 (here written as a parametrized specification with parameter *data* instead of the special algebra $N$ used in Example 12.6), and $\bot$ instead of 0).

IMPORT introduces a rigid specification *Array [nat] of data* (here we use the built-in data type CARDINAL) with sort *array*, constant *initarray* of type $\rightarrow$ *array*, and functions *assign* of type *array nat data* $\rightarrow$ *array* and *read* of type *array nat* $\rightarrow$ *data*. Its axioms are:

$$\text{assign(assign}(A,I,D),I,E) = \text{assign}(A,I,E)$$
$$\neg I = J \rightarrow \text{assign(assign}(A,I,D),J,E)) = \text{assign(assign}(A,J,E),I,D)$$

$$\text{read(initarray},I) = \bot$$
$$\text{read(assign}(A,I,D),I) = D$$
$$\neg I = J \rightarrow \text{read(assign}(A,I,D),J) = \text{read}(A,J)$$

$$A_1 = A_2 \leftrightarrow \forall I \ \text{read}(A_1,I) = \text{read}(A_2,I)$$

The reader should attempt to show that these axioms indeed are a rigid specification of *Array [nat] of data*.

Finally, we present the IMPLEMENTATION part:

REP(stack) is RECORD arr:array ; ptr: nat END
(using built-in data type constructor RECORD)

```
PROCEDURE P_emptystack(out Y_emptystack: REP(stack));
BEGIN
    Y_emptystack.arr:=initarray; Y_emptystack.ptr:=0
END

PROCEDURE P_push(in D:data, A:REP(stack); out Y_push: REP(stack));
BEGIN
    Y_push.arr:=assign(A.arr,A.ptr,D); Y_push.ptr:=A.ptr+1
END

PROCEDURE P_pop(in A:REP(stack); out Y_pop: REP(stack));
BEGIN
    Y_pop.arr:=A.arr; Y_pop.ptr:=A.ptr-1
END

PROCEDURE P_top(in A:REP(stack); out Y_top:data);
BEGIN
    IF A.ptr=0 THEN Y_top:=⊥
    ELSE Y_top:=read(A.arr,A.ptr-1)
    END
END

PROCEDURE stack-Equal(in A,B: REP(stack); out B: boole);
BEGIN
    IF A.ptr=B.ptr THEN B:=true
    ELSE B:=false
    END;
    I:=0;
    WHILE I<A.ptr ∧ B=true DO
        IF ¬read(A.arr,I)=read(B.arr,I) THEN B:=false
        END;
        I:=I+1
    END
END
```

A short look at the implementing procedures shows that indeed every pair (A,ptr) with a natural number ptr and a parameter generated array A, that is, array A with the property that at most a finite number of entries are different from the exceptional value $\perp$, are used for the representation of EXPORT data.

## 18.2    Formal definition of module syntax

In this section, a **specification** is a triple $(S,\Sigma,Ax)$, with a many-sorted signature $(S,\Sigma)$ and a set Ax of universal $(S,\Sigma)$-formulas. A triple $(S',\Sigma',Ax')$ is called an **extension** of a specification $(S,\Sigma,Ax)$ iff $(S\cup S',\Sigma\cup\Sigma',Ax\cup Ax')$ is again a specification and S and S', as well as $\Sigma$ and $\Sigma'$ are disjoint.

A **module** M is a 4-tuple

(PARAMETER,IMPORT,EXPORT, IMPLEMENTATION)

with the following properties:

- PARAMETER is a specification $(S_{par},\Sigma_{par},Ax_{par})$.
- EXPORT is an extension $(S_{exp},\Sigma_{exp},Ax_{exp})$ of $(S_{par},\Sigma_{par},Ax_{par})$.
- IMPORT is an extension $(S_{imp},\Sigma_{imp},Ax_{imp})$ of $(S_{par},\Sigma_{par},Ax_{par})$.
- IMPLEMENTATION consists of a **representation part** REP, a **program part** PROC and an **identification part** IDENTIFICATION.

  - REP is a function from $S_{par}\cup S_{imp}$ to $S_{par}\cup S_{imp}$ such that REP(s)=s for all $s\in S_{par}$.
  - PROC contains for each f in $\Sigma_{exp}$ of type $f{:}s_1s_2...s_n\rightarrow s$ a procedure $P_f$ over $(S_{par}\cup S_{imp},\Sigma_{par}\cup\Sigma_{imp})$ with head

    $P_f(\text{in } Y_1{:}REP(s_1);...;Y_n{:}REP(s_n); \text{ out } Y_f{:}REP(s))$.

    Likewise, PROC contains for each relation symbol r in $\Sigma_{exp}$ of type $r{:} s_1s_2...s_n$ a procedure $P_r$ over $(S_{par}\cup S_{imp},\Sigma_{par}\cup\Sigma_{imp})$ with head

    $P_r(\text{in } Y_1{:}REP(s_1);...;Y_n{:}REP(s_n); \text{ out } Y_r{:}boole)$.

  - IDENTIFICATION contains for *some* of the sorts s in $S_{exp}$ a procedure s-Equal over $(S_{par}\cup S_{imp},\Sigma_{par}\cup\Sigma_{imp})$ with head

    $s{\text -}Equal(\text{in } Y_1,Y_2{:}REP(s); \text{ out } B{:}boole)$.

PROC$\cup$IDENTIFICATION forms a procedure environment.

The definition is not yet complete. One thing that is missing is a **termination condition** for the procedures occurring in our module. In order to state this requirement, we need some more terminology:

- An $(S_{par} \cup S_{imp}, \Sigma_{par} \cup \Sigma_{imp})$-algebra is called an **import algebra**. A model of $Ax_{par} \cup Ax_{imp}$ is called an **import model**. An $(S_{par} \cup S_{imp}, \Sigma_{par} \cup \Sigma_{imp})$-term is called **an import term**. Likewise, we introduce the concepts of **export algebra, export model,** and **export term**.

- For an import or export algebra $A$, the elements of the domains $A_p$, for $p \in S_{par}$, are called **parameter objects** of $A$. An import or export algebra $A$ is called **parameter generated** iff all of its elements can be obtained from parameter objects of $A$ by iterated application of the functions of $A$. The **parameter generated subalgebra** of $A$ is the subalgebra of $A$ whose domains consist of the parameter-generated objects of $A$.

Using this terminology we may state the missing **termination condition**:

- For every import algebra $A$ and operation symbol f in $\Sigma_{exp}$ execute the corresponding procedure $P_f$ over $A$ obtaining a partial function $f_A$. Functionalities are related as follows:

$$f : s_1 s_2 \ldots s_n \to s$$
$$P_f(\text{in } Y_1 : REP(s_1); \ldots; Y_n : REP(s_n); \text{ out } Y_f : REP(s))$$
$$f_A : A_{REP(s_1)} \times \ldots \times A_{REP(s_n)} \to A_{REP(s)}$$

- The same is done for every relation symbol r in $\Sigma_{exp}$ resulting in a boolean valued function $r_A$:

$$r : s_1 s_2 \ldots s_n$$
$$P_r(\text{in } Y_1 : REP(s_1); \ldots; Y_n : REP(s_n); \text{ out } Y_r : \text{boole})$$
$$r_A : A_{REP(s_1)} \times \ldots \times A_{REP(s_n)} \to \{\text{true,false}\}$$

- Finally, every procedure of the identification part is executed over $A$ resulting in a boolean valued function s-Equal$_A$:

$$\text{s-Equal(in } Y_1, Y_2 : REP(s); \text{ out } B : \text{boole})$$
$$\text{s-Equal}_A : A_{REP(s)} \times A_{REP(s)} \to \{\text{true,false}\}$$

Now it is required that the restriction of the obtained partial functions $f_A$, $r_A$ and s-Equal$_A$ to the parameter generated subalgebra of $A$ are total functions.

## Remark

The termination condition deserves some comment. It was required that the interpretation of procedures over an import model leads to a function that is defined for all parameter generated inputs. Why is termination not required for arbitrary inputs? Simply, since such a strong requirement would be indeed too strong. As we proved in Chapter 17, a program a that terminates over every model A of a first-order axiom system Ax on every input from A terminates within a constant number c of interpreter steps uniformly over every model A of Ax and every input from A. Such a program can be unwound into an equivalent loop-free program. Contrary to this uninteresting 'trivial halting behaviour', procedures may well exhibit a non-uniform halting behaviour on the parameter generated subalgebra of an import algebra, owing to the availability of term representation as a means of counting down in a program.

## 18.3    Module semantics and correctness

Let M be a module. The meaning of M is an operation $\text{Sem}_M$ that assigns to each import model $A$ of M an export algebra $\text{Sem}_M(A)$. The construction of $\text{Sem}_M(A)$ proceeds in 3 steps.

### Step 1: Interpretation of procedures and information hiding

From $A$ we define an export algebra $B$ as follows:

- For $s \in S_{par} \cup S_{exp}$ we define

$$B_s = \{x \in A_{REP(s)} | \ x \text{ is parameter generated over } A \ \}$$

In particular, $B_s = A_s$ for $s \in S_{par}$.

- For every function symbol $f: s_1 s_2 \ldots s_n \to s$ and every relation symbol $r: s_1 s_2 \ldots s_n$ in $\Sigma_{par} \cup \Sigma_{exp}$ we define

$$f_B = \text{the restriction of } f_A \text{ to } B_{s_1} \times \ldots \times B_{s_n}$$

$$r_B = \{(a_1, \ldots, a_n) \in B_{s_1} \times \ldots \times B_{s_n} | \ r_A(a_1, \ldots, a_n) = \text{true}\}$$

In particular, for a function symbol $f: s_1 s_2 \ldots s_n \to s$ and a relation symbol $r: s_1 s_2 \ldots s_n$ in $\Sigma_{par}$ we know that $f_B = f_A$ and $r_B = r_A$.

**Step 2: Reachability**

From $B$ we define next an export algebra $C$ as follows:

- For a sort $s \in S_{par} \cup S_{exp}$ we define

    $reachable_s(A) = \{x \in B_s \mid x$ is a parameter-generated over $B\}$

    Thus, $reachable_s(A)$ consists of all elements of $A_{REP(s)}$ which can be generated from the parameters of $A$ by the procedures of the module's implementation part. In particular, $reachable_s(A) = A_s$ for $s \in S_{par}$.

- Then we define $C$ as the subalgebra of algebra $B$ with domains $C_s = reachable_s(A)$, for every sort $s \in S_{par} \cup S_{exp}$.

**Step 3: Identification**

From $C$ we define our desired export algebra $Sem_M(A)$ as follows:

- For every sort $s \in S_{exp}$ such that the identification part of M contains a procedure s-Equal(x,y;B), define a 2-ary relation $\sim_s$ on $C_s$ by

    $x \sim_s y$ iff $s\text{-}Equal_A(x,y) = true$

- For the remaining sorts $s \in S_{par} \cup S_{exp}$ we define $x \sim_s y$ iff $x = y$.

- Let $\sim$ be the family of relations $(\sim_s)_{s \in S_{par} \cup S_{exp}}$. If $\sim$ is a congruence relation on algebra $C$, we say that $Sem_M(A)$ **is defined**. In this case we define $Sem_M(A)$ to be the quotient algebra of $C$ modulo $\sim$.

**Definition 18.1:** A module M is called **correct** iff, for every import model $A$, the algebra $Sem_M(A)$ is defined and is an export model.

## 18.4     Expressing module correctness in Hoare logic

### 18.4.1     Outline of the construction

Let a module M be given. *Uniformly in M,* we construct a finite set $Correct_M$ consisting of partial correctness assertions in the import signature

$(S_{par} \cup S_{imp}, \Sigma_{par} \cup \Sigma_{imp})$ such that the following holds:

$$M \text{ is correct iff } Ax_{com} \cup Ax_{par} \cup Ax_{imp} \models \text{Correct}_M \qquad (18.1)$$

Note that emphasis is put on uniformity. The pure existence of $\text{Correct}_M$ is trivial. We construct $\text{Correct}_M$ in such a way that for all import models $A$ the following holds:

$$\text{Sem}_M(A) \text{ is defined and } \text{Sem}_M(A) \models Ax_{exp} \text{ iff } A \models \text{Correct}_M \qquad (18.2)$$

Looking at this latter equivalence we recognize that formula set $\text{Correct}_M$ must express two different properties of import models $A$ of M, namely:

(1)    Validity of all export axioms in $\text{Sem}_M(A)$.

(2)    The property that the relation $\sim$ resulting from the identification step in the definition of $\text{Sem}_M(A)$ indeed defines a congruence relation.

Concerning (2), we must, for example, express symmetry of $\sim_s$ for every sort $s \in S_{exp}$ such that the identification part of M contains an identifying procedure s-Equal(x,y;B). This means showing that for all $a_1, a_2 \in \text{reachable}_s(A)$, if s-Equal$_A(a_1, a_2)$=true then s-Equal$_A(a_2, a_1)$=true. This statement is to be expressed as a partial correctness assertion over $A$. The only difficulty in doing so is the *restriction of the universal quantification to reachable elements*.

Concerning (1), consider one of the axioms at the export interface, say a closed universal formula $\forall X_1 \dots \forall X_k \varphi$ from $Ax_{exp}$ with quantifier free formula $\varphi$. Then $\text{Sem}_M(A) \models \forall X_1 \dots \forall X_k \varphi(X_1, \dots, X_k)$ means:

$$\text{Sem}_M(A) \models \varphi(a_1, \dots, a_k), \text{ for all reachable elements } a_1, \dots, a_k \qquad (18.3)$$

Statement (18.3) is to be expressed as a partial correctness assertion over $A$. Again, we have to cope with the restriction to reachable elements.

This is in fact the only difficulty that must be overcome. Validity of the atomic subformulas of $\varphi$ is simply simulated by appropriate calls of the procedures that implement the functions and relations of the export interface. In particular, equations at the export interface are simulated by appropriate calls to the procedures from the identification part.

## 18.4.2    Reachability

We first discuss the restriction of the occurring universal quantifiers to reachable elements. This restriction will be realized by using partial correctness assertions of the form

$$\{true\} \text{ CALL Reachable}_{s_1}(Y_1) ; \ldots; \text{CALL Reachable}_{s_k}(Y_k) ; \alpha \{\psi\}$$

with variables $Y_1,\ldots,Y_k$ of sort $REP(s_1),\ldots,REP(s_k)$ (corresponding to variables $X_1,\ldots,X_k$ of sort $s_1,\ldots,s_k$ occurring in the considered formula) and calls of a procedure Reachable$_s$(Y) which terminates over an arbitrary import model $A$ exactly for the reachable elements of $A$ of sort s as actual parameters. Then, the semantics of partial correctness assertions fairly mirrors the restriction of universal quantifiers to reachable elements by termination of programs reachable(Y).

> **Definition 18.2:**   A module M **allows implementation of reachability** iff for every sort $s \in S_{exp}$ there exists a procedure with head Reachable$_s$(in Y:REP(s)) that terminates over every import model $A$ of M on exactly the elements of reachable$_s(A)$.

Unfortunately, it depends on the module's import interface whether such procedures reachable$_s$ exist. Since this problem is rather intricate we shall postpone its discussion until Subsection 18.4.5. There we will give sufficient conditions for a module M to allow implementation of reachability. In the last section of this chapter we will finally present an alternative attempt to cope with reachable elements that is better suited for practical verification of concrete modules.

## 18.4.3   Expressing the congruence property

*Lemma 18.1*
Given a module M that allows implementation of reachability, we may construct a finite set Congruence$_M$ of partial correctness assertions such that for every import model $A$ the following are equivalent:

(1)     $A \models$ Congruence$_M$
(2)     Sem$_M(A)$ is defined

*Proof*     Let reachable$_s$(Y) be the procedure implementing reachability, for $s \in S_{exp}$. We have to express reflexivity, symmetry, transitivity and compatability with export functions and relations of the relation $\sim$ on reachable elements defined in the identification step as partial correctness assertions. We treat two examples, leaving the remaining cases as simple exercises to the reader.

*Reflexivity*    For every sort s∈$S_{exp}$ with an identifying procedure s-Equal take the partial correctness assertion

{true} CALL Reachable$_s$(X); CALL s-Equal(X,X,B) {B=true}.

*Compatibility with a function* f:$s_1s_2p{\to}s$ in $\Sigma_{exp}$    We consider the case that s,$s_1$,$s_2\in S_{exp}$, p∈$S_{par}$, and assume that there exist identifying procedures for sorts s and $s_1$, but no such procedure for sort $s_2$. Take the following partial correctness assertion:

{true}
    CALL Reachable$_{s_1}$($X_1$); CALL Reachable$_{s_2}$($X_2$);

    CALL Reachable$_{s_1}$($Y_1$); CALL Reachable$_{s_2}$($Y_2$);

    CALL $s_1$-equal($X_1$,$Y_1$,$B_1$);

    CALL $P_f$($X_1$,$X_2$,D,U); CALL $P_f$($Y_1$,$Y_2$,E,V) ;

    CALL s-Equal(U,V,B)
{($B_1$=true ∧ $X_2$=$Y_2$ ∧ D=E) → B=true}.

## 18.4.4    Simulating export formulas

Instead of a formal presentation of the way in which validity of export formulas over $Sem_M(A)$ is reduced to validity of appropriate partial correctness assertions over $A$, we treat an example. Consider the following export formula φ:

f(c)=g(f(X)) ∧ ¬(r(h(D),X) ∨ q(f(X)) ∨ k(X)=X)

with c,f,g,k,r from $\Sigma_{exp}$, and h,q from $\Sigma_{par}$. Assume that the second equality symbol is equipped with a corresponding procedure Equal(X,Y,B) in the identification part of M, whereas the first one is not. Assume that X is a variable of sort from $S_{exp}$ and D is a variable of parameter sort. The translation of φ into a partial correctness assertion is as follows (variable Y corresponds to X):

---

*simulates*

---

{true}
    CALL Reachable(Y);                    *restriction of quantifier ∀X*

CALL $P_c(Y_c)$;                                     *evaluation of c*

CALL $P_f(Y_c, Y_{f(c)})$;                           *evaluation of f(c)*

CALL $P_f(Y, Y_{f(X)})$;                             *evaluation of f(X)*

CALL $P_g(Y_{f(X)}, Y_{g(f(X))})$;                   *evaluation of g(f(X))*

CALL $P_r(h(D), Y, Y_{r(h(D),X)})$;                  *validity of r(h(D),X)*

CALL $P_k(Y, Y_{k(X)})$;                             *evaluation of k(X)*

CALL $Equal(Y_{k(X)}, Y, Y_{k(X)=X})$                *validity of k(X)=X*

$\{Y_{f(c)} = Y_{g(f(X))} \wedge \neg (Y_{r(h(D),X)}) = \text{true} \vee q(Y_{f(X)}) \vee Y_{k(X)=X} = \text{true})\}$

---

Of course, we could have shifted the evaluation of $\varphi$ completely into the program of the above partial correctness assertion, coming up with a postcondition $Y_\varphi = \text{true}$. The strategy used above, shifting into the program part only what is absolutely necessary, is better suited for practical verification work, since Hoare logic works on the level of program structure.

On the basis of the example treated above the reader may readily prove the following lemma.

### Lemma 18.2

Given a module M allowing tthe implementation of reachability, we can construct for every closed formula $\forall X_1 ... \forall X_n \forall D_1 ... \forall D_m \varphi$ in $Ax_{exp}$ with quantifier free formula $\varphi$, parameter variables $D_1, ... , D_m$ and non-parameter variables $X_1, ... , X_n$ of sorts $s_1,...,s_n \in S_{exp}$ a partial correctness assertion $\text{correct}_M(\varphi)$ of the form

$\{\text{true}\}$ CALL $\text{Reachable}_{s_1}(Y_1)$ ;...; CALL $\text{Reachable}_{s_n}(Y_n)$ ; $\alpha$ $\{\psi\}$

such that for all import models $A$ the following are equivalent:

(1)     $\text{Sem}_M(A) \models \varphi$

(2)     $A \models \text{correct}_M(\varphi)$

Here, $\alpha$ transforms subterms and non-parameter atomic subformulas of $\varphi$ into simulating procedure calls that pass back boolean values, and $\psi$ combines these boolean values into a propositional formula that mirrors the propositional structure of $\varphi$.

### Theorem 18.1

Given a module M allowing the implementation of reachability, we can construct a finite set $Correct_M$ of partial correctness assertions in the import signature such that for all import models $A$ the following are equivalent:

(1)     $Sem_M(A)$ is defined and is an export model

(2)     $A \models Correct_M$

*Proof*   Define $Correct_M = Congruence_M \cup \{correct_M(\varphi) \mid \varphi \in Ax_{exp}\}$.     ∎

### Corollary

Given a module M allowing the implementation of reachability, we can construct a finite set of partial correctness assertions $Correct_M$ in the import signature such that the following are equivalent:

(1)     M is correct

(2)     $Ax_{par} \cup Ax_{imp} \models Correct_M$.

---

**Exercise 18.1**  Discuss how built-in data types together with rigid specifications fit into the constructions above.

---

## 18.4.5   Modules allowing implementation of reachability✂

We now consider the property of modules allowing implementation of reachability. It will be shown that two simple and quite natural conditions of a module's import interface, namely **decompositionality of structured objects** and the availability of counters, suffice to guarantee this property. These conditions will be introduced next.

**Definition 18.3:**   Let M be a module. We say that the import interface of M **allows decomposition of structured objects**, iff for every import term $t(D_1,...,D_k)$ of a sort $s \in S_{imp}$, all of whose variables $D_1,...,D_k$ are variables of parameter sorts, there exist import terms $\Pi_1(Z),...,\Pi_k(Z)$ of parameter sort (with a single variable Z of sort s) such that

$$Ax_{par} \cup Ax_{imp} \models \bigwedge_{i=1,...,k} \bigvee_{j=1,...,k} D_i = \Pi_j(t(D_1,...,D_k)) \wedge$$

$$\bigwedge_{j=1,...,k} \bigvee_{i=1,...,k} D_i = \Pi_j(t(D_1,...,D_k))$$

Let us look at some examples. The usual specification of lists possesses the property above; for example, for a list L with two data elements, these elements can be accessed by terms car(L) and car(cdr(L)). The usual specification of sets with operations emptyset, insert, delete and elementtest does not possess this property, since there is no possibility of accessing the elements of a set with these operations. But if we extend the specification by a function minelement selecting the minimal element of a set (w.r.t. a linear order), then decompositionality of structured objects is fulfilled: we may access the data elements of a set M containing two elements by terms minelement(M) and minelement(delete(minelement(M),M)).

Let M be a module such that its import interface allows decomposition of structured objects. This means that from an import term $t(D_1,...,D_k)$ of sort $s \in S_{imp}$, all of whose variables $D_1,...,D_k$ are parameter variables, the *set* $\{D_1,...,D_k\}$ of parameters used to build $t(D_1,...,D_k)$ can be recovered using 'access paths' $\Pi_1(Z),...,\Pi_k(Z)$, uniformly in all import models. It does not mean that we can find a definite access path $\Pi_i(Z)$ for each $D_i$, for i=1,...,n. As an example, consider the set term insert(insert(emptyset,$D_1$),$D_2$). The access path minelement(Z) accesses one of $D_1$ and $D_2$, but we cannot definitely say which one, since this depends on the order of $D_1$ and $D_2$. Apparently, the access path minelement(delete(Z,minelement(Z))) accesses the other one of $D_1$ and $D_2$. It should also be pointed out that we do not require that the terms $\Pi_1(Z),...,\Pi_k(Z)$ can be effectively computed from $t(D_1,...,D_k)$; all what is needed is their existence.

### Lemma 18.3

Let M be a module such that its import interface allows decomposition of structured objects and *A* be an import model of M such that $Sem_M(A)$ is defined. Then $x \in$ reachable$_s(A)$ iff x can be obtained from the set *par(x)* consisting of all parameter objects $val_A(\Pi(x))$, for an import term $\Pi(Z)$ of parameter type and a variable Z of type REP(s), by iterated application of M's procedures.

*Proof*    Assume that x is an element of reachable$_s(A)$. This means that x can be obtained from certain parameter elements $d_1,...,d_k$ by a sequence of procedure calls of the procedures from M's implementation part. Since the called procedures operate on the considered import model *A*, we conclude that x is of the form $val_A(r(d_1,...,d_k))$, for an import term $r(D_1,...,D_k)$ of sort s containing only variables $D_1,...,D_k$ of parameter sort. Applying the definition of decompositionality, we conclude that the used parameter values $d_1,...,d_k$ are of the form $val_A(\Pi(x))$, where $\Pi(Z)$ is an import term of parameter sort with a single variable $Z$ of sort s. ∎

This 'parameter-free representation' of reachable elements introduced in Lemma 18.7 will play an important role in the following construction. Besides this decompositionality property we need a further minor condition to express correctness as a partial correctness assertion. This concerns the minimal amount of counting ability present in a module's import interface.

> **Definition 18.4:** Let M be a module. We say that M **contains counters** iff there is a ground import term $t_0$ and an import term $t(X)$ with exactly one variable, both $t_0$ and $t(X)$ of the same import sort s, such that the following holds (writing $t^i(t_0)$ for $t(t(...(t(t_0))...))$, with i occurrences of term t) :
>
> $$Ax_{par} \cup Ax_{imp} \models \neg t^i(t_0)=t^j(t_0), \text{ for all natural numbers } i<j.$$

Thus, given a module with counters, an isomorphic copy of $(\mathbb{N},0,succ)$ can be defined within every import model, *uniformly* for every import model, by a single pair of terms $(t_0,t(X))$ as above. In concrete examples, it is usually the case that the stronger formulas $t(X)=t(Y) \to X=Y$ and $\neg t_0=t(X)$ follow from $Ax_{par} \cup Ax_{imp}$.

### Lemma 18.4 (Characterization of reachable elements)

> Let M be a module such that it contains counters and its import interface allows decomposition of structured objects. Then, M allows implementation of reachability.

*Proof*   For every sort $s \in S_{exp}$ we must construct a procedure $Reachable_s(Y_s)$ with a variable $Y_s$ of sort REP(s), such that for every import model $A$ such that $Sem_M(A)$ is defined, and for every element $x \in A_{REP(s)}$ the following are equivalent:

(1)     CALL $Reachable_s(x)$ terminates over $A$.

(2)     $x$ is an element of set $reachable_s(A)$.

According to Lemma 18.3, $x$ is an element of set $reachable_s(A)$ iff $x$ can be obtained from a sequence of parameter objects from set $par(x)$ by a sequence of calls over $A$ of procedures from M's IMPLEMENTATION part. Remember that every element of $par(x)$ is described by some access path $\Pi(Z)$. Access paths (being import terms) as well as sequences of procedure calls are syntactical objects that may be encoded by counters.

As long as element x has not been found, procedure $Reachable_s(Y)$ systematically generates every combination of a sequence of procedure calls ProcSeq and appropriate access paths $\Pi_1(Z),\ldots,\Pi_k(Z)$ (both encoded by counters), executes in parallel all these sequences of procedure calls ProcSeq with $\Pi_1(x),\ldots,\Pi_k(x)$ as actual parameters, and compares the computed values with x. How is the execution of ProcSeq with actual parameters $\Pi_1(x),\ldots,\Pi_k(x)$ implemented on a given import model? As follows: we *symbolically execute* ProcSeq with inputs $\Pi_1(x),\ldots,\Pi_k(x)$ up to the first test. 'Symbolic execution' means formal execution on the basis of terms without any evaluation of terms. Since WHILE programs form a universal programming language over algebra $N$, such a symbolic execution is possible using a standard Godel numbering via counters of terms, procedures, and so on, and WHILE programs as the implementation language, but only up to the first test. It is only the execution of such tests that cannot be done on the basis of counters and WHILE programs alone. What we need is the possibility of evaluating encoded import terms given values for the parameter variables in the considered term. This is finally provided by the following procedure Eval.

PROCEDURE
    Eval( in   C: code of a term $t(Z_1,\ldots,Z_m)$,

               $A_1,\ldots,A_m$: import algebra elements;

      out Y: algebra element);
BEGIN
    Analyse C (algebra-independent, done via counters);
    IF C encodes the variable $Z_i$ THEN $Y:=A_i$
    ELSE
        IF C encodes term $f(t_1(Z_1,\ldots Z_m),\ldots,t_n(Z_1,\ldots Z_m))$
        with import function symbol f and subterms
        $t_1(Z_1,\ldots Z_m),\ldots,t_n(Z_1,\ldots Z_m)$ with codes $C_1,\ldots,C_n$
        THEN
            CALL Eval($C_1,A_1,\ldots A_m,Y_1$); ... ;
            CALL Eval($C_n,A_1,\ldots A_m,Y_n$);
            $Y:=f(Y_1,\ldots,Y_n)$
        END
    END
  END

Whenever required, a call of procedure Eval is build into the symbolic execution described above. ∎

## 18.5    Practical verification of modules

We have shown that correctness of a module M with counters, whose import interface allows decomposition of structured objects, is equivalent to the assertion that

$$Ax_{par} \cup Ax_{imp} \models Correct_M$$

This means that this latter assertion has to be shown. Remember that it was stated in a lemma after the definition of module semantics that only *parameter generated models* of $Ax_{par} \cup Ax_{imp}$ (that is, models whose elements may be represented by import terms with arbitrary parameter values admitted) must be considered in order to show the assertion above.

Looking back at our discussion in Chapter 16 on the provability of partial correctness assertions from a set of axioms, we must expect successive extensions of the underlying signature and axiom set to be necessary to prove the desired claim $Ax_{par} \cup Ax_{imp} \models Correct_M$. Anticipating and collecting all the required extensions we introduce here a set of predicate logic formulas P which we require to fulfil the following properties:

- Every parameter generated model of $Ax_{par} \cup Ax_{imp}$ can be expanded to a model of $P \cup Ax_{par} \cup Ax_{imp}$.

- $P \cup Ax_{par} \cup Ax_{imp} \vdash_{HC} Correct_M$

   Here, because of our restriction to parameter generated models, we may apply *structural induction on import terms* (what this precisely means is clarified later).

Having shown these two claims, we are done. Let us now take a closer look at the formulas in the set $Correct_M$. Such a formula is a partial correctness assertion of the form

$$\{true\} \ CALL \ Reachable_{S_1}(Y_1); \dots; CALL \ Reachable_{S_k}(Y_k); \ \alpha \ \{\psi\}$$

with a program $\alpha$ and postcondition $\psi$, that are obtained in a quite straightforward manner from module M. Verifying a partial correctness assertion as above by Hoare's calculus requires the invention of an intermediate formula $\xi$ such that the following partial correctness assertions can be derived in Hoare's calculus from $P \cup Ax_{par} \cup Ax_{imp}$:

- $\{true\} \ CALL \ Reachable_{S_1}(Y_1); \dots; CALL \ Reachable_{S_k}(Y_k) \ \{\xi\}$

- $\{\xi\} \alpha \{\psi\}$

The choice of $\xi$ is done in a special manner. We assume that axiom set P contains a definition of a certain predicate $p_s(Y)$, for every $s \in S_{par} \cup S_{exp}$, with a variable Y of sort REP(s). For $s \in S_{par}$ we assume that $p_s(Y)$ is defined as a valid formula true.

Now we propose a special form for the intermediate formula $\xi$, namely as a conjunction $p_{s_1}(Y_1) \wedge ... \wedge p_{s_n}(Y_n)$.

On the basis of this choice we have to derive the following partial correctness assertions from $P \cup Ax_{par} \cup Ax_{imp}$:

(1)    {true}

   CALL Reachable$_{s_1}(Y_1)$ ; ... ; CALL Reachable$_{s_k}(Y_k)$

   $\{p_{s_1}(Y_1),...,p_{s_n}(Y_n)\}$

(2)    $\{p_{s_1}(Y_1),...,p_{s_n}(Y_n)\} \; \alpha \; \{\psi\}$

Proof of (2) causes no difficulties and is shown along the lines discussed in previous chapters (Hoare's calculus for recursive programs). (1) is treated differently, as the procedures Reachable$_s$ constructed in Subsection 18.4.5 are of a purely theoretical nature, completely unsuited to practical usage. An alternative approach is obtained if we recognize that (1) expresses that predicate $p_{s_i}$ defines a superset of reachable$_{s_i}$, for i=1,...,k. Bearing in mind the inductive definition of reachable elements as those that can be generated from the parameter objects by application of the implementing procedures of module M, we may replace the considered claim by the following **representation invariant** property:

(3)    For every procedure in the module's implementation part

   $P_f(in \; Y_1:REP(s_1),...,Y_n:REP(s_n); \; out \; Y_f:REP(s))$

   derive the following pca from $P \cup Ax_{par} \cup Ax_{imp}$:

   $\{p_{s_1}(Y_1),...,p_{s_n}(Y_n)\} \; CALL \; P_f(Y_1,...,Y_n,Y_F) \; \{p_s(Y_f)\}$

(3) states that validity of predicates $p_s(Y)$ is invariant under application of the procedures of our module's implementation part. For this reason, predicates $p_s(Y)$ are called in the literature **representation invariants** of a module. (They are indeed comparable to loop invariants for WHILE statements.)

Note that, for certain function symbols f, some or even all of the variables $Y_1,...,Y_n$ occurring in (3) may be of parameter type. Since $p_s(Y)$ is true for every $s \in S_{par}$, (3) indeed implies that predicate $p_s(Y)$ defines a superset of the set of reachable elements of sort s, for every $s \in S_{par} \cup S_{exp}$.

With property (3) we have again arrived at a simple Hoare like task.

Again, induction on import terms may be applied to show (3).

One further note on the role of predicates $p_s(Y)$ is in order. Predicates $p_s(Y)$ capture the idea that a programmer may have about the concrete representation of export objects by import objects. So far, this idea has gone into the definition of the module's procedures and has been hidden there. It requires a lot of work to recover this information from the implementation part by means of the programs $\text{Reachable}_s(Y)$. As an example, in the following verification of our module implementing finite sets by sorted queues, $p_{set}(Q)$ will just be the predicate $sorted(Q)$ with an appropriate definition of what sorted queues are.

It should also be pointed out that the work invested in Subsection 18.4.5 for determining reachability programs has been by no means in vain: it has allowed us to show *uniform* expressibility of module correctness in the language of Hoare logic, whereas the approach sketched in this section depends heavily on the module under consideration.

We now summarize the approach introduced here.

## 18.5.1   Verification strategy for module correctness

### Step 1: Auxiliary predicates introduced by a conservative extension

Introduce new function and predicate symbols, among them a unary predicate symbol $p_s(Y)$ with $Y$ of sort REP(s), for every $s \in S_{exp}$, and describe the intended meaning of the new symbols by an axiom system P. For $s \in S_{par}$ define $p_s(Y)$ to be a valid formula true. Show that:

Every parameter generated model $A$ of $\text{Ax}_{par} \cup \text{Ax}_{imp}$ can be expanded to a model $B$ of $P \cup \text{Ax}_{par} \cup \text{Ax}_{imp}$.

### Step 2: Representation invariants

Show that the family of predicates $(p_s(Y_s))_{s \in S_{par} \cup S_{exp}}$ introduced in Step 1 is invariant under application of the procedures of the module's body. Formally this means deriving from $P \cup \text{Ax}_{par} \cup \text{Ax}_{imp}$ in Hoare's calculus the pca

$$\{p_{s_1}(Y_1), \ldots, p_{s_n}(Y_n)\} \text{ CALL } P_f(Y_1, \ldots, Y_n, Y_f) \{p_s(Y_f)\}$$

for every $f: s_1 \ldots s_n \to s$ in $\Sigma_{exp}$. *Structural induction over import terms* (see below) may be applied.

**Step 3: Simulation of export specification**

For every partial correctness assertion

$$\{true\} \text{ CALL Reachable}_{s_1}(Y_1);\dots;\text{CALL Reachable}_{s_k}(Y_k); \alpha \{\psi\}$$

occurring along the lines of Subsections 18.4.3. and 18.4.4 as a pca simulating an export axiom or a congruence relation axiom derive from $P \cup Ax_{par} \cup Ax_{imp}$ in Hoare's calculus together with structural induction the partial correctness assertion

$$\{p_{s_1}(Y_1),\dots,p_{s_n}(Y_n))\} \alpha \{\psi\}$$

**Structural induction on import terms**

Let $\varphi=(\varphi_s(X_s))_{s \in S_{par} \cup S_{imp}}$ be a family of formulas over the import signature with $X_s$ of sort s. Assume that $\varphi_s(X_s)$ is a valid formula *true*, for every $s \in S_{par}$. Then the **structural induction formula** with respect to $\varphi$ is the following formula Ind($\varphi$), where $S_{imp}=\{s_1,\dots,s_m\}$ and $\Sigma_{imp}=\{f_1,\dots,f_n\}$:

$$(closed(\varphi,f_1)\wedge\dots\wedge closed(\varphi,f_n)) \to (\forall X_{s_1} \varphi_{s_1}(X_{s_1})\wedge\dots\wedge\forall X_{s_m} \varphi_{s_m}(X_{s_m}))$$

with formulas

$$closed(\varphi,f)=\forall X_{r_1}\dots\forall X_{r_m}((\varphi_{r_1}(X_{r_1})\wedge\dots\wedge\varphi_{r_k}(X_{r_k})) \to \varphi_r(f(X_{r_1},\dots,X_{r_k}))$$

for every function symbol f: $r_1\dots r_k \to r$ in $\Sigma_{imp}$.

Since every parameter generated import algebra is obviously a model of Ind($\varphi$), we may assume that all such formulas Ind($\varphi$) are part of our conservative extension P from Step 1.

As an example, for an import interface with function symbols [] and [D|L] as we used in in prior examples, Ind($\varphi$) looks as follows:

$$(\varphi([]) \wedge \forall D \forall L (\varphi(L)\to\varphi([D|L]))) \to \forall L \varphi(L)$$

## 18.6    A concrete module verification

Let us verify the module from Section 18.1 implementing finite sets of data elements by queues of data elements in ascending order. Although the reader could state, with some justification, that 'it is obvious that this module is correct', we will nevertheless work out all the steps required to show the correctness of this module. The reason is to give the reader an impression of the amount of work which has to be done, work which, in part, seems to be so simple (for example the lemmas on the underlying data structures), that one could imagine automatization to be possible and helpful.

Of course, we do not expect a system that would imagine axiomatic extensions or loop invariants. It will be the task of a user in an interactive system to give these components. But then, further processing of the asserted programs, in particular the generation of the lemmas to be proved and the attempt to derive the simpler ones automatically using induction and a theorem prover, does not seem to be unrealistic. The harder lemmas should be passed back to the user for further examination. We collect the axioms of the parameter and import parts that are used in the verification:

| | |
|---|---|
| (Ax0) | Standard axiomatization of BOOLEAN |
| (Ax1) | $\neg D < D$ |
| (Ax2) | $D < E \lor D = E \lor E < D$ |
| (Ax3) | $(D < E \land E < F) \rightarrow D < F$ |
| (Ax4) | $\neg [D|L] = []$ |
| (Ax5) | $\text{first}([]) = \bot$ |
| (Ax6) | $\text{first}([D]) = D$ |
| (Ax7) | $\text{first}([E,D|L]) = \text{first}([D|L])$ |
| (Ax8) | $\text{rest}([]) = []$ |
| (Ax9) | $\text{rest}([D]) = []$ |
| (Ax10) | $\text{rest}([E,D|L]) = [E|\text{rest}([D|L])]$ |

### First step

It is obvious how to define the required representation invariant predicate $p_{\text{set}}(L)$, with L of sort queue. We call it sorted(L) and define it by the following axiom system:

| | |
|---|---|
| (Ax11) | $\text{sorted}([])$ |
| (Ax12) | $\text{sorted}([D])$ |
| (Ax13) | $\text{sorted}([D,E|L]) \leftrightarrow (D < E \land \text{sorted}([E|L]))$ |

For later use within loop invariants, we introduce a 2-ary, infixedly noted relation symbol $D \in L$ and define it by

| | |
|---|---|
| (Ax14) | $D \in [E|L] \leftrightarrow (D = E \lor D \in L)$ |

(Ax15)          $\neg D \in []$

as well as relations $L_1 < L_2$ and $D < L$ and a concatenation function written $L \bullet M$ with the following definitions (for simplicity, $<$ is used in a polymorphic manner, for $D<E$, $D<L$ and $L_1<L_2$; which one is meant in a concrete case can always be seen from the context):

(Ax16)          $L_1 < L_2 \leftrightarrow \forall D\ \forall E\ (D \in L_1 \wedge E \in L_2 \rightarrow D < E)$

(Ax17)          $D < L \leftrightarrow \forall E\ (E \in L \rightarrow D < E)$

(Ax18)          $[] \bullet M = M$

(Ax19)          $[D|L] \bullet M = [D|L \bullet M]$

Let us show that every parameter generated model of (Ax1)-(A10) can be extended to a model of (Ax1)-(Ax19). First of all, remember that (Ax1)-(Ax10) is a rigid specification of the standard algebra *Queue(P)* of queues over a parameter algebra *P*. An extension of *Queue(P)* to a model of (Ax1)-(Ax19) is trivially obtained. The reason is, that (Ax11)-(Ax15) and (Ax18)-(Ax19) are recursive definitions over a free term structure (thus immediately realizable over *Queue(P)*, and (Ax16), (Ax17) are even explicit definitions realizable over *Queue(P)* by interpretation. (Of course, various other definitions of $<$, concatenation, and so on, are possible, serving the same purpose as the ones above.)

## Second step

Here, the following partial correctness assertions must be derived (eventually using induction over import terms) from (Ax0)-(Ax19):

$$\{\text{true}\}\ \text{CALL}\ P_{\text{emptyset}}(L_1)\ \{\text{sorted}(L_1)\}$$
$$\{\text{sorted}(L)\}\ \text{CALL}\ P_{\text{insert}}(L,D,L_1)\ \{\text{sorted}(L_1)\}$$
$$\{\text{sorted}(L)\}\ \text{CALL}\ P_{\text{delete}}(L,D,L_1)\ \{\text{sorted}(L_1)\}$$

Anticipating the work to be done in the third step, we derive the following even stronger partial correctness assertions. This will prevent us from repeating verification work:

(a)     $\{\text{true}\}$

$\quad$ CALL $P_{\text{emptyset}}(Y)$

$\{\text{sorted}(Y), Y=[]\}$

(b)     $\{\text{sorted}(L)\}$

$\quad$ CALL $P_{\text{insert}}(L,D,Y)$

$\{\text{sorted}(Y), \forall E(E \in Y \leftrightarrow (E=D \vee E \in L))\}$

(c)     $\{\text{sorted}(L)\}$

$\quad$ CALL $P_{\text{delete}}(L,D,Y)$

$\{ sorted(Y),\ \forall E(E\in Y \leftrightarrow (\neg E=D \wedge E\in L))\}$

Likewise, we deal with the remaining procedures in our module's BODY.

(d)    $\{ sorted(L)\}$

      CALL $P_{element}(D,L,B)$

    $\{(D\in L\rightarrow B=true),\ (\neg D\in L\rightarrow B=false)\}$

(e)    $\{ sorted(L)\}$

      CALL $P_{max}(L,D)$

    $\{(L=[]\wedge D=\bot)\vee(D\in L\wedge\forall E(E\in L\rightarrow(E=D\vee E<D)))\}$

Let us now verify (a)-(e). To do this, we equip the procedure heads of procedure $P_{emptyset}$ with the corresponding assertions, for example:

PROCEDURE
    $\{ sorted(L)\}$ $P_{insert}(L,D,Y)$ $\{ sorted(Y),\ \forall E(\ E\in Y\leftrightarrow(E=D\vee E\in L))\}$

and apply the deduction rule associated with these asserted heads to derive the corresponding asserted procedure bodies. The required lemmas are collected and proved at the end of this section. The following are the correctly asserted procedure bodies w.r.t the asserted procedure heads from (a)-(e) (the numbers on the right of the symbol ☞ indicate subsequent lemmas):

For (a):    $\{ true\}$

        $Y:=[]$

    $\{ sorted(Y),Y=[]\}$ ☞(1)

For (e):    $\{ sorted(L)\}$

        IF L=[] THEN

            $\{L=[],\ sorted(L)\}$

            $D:=\bot$

            $\{(L=[]\wedge D=\bot)\vee(D\in L\wedge\forall E(\ E\in L\rightarrow(E=D\vee E<D)))\}$ ☞(23)

        ELSE

            $\{\neg L=[],\ sorted(L)\}$

            $D:=first(L)$

            $\{(L=[]\wedge D=\bot)\vee(D\in L\wedge\forall E(\ E\in L\rightarrow(E=D\vee E<D)))\}$ ☞ (0),(25)

        END

    $\{(L=[]\wedge D=\bot)\vee(D\in L\wedge\forall E(\ E\in L\rightarrow(E=D\vee E<D)))\}$

For (b):      {sorted(L)}

L':=L; Y:=[] ;

{sorted(L'), sorted(Y), L'<Y, D<Y, L'•Y=L} ☞(1),(2),(3),(4)

WHILE ¬L'=[] ∧ D<first(L') DO

{¬L'=[], D<first(L'), sorted(L'), sorted(Y), L'<Y, D<Y, L'•Y=L}

Y:=[first(L')|Y]); L':=rest(L')

{sorted(L'), sorted(Y), L'<Y, D<Y, L'•Y=L} ☞(5),(6),(7),(8),(9)

END ;

{L'=[] ∨ ¬D<first(L'), sorted(L'), sorted(Y), L'<Y, D<Y, L'•Y=L}

IF L'=[] THEN

{L'=[], sorted(L'), sorted(Y), L'<Y, D<Y, L'•Y=L}

Y:=[D|Y]

{sorted(Y), ∀E( E∈ Y ↔ (E=D ∨ E∈ L)) } ☞(10),(11),(12)

ELSE

{¬D<first(L'), sorted(L'), sorted(Y), L'<Y, D<Y, L'•Y=L}

IF first(L')<D THEN

{¬D<first(L'),first(L')<D, sorted(L'),

sorted(Y), L'<Y, D<Y, L'•Y=L }

Y:=[D|Y]

{sorted(L'), sorted(Y), L'<Y, ∀E( E∈ L'•Y ↔ (E=D ∨ E∈ L))}

☞(10),(13),(14)

END;

{sorted(L'), sorted(Y), L'<Y, ∀E( E∈ L'•Y ↔ (E=D ∨ E∈ L))} ☞(15)

WHILE ¬L'=[]  DO

{¬L'=[],sorted(L'), sorted(Y), L'<Y,

∀E( E∈ L'•Y ↔ (E=D ∨ E∈ L))}

Y:=[first(L')|Y]; L':=rest(L')

{sorted(L'), sorted(Y), L'<Y, ∀E( E∈ L'•Y ↔ (E=D ∨ E∈ L))}

☞(5),(6),(7),(16)

END ;

{L'=[],sorted(L'), sorted(Y), L'<Y, ∀E( E∈ L'•Y ↔ (E=D ∨ E∈ L))}

{sorted(Y), ∀E( E∈ Y ↔ (E=D ∨ E∈ L))} ☞(12)

END

{sorted(Y), ∀E( E∈ Y ↔ (E=D ∨ E∈ L))}

For (c):    {sorted(L)}

$$L':=L; \ Y:=[];$$

{sorted(L'), sorted(Y), L'<Y, D<Y, L'•Y=L}  ☞(1),(2),(3),(4)

$$\text{WHILE } \neg L'=[] \wedge D<first(L') \text{ DO}$$

{¬L'=[], D<first(L'), sorted(L'), sorted(Y), L'<Y, D<Y, L'•Y=L}

$$Y:=[first(L')|Y]); \ L':=rest(L')$$

{sorted(L'), sorted(Y), L'<Y, D<Y, L'•Y=L}  ☞(5),(6),(7),(8),(9)

$$\text{END };$$

{L'=[]∨¬D<first(L'), sorted(L'), sorted(Y), L'<Y, D<Y, L'•Y=L}

$$\text{IF } \neg L'=[] \wedge D=first(L') \text{ THEN}$$

{¬L'=[] ∧D=first(L'),sorted(L'), sorted(Y), L'<Y, L'•Y=L}

$$L':=rest(L')$$

{sorted(L'), sorted(Y), L'<Y, ∀E( E∈ L'•Y ↔ (¬E=D ∧ E∈ L))}
☞( 5),(17),(18)

$$\text{END };$$

{sorted(L'), sorted(Y), L'<Y, ∀E( E∈ L'•Y ↔ (¬E=D ∧ E∈L))}  ☞(19)

$$\text{WHILE } \neg L'=[] \text{ DO}$$

{¬L'=[],sorted(L'),sorted(Y), L'<Y, ∀E( E∈ L'•Y ↔ (¬E=D ∧ E∈L))}

$$Y:=[first(L')|Y]; \ L':=rest(L')$$

{sorted(L'),sorted(Y), L'<Y, ∀E( E∈ L'•Y ↔ (¬E=D ∧ E∈L))}
☞(5),(6),(16)

$$\text{END}$$

{L'=[], sorted(L'),sorted(Y), L'<Y, ∀E( E∈ L'•Y ↔ (¬E=D ∧ E∈L))}

{sorted(Y), ∀E( E∈ Y ↔ (¬E=D ∧ E∈ L))}  ☞(12)

For (d):    {sorted(L)}

$$L':=L;$$

{sorted(L'), D∈ L ↔ D∈ L'}

$$\text{WHILE } \neg L'=[] \wedge D<first(L') \text{ DO}$$

{¬L'=[], D<first(L'), sorted(L'), D∈ L ↔ D∈ L'}

$$L':=rest(L')$$

{sorted(L'), D∈ L ↔ D∈ L'}  ☞(5),(20)

END;

$\{L'=[] \vee \neg D<first(L'), sorted(L'), D\in L \leftrightarrow D\in L'\}$

IF $L'=[] \vee first(L')<D$ THEN

   $\{L'=[] \vee first(L')<D, sorted(L'), D\in L \leftrightarrow D\in L'\}$

   B:=false

   $\{\neg D\in L, B=false\}$ ☞(21),(22)

   $\{(D\in L\rightarrow B=true), (\neg D\in L\rightarrow B=false)\}$☞(23)

ELSE

   $\{\neg L'=[] \wedge D=first(L'), sorted(L'), D\in L \leftrightarrow D\in L'\}$

   B:=true

   $\{D\in L, B=true \}$ ☞(24)

   $\{(D\in L\rightarrow B=true), (\neg D\in L\rightarrow B=false)\}$☞(23)

END

$\{(D\in L\rightarrow B=true), (\neg D\in L\rightarrow B=false)\}$

## Third step

The following partial correctness assertions mirroring the export equations must be derived from (Ax0)-(Ax19) (with induction allowed):

(f)     $\{sorted(L)\}$
         CALL $P_{insert}(L,D,L_1)$;
         CALL $P_{insert}(L_1,E,L_2)$;
         CALL $P_{insert}(L,E,L_3)$;
         CALL $P_{insert}(L_3,D,L_4)$
         $\{L_2=L_4\}$

(g)     $\{sorted(L)\}$
         CALL $P_{insert}(L,D,L_1)$;
         CALL $P_{insert}(L_1,E,L_2)$
         $\{L_1=L_2\}$

(h)     $\{sorted(L)\}$
         CALL $P_{insert}(L,D,L_1)$;
         CALL $P_{emptyset}(L_2)$
         $\{ \neg L_1=L_2\}$

(i)     $\{true\}$
         CALL $P_{emptyset}(L_1)$;
         CALL $P_{delete}(L_1,D,L_2)$
         $\{L_1=L_2\}$

(j)     $\{sorted(L)\}$
         CALL $P_{insert}(L,D,L_1)$;
         CALL $P_{delete}(L_1,D,L_2)$;
         CALL $P_{delete}(L,D,L_3)$;
         $\{L_2=L_3\}$

(k)     $\{sorted(L)\}$
         CALL $P_{insert}(L,E,L_1)$;
         CALL $P_{delete}(L_1,D,L_2)$;
         CALL $P_{delete}(L,D,L_3)$;
         CALL $P_{insert}(L_3,E,L_4)$
         $\{\neg D=E\rightarrow L_2=L_4\}$

(l)    {true}
    CALL $P_{emptyset}(L)$;
    CALL $P_{element}(D,L,B)$
    {B=false}

(m)    {sorted(L)}
    CALL $P_{insert}(L,E,L_1)$;
    CALL $P_{element}(D,L_1,B_1)$;
    CALL $P_{element}(D,L,B)$;
    {$B_1$=true$\leftrightarrow$(D=E$\lor$B=true)}

(n)    {sorted(L)}
    CALL $P_{insert}(L,E,L_1)$;
    CALL $P_{element}(D,L_1,B_1)$;
    CALL $P_{element}(D,L,B_2)$
    {$\neg$D=E$\rightarrow$$B_1$=$B_2$}

(o)    {true}
    CALL $P_{emptyset}(L_1)$;
    CALL $P_{max}(L_1,D)$;
    {D=$\perp$}

(p)    {true}
    CALL $P_{emptyset}(L_1)$;
    CALL $P_{insert}(L_1,D,L_2)$;
    CALL $P_{max}(L_2,D_1)$
    {$D_1$=D}

(q)    {sorted(L)}
    CALL $P_{insert}(L,D,L_1)$;
    CALL $P_{insert}(L_1,E,L_2)$;
    CALL $P_{insert}(L_1,E,L_2)$;
    CALL $P_{max}(L_2,D_1)$;
    CALL $P_{max}(L_1,D_2)$;
    { $\neg$D<E$\rightarrow$$D_1$=$D_2$}

(r)    {sorted(L)}
    CALL $P_{insert}(L,D,L_1)$;
    CALL $P_{insert}(L_1,E,L_2)$;
    CALL $P_{max}(L_2,D_1)$;
    CALL $P_{insert}(L,E,L_3)$;
    CALL $P_{insert}(L,E,L_3)$;
    CALL $P_{max}(L_3,D_2)$;
    {D<E$\rightarrow$$D_1$=$D_2$}

Fortunately, (f)-(r) can be deduced from (a)-(e) without going into the details of the programs again. Pure predicate logic deductions using the 'universal' lemma (27) below suffice to obtain them. (The reason, of course, is that our treatment of procedures with suitable proof rules already contains a lot of modularity, with respect to both programming and verification.)

Let us verify (f). The verification of the remaining partial correctness assertions follows exactly the same lines and is left to the reader. Remember that in parameters of a procedure remain unchanged; so, any precondition concerning only in parameters being true before a procedure call remains true after the procedure call.

The following is a correctly asserted program:

{sorted(L)}
    CALL $P_{insert}(L,D,L_1)$;
{sorted(L), sorted($L_1$), $\forall X( X \in L_1 \leftrightarrow (X=D \lor X \in L))$}

CALL $P_{insert}(L_1,E,L_2)$;

{sorted(L), sorted($L_1$), sorted($L_2$),

$\forall X(X \in L_1 \leftrightarrow (X=D \lor X \in L))$, $\forall X(X \in L_2 \leftrightarrow (X=E \lor X \in L_1))$}

CALL $P_{insert}(L,E,L_3)$;

{sorted(L), sorted($L_1$), sorted($L_2$), sorted($L_3$), $\forall X(X \in L_1 \leftrightarrow (X=D \lor X \in L))$,

$\forall X(X \in L_2 \leftrightarrow (X=E \lor X \in L_1))$, $\forall X(X \in L_3 \leftrightarrow (X=E \lor X \in L))$}

CALL $P_{insert}(L_3,D,L_4)$

{sorted(L), sorted(L1), sorted(L2), sorted(L3), sorted(L4),

$\forall X(X \in L_1 \leftrightarrow (X=D \lor X \in L))$, $\forall X(X \in L_2 \leftrightarrow (X=E \lor X \in L_1))$,

$\forall X(X \in L_3 \leftrightarrow (X=E \lor X \in L))$, $\forall X(X \in L_4 \leftrightarrow (X=D \lor X \in L_3))$}

{sorted($L_2$), sorted($L_4$), $\forall X(X \in L_2 \leftrightarrow (X=E \lor (X=D \lor X \in L)))$,

$\forall X(X \in L_4 \leftrightarrow (X=D \lor (X=E \lor X \in L)))$ }

{sorted($L_2$), sorted($L_4$), $\forall X(X \in L_2 \leftrightarrow X \in L_4)$} ☞(27)

{$L_2=L_4$}

The lemmas used are:

(0)     $\neg L=[] \rightarrow first(L) \in L$

(1)     sorted([])

(2)     $L<[]$

(3)     $D<[]$

(4)     $L \bullet []=L$

(5)     sorted(L)$\rightarrow$sorted(rest(L))

(6)     (sorted(Y) $\land$ L<Y $\land$ $\neg L=[]$)$\rightarrow$sorted([first(L)|Y])

(7)     (sorted(L) $\land$ $\neg L=[]$ $\land$ L<Y)$\rightarrow$rest(L)<[first(L)|Y]

(8)     (D<Y $\land$ D<E)$\rightarrow$D<[E|Y]

(9)     $\neg L=[]$$\rightarrow$LY=rest(L)$\bullet$[first(L)|Y]

(10)    (sorted(Y) $\land$ D<Y)$\rightarrow$sorted([D|Y])

(11)    $\forall E(E \in [D|Y] \leftrightarrow (E=D \lor E \in Y))$

(12)    $[] \bullet Y=Y$

(13)    (L<Y $\land$ $\neg L=[]$ $\land$ sorted(L) $\land$ first(L)<D)$\rightarrow$L<[D|Y]

(14)    $L' \bullet Y=L \rightarrow \forall E(E \in L' \bullet [D|Y] \leftrightarrow (E=D \lor E \in L))$

(15)    ($\neg L'=[]$ $\land$ $\neg D<first(L')$ $\land$ $\neg first(L')<D$ $\land$ $L'Y=L$)$\rightarrow$

            $\forall E(E \in L' \bullet Y \leftrightarrow (E=D \lor E \in L))$

(16)    $\neg L=[]$$\rightarrow$$\forall E(E \in L \bullet Y \leftrightarrow E \in rest(L) \bullet [first(L)|Y])$

(17)    $L<Y \rightarrow rest(L)<Y$

(18)    ($L' \bullet Y=L$ $\land$ D=first(L') $\land$ sorted(L') $\land$ sorted(Y) $\land$ L'<Y)

$\rightarrow \forall E(\ E \in rest(L') \bullet Y \leftrightarrow (\neg E=D \wedge E \in L))$

(19)    $(L'=[] \vee \neg D=first(L')) \wedge (L'=[] \vee \neg D<first(L')) \wedge$
      $sorted(L') \wedge D<Y \wedge L' \bullet Y=L) \rightarrow \forall E(E \in L' \bullet Y \leftrightarrow (\neg E=D \wedge E \in L))$

(20)    $(\neg L'=[] \wedge D<first(L')) \rightarrow (D \in L' \leftrightarrow D \in rest(L'))$

(21)    $((L'=[] \vee first(L')<D) \wedge sorted(L') \wedge (D \in L \leftrightarrow D \in L')) \rightarrow \neg D \in L$

(22)    $\neg D \in []$

(23)    propositional logic

(24)    $((\neg L'=[] \wedge D=first(L')) \wedge (D \in L \leftrightarrow D \in L')) \rightarrow D \in L$

(25)    $(\neg L=[] \wedge sorted(L)) \rightarrow \forall E(E \in L \rightarrow (E=first(L) \vee E<first(L)))$

(26)    $sorted([D|L]) \rightarrow (D<L \wedge sorted(L))$

(27)    $(sorted(L) \wedge sorted(M)) \rightarrow (L=M \leftrightarrow \forall E(\ E \in L \leftrightarrow E \in M\ ))$

---

**Exercise 18.2**   Verify lemmas (0)-(27):

(a) Give informal (mathematical) proofs on the basis of an intuitive understanding of the predicates and functions involved.

(b) Give formal proofs using structural induction.

(c) If you have time available and already have some experience with the implementation of AI-systems, write an induction verifier and let it prove the lemmas. Be aware of the fact that this is not a three-day project.

**Exercise 18.3**   Verify the other modules from Section 18.1.

**Exercise 18.4**   Write and verify a module providing the user with the quicksort algorithm.

**Exercise 18.5**   Write and verify a module implementing sets by binary search trees (it is assumed that there is a linear order on data objects). Binary search trees are trees with the property that, for all nodes, all labels of the left subtree are less than the label of the parent node, and all labels of the right subtree are greater than the label of the parent node.

**Exercise 18.6**   Assume that you have developed modules implementing sets by sorted queues and queues by arrays respectively. Further assume that you want to combine these modules and build a runnable program. What has to be shown in order to prove correctness of this program?

---

# Historical remarks and recommended reading for Part IV

Hoare (1969) presented his axiomatic proof system for programs partially based on the ideas of Floyd (1967) on inductive assertions. His work was influential and dominates the verification field to this day. Research has taken several directions which we want to discuss briefly in the following; a good overview can be found in Apt (1981). A good presentation of the verification field is Loeckx and Sieber (1987).

Naturally, completeness issues were thoroughly discussed after presentation of Hoare's calculus for WHILE programs. It was Cook who in 1978 introduced the notion of expressiveness and gave a completeness proof for expressive structures. Wand (1978) presented the first concrete algebra with incomplete Hoare logic. Further works in the field of completeness of Hoare's calculus for WHILE programs are Bergstra and Tucker (1982a) and (1982b), Bergstra et al. (1982), Antoniou and Sperschneider (1984), Bergstra and Tiuryn (1984) and Hortala-Gonzales et al. (1988).

Another line of research was concerned with extending Hoare's calculus to further programming constructs. Works in this field are Lipton (1977) and de Bakker (1980). There are, however, some limitations on existence of complete Hoare logics for some constructs. The reader may find results on this topic not discussed here in Clarke (1979) and (1985). Application of Hoare logic to verification of modules as discussed here is new; some discussion and examples may be found in Antoniou (1989) and Antoniou and Sperschneider (1989).

Hoare logic can also be used to provide axiomatic programming language semantics merely consisting of the calculus' axioms and deduction rules. More on this field can be found in Halpern and Meyer (1981), Bergstra and Tucker (1982c) and Bergstra and Tiuryn (1984).

# Appendix
# Basic Mathematical Concepts

## Set theoretic notions

| | |
|---|---|
| $x \in M$ | x is an element of M |
| $\{x \in M \mid p(x)\}$ | set of all elements $x \in M$ with property p |
| $\varnothing$ | empty set |
| $M \subseteq N$ | M is a subset of N |
| M-N | set difference, that is $\{x \in M \mid x \notin N\}$ |
| $M \cup N$ | union of M and N |
| $M \cap N$ | intersection of M with N |
| $\bigcup_{i \in I} M_i$ | set of all x such that there is some $i \in I$ with $x \in M_i$ |
| $\bigcap_{i \in I} M_i$ | set of all x such that for all $i \in I$, $x \in M_i$ |
| $M_1 \times ... \times M_n$ | Cartesian product = |
| | $\{(a_1,...,a_n) \mid a_i \in M_i$, for i=1,...,n$\}$ |
| $M^n$ | $M \times ... \times M$, n times set M, for n>0 |
| $\mathbb{N}$ | set of natural numbers, including 0 |

We use ordinal and cardinal numbers only in the notion of categoricity of axiom systems in Chapter 5; the reader not familiar with ordinal and cardinal numbers should consult Chang and Keisler (1973) or omit these sections.

## Relations

A *relation* R on sets $M_1,...,M_n$ is a subset of $M_1 \times ... \times M_n$. An *n-ary relation* R on M is a subset of $M^n$, n is called the *arity* of R. For a 2-ary relation R we often use infix notation and write xRy instead of $(x,y) \in R$. A 2-ary relation R on M is called

| | |
|---|---|
| *reflexive* | iff xRx, for all $x \in M$ |
| *symmetric* | iff xRy implies yRx, for all $x,y \in M$ |
| *transitive* | iff xRy and yRz imply xRz, for all $x,y,z \in M$ |

| | |
|---|---|
| *equivalence relation* | iff R is reflexive, symmetric and transitive |
| *irreflexive* | iff not xRx, for all $x \in M$ |
| *antisymmetric* | iff xRy and yRx imply x=y, for all $x,y \in M$ |
| *partial ordering* | iff R is irreflexive and transitive |
| *linear ordering* | iff R is a partial ordering and, for all $x,y \in M$, xRy or x=y or yRx |
| *well-founded* | iff there does not exist a sequence $(x_i)_{i \in \mathbb{N}}$ with $x_{i+1}Rx_i$, for all $i \in \mathbb{N}$ |

The *identity relation* $=_M$ on M is the relation $\{(x,x) \mid x \in M\}$. For relations R on L×M and S on M×N we define the *composition* R∘S of R and S as the relation $\{(x,y) \mid$ there exists an element z such that xRz and zSy$\}$. For a 2-ary relation R on M we define the *powers* $R^n$ of R as follows: $R^0$ is $=_M$, $R^{n+1}$ is $R^n \circ R$. The *transitive closure* $R^+$ of R is the union of all $R^n$, for n>0. The *reflexive and transitive closure* $R^*$ of R is the union of all $R^n$, for n≥0. More explicit definitions of these notions are:

$xR^ny$ iff there are $x_0,x_1,...,x_n$ with $x=x_0$, $x_iRx_{i+1}$ for all i<n, and $x_n=y$

$xR^+y$ iff there are n>0 and $x_0,x_1,...,x_n$ with $x=x_0$, $x_iRx_{i+1}$ for all i<n, and $x_n=y$

$xR^*y$ iff x=y or $xR^+y$

## Functions

A *partial function* f from set A to set B, written f:A→B, is a triple (f,A,B) such that f is a relation on A×B with the property that for every $a \in A$ there is at most one $b \in B$ with $(a,b) \in f$. f is called the *graph* of function f:A→B. We define f(a) to be the unique b with $(a,b) \in f$, if b exists. If such a b does not exist we say that f(a) is undefined. The *domain* of f is the set dom(f)={a | f(a) is defined}. The *range* of f is the set rg(f)={f(a) | a ∈ dom(f)}. f:A→B is called a *total function* on A iff dom(f)=A. The *identity function* on A is the function $id_A$:A→A with $id_A$(a)=a, for all $a \in A$.

Let f:A→B and g:B→C be total functions. Then the *composition* of g and f is the function g∘f:A→C with g∘f(a)=g(f(a)). For a total function f:A→A we define the *iterations* $f^n$:A→A of f by $f^0=id_A$ and $f^{n+1}=f \circ f^n$. Function f:A→B is called *injective* iff, for all $a \in A$ and $a' \in A$, f(a)=f(a') implies a=a'. f is called *surjective* iff for all $b \in B$ there is an $a \in A$ such that f(a)=b. f is *bijective* iff f is injective and surjective. Thus a function f:A→B is bijective iff there exists a unique function $f^{-1}$:B→A, called the *inverse* of f, with $f \circ f^{-1}=id_B$ and $f^{-1} \circ f=id_A$.

Given a total function f:A→B and L⊆A and M⊆B, we define the *image* of L under f as the set f(L)={f(a)|a∈L}, the *inverse image* of M under f as the set f⁻¹(M)={a∈A|f(a)∈M}, and the *restriction* of f to L as the function f|$_L$:L→B with f|$_L$(a)=f(a), for all a∈L.

Given two sets I and M, an I-indexed *sequence* is a function a:I→X, usually denoted by $(a_i)_{i \in I}$. I is called *index set*. If I is the set $\mathbb{N}$ of natural numbers, then $(a_i)_{i \in I}$ is called a *countable sequence*, whereas if I is some set {0,1,...,n} then $(a_i)_{i \in I}$ is called a *finite sequence*.

A set M is called *countably infinite* iff there is a bijective function f:$\mathbb{N}$→M. M is called *countable* iff it is finite or countably infinite.

## Strings and trees

Let S be an arbitrary set (often called *alphabet*, or set of *symbols*). A finite sequence of symbols in S is called a *string* over S. S* is the set of all strings over S. Strings are written in the form $s_1...s_n$ with symbols $s_1,...,s_n \in$ S. n is called the *length* of the string. The unique string of length 0 is called the *empty string* and is denoted by ε. Two strings can be *concatenated* by joining the two sequences: concatenation of $u=s_1...s_n$ and $v=t_1...t_m$ yields the string $uv=s_1...s_n t_1...t_m$. A string v is called a *prefix* of string u iff there is a string w such that u=vw. v is a *proper prefix* of u if additionally w≠ε. v is called a *suffix* of u iff there is a string w such that u=wv. v is a *substring* of u iff there are strings x,y such that u=xvy.

*Trees* are a data structure which is extensively used in this book. Trees are introduced as a special form of directed graphs with nodes and edges as follows:

(1)    A single node by itself is a tree. This node is also the tree's *root*.

(2)    Let n be a node and $T_1,...,T_k$ be trees with roots $n_1,...,n_k$ respectively. Then the following graph is also called a tree:

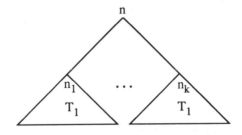

In this tree n is the *root* and $T_1,...,T_k$ the *subtrees* of the root. Nodes $n_1,...,n_k$ are called *children* of node n, n is the parent of $n_1,...,n_k$. Note that there is an order on the children of each node.

It is often convenient to include among trees the *empty tree* $\lambda$ consisting of no nodes. Also, nodes and edges are often associated with *labels*, in general words over a specific alphabet. Nodes without children are called *leaves*. A *path* from node m to node n is a sequence $n_0,...,n_k$ with $n_0=m$, $n_k=n$ and $n_i$ is parent of $n_{i+1}$, for i=0,...,k-1. k is called the path's *length*. The *height* of a tree T is the maximal length of paths connecting nodes of T (or equivalently, the maximal length of a path from the root to a leaf). Trees with the property that each node is either a leaf or has exactly two children are called *binary trees*.

Binary trees are widely used in computer science. It is often desirable to visit all nodes of a tree in a systematic order. The three most important possibilities for this task are preorder, inorder and postorder:

(1)    If tree T is empty, the empty list is the preorder, inorder and postorder listing of nodes of T.

(2)    If T consists solely of one node, then the list containing exactly this node is the result in all three cases.

(3)    If T is a tree with root n and subtrees $T_1,...,T_k$, then

    (a)    $preorder(T) = n,preorder(T_1),...,preorder(T_k)$

    (b)    $inorder(T) = inorder(T_1),n,...,inorder(T_k)$

    (c)    $postorder(T) = postorder(T_1),...,postorder(T_k),n$

Consider as an example the following tree T:

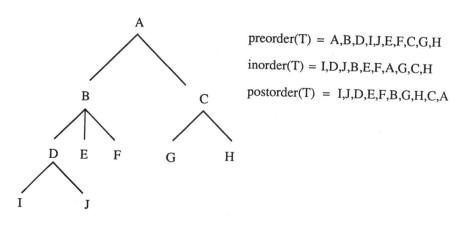

$preorder(T) = A,B,D,I,J,E,F,C,G,H$

$inorder(T) = I,D,J,B,E,F,A,G,C,H$

$postorder(T) = I,J,D,E,F,B,G,H,C,A$

The definition of trees given above can be readily extended to admit infinite paths as well as nodes with an infinite number of children. A tree with a finite number of nodes is called a finite tree.

### *König's Lemma*

A tree T such that every node of T has a finite number of children (the number may depend on the node) is finite iff it contains only paths of finite length.

*Proof*  If T is finite, then it obviously possesses only paths. Now let T be infinite and let $k_0$ be T's root. Since T is infinite, there is one node $k_1$ among the children of $k_0$ that is root of an infinite subtree of T. Among the children of $k_1$ there is a node $k_2$ that is root of an infinite subtree of $k_1$. This process does not end and yields an infinite path $(k_i)_{i=0,1,...}$ of T.                 ■

# Mathematical induction

Mathematical (or complete) induction on the set of natural numbers is a very important and broadly used proof method. Owing to the relation between natural numbers and countable sets, induction is more generally used for proving properties of sets with elements of 'unbound length'. In fact, induction (in its variants) is the major proof tool in logic and program verification.

## Induction principle I

In order to prove property P(n), for all $n \in \mathbb{N}$, show:

- P(0) (*Induction base*)
- For all $n \in \mathbb{N}$, P(n) implies P(n+1)  (*Induction step*)

## Induction principle II

In order to prove property P(n), for all $n \in \mathbb{N}$, show:

- For all $n \in \mathbb{N}$, if P(m) holds for all  m<n then P(n) (*Induction step*)

(Note that the induction base of induction principle I is present in induction principle II in hidden form for n=0.)

## Inductively defined domains

Given a set $\Omega$, a subset Init of $\Omega$ and a set Func of functions (of different arities) on $\Omega$, we often define a subset $\Delta$ of $\Omega$ as follows:

- *Initialization*  Every element of Init is an element of $\Delta$.

- *Closure*    If $f:\Omega^n \to \Omega$ is a function in Func and $\delta_1,...,\delta_n$ are already obtained elements of $\Delta$, then $f(\delta_1,...,\delta_n)$ is also an element of $\Delta$.

The definition of set $\Delta$ is not yet complete. Note that $\Omega$ itself taken as $\Delta$ fulfils the two 'closure conditions' of the definition above. But $\Omega$ is not the intended subset. What is missing is the requirement that we search for

- The *least subset* $\Delta$ of $\Omega$ with the two properties above

Here the phrase 'least subset' can be made precise in different ways, either by saying that $\Delta$ is the intersection of all subsets $\Delta'$ that contain Init and are closed under the functions of Func (this intersection then also contains Init and is closed under the functions of Func), or by saying that the elements, and *only those*, obtainable by the two closure conditions above are elements of $\Delta$. For a subset $\Delta$ of $\Omega$ that is inductively defined as above, a generalized induction principle can be used to show that all elements of $\Delta$ fulfil property P.

## Induction principle III for inductively defined domains

Let $\Delta$ be an inductively defined domain as above. In order to prove property $P(\delta)$ for all $\delta \in \Delta$, show:

- $P(\delta)$, for all $\delta \in$ Init (*Induction base*)
- For all n-ary functions f from Func and $\delta_1,...,\delta_n \in \Omega$, if $P(\delta_1),...,P(\delta_n)$ then $P(f(\delta_1,...,\delta_n))$ (*Induction step*)

# Bibliography

Antoniou, G. (1989). *Über die Verifikation modularer Programme*. Dissertation, Fachbereich Mathematik/Informatik, Universität Osnabrück

Antoniou, G. and Sperschneider, V. (1984). *Incompleteness of Hoare's calculus over simple data structures*. Interner Bericht 3/84 der Fakultät für Informatik, Universität Karlsruhe

Antoniou, G. and Sperschneider, V. (1989). On the verification of modules. In *Proc. CSL' 89*, Springer LNCS

Apt, K. R. (1981). Ten years of Hoare's logic: A Survey-Part I. *ACM TOPLAS*, 3, 431-483

Apt, K. R. and van Emden, M. H. (1982). Contributions to the theory of logic programming. *J. ACM*, 29(3), 841-862

Bachmair, L., Dershowitz, N. and Plaisted, D.A. (1989). Completion without failure. In: *Resolution and Equations in Algebraic Structures 2* (Ait-Kaci, H. and Nivat, M., eds.), pp. 297-367. Boston: Academic Press

de Bakker, J. W. (1980). *Mathematical Theory of Program Correctness*. London: Prentice-Hall

Bergstra, J. A., Chmielinska, A. and Tiuryn, J. (1982). Another incompleteness result for Hoare's logic. *Inform. and Control*, 52, 159-171

Bergstra, J. A., Heering, J. and Klint, P. (1987). *Module Algebra*. Technical Report 1986, Centrum voor Wiskunde en Informatica CWI

Bergstra, J. A. and Klop, J. W. (1984). Proving program inclusion using Hoare's logic. *Theor. Comp. Science*, 30, 1-48

Bergstra, J. A. and Tiuryn, J. (1984). PC-compactness, a necessary condition for the existence of sound and complete logics of partial correctness. In: *LNCS 164*, pp. 45-56. Berlin, Heidelberg, New York: Springer

Bergstra, J. A., Tiuryn, J. and Tucker, J. V. (1982). Floyd's principle, correctness theories and program equivalence. *Theor. Comp. Sci.*, 17, 113-149

Bergstra, J. A. and Tucker, J. V. (1979). *A characterization of computable data types by means of a finite equational specification method*. Techn. Rep., Math. Centrum, Amsterdam

Bergstra, J. A. and Tucker, J. V. (1982a). Expressiveness and the completeness of Hoare's logic. *J. Comp. Sys. Sci.*, 25, 267-284

Bergstra, J. A. and Tucker, J. V. (1982b). Some natural structures that fail to possess a sound and decidable Hoare-like Logic for their while-programs. *Theor. Comp. Science*, 17, 303-315

Bergstra, J. A. and Tucker, J. V. (1982c). The axiomatic semantics of programs based on Hoare's logic. *Acta Inform.*, 21, 293-320

Bibel, W. (1982). *Automated Theorem Proving*. Braunschweig: Vieweg

Bidoit, M., Gaudel, M.-C. and Mauboussin, J. (1987). *How to make algebraic specifications more understandable - An experiment with the PLUSS specification language.* Technical Report 343, Univ. Paris-Sud

Birkhoff, G. (1935). On the structure of abstract algebras. *Proc. Cambridge Philos. Society*, 31, 433-454.

Bläsius, K.-H. and Bürckert, H.-J. (1987). *Deduktionssysteme.* München: Oldenbourgh Verlag

Börger, E. (1985). *Berechenbarkeit, Lomplexität, Logik.* Braunschweig: Vieweg

Bossi, A. and Cocco, N. (1989). Verifying correctness of logic programs, *Tapsoft 89*, pp. 96-110, LNCS 352, Berlin, Heidelberg, New York: Springer

Boyer, R. S. and Moore, J. S. (1979). *A Computational Logic.* New York: Academic Press

Bratko, I. (1990). *Prolog Programming for Artificial Intelligence* 2nd edn. Wokingham: Addison-Wesley

Buchberger, B. and Loos, R. (1982). Algebraic Specification. In: *Computer Algebra - Symbolic and Algebraic Computation* (Buchberger, B., Collins, G. and Loos, R., eds.), pp. 11-43. Wien: Springer

Buchberger, B. (1987). History and basic features of the critical-pair/completion procedure. *J. Symbolic Computation*, 3, 3-38

Büchi, J. (1960). Weak second-order arithmetic and finite automata. *Zeitschrift für mathematische Logik und Grundlagen der Mathematik*, 6, 66-92

Bundy, A. (1983). *The Computer Modelling of Mathematical Reasoning.* New York: Academic Press

Burstall, R. M. and Goguen, J. A. (1977). Putting theories together to make specifications. *Proc. IJCAI 77*

Burstall, R. M. and Goguen, J. A. (1980). Semantics of CLEAR, a specification language. In: *Abstract Software Specifications* (Bjorner, D. ed.), pp. 292-332. Springer LNCS 86

Burt, A. D., Hill, P. M. and Lloyd, J. W. (1990). *Preliminary Report on the Logic Programming Language Goedel.* Technical Report TR-90-02, Department of Computer Science, Univ. of Bristol

Campbell, J. A. (1984). *Implementations of Prolog.* Chichester: Ellis Horwood

Ceri, S., Gottlob, G. and Tanca, L. (1989). What you always wanted to know about Datalog (and never dared to ask). In: *IEEE Trans. on Knowledge and Data Engineering, Vol. 1*, pp. 146-166

Chang, C.-C. and Keisler, H. J. (1973). *Model Theory.* Amsterdam: Elsevier North-Holland

Chang, C.-L. and Lee, R. C. (1973). *Symbolic Logic and Mechanical Theorem Proving.* New York: Academic Press

Church, A. (1936). A note on the Entscheidungsproblem. *J. Symb. Logic*, 1, 40-41.

Clarke, E. M. (1979). Programming language constructs for which it is impossible to obtain good Hoare-like axioms. *J. ACM*, 26, 129-147

Clarke, E. M. (1985). The characterization problem for Hoare logics. In: *Mathematical Logic and Programming Languages* (Atiyah, M., Hoare, C. A. and Shepherdson, J. C., eds.). Englewood Cliffs NJ: Prentice-Hall

Clocksin, W. F. and Mellish, C. S. (1981). *Programming in PROLOG*, Berlin, Heidelberg, New York: Springer

Colmerauer, A., Kanoui, H., Roussel, P. and Pasero R. (1973). *Un Système de Communication Homme-Machine en Français*. Groupe de Recherche en Intelligence Artificielle, Université d'Aix-Marseille

Cook, S. A. (1978). Soundness and completeness of an axiom system for program verification. *SIAM Journ. Comput.*, 7, 70-90; Corrigendum 10 (1981), 612.

Davis, M. and Putnam, H. (1960). A computing procedure for quantification theory. *J. ACM*, 7, 201-215

Dershowitz, N. (1989). Completion and its applications. In: *Resolution and Equations in Algebraic Structures 2* (Ait-Kaci, H. and Nivat, M., eds.), pp. 297-367. Boston: Academic Press

Ebbinghaus, H.-P., Flum, J. and Thomas, W. (1978). *Einführung in die mathematische Logik*. Darmstadt: Wissenschaftliche Buchgesellschaft

Ehrig, H., Kreowski, H.-J., Mahr, B. and Padawitz, P. (1982). Algebraic implementation of abstract data types. *Theor. Comp. Sci.*, 20, 209-263

Ehrig, H. and Mahr, B. (1985). *Fundamentals of Algebraic Specification Vol 1*. Berlin, Heidelberg, New York: Springer

Ehrig, H. and Mahr, B. (1990). *Fundamentals of Algebraic Specification Vol 2*. Berlin, Heidelberg, New York: Springer

Ehrig, H. and Weber, H. (1985). Algebraic Specifications of Modules. In: *Proc. IFIP Work Conf.: The Role of Abstract Models in Programming* (Wien), Amsterdam: North-Holland

Fey, W. (1986). *Introduction to Algebraic Specification in ACT TWO*. Research Rep. 86-13, Dep. of Comp. Sci., TU Berlin

Floyd, R. W. (1967). Assigning meanings to programs. In *Proc. Symp. on Appl. Math. 19*, American Mathematical Society, New York

Frege, G. (1879). *Begriffsschrift, eine der arithmetischen nachgebildete Formelsprache des reinen Denkens*. Halle a.S.: Nebert Verlag

Futatsugi, K., Goguen, J., Jouannaud, J. P. and Meseguer, J. (1985). Principles of OBJ2. In: *Proc. 12th ACM Symposium on the Principles of Progr. Languages*, pp. 52–66. ACM

Gallaire, H. and Minker, J., eds. (1978). *Logic and Data Bases*. New York: Plenum Press

Gallaire, H., Minker, J. and Nicolas, J.-M. (1984). Logic and data bases: a deductive approach. *Computing Surveys*, 16(2), 153-185

Gallier, J. H. (1986). *Logic for Computer Science - Foundations of Automatic Theorem Proving*. New York: Wiley

Gaudel, M.-C. (1984). *A first introduction to PLUSS*. Technical Report, Univ. Paris-Sud

Genesereth, M. R. and Nilsson, N. J. (1987). *Logical Foundations of Artificial Intelligence*. Los Altos CA: Morgan Kaufmann

Gödel, K. (1930). Die Vollständigkeit der Axiome des logischen Funktionenkalkül. *Monatshefte für Mathematik und Physik*, 37, 349-360

Gödel, K. (1931). Über formal unentscheidbare Sätze der Principia Mathematica und verwandter Systeme. *Monatshefte für Mathematik und Physik*, 38

Goguen, J. A. and Meseguer, J. (1984). Equality, types, modules and (why not?) generics for logic programming. *J. Logic Programming*, 1(2), 179–210

Goguen, J. A., Thatcher, J. W. and Wagner, E. G. (1976). *An initial algebra approach to the specification, correctness and implementation of abstract data types*. IBM Research Rep. RC 6487

Goguen, J. A., Thatcher, J. W., Wagner, E. G. and Wright, J. B. (1977). Initial algebra semantics and continuous algebras. *J. ACM*, 24, 68-95

Gries, D. (1981). *The Science of Programming*. New York: Springer

Guttag, J. V. and Horning, J. J. (1978). The algebraic specification of abstract data types. *Acta Informatica*, 10(1), 27-52

Halpern, J. and Meyer, A. (1981). Axiomatic definitions of programming languages II. In: *Proc. 8th POPL Conf.*, pp. 139-148

van Heijenoort, J. (1967). *From Frege to Gödel - A Source Book in Mathematical Logic*. Harvard University Press

Herbrand, J. (1930). Investigations in proof theory. In: *From Frege to Goedel: A Source Book in Mathematical Logic 1879–1931* (v. Heijenoort, ed.), pp. 525-581. Harvard University Press

Hoare, C. A. (1969). An axiomatic basis for computer programming. *Comm. ACM*, 12, 567-580

Hoare, C. A. (1972). Proof of correctness of data representations. *Acta Inf.*, 1, 271-281

Hogger, C. (1984). *Introduction to Logic Programming*. London: Academic Press

Hortala-Gonzalez, M. T., Lucio-Carrasco, F. and Rodriguez-Artalejo, M. (1988). Some general incompleteness results for partial correctness logics. *Inform. and Comput.*, (1988), 22-42

Jounnaud, J. P. and Kirchner, H. (1986). Completion of a set of rules modulo a set of equations. *SIAM J. Comput.*, 15(4), 1155-1194

Kaplan, S (1984). Conditional rewrite rules. *Theor. Computer Science*, 33, 175-193

Kfoury, A. J. and Urzyczyn, P. (1985). Necessary and sufficient conditions for the universality of program formalisms. *Acta Inform.*, 22, 347-377

Kleene, S. C. (1952). *Introduction to Metamathematics*. Amsterdam: North-Holland

Kleene, S. C. (1967). *Mathematical Logic*. New York: Wiley

Knuth, D. E. and Bendix, P. B. (1967). Simple word problems in universal alge-bras. In: *Proc. of the Conf. on Computational Problems in Abstract Alge-bras* (Leech, J. ed.), pp. 263-298. Oxford: Pergamon

Kowalski, R. A. (1975). A proof procedure using connection graphs. *J. ACM*, **22**(4)

Kowalski, R. A. and van Emden, M. H. (1976). The semantics of predicate log-ic as a programming language. *J. ACM*, **23**(4), 733-742

Kowalski, R. A. (1979a). *Logic for Program Solving*. New York: North-Hol-land

Kowalski, R. A. (1979b). Algorithm = Logic + Control. *J. ACM*, 22, 424-436

Lipton, R. J. (1977). A necessary and sufficient condition for the existence of Hoare logics. In: *Proc. 18th IEEE Symp. on Found. of Comp. Sci.*, pp. 1-6, Providence

Lloyd, J. W. (1987). *Foundations of Logic Programming*, 2nd edn. Berlin, Heidelberg, New York: Springer

Loeckx, J. and Sieber, K. (1987). *The Foundations of Program Verification*, 2nd edn. Chichester: Wiley

Loveland, D. W. (1978). *Automated Theorem Proving: A Logical Basis*. New York: Elsevier North-Holland

Manna, Z. and Waldinger, R. (1985). *The Logical Basis for Computer Program-ming*. Reading MA: Addison-Wesley

Mendelson, E. (1964). *Introduction to Mathematical Logic*. Princeton NJ: Van Nostrand

Minker, J., ed. (1988). *Foundations of Deductive Databases and Logic Pro-gramming*. Los Altos CA: Morgan Kaufmann

Nilsson, N. J. (1982). *Principles of Artificial Intelligence*. Berlin, Heidelberg, New York: Springer

Oberschelp, A. (1962). Untersuchungen zur mehrsortigen Quantorenlogik. *Math. Annalen*, 145, 297-333

Olderog, E.-R. (1981). Sound and complete Hoare-like calculi based on copy-rules. *Acta Informatica*, 16, 161-197

Parnas, D.C. (1972). A technique for software module specification with exam-ples. *CACM* **15**(5), pp. 330-336

Peano, G. (1889). *Arithmetices principia, nova methodo exposita*. Turin

Richter, M. M. (1978). *Logikkalküle*. Stuttgart: Teubner

Robinson, J. A. (1965). A machine oriented logic based on the resolution prin-ciple. *J. ACM*, **12**(1), 23–41

Robinson, J. A. (1979). *Logic: Form and Function*. New York: Elsevier North-Holland

Robinson, G. A. and Wos, L. (1969). Paramodulation and theorem-proving in first-order logic with equality. *Machine Intelligence*, 4, 135-150

Rogers, H. (1967). *Theory of Recursive Functions and Effective Computabili-*

*ty.* New York: McGraw-Hill

Russell, B. (1946/61). *History of Western Philosophy.* Unwin University Books

Schöning, U. (1987). *Logik für Informatiker.* Mannheim: BI Wissenschaftsverlag

Sheperdson, J. C. (1985). Negation as failure II. *J. Logic Programming,* 2(3), 185-202

Sheperdson, J. C. (1987). Negation in logic programming. In: *Foundations of Deductive Databases and Logic Programming* (Minker J., ed.). Los Altos CA: Morgan Kaufmann

Shoenfield, J. R. (1967). *Mathematical Logic.* Reading MA: Addison-Wesley

Siekmann, J. (1984). Universal unification. In: *LNCS 170,* pp. 1-42. Berlin, Heidelberg, New York: Springer

Siekmann, J. and Stephan, W. (1976). *Completeness and Soundness of the Connection Graph Proof Procedure.* Interner Bericht 7/76, Inst. f. Informatik I, Universität Karlsruhe.

Siekmann, J. and Wrightson, G., eds. (1983a). *Automation of Reasoning; Classical Papers on Computational Logic 1957-1966.* Berlin, Heidelberg, New York: Springer

Siekmann, J. and Wrightson, G., eds. (1983b). *Automation of Reasoning; Classical Papers on Computational Logic 1967-1970.* Berlin, Heidelberg, New York: Springer

Smolka, G., Nutt, W. and Goguen, J.A. (1989). Order-sorted equational computation. In: *Resolution and Equations in Algebraic Structures 2* (Ait-Kaci, H. and Nivat, M., eds.), pp. 297-367. Boston: Academic Press

Smullyan, R. M. (1968). *First Order Logic.* Berlin, Heidelberg, New York: Springer

Sterling, L. and Shapiro, E. Y. (1986). *The Art of Prolog.* Cambridge MA: MIT Press

Stickel, M. (1985). Automated deduction by theorem resolution. *J. Autom. Reas.,* 1(4), 333-356.

Takeuti, G. (1975). *Proof Theory.* Amsterdam: Elsevier North-Holland

Thayse, A., ed. (1988). *From Standard Logic to Logic Programming.* Chichester: Wiley

Thayse, A., ed. (1989). *From Modal Logic to Deductive Databases.* Chichester: Wiley

Walther, C. (1987). *A Many-Sorted Calculus Based on Resolution and Paramodulation.* London: Pitman & Los Altos: Morgan Kaufmann

Wand, M. (1978). A new incompleteness result for Hoare's system. *J. ACM,* 25, 168-175.

Warren, D. H. D. (1983). *An Abstract Prolog Instruction Set.* Techn. Rep., SRI International

Wos, L., Overbeek, R., Lusk, E. and Boyle, J. (1984). *Automated Reasoning-Introduction and Applications.* Englewood Cliffs NJ: Prentice-Hall

# Index